IMPROVEMENT OF BUILDINGS STRUCTURAL QUALITY BY NEW TECHNOLOGIES

OUTCOME OF THE COOPERATIVE ACTIVITIES

COST Action C 12

Improvement of Buildings Structural Quality by new Technologies
Outcome of the Cooperative Activities

Final Scientific Report - 2004

Edited by
Christian Schaur (Austria)
Federico Mazzolani (Italy)
Gerald Huber (Austria)
Gianfranco de Matteis (Italy)
Heiko Trumpf (Germany)
Heli Koukkari (Finland)
Jean-Pierre Jaspart (Belgium)
Luis Bragança (Portugal)

LEIDEN/LONDON/NEW YORK/PHILADELPHIA/SINGAPORE

Published by: A.A. Balkema Publishers, Leiden, The Netherlands a member of
Taylor & Francis Group plc
www.balkema.nl and www.tandf.co.uk

ISBN: 04 1536 610 0

Printed and bound in Great Britain by Antony Rowe Ltd, Eastbourne

COST Office
Urban Civil Engineering

Contact:
Mr. Jan SPOUSTA, Science Officer
COST Transport & COST Urban Civil Engineering
ESF COST Office
Av. Louise 149
P.O.Box 12
B-1050 BRUSSELS
Belgium
phone +32 2 533 38 32
jspousta@cost.esf.org

What is COST

COST- the acronym for European **CO**operation in the field of **S**cientific and **T**echnical Research- is the oldest and widest European intergovernmental network for cooperation in research. Established by the Ministerial Conference in November 1971, COST is presently used by the scientific communities of 35 European countries to cooperate in common research projects supported by national funds. The funds provided by COST - less than 1% of the total value of the projects - support the COST cooperation networks (COST Actions) through which, with only around €20 million per year, more than 30.000 European scientists are involved in research having a total value which exceeds €2 billion per year. This is the financial worth of the European added value which COST achieves. A "bottom up approach" (the initiative of launching a COST Action comes from the European scientists themselves), "à la carte participation" (only countries interested in the Action participate), "equality of access" (participation is open also to the scientific communities of countries not belonging to the European Union) and "flexible structure" (easy implementation and light management of the research initiatives) are the main characteristics of COST. As precursor of advanced multidisciplinary research COST has a very important role for the realisation of the European Research Area (ERA) anticipating and complementing the activities of the Framework Programmes, constituting a "bridge" towards the scientific communities of emerging countries, increasing the mobility of researchers across Europe and fostering the establishment of "Networks of Excellence" in many key scientific domains such as: Physics, Chemistry, Telecommunications and Information Science, Nanotechnologies, Meteorology, Environment, Medicine and Health, Forests, Agriculture and Social Sciences. It covers basic and more applied research and also addresses issues of pre-normative nature or of societal importance.

For more information visit http://cost.cordis.lu/src/home.cfm

COST ACTION C12

Foreword

The COST C12 Action has been launched in 2000 and all along its whole duration, about 100 scientists from 21 of the 35 COST member countries contributed to its success.

During four years, efforts have been devoted to the achievement of the main objectives of the Action which may be summarised as follows:
- to develop, combine and disseminate new technical engineering technologies;
- to improve the quality of urban buildings;
- to propose new technical solutions to architects and planners;
- to reduce the disturbances of the construction process in urban areas, and finally improve the quality of living in the urban habitat.

The present publication reflects the outcome of these cooperative activities. It includes, in a format of separate datasheets, all the knowledge and expertise acquired during the action. These datasheets present in a direct and synthetic way basic, but also more practical key information on various topics relevant to the three following thematic fields:
- mixed building technology;
- structural integrity under exceptional actions;
- urban design.

On behalf of the scientific committee in charge of the edition of the present book, I would like to thank warmly all those who contributed actively to the Action and more particularly the authors of the datasheets. The quality of their work and their involvement all along the Action have been highly appreciated. Special gratitude is also addressed to Mr I. Samaras from COST offices and Mr J. Spousta and I. Silva-Ballesteros from ESF (European Science Foundation) as well as to Mr L. Bijnsdorp from BALKEMA for their support and help, respectively in administrative and publication matters.

Finally, the financial support from the European Commission has to be underlined.

Jean-Pierre Jaspart
C12 Chairman

COST ACTION C12

COST C12 Management Committee

Chairman: Jean-Pierre Jaspart
Vice-Chairman :   Gerald Huber

Chairmen of Working Groups

WG1 "Mixed Building Technology":   Christian Schaur
WG2 "Structural Integrity under Exceptional Actions":   Federico Mazzolani
                                            Gianfranco De Matteis
WG3 "Urban Design":   José Mendes
                                               Luis Bragança
                                               Heli Koukkari

National representatives

Country	National representatives		E-mail addresses
	Family name	First name	
Austria	Huber	Gerald	geraldhuber@telering.at
	Schaur	Christian	schaur@tirol.com
Belgium	Jaspart	Jean-Pierre	jean-pierre.jaspart@ulg.ac.be
	Maquoi	René	r.maquoi@ulg.ac.be
Czech Rep.	Wald	Frantisek	wald@fsv.cvut.cz
Denmark	Engelmark	Jesper	je@byg.dtu.dk
Finland	Leino	Tapio	tapio.r.leino@vtt.fi
France	Muzeau	Jean-Pierre	muzeau@cust.univ-bpclermont.fr
	Terrin	Jean Jacques	jean-jacques.terrin@utc.fr
F.Y.R. of Macedonia	Gramatikov	Kiril	gramatikov@gf.ukim.edu.mk
	Bozinovski	Zivko	zivko@pluto.iziis.ukim.edu.mk
Germany	Haller Menzel	Peer Ralf (deputy)	haller@bbhu01.bau.tu-dresden.de menzel@bbhu01.bau.tu-dresden.de

	Trumpf	Heiko	tru@stb.rwth-aachen.de
Greece	Baniotopoulos	Charalambos	ccb@civil.auth.gr
	Perdikaris	Filippos	filperd@civ.uth.gr
Ireland	Broderick	Brian	bbrodrck@tcd.ie
Italy	Mazzolani	Federico	fmm@unina.it
	Zandonini	Riccardo	riccardo.zandonini@ing.unitn.it
Latvia	Rocens Serdjuks	Karlis Dimitrijs (deputy)	rocensk@latnet.lv
Lithuania	Kudzys	Antanas	asi@asi.lt
	Vektaris	Bronius	asi@asi.lt
Netherlands	Gresnigt	A.M. Nol	a.m.gresnigt@citg.tudelft.nl
	Blok	Rijk	r.blok@bwk.tue.nl
Poland	Kozlowski	Aleksander	kozlowsk@ewa.prz.rzeszow.pl
Portugal	Silva	Luis	luis_silva@gipac.pt
	Mendes Braganca	José Luis (deputy)	jmendes@civil.uminho.pt braganca@civil.uminho.pt
Romania	Dinu	Florea	florin@constructii.west.ro
	Dubina	Dan	dubina@constructii.west.ro
Slovenia	Bosiljkov	Vlatko	bosiljko@zag.si
	Fischinger	Matej	matej.fischinger@ikpir.fagg.uni-lj.si
Spain	Cancela Rey Serra	Dolores Javier (deputy)	mdcancela@mfom.es jserra@mfom.es
Switzerland	Hamm	Jan	jan.hamm@hsb.bfh.ch
United Kingdom	Mottram	James Toby	jtm@eng.warwick.ac.uk

COST ACTION C12

Table of contents

COST ACTION C12

General introduction

From 1991 to 1999, in the frame of the C1 COST Action devoted to the "Control of the semi-rigid behaviour of civil engineering structural connections", European scientists developed unified design approaches for the conceptual design and the calculation of structural joints in civil engineering structures and so contributed to the recent improvements of the European codes (Eurocodes 3 and 4) in this field.

At the end of this quite successful action, needs for further developments were identified and it was recognised that there was, throughout Europe, a special need for higher quality in the field of urban buildings. The development of new and suitable strategies for architects, sociologists, urban planners, local authorities and engineers were required and, until that time, civil engineering aspects had been somewhat disregarded in this process.

Analysing the industries associated with structures of steel, concrete, steel-concrete composite, timber, polymeric composite and structural glass it is found that they were operating more or less independently. Synergy-effects of the different construction methods were seen as the future to widen the creative possibilities of architects and planners, and to improve urban building technology.

New building technologies had been developed in many countries in Europe and these had not yet been transferred into practice so as to achieve higher quality, economy and safety in urban buildings as well as a lower impact on urban environment. As a result, a number of technical fields of development were identified as being appropriate to this transfer; amongst them:

· the Mixed Building Technology (MBT)

· the Structural Integrity of urban buildings under exceptional actions (SI).

Mixed Building Technology (MBT) aims at integrating structural members of different construction methods into a single building. The situation for the development of MBT in Europe was excellent and fundamental connection problems had been successfully solved, both for different material technologies in isolation and in combination, within the COST C1 Action. Furthermore, unified safety and action concepts for most construction methods had just been agreed at the European level and implemented in the structural Eurocodes, published by CEN.

The effect of normal actions on buildings is extensively covered by the existing Eurocodes. Many structures are, however, exposed to certain extreme non-initially expected actions arising from natural and human-made hazards, such as: fire, earthquakes, wind storms, explosions, exceptional snow loading. Robustness is the capability of the building to remain stable even when one of its parts is heavily damaged, for instance because of a local explosion on one storey or the impact of a truck against the structure. Robustness is a strong requirement as far as human integrity of inhabitants and neighbours is of concern and no design guidelines are provided by the structural Eurocodes.

Furthermore there was a need for an appropriate quality management system in these fields. This is crucial to assure the desired quality level in practice. The introduction of manufacturing techniques, thus enabling prefabrication in the factory instead of construction works on site, is also needed to provide better quality structural elements which meet tighter dimensional tolerances, etc.

But the smooth and efficient transfer of these new technologies within the specific field of urban habitat requires a strong link between all the partners involved in the construction business, such as urban planners, local authorities, architects, engineers, …One could speak about "Urban Design (UD)" using MBT. Expertise on each of these topics was existing locally in many European countries, but no integrated knowledge combining MBT, SI and UD was available, which could lead to new urban structures meeting the imposed structural requirements and the highest non-structural demands of the people using them. In this context the COST scheme was an ideal platform to have a multidisciplinary Action allowing to integrate these new technologies in the field of urban buildings and to disseminate understanding, knowledge and a unified technical vocabulary throughout Europe.

Two coordinated COST Actions were set up in 2000 with the main objective to develop, combine and disseminate new engineering technologies, to improve the quality of urban buildings, to propose new technical solutions to architects and planners, to reduce the disturbances of the construction process in urban areas and finally to improve the quality of living in the urban habitat:

- COST C12 *"Urban Buildings – Quality by New Technologies"*
  dealing with:
  - Mixed Building Technology (MBT);
  - Structural Integrity under exceptional loading (SI);
  - Urban Design (UD);
- COST C13 *"Glass and Interactive Building Envelopes"*.

So-called Memorandum of Understanding (MoU) describing the backgrounds, the objectives and the work planning of these two actions have been drafted and used as a reference all along their duration. They can be found on the COST web site: http://cordis.lu/scr/home.cfm.

As far as COST C12 is concerned, scientists from 21 European countries put a lot of effort into it from 2000 to 2004 so as to turn the initial objectives stated in the MoU into reality.

The present final report is the outcome of these activities. It is not aimed at covering all the detailed information gathered by the C12 members during four years, but much more at providing designers, architects, planners and code writers with *(i)* key data on all the addressed topics and *(ii)* expert contacts in the respective fields. To achieve this goal, a standardised "datasheet" format of presentation has been adopted.
Prevailing rules for the drafting of these datasheets may be summarised as follows :
- Use of "bullet points" so as to get an "easy-to-consult" publication;
- brief state-of-the-art presentation of the addressed topic;

- report of the suggested solutions or recommendations;
- references to available papers or reports where more detailed information may be found;
- Inclusion of a list of contact persons to whom further requests could be addressed by interested readers. Alternatively, contacts may also be directly established with all the COST C12 National Representatives listed before.

Despite this sake for conciseness, the presented material is rather voluminous. It has however to be clearly stated that it only represents the outcome of the cooperative works within COST C12 and has therefore not to be considered as exhaustive.

Anyway the authors have the feeling that they have done a useful piece of work and hope that the present book will represent a valuable source of information for the various protagonists in the urban and engineering construction fields.

Chapter I

Structural Building Quality

COST ACTION C12

Chapter I - Structural Building Quality

Introduction

I.1 Datasheets on General Methodology
I.1.1 Performance based design
I.1.2 Probabilistic analysis
I.1.3 Risk analysis
I.1.4 Optimisation

I.2 Datasheets on Exceptional Actions
I.2.1 Explosions
I.2.2 Earthquake
I.2.3 Landslides
I.2.4 Fire
I.2.5 Mining subsidence

I.3 Datasheets on Structural Integrity
I.3.1 Seismic actions
    I.3.1.1 New buildings
    I.3.1.2 Existing buildings
I.3.2 Fire actions
I.3.3 Explosion actions
I.3.4 Impact actions

I.4 Datasheets on Robustness
I.4.1 Members
I.4.2 Connections

Introduction to Chapter I

The main peculiarity of WG 2 "*Structural integrity of buildings under exceptional actions*" consists in being related to a very wide and multi-disciplinary area. On one hand, it deals with the definition of exceptional actions on buildings and, on the other hand, it has to face the determination of the structural performance due to these actions. The analysis of relevant structures has to be developed always beyond the elastic limit and also in dynamic range, and has to be related to many types of buildings present in the urban habitat, old and new, made of different constructional materials and technological systems. An important aspect is, therefore, the examination of a variety of both traditional and innovative technologies, which are referred to several building materials (like reinforced concrete, steel, masonry, aluminium, stainless steel, timber), for analysing their behaviour under exceptional situations produced by catastrophic events. As a consequence, the WG2 activity has been differentiated accordingly, involving specialised experts in the working teams dealing with several aspects of the so-called structural engineering field. This provided an important interdisciplinary character in the scientific expertise present within the group, but at the same time it was unavoidable to have a limited number of specialists for each topic in relation with the potential knowledge of the participating Countries.

The first step of the activity has been devoted to clearly define the main related topics that it was decided to cover in the working group. The Seminar held in Lisbon on April 2002 was the first milestone, where a preliminary consensus on both the general definitions of the relevant keywords (structural integrity, robustness, exceptional actions etc,) and the general scientific programme to be developed was reached. At that time, it was immediately recognised that it was not possible to cover every aspects of this so large field, mainly due to the lack of knowledge in some specific field in relation to the actual expertise of the representatives of the European Countries who decided to participate to this Action. In particular, from the point of view of exceptional actions, it immediately appeared that earthquake and fire were very popular in all the Countries, but, contrarily, other actions, like gas explosions, impacts, landslides, volcanic explosions etc., could not be satisfactorily covered because they did not represent the current research topic of the attending members. It has been observed that even earthquake and fire could not be considered as "exceptional" actions, when their effect has been correctly considered in the design process. Also wind can produce exceptional effects under form of typhoons and hurricanes, but it has been excluded from the WG2 activity not only because these extreme situations are not very common in Europe, but also because there was contemporary another COST action devoted to this subject.

The consequence of exceptional actions on the building performance was the second step to be faced. It required the identification of a proper unitary methodology of analysis which has to be necessarily complicated, since it has to include the inelastic and dynamic behaviour of material and structures, without ignoring both probabilistic and risk aspects. The out-put of the general analysis allows a judgment of the "structural integrity" of a building. In this perspective, the concept of structural robustness was introduced like a measure of the available capacity in terms of the triad "strength, stiffness and ductility", which allows the structures to possess enough structural integrity to survive under exceptional situations of loading.

In order to cover these main aspects of the whole subject, it was decided to organise the output of the activity developed in the current project according to four main chapters" (1) General methodology, (2) Exceptional actions, (3) Structural integrity" and (4) Robustness. The datasheets produced by WG2 and selected to be published in this volume in number of 49 are treated in a unitary way and are distributed along these four chapters in order to cover there the main involved aspects, but without any intention to exhaust all possible topics. In any case, also with the above mentioned limits, we are convinced that our contribution will be useful for researchers and engineers, providing a kind of state-of-the-art in this quite complex field.

Federico M. Mazzolani   Gianfranco De Matteis
WG2 Chairman   WG2 Vice-chairman

PERFORMANCE BASED (SEISMIC) ENGINEERING (PBSE)

Description

- By nature engineers have been interested in the performance of their structures and products. However, performance has been typically addressed in rather general terms, and seldom explicitly quantified with appropriate engineering demand parameters. This has been particularly true in earthquake engineering, where appropriate level of seismic design forces had been defined predominantly empirically. It has been simply known that standard structural systems survive "standard" earthquakes, if they were designed for specific level of design seismic forces. There has been little interest in the relation of such design values and actual physical behaviour of the structure. By definition such empirically based procedures can not be used for innovative structural systems, addressed within the COST 12 action. Moreover, even for standard systems, designers have been predominantly interested in survival of the structure in the case of a major earthquake, with acceptably low probability of occurrence (i.e. 10% in 500 years). It has been further assumed that so designed structure will experience acceptable level of damage at more frequent weaker earthquakes. Typically, neither "survival" nor "acceptable damage" has been quantified

- Some recent seismic events (in particular Northridge 1994, Kobe 1995 as well as Bovec, Slovenia 2004) clearly demonstrated serious deficiencies of standard seismic design procedures. First of all it was observed that engineers and stakeholders (i.e. owners, tenants, public authorities and media) talk different languages. While engineers have been interested predominantly in survival, stakeholders have been as much interested in direct as well as indirect damage. And damage was indeed massive (i.e. 30 billion USD in Northridge and even 200 billion USD in Kobe, both relatively low magnitude events). In Bovec magnitude was extremely low (M = 4.9), but due to strong amplifications in glacial deposits the local intensity was high with maximum ground acceleration as much as 0.5g, which can be considered as typical example of exceptional load as defined within COST 12 action. However, public opinion and media were not willing to accept any damage at this low magnitude event. Moreover, it is more and more evident that public opinion is not willing to accept any damage to engineered structures at any (even very strong) earthquake.

- In short the performance foreseen by engineers and that expected by stakeholders is very much different. The present code design procedures can not bridge this gap. They operate with vague terms and imprecise definitions as "moderate actions", "some damage" or "less safe than" based on certain "engineering feeling" typically not understood by users. As such these procedures often cannot satisfy (a) the owner's desire for understanding cost and benefits of particular structural solution, (b) the society's needs for informed decision making and (c) the designer's desire for understanding of the procedure (Krawinkler 1997).

- Therefore there is an urgent need for new innovative methodologies for seismic design and assessment, capable of delivering facilities of predictable and quantified performance for multiple performance objectives. Such methodologies commonly referred as Performance Based /Seismic/ Engineering (PB/S/E) have been recently developed predominantly for seismic applications, but they have been subsequently extended to other engineering applications (i.e. fire design after Sept. 11 attacks).

- Several methodologies related to PBSE have been developed. This contribution addresses only three relatively simple procedures, which have already got application in practice and codes:
    - Performance assessment based on deterministic simplified inelastic analysis.
    - Cornell's approach for probabilistic seismic performance assessment (simplified probabilistic procedure based on total probability theorem) – basic PEER methodology
    - Enhanced PEER/ATC-58 methodology (enhanced Cornell's approach, which seems to form the basis of the new generation of seismic codes in the USA)
- While such new methodologies represent considerable step forward in the design, the well-known standard structural systems, for which enough empirical evidence is available, could still be analysed using standard procedures in the present codes. However, there is no alternative in the case of innovative systems or exceptional loads. By definition, there is no empirical evidence in such cases. Therefore the issue of innovative general design procedures forms an integral part within the topics addressed by the COST–12 action.

General definitions

- *Performance based engineering (PBE)*
  PBE is defined as integrated effort of design, construction and maintenance needed to produce engineered facilities of predictable performance for multiple performance objectives.

- *Performance based seismic engineering (PBSE)*
  PBSE is PBE related to seismic design situation.

- *Performance based seismic assessment (PBSA)*
  Assessment of the seismic performance of the already designed structure or existing structure.

- *Performance based seismic design (PBSD)*
  Design of the structure, driven by the set performance objectives from the very beginning of the design process.

- *Design performance objectives*
  A design performance objective is an expression of the desired performance level for the building for each earthquake design level. According to this definition, design performance objective couples expected or desired performance levels with levels of possible seismic hazards. This is illustrated for example by the Design Performance Objective Matrix (Figure 1), which has been proposed for buildings by the Vision 2000 Committee and is nowadays widely used by the earthquake engineering community.

- *Minimum and enhanced performance objectives*
  Individual owners and tenants may desire on a case-by-case basis to design their facilities for better performance or lower risk than the minimum objectives recommended in the design performance objective matrix. These design performance objectives which result in less damage at one or more earthquake design levels than the minimum recommended are termed *"Enhanced Objectives."*

- *Performance level*
  Performance level is an expression of the maximum desired extent of damage to a facility given that a specific earthquake design level affects it. From the entire

spectrum of all possible states four discrete states were chosen by Vision 2000 Committee and have been frequently used by researchers and engineers as well as in the codes: *Fully Operational or Serviceable* (facility continues in operation with negligible damage); *Operational or Functional* (facility continues in operation with minor damage and minor disruption in non-essential services); *Life Safety* (life safety is substantially protected, damage is moderate to extensive); and *Near Collapse or Impending Collapse* (life safety is at risk, damage is severe, and structural collapse is prevented).

However, although providing important development in earthquake engineering, these definitions have been sometimes found rather vague, unspecified, non-quantified and first of all difficult to understand by non-engineering community.

In recent developments (ATC-58, PEER, Hamburger, 2004) an effort has been made to replace them with *3D concept*, where 3D stands for (deaths, downtime and dollars). It is tried to made quantified probabilistic estimate of these three quantities easily understood by stakeholders given that a specific earthquake level is chosen.

Earthquake Performance Level

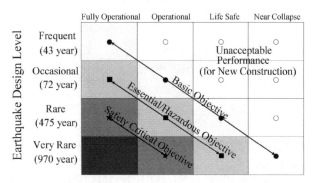

Figure 1. Seismic performance design objective matrix (Vision 2000).

Basis of the methods

- A. Simplified inelastic assessment procedures

  For the time being, the most rational analysis and performance evaluation methods for practical applications seem to be simplified non-linear procedures, which combine the non-linear static (pushover) analysis of a relatively simple mathematical model and the response spectrum approach. Although general concepts have been known for decades (i.e. Freeman et al. (1975), Fajfar & Fischinger (1989)), only in recent years, a breakthrough of these procedures has been observed. They have become widely used and they have been implemented into a number of modern guidelines and codes, including Eurocode 8 (2003).

  Most methods are formulated in the acceleration - displacement (AD) format. In this format, the capacity of a structure is directly compared with the demands of earthquake ground motion on the structure. The graphical presentation makes possible a visual interpretation of the procedure and of the relations between the basic quantities controlling the seismic response. The capacity of the structure is represented by a force - displacement curve, obtained by non-linear static (pushover) analysis. The base shear forces and roof displacements are converted to the spectral accelerations and spectral displacements of an equivalent single-degree-of-freedom (SDOF) system, respectively. These spectral values define the capacity diagram. The intersection of the capacity curve and the demand spectrum provides an estimate of the inelastic acceleration (strength) and displacement demand.

Typical representative of such procedures is the N2 method (Fajfar 2002), which has been incorporated into the latest version of Eurocode 8. "N" stands for Non-linear and "2" for two mathematical models – a SDOF and a MDOF model. Summary of the method is in Table 1.

Table 1. Summary of the N2 method

I. Data
   a) Structure (including moment rotation relationships)
   b) Elastic acceleration spectrum S_{ae}

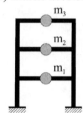 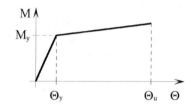 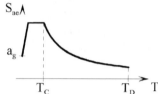

II. Demand spectra in ADRS format
   a) Determine elastic spectrum: $S_{de} = (T^2/4\pi^2)\, S_{ae}$
   b) Determine inelastic spectra for constant ductilities
      $S_a = S_{ae}/R_\mu,$ $S_d = (\mu/R_\mu)\, S_{de}$
      $R_\mu = (\mu - 1)\,(T/T_C) + 1$ $T < T_C$
      $R_\mu = \mu$ $T \geq T_C$

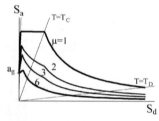

III. Pushover analysis
   a) Assume displacement shape $\{\Phi\}$
   b) Determine vertical distribution of lateral forces : $\{P\} = [M]\,\{\Phi\}$, $P_i = m_i\,\Phi_i$
   c) Determine base shear (V) – top displacement (D_t) relationships

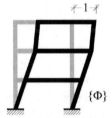 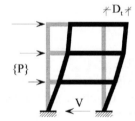 

IV. Equivalent SDOF model
   a) Transform MDOF quantities (Q) to SDOF quantities (Q^*)

$$Q^* = Q/\Gamma, \quad \Gamma = \left(\sum m_i\, \Phi_i\right) / \left(\sum m_i\, \Phi_i^2\right)$$

   b) Assume an approximate elasto-plastic force – displacement relationship
   c) Determine mass m^*, strength F_y^*, yield displacement D_y^*, and period T^*

 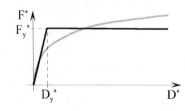

$$m^* = \sum m_i \Phi_i$$
$$k^* = F_y^* / D_y^*$$
$$T^* = 2\pi \sqrt{m^*/k^*}$$

   d) Determine capacity curve (acceleration versus displacement) $S_a = F^*/m^*$

V. Seismic demand for SDOF model
a) Determine reduction factor R_μ
b) Determine displacement demand $S_d = D^*$

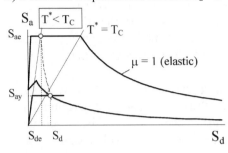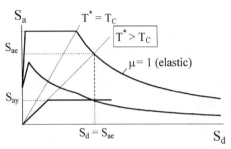

$$R_\mu = S_{ae}/S_{ay}$$

$$S_d \equiv D^* = S_{de}\,[1+(R_\mu - 1)\,(T_C/T^*)]\,/\,R_\mu \qquad\qquad S_d \equiv D^* = S_{de}$$

c) If necessary, modify displacement demand

VI. Global seismic demand for MDOF model
a) Transform SDOF displacement to the top displ. of the MDOF model: $D_t = \Gamma D^*$

VII. Local seismic demands (the results obtained in step III. can be used)
a) Perform pushover analysis of MDOF model up to the top displacement D_t
b) Determine local quantities (e.g. storey drifts, rotations Θ), corresponding to D_t

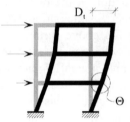

VIII. Performance evaluation (Damage analysis)
a) Compare local and global seismic demands with the capacities for the relevant performance level

- B. Cornell's approach for probabilistic seismic performance assessment (basic PEER methodology)

At Stanford University the foundation of an approach for probabilistic seismic performance assessment was developed (Cornell & Krawinkler 2000). It represents a generic structure for coordinating, combining and assessing the many considerations implicit in performance-based seismic assessment and design. Recent research in Pacific Earthquake Engineering Research (PEER) Center as well as in some other institutions, including University of Ljubljana, takes advantage of this formulation.

The proposed methodology in its generalized form presumes that the basis for assessing adequacy of the structure or its design will be a certain key Decision Variable, *DV*, such as the annual earthquake loss and/or the exceedance of one or more limit states (e.g., collapse). These can only be predicted probabilistically. Therefore the specific objectives of engineering assessment analyses are quantities such as λIcoll, the mean annual frequency (MAF, which is approximately equal to the annual probability for the small values of interest here) of collapse, or a similar quantity related to loss express in money terms.

Then the practical and natural analysis strategy involves the expansion and/or "disaggregation" of the MAF, *λ(DV)*, in terms of a structural Damage Measure, *DM*, and ground motion Intensity Measure, *IM*, which we can write symbolically as:

$$\lambda(DV) = \iint G(DV|DM)\,dG(DM\,|\,IM)\,d\lambda(IM) \qquad\qquad (1)$$

Here $G(DV|DM)$ is the probability that the decision variable exceeds specified values given (i.e., conditional on knowing) that the engineering Damage Measure (e.g., the maximum interstory drift) are equal to particular values. Further, $G(DM|IM)$ is the probability that the Damage Measure exceeds these values given that the Intensity Measure (such as spectral acceleration at the fundamental mode frequency or peak ground acceleration) equal particular values. Finally $\lambda(IM)$ is the MAF of the Intensity Measure.

Note that $\lambda(IM)$ is in fact just a seismic hazard curve, the determination of which is conventional Probabilistic Seismic Hazard Analysis (PSHA). Secondly, the estimation of $G(DM|IM)$ is the objective of linear and nonlinear dynamic analysis or seismic "demand analysis" under accelerograms with a given intensity level. Examples of $G(DV|DM)$ when the DV is a binary damage state indicator variable are various "fragility curves". In the case of the more recent HAZUS software it is the likelihood of a given discrete loss state.

The generic equation (1) is an example of the Total Probability Theorem of probability theory. It "de-constructs" the assessment problem into the three basic elements of hazard analysis, dynamic demand prediction, and failure or loss estimation, by introduction of the two "intermediate variables", DM and IM. Then it re-couples the elements via integration over all levels of the selected intermediate variables.

Using the above formulation, a methodology that focuses on specific performance levels (which may range from continuous operation to collapse prevention) and an acceptable annual probability of exceeding these levels (the "performance objective" approach advocated in present guidelines) can be formulated.

Typical result of this methodology is given in the form of so-called IDA (Incremental Dynamic Analysis) curves. This is a set of curves (based on non-linear dynamic analyses) which link maximum value of the engineering demand parameter – EDP (i.e. maximum story drift) versus intensity measure – IM (i.e. peak ground acceleration) of a set of chosen accelerograms. Example IDA curves (Figure 2) are taken from paper (Fischinger et. al., 2003), which is suggested as further reading.

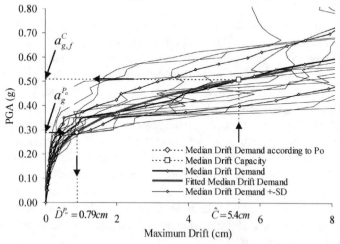

Figure 2. IDA curves representing maximum drift versus peak ground acceleration for different accelerograms, median drift demand and fitted median drift demand.

- C. Enhanced PEER (ATC-58) methodology

Although the basic PEER procedure actually initiated a totally new advanced concept in EE, several shortcomings have been recognized. Current procedures evaluate performance on the basis of local rather than global behaviour, do not adequately characterize the performance of non-structural components and systems, provide no guidance on how to proportion a structure, other than by iterative trial and error procedures, are of unknown reliability and are tied to performance levels that do not directly address the needs of the decision makers who must select the appropriate performance criteria for specific projects. In September 2001, the Federal Emergency Management Agency in the USA contracted with the Applied Technology Council to develop next-generation performance-based seismic design guidelines intended to address these shortcomings. The resulting 10-years ATC-58 project (primarily based on modified and extended PEER methodology) will culminate with the publication of next-generation performance-based seismic design guidelines for buildings as well as companion publications intended to assist decision makers in using and taking advantage of performance-based approaches. Though primarily intended to address seismic design, substantial efforts are being made to ensure that the guidelines are developed compatibly with parallel efforts to develop performance-based design criteria for resistance to other extreme loads including fire and blast.

The project is still in the initial phase and the reader is suggested to refer to papers by Moehle & Deierlein (2004) and Hamburger (2004) for more details. Nevertheless, two important changes in the methodology (regarding performance objectives and generic formula of the procedure) will be additionally addressed here.

It is the goal of the ATC-58 project is to utilize performance objectives that are quantifiable and predictable, as well as meaningful and useful for the decision makers who must select or approve the objectives used as a basis for design. Decision makers are a disparate group that includes building developers, facility managers, risk managers, lenders, insurers, public agencies, regulators and individual members of the public. Each of these decision makers view seismic performance from a different perspective and select desired performance using different decision making processes. Regardless of the specific process used, selection of appropriate performance will involve development of an understanding of the risk associated with a given choice and the resources that must be invested to reduce this risk beyond certain thresholds. To facilitate this process, in the ATC-58 project performance objectives are expressed as the risk of incurring three specific kinds of earthquake induced loss. These include direct financial losses (dollars) associated with the cost of repairing or replacing damaged facilities, earthquake-induced life losses (deaths and serious injuries) and lost use of facilities (downtime) while they are being repaired, replaced or otherwise restored to service. Different decision makers characterize these risks and determine acceptable levels of risk in different ways. Therefore the next-generation guidelines permit alternative methods of stating performance objectives including: expected losses, given the occurrence of a specific earthquake scenario, annualized losses, or the expected loss over a given period of years, each expressed together with a level of confidence associated with the estimate. Examples include Probable Maximum Losses (estimates of the 90% confidence level loss given a specific event), average annual loss (the mean loss per year averaged over many years) or the 500-year loss (that loss which has a 10% chance of being exceeded in 50 years). Many other similar means of expression will be accommodated (Hamburger, 2004). This approach replacing »standard« performance objectives (i.e. IO – imediate occupancy; LS – life save and CP – collapse prevention) is illustrated in Figure 3.

Another important change has been applied to generic formula in basic PEER procedure (1), where the damage measure (DM) – intensity measure (IM) relation has been further decomposed into engineering demand parameter (EDP) – intensity measure (IM) relation and damage measure (DM) – engineering demand parameter (EDP) relation, where EDP is defined as physical seismic response quantity (i.e. story drift, dissipated energy, …)

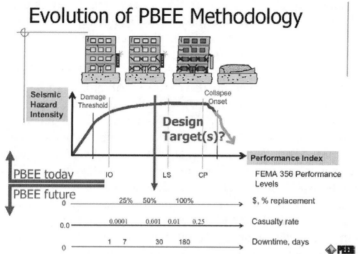

Figure 3. A visualization of performance-based earthquake engineering (after Moehle & Dieirlein, 2004).

Examples of application

- Due to space limitations only references to relevant, most characteristic publications are given in this section:
  (Fajfar, 2000) related to "Simplified inelastic assessment procedures".
  (Fischinger et. al., 2002), (Fajfar & Krawinkler eds., 1997), (Fajfar & Krawinkler eds., 2004) related to "Cornell's approach for probabilistic seismic performance assessment (basic PEER methodology)".
  (Fajfar & Krawinkler eds.,2004) related to "Enhanced PEER - ATC-58 methodology".

References

- Cornell C. A. & Krawinkler H. 2000. Progress and challenges in seismic performance assessment. *PEER Center News* (4)1:1-3.
- Eurocode 8 (2003): Design of structures for earthquake resistance, Part 1: General rules, seismic actions and rules for buildings, prEN 1998-1, *European Committee for Standardization.*
- Fajfar P. & Fischinger M. 1989. A method for non-linear seismic analysis of regular buildings. *Proceedings of the 9th WCEE*, Tokyo/Kyoto, Maruzen, Tokyo, Vol. V, 111-116.
- Fajfar P. & Krawinkler H. (editors) 1997. Seismic design methodologies for the next generation of codes. *Proceedings of the International workshop on seismic design methodologies for the next generation of codes, Bled/Slovenia, 24-27 June 1997,* Balkema, Rotterdam.
- Fajfar P. 2000. A nonlinear analysis method for performance-based seismic design.

Earthquake Spectra 16(3), 573-92.

- Fajfar P. & Otani S. (editors) 2002. Proceedings of Slovenia – Japan workshops on performance based seismic design methodologies, *Ljubljana/Slovenia, 2-3 October 2000 and 1-2 October 2001*.

- Fajfar P. & Krawinkler H. (editors) 2004. Proceedings of the International workshop on performance based seismic design – concepts and implementation, *Bled/Slovenia, 28 June – 1 July 2004, PEER publication*, in print.

- Fischinger M., Fajfar P., Dolsek M., Zamfiresku D., Stratan A., Leino T., Vacareanu R., Cornea T., Lungu D., Wolinski S., Alarcon E., 2003. General methodologies for evaluating the structural performance under exceptional loadings. *In the proceedings of the international seminar Improvement of buildings' structural quality by new technologies, Lisbon/Portugal 19-20 April 2002*, Luis Simones da Silva & J. F. G. Mendes (editors), *COST C12 publication*, EUR 20728.

- Freeman S. A., Nicolleti J. P. & Tyrell J. V. 1975. Evaluations of existing buildings for seismic risk – A case study of Pudget Sound Naval Shipyard, Bremerton, Washington. *Proceedings of the 1st USNCEE*, EERI, Berkeley, CA, 113-122.

- Hamburger R., 2004. Development of next generation performance-based seismic design guidelines for buildings. In P. Fajfar & H. Krawinkler (editors), *International workshop on performance based seismic design – concepts and implementation, Bled/Slovenia, 28 June – 1 July 2004, PEER publication*, in print.

- Kabeyasawa, T. & Moehle J. P. (editors), 2000, 2001, 2002, 2003, 2004. Proceedings of the U.S. – Japan workshops on performance-based earthquake methodology for reinforced concrete building structures. *1st workshop, 1-3 September 1999, Maui, Hawai; 2nd workshop, 11-13 September 2000, Sapporo; 3rd workshop, 16-18 August 2001, Seattle; 4th workshop, 22-24 October 2002, Toba, Mie; 5th workshop, 9-11 September 2003, Hakone*.

- Kabeyasawa T., 2004. Development of performance-based design methodologies. Research report, *US-Japan Cooperative Research in Earthquake Disaster Mitigation*, Theme 2-1.

- Krawinkler, H., 1997. Research issues in performance based seismic engineering. In P. Fajfar & H. Krawinkler (editors), *Seismic design methodologies for the next generation of codes, Bled/Slovenia 1997*, Balkema, Rotterdam, pp. 47-58.

- Moehle J. & Deierlein G. G:, 2004. A framework methodology for performance-based earthquake engineering. In P. Fajfar & H. Krawinkler (editors), *International workshop on performance based seismic design – concepts and implementation, Bled/Slovenia, 28 June – 1 July 2004, PEER publication*, in print.

Contacts

- Matej Fischinger
  - Address: Faculty of Civil and Geodetic Engineering, University of Ljubljana, Jamova 2, Ljubljana Slovenia
  - E-mail: Matej.Fischinger@ikpir.fgg.uni-lj.si
  - Phone: +386 1 4768593
  - Fax: +386 1 4250693

The Web address http://bled2004.ikpir.com/participants.htm contains contacts for 49 participants of the International workshop on performance based seismic design – concepts and implementation, Bled/Slovenia, 28 June – 1 July 2004.

GENERAL METHODOLOGY FOR PROBABILISTIC ANALYSIS

Description

- Because of uncertainties in the building process absolute safe structures cannot be achieved and consequently building structures must be designed and verified to serve their function with a finite probability of failure and serviceability.
- Probabilistic methods can be used for structural design and assessment when the set of random events and variables can be identified.
- Three levels of probabilistic methods can be distinguished:
  - semi-probabilistic (Level I; Partial Factors Method),
  - simplified probabilistic (Level II; First Order Reliability Method FORM and Second Order Reliability Method SORM),
  - full probabilistic (Level III).
- In both the simplified probabilistic and full probabilistic methods the measure of reliability is identified with the survival probability $P_s = 1 - P_{fd}$, where P_{fd} is the target failure probability which depends on various design situations, failure modes and reference periods.
- In the simplified probabilistic analysis, an alternative reliability measure is usually defined by the reliability index β which is related to P_f by: $P_f = \Phi(-\beta)$, where Φ is the cumulative distribution function (CDF) of the standardised Gaussian distribution.
- The exact analytical methods for the full probabilistic analysis are too sophisticated and difficult for effective practical applications, and the direct numerical integration methods are realistic when the failure function includes less than four or five random variables and the area of integration is known in advance.
- Application of the probabilistic simulation methods enable to obtain approximate solutions and make possible direct insight into structural reliability, basic variables and decisive parameters.
- Nevertheless, the simulation-based methods for probabilistic analysis should not be considered as a replacement to analytical methods, but as a rational and efficient alternative to current procedures employed by civil engineers.

Basis of the method

- Three conventions of probability-based design can be distinguished (ISO 2394, 1998, prEN 1990, 2002, Murzewski, 1998):
  - target probability specified for a realiability class of failure should be equal for all structural elements: P_{fd} = const.,
  - target probabilities should be proportional to the sensitivity factors related to action effects and resistance,
  - probability of failure for each structural element should be minimal: P_{fd} = min.
- The probability of structural failure in the case of time-invariant problems can be calculated

from: $P_f = \int_{FA} f_{\vec{x}}(\vec{x})d\vec{x} \leq P_{fd}$, where FA is the failure domain, $f_{\vec{x}}(\vec{x})$ is the joint probability density function of random variables $\vec{X}$.

- Calculation of failure probabilies in cases of time-variant problems (namely in cases of overload failure and cumulative failures) when time-dependent quantities need to be represented by stochastic processes, is concerned with transformation into time-invarinat problems or with simulation methods.
- The analytical method may be refined by expressing all random variables of arbitrarily continuous probability distributions in Gaussian space of random variables and using appropriate linear approximation procedures when the surface of failure is nonlinear (Hohenbichler and Rakwitz, 1981, Ranganathan, 1990).
- The direct numerical integration methods are realistic when the failure function includes less than four or five random variables and the failure domain is known in advance.
- Numerous models of structural behaviour are formulated in an implicit way, so that the failure domain (area of integration) is not known in advance and these cases there is possibility to check it by means of random simulation methods.
- Simulation methods can be divided into (ISO 2394, 1998):
  - direct Monte Carlo simulation with sampling density taken as the original probability density,
  - importance sampling where the Monte Carlo techique is applied with a fictitious density function close to the design point,
  - adaptive sampling in which importance sampling is applied with successive updating of the density function,
  - directional simulation and axis orthogonal simulation.

Application to structural analysis

- Probabilistic methods can be applied to the design of new structures, the evaluation of existing ones and for calibration of reliability or safety measures of semi-probabilistic, empirical and assisted by testing methods.
- Several main steps of probabilistic design or analysis of any structure may be distinguished:
  - describe a structure,
  - identify the relevant basic variables for each specific limit state, i.e. the variables which characterize actions and environmental influences, properties of materials and soils, geometrical properties (prEN 1990, 2002, JCSS Model Code, 2001),
  - select the target probability of failure and decommissioning of a structure (prEN 1990, 2002, ISO 2394, 1998),
  - select a method of structural design or evaluation,
  - perform the standardized design or assessment calculations for selected loads and resistance models (e.g. using a random simulation method),
  - compare results with assumptions.
- Most building structures consist of a system of interconnected components and members and therefore, it is important to distinguish between the failure probability of each component and the failure probability of the entire system (Nowak and Collins, 2000).
- Application of the direct Monte Carlo simulation method can be used when structural dimensioning, evaluation of resistance or load effect with assumed in advance probability of failure is considered (Figures 1 and 2).

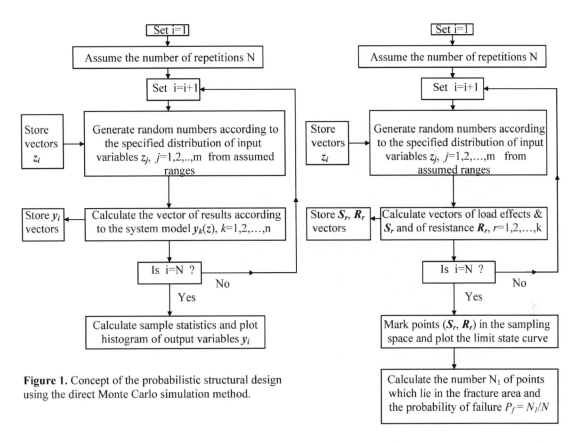

Figure 1. Concept of the probabilistic structural design using the direct Monte Carlo simulation method.

Figure 2. Concept of the probabilistic evaluation using the direct Monte Carlo simulation method.

Example of application

▪ The failure probability of the simply supported braced steel element shown in Figure 3 exposed to the axial load N and the bending load P is calculated using direct Monte Carlo simulation.

▪ The second order theory and elastic range of the response to the loading are considered, and the performance function is given as follows:

$$g = R - S = f_y - \{1.4N/A + [(N/2k)(\sin(2.5k)/\cos(2.5k) + PL/4]/W\}, \quad k = \sqrt{N/EI}.$$

▪ Random variables considered in the analysis, their probability distributions and statistical parameters are defined as: the steel yield stress $f_y \Rightarrow$ LN (486, 29.2) MPa, the modulus of elasticity $E \Rightarrow$ N (200, 13.4) GPa, the axial load $N \Rightarrow$ N (280, 28) kN, the bending load $P \Rightarrow$ Gamma (150, 25) kN, length of the element $L = 2.50$ m (deterministic), cross-sectional characteristics of the element (hot rolled I section): area $A_s \Rightarrow$ N (149, 4.9) cm^2, elastic section modulus $W \Rightarrow$ N (1680, 47.5) cm^3, moment of inertia $I \Rightarrow$ N (25170, 712) cm^4.

for direct Monte Carlo simulation (Marek and Gustar, 1999), and the total number of 500000 simulation steps were applied.

• The output of AntHill program is given in Figure 3, and the failure probability of the considered steel element is $P_f = 1.6 \times 10^{-5}$.

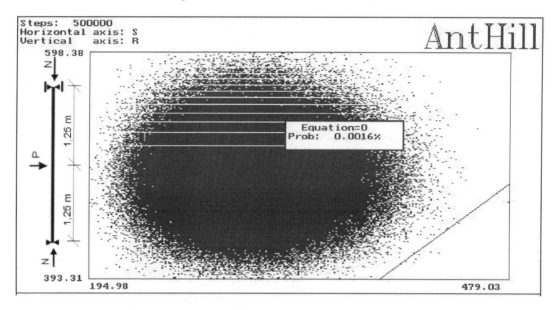

Figure 3. Steel element and its loading. AntHill program output

References

• ISO 2394 : 1998 (E), International Standard. *General principles on reliability for structures*, ISO, Geneve, Switzerland.
• EN 1990 : 2002 (E). *Eurocode – Basis of structural design*, CEN, Brussels, Belgium.
• Murzewski J., 1998. *A more fundamental reassessment of probabilistic design method*, Proceedings of Eight IFIP WG7.5 Working Conference on Reliability and Optimization of Structural Systems. The University of Michigan, Ann Arbor, Michigan, USA, pp. 39-50.
• Hohenbichler M., Rackwitz R., 1981. *Non-normal dependent vectors in structural reliability*, Journal Eng. Division ASCE, 107, pp. 1127-1238.
• Ranganathan R., 1990. *Reliability analysis and design of structures,* McGraw – Hill, New York.
• Joint Committee on Structural Safety, 2001. *Probabilistic Model Code*. JCSS Internet Site: http://www.iabse.ethz.ch/lc/jcss.html.
• Nowak A.S, Collins K.R., 2000. *Reliability of Structures,* McGraw – Hill, Int. Edition.
• Marek P., Gustar M., 1999, *Monte Carlo simulation programs for PC. AntHill, M-Star, Res-Com, DamAc, LoadCom,* ARTech Prague, Czech Republic.

Contacts

• Assoc. Prof. Szczepan Wolinski
Address: Faculty of Civil and Environmental Engineering,
Rzeszow University of Technology,
Ul. Poznanska 2,   35-040 Rzeszow, Poland
e-mail: szwolkkb@prz.rzeszow.pl

COST C12 - WG2

Datasheet I.1.2.2

Prepared by: A. Kudzys

Institute of Architecture and Construction of Kaunas University of
Technology Kaunas, Lithuania

AUTOSYSTEMS IN PROBABILITY-BASED DURABILITY PREDICTION

Definition

- Durability prediction analysis for load-carrying structures must be considered as a prediction of the technical service life of members as autosysems representing multicriteria failure mode and characterizing by stochastically dependent conditional elements in series, parallel or mixed connections.

- Technical service life as main durability parameter of autosystems is the random period of time at which they can actually perform according to the requirements based on intended purpose without major repairs of members at a preset target reliability index P_{min}.

- Since technical service life of irreparable autosystems is less or greater than this parameter of member components the reliability index P_{min} must be associated not only with the reliability classes but also with reliability peculiarities of autosystems.

- Homogeneous in nature but different in structural reliability properties autosystems need to different approaches in design and residual technical service life predictions.

Basis of the methodology

- A wide range of applied reliability issues of deteriorating material resistances due to steel corrosion, concrete degradation or wood putrefaction caused by random aggressive environmental conditions or other adverse circumstances can be neither formulated nor solved by deterministic analysis method [CEB Bulletin 238, 1997, Leira et al., 1999].

- Long-term quality of deteriorating load-carrying structures of buildings, construction and engineering works may be closely defined only by probabilistic-based durability prediction parameters using the available probabilistic methodology, life cycle assessment concepts and methods [Kudzys, 1992; Internation Standard ISO 2394, 1998; Melchers, 1999].

- Structural durability prediction analysis is based on consideration of material mechanical properties as multi-axial stress behaviours, action duration effects and long-term probabilistic responses of members to mechanical, physical and chemical influences [CEB Bulletin 238, 1997; Vrowenwelder, 2002; COST Action C12, 2003].

- Joint Committee on Structural Safety, 2001, presents probability-based durability formulation taking into account stationary and non-stationary processes of material deteriorations, external actions and loads dedicating failure probability calculations.

- It is expedient to divide life cycles of deteriorating structural members into initiation and propagation periods of mechanical, physical or chemical degradations [Kudzys, 1992; Joint Committee on Structural Safety, 2001].

Application concepts and methods

- To durability prediction problems belong an analysis on durability not only of deteriorating structural members but also not-deteriorating ones for which a failure probability is time-dependent random value due to random fluctuations of gravity and lateral (horizontal) extreme actions.

- Contrary to traditional structural safety analysis, it is expedient to consider durability prediction issues not for cross-sections, connections and other components of load-carrying structures but for whole repairable or replaceable members representing by autosystems of components.

- Predicted durability parameters of load-carrying structures are in close relation to probabilistic reliability indices which minimum values P_{min} must be calibrated taking into account not only member working classes [European Committee for Standardization, 2002] but also time-dependent both mechanical properties and economic optimization factors of autosystems.

- Necessity to base a prediction of the technical service life of structural members on autosystem models is illustrated in Figure 1, in which t_1, t_{sa}, t_{pa} and t_{ma} are the values of the technical service life of single elements, series, parallel and mixed autosystems, respectively.

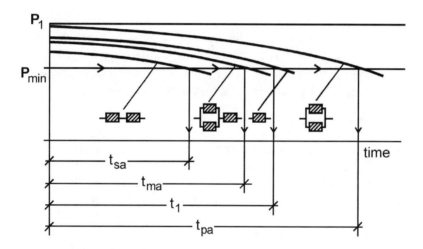

Figure 1. *Effect of autosystem type on the technical service life of structural members*

- Technical service life values of autosystems exposed to permanent and variable actions may be calculated by Monte Carlo simulation and numerical integration methods using the concepts of the analytical method of transformed conditional probability (TCPM) and taking into account peculiarities of initial and propagation deterioration periods (Figure 2).

- Technical service life values t_l and t_r of irreparable members may be calculated from failure criteria conditions, respectively, $\mathbf{P}(T \geq t_l) = \mathbf{P}_{min}$ and $\mathbf{P}(T \geq t_r) = \mathbf{P}_{min}$, where T is the lifetime of autosystems (as random value).

- Residual service life values of existing structures must be revised taking into account service-proven actions and truncated resistance and lifetime distributions autosystems.

- Autosystem design methodology admit to make more exactly the technical service life of members and their rational design working life for which a member is to be used for its intended purpose with balancing structural reliability.

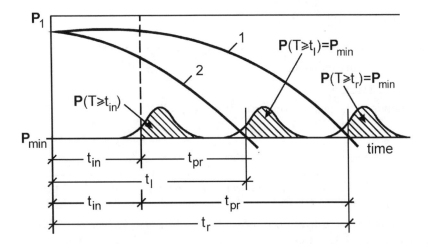

Figure 2. *Reliability functions versus the time for autosystems subjected to permanent (1) and variable actions: t_{in} and t_{pr} are the initiation and propagation periods of degradation process*

Realization of the application

- A reinforced concrete beam (Figure 3) is subjected to random permanent *g* and reiterated peak variable *q* actions.

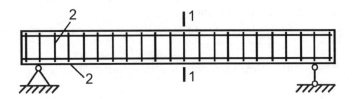

Figure 3. *Normal (1) and oblique design sections of the reinforced concrete beam*

- Failure criteria for normal (1) and oblique (2) sections

$$Z_1(t) = g_1(\boldsymbol{\theta}_1, \mathbf{X}(t)) = 0$$
$$Z_2(t) = g_2(\boldsymbol{\theta}_2, \mathbf{X}(t)) = 0$$

where θ is the vector of additional random variables; $\mathbf{X}(t)$ is the process of basic ones.

- Reliability margin function analysis

$$Z_1(t) = \theta_{R1} R_1(t) - \theta_g M_g - \theta_q M_q(t)$$
$$Z_2(t) = \theta_{R2} R_2(t) - \theta_g V_g - \theta_q V_q(t)$$

where $R_1(t)$ and $R_2(t)$ are section resistances; M_g, M_q and V_g, V_q are bending moments and transverse forces.

- Series autosystem reliability index analysis

$$\mathbf{P}(T \geq t_r) = \mathbf{P}(Z_1(t) > 0) \cdot \mathbf{P}(Z_2(t) > 0 | Z_1(t) > 0) =$$
$$= \mathbf{P}(Z_1(t) > 0) \cdot \mathbf{P}(Z_2(t) > 0)\{1 + \rho_{12}[1/\mathbf{P}(t) - 1]\}$$

where ρ_{12} is the coefficient of correlation of reliability margin functions; $\mathbf{P}(t)$ is the greater value of the reliability indices.

- Technical service life prediction

The value t_r corresponds to the design condition $\mathbf{P}(T \geq t_r) = \mathbf{P}_{min}$.

References

- CEB Bulletin 238, 1997. *New approach to durability design*. Printed by Sprint-Druck, Stuttgart, Deutschland.

- European Committee for Standardization (CEN), 2002. EN 1990, *Eurocode – Basis of structural design*, Brussels, Belgium.

- COST Action C12 (2003). *Improvement of buildings structural quality by new technologies*. Proceedings of the international seminar Lisabon 2002, EUR 20728, Luxembourg.

- International Standard ISO 2394:1998 E, 1998. *General principles on reliability for structures*, Switzerland.

- Kudzys, A., 1992. *Probability Estimation of Reliability and Durability of Reinforced Concrete Structures*, Vilnius technical university, Vilnius, Lithuania.

- Leira, B., Lindgräd, A., Nesje, A., Sund, E., Saegrov, S., 1999. Degradation analysis by statistical methods. *Durability of Building Materials and Components 8*, Edited by Lacasse, M.A. and Vanier, D.J., Ottawa, Canada.

- Melchers, R.E., 1999. Structural Reliability Analysis And Prediction. John Wiley and Sons, England.

- Vrouwenvelder, A.C.W.M., 2002. Developments towards full probabilistic design codes. *Structural Safety*, N° 24.

Contacts

- Prof., Habil. Dr. Antanas Kudzys
  Address:
  > Institute of Architecture and Construction of
  > Kaunas University of Technology.
  > Tunelio 60, 44405 Kaunas, Lithuania.
  > E-mail: asi@asi.lt, Fax: (+370 5) 2340123

COST C12 - WG2

Datasheet I.1.2.3

Prepared by: B. Snarskis[1], A. Galvonaité[2], R. Šimkus[1], V.Doveika[1]

[1] Institute of Architecture and Construction of the Kaunas University of Technology, Kaunas, Lithuania

[2] Lithuanian Agency of Hydrology and Meteorology, Vilnius, Lithuania

SOME METHODOLOGICAL PROBLEMS OF NORMATIVE REGULATIONS CONCERNING THE LEVELS OF WIND AND SNOW LOADS
(exemplified by analysis of Lithuanian data and normatives)

Description

- The directions of Eurocodes governing the choice of characteristic and design values of wind and snow loads (especially, directions given in [1]) constitute one of the foundations for structural design and, first of all, for elaboration of various special-purpose normatives. Nevertheless, there are serious methodological problems of interpretation and implementation of these directions formulated somewhat abstractedly and aimed mainly at rather specific conditions of application.

- This datasheet presents a brief overview of some such problems based on analysis done by the authors when examining the normatives recently adopted in Lithuania (the so-called 'Technical Regulations of Construction' [Lith. *Statybos techniniai reglamentai - STR*], including *STR 2.05.04:2003* [2],which deals with snow and wind loads) and analysing the pertinent climatological data. Our main objective here is, however, not a critical revision of these local regulations (although such a revision is actually a necessity) but <u>general methodological remarks</u> following from our analysis and from some unqestionable theoretical considerations. (Nevertheless, some particular <u>quantitative results relating to the local Lithuanian regulations</u> and based on our calculations[1] will be presented in the closing fragments of this datasheet as an illustration of practical meaning of the considerations put forward here.)

- It is hard to formulate immediately all conclusions to be presented here – and the presentation of them will take allmost all of the remaining part of this paper – but in a very concise exposition our main assertions may be outlined as follows.

- Firstly, we emphasize the necessity of a careful taking into account the proper measure of uncertainty associated with the random parameter that is characterized (e.g., the coefficient of variation of the snow or wind load – or any other similar statistical indicator); and, particularly, the necessity of taking into account the «spatial» («territorial») component of variance alongside the more obvious anlysis of the «local temporal» variance is stressed.

- Second, we call attention to the possible paramount importance of wind gusts for the total wind loads – a circumstance occurring at least in some quite not unusual cases but being, in all likelihood, underrated rather frequently.

- These statements, however, stay in need of more comprehensive explanation.

---

[1] A corresponding study is now conducted in the Institute of architecture and construction of KTU.

The data used in this study may be defined briefly as gathered from 19 meteorological stations in 1970-2002. (Minor adjustments of this statement would be appropriate depending on particular tasks of analysis and some gaps in observations; however, such accuracy is not strongly needed here.) Supplementing of data with older information is envisaged and realized gradually – but it takes long because of difficulties of handling the old archive materials. So this study is not completed yet; nevertheless some primary results seem to be worth of presenting them already now.

Considerations relating to distinction between various components of uncertainty in statistical models of atmospherical actions

- The Eurocodes describe the characteristic values of wind and snow loads as the quantiles of annual maxima of corresponding actions with probability 0.02 of excess. Nearly the same requirement, in the sense of its intended practical meaning, is presented in *STR 2.05.04* adopted now in Lithuania (except for somewhat different formal definition and for some slight misuderstanding in minor details that are not worthy of treatment here). It might appear at first sight as a very clear-cut definition but in actuality it is not free of problems, mainly in probabilistic interpretation of 'corresponding actions' mentioned above.

- First (but not foremost), problems arise not infrequently when the influence of specific physical circumstances at the given site (such as configuration of the building, possibility of thawing of the snow on the roof caused by heat transfer from within the building, etc.) must be measured quantitatively. However, it seems likely that we may omit such circumstances here since auxiliary formulae appearing in [1] and other normative literature allow mostly quite agreeable account of these circumstances – perhaps with only not dangerous distortions in the picture of uncertainties and random deviations that is to be considered when reliability requirements are chosen. A question of prominent importance, however, remains latent and calling for a more attention in this statistical scheme – namely, a question about a <u>correct probabilistic and statistical definition of the random variable mentioned as '*annual maximum of the load*'</u>. In other words, the question is: what is the amount of probabilistic uncertainty of such a variable (and, respectively, of its statistical variance as reflected in observations) that must be taken into account when analysing the distribution of such variable and estimating its quantiles?

- There are various interesting aspects of this question – but of special note are problems connected with uncertainties in the probabilistic model of the variable under consideration (maximum of the load) resulting not only from variability of actions <u>at a strictly defined location</u> (for which a more or less representative sample of data is presumably available) but also from the fact that the structure being considered is to function <u>in a location which we are not able to characterize so definitely</u> [2]. And we put forth here the assertion that in typical cases the variance of the analysed variable (here – the random level of an annual maximum of the load) must be viewed as including at least <u>two components of variance</u>:
    - (1) the *temporal* component – i.e., the component of uncertainties *related with random properties of a chosen time interval* (usually of a chosen <u>annual</u> period and/or a sequence of such periods);
    - (2) the *spatial* component – i.e., the component of uncertainties *related with random properties of a site* (and/or a subset of sites chosen from the general set of sites taken into account).

Practically, in analysis of data samples, the first component may be observed in comparison of periods surveyed, provided that in any of such confrontations the place of observations is uniformly fixed; and the second may be observed in comparison of all sites treated by official regulations as belonging <u>to the same region</u> – provided that

---

[2] In many cases these uncertainties spring up simply because of scantiness of climatological information related to the individual site, – but sometimes even because the localization of the site itself is not determined yet (as it happens when a design is used repeatedly in numerous places).

in any of such confrontations the time of observations is uniformly fixed.

- In reality, neglection of the «spatial» uncertainties in practical statistical investigations of atmospheric loads may be observed in numerous publications where probability distribution of the load is straightforwardly identified with some empirical distribution obtained at the nearest meteorological station – so that special possibilities of the site targeted in such practical application are totally ignored. Nevertheless this simple-minded assumption is reasonable quite rarely – maybe only if a representative sample of data has been obtained in a rather close proximity to the site for which the design values are to be determined. It should be realized here that substantial differences of load data (comparable even with the scale of the load variance throughout all the region for which the normative load levels are defined uiformly) can be observed sometimes even under conditions when the comparison refers to a pair of localities separated by a distance of several kilometres in all. Therefore we may not rely upon such simplifications.

Considerations relating to the treatment of data describing the wind velocities

- We must also notice the problematical character of the fact that in Eurocodes (and as well in the *STR*), when the observational data are treated, the characteristic and design values of wind loads are related mainly to measurements of "moving" averages of the wind velocity – with averaging on time intervals as long as 10 minutes. The action of more intensive momentary gusts is to some extent taken into account, but the inclusion of parameters referring to gust velocities is only indirect and so organized that the long-term effects are usually prevailing in the end results (at least when the most compli-cated methods of dynamical calculations are not applied and simplified normative recommendations are followed). Such construction of norms is likely to be sound when rather tall and relatively not very stiff buildings with rather prolonged oscillatory motions are kept in mind – but for relatively simple and stiff structures of low buildings this approach is somewhat misleading since even single gust can make a fatal impact.
- The analysis of wind observations conducted by us till now is based mainly on data representing maximal velocities in gusts. Therefore it is difficult at this point to compare our conclusions with the recommendations of Eurocodes or the STR – though such a comparison is not meaningless however. (By the way, analysis of data for "moving averages" of wind velocities is being performed by us currently.) The problems mentioned here (and, moreover, some problems associated with a possibility of certain provisional modifications of initial data to account for particular effects of local environment at meteorological stations) are the main causes of the fact that in this publication most attention is concetrated on snow (not wind) loads. Nevertheless the results of statistical analysis of wind velocities obtained so far and briefly presented below seem to indicate definitely that the gust velocities are significantly more dangerous than it is believed to be on the grounds of norms adopted now.

Some remarks anticipating the review of statistical results (and in our opinion deserving notice from the standpoint of other similar investigations)

- The properties of empirical distributions of sample data are such that approximative description of these distributions is, as a rule, quite good when the lognormal model is

31

taken as the «theoretical» model [3].

- Having adopted this scheme of approximative description, let us recollect briefly some important properties of lognormal variables and characteristics of their distributions. (We hope that these remarks, though not new in their origins, can be interesting for authors of other similar investigations. Furthermore, they are necessary for the following presentation of our statistical illustrations.)

- Firstly, if a numerical random variable Z has lognormal distribution, then its logarithm $L := \log_a Z$ is distributed normally – and it is easy to understand that the choice of the base of logarithms (a) results here only in scale transformation of the variable L . It is, moreover, also evident that in efforts to analyse such variable Z and to give its mathematical description it may be advantageous to use such variable L as a convenient auxiliary tool since the properties of normal distributions are very helpful and well explored. To be more specific, in doing so there is good reason to define logarithms mentioned just above as <u>natural</u> logarithms $\ln Z$ (with the base of logarithms equal to $e=2.71...$) [4]. Hereafter we shall keep in mind just this way of using the auxiliary variable (*logarithmic image*) L associated with a given supposedly lognormal Z (that is to be described) by logarithmic transformation (and being supposedly normal).

- It is needless to say that for description of lognormal random variable Z all usual parameters of its distribution such as mean value μ_Z , median η_Z , mean square deviation [standard deviation] σ_Z , coefficient of variation υ_Z , etc. (or, perhaps, sample estimates of these parameters for a sample $\mathbf{Z}=(z_1, z_2, ..., z_n)$ presenting the observed values of Z – if empirical distributions and statistical estimations are to be considered explicitly rather than a probabilistic model adopted as a result) may be employed in a quite traditional way. Nevertheless there are grounds for use of some other parameters more specific to lognormal model – in particular, some parameters of the normal image L (μ_L, η_L , σ_L ,...) similar to the parameters of Z mentioned above (or some sample characteristics of L from the sample $\mathbf{L}=(\ln z_1, \ln z_2, ..., \ln z_n)$). Of special interest here are two such parameters:
  (1) the *mean* μ_L of the distribution of L (or an estimate of such a mean if statistical estimations are presented explicitly), value of which obviously corresponds to the median η_Z of the original variable ($\eta_Z = \exp \mu_L$) if the assumption of lognormality is really applicable to Z;
  (2) the *mean square deviation* σ_L of the distribution of L (or an estimate of such a parameter).

The interpretation of these parameters is quite traditional and does not require any explanation; but it might be well to point out the practical consequences following from such treatment of the model (when the relation between Z and its normal image L is put to use). One of these consequences is that we arrive at using the <u>median</u> of Z (or, if we are adhered to strictly statistical point of view, – the estimate of this median) in a role closely resembling the usual role of mean values or sample averages

---

[3] We do not advocate thereupon that this model has been supposedly established as undoubtedly superior above other possible models. We state only that the approximation works well at least to the level of quantiles covered here – without manifestation of any evident systematic deviations. In addition, this model is very handy by its mathematical features.

[4] Among other motives, it is pertinent to consider here the fact that such description of lognormal variable leads to a situation where the «logarithmical image» L of the original lognormal variable Z is distributed so that mean square deviation σ_L of L is asymptotically equivalent (as its value decreases) to the coefficient of variation υ_L of the original variable Z. This is doubtlessly a great convenience for an easy and clear understanding of the values assigned to the parameters thought of here.

(although the difference between the two is worthy of consideration and generally is not negligeable [5]). The second consequence is the wide use of the *mean square deviation of the logarithmic image* [of the variable under consideration, – i.e., use of σ_L for the given Z] – use of a characteristic that is, as it has been indicated before, similar to υ_Z, although not identical to it.

- Bearing in mind the importance and frequent use of σ_L we shall abbreviate its denomination ('*mean square deviation of the logarithmic image*') with brief notation '*m.sq.d.l.i.*'.

Some preliminary results referring to the Lithuanian statistical data as an example of the approach presented above

- Let us begin, firstly, with the data on <u>snow loads</u>.

- When the random variable being analysed is defined as a <u>maximum for a winter season amount of water per unit surface</u> (observed at some site which is, however, not fixed *a priori* and is treated so far as a variable of the model) – let us denote it traditionally as s and express its values in kg/m^2 – and the results are concurrently surveyed for all the analysed sites <u>without regional divisions of the country</u> (although grouping of data by individual sites and confrontation of results from various sites is of course a requisite perpetually), the following results are obtained.

- The most typical estimates of the <u>median</u> value η_s of s are, in a review of data for fixed sites or for single winters, about 45…50 kg/m^2 – however, with substantial variation from one site to another (from approximately 33 to 73 kg/m^2) and in comparing of various winters (from very low values as some winters are almost snowless, and to about 122 kg/m^2 in 1995/96). The estimate of median for the joined sample of all observations is about 46.0…46.3 kg/m^2. (Slightly different results can be obtained depending on one of possible methods of treating data.)

- Probably yet more interesting are the estimates of the <u>m.sq.d.l.i.</u> $\sigma_{\ln s}$ of s – the measures of its variance as observed on variuous grouped samples. These are about 0.56 for the «local» samples (taken at fixed sites and representing the <u>«serial»</u> variation of the variable) – and about 0.35 for the «territorial» samples (gathered in a fixed time period, usually a winter, and representing the <u>«spatial»</u> component of variance). Thus the «local serial» component of variance is distinctly prevailing over the «spatial» («territorial») component; however, both are great (greater than almost all analogous characteristics of variance defined for other factors of reliability essential in stuctural design) so that <u>no one of them</u> may be neglected.

- The most necessary characteristic of variance here is, naturally, a characteristic in some way summarizing the tendencies of «serial» and «territorial» variance marked above. Such a combined estimate may be obtained assuming that influences of «serial» and «territorial» variance are mutually independent in the probabilistic sense and expressed multiplicatively by respective factors – random multipliers. Then the overall measure of variance may be represented by m.sq.d.l.i., equal to approximately 0.65 or 0.66 (again with possibility of slightly different results under various assumptions as to the treatment of data).

- If we were to establish the characteristic values s_k of snow load <u>without any regional</u>

---

[5] More precisely, for a lognormal variable Z the relation between η_Z and μ_Z is given by formula
$$\mu_Z = \eta_Z \left(1+\upsilon_Z^2\right)^{\frac{1}{2}}.$$

divisions and expressing the load in terms of water content (as is the case in the remarks above), values of s_k about $\eta_s \exp(2.054\sigma_{\ln s}) \cong 177$ kg/m^2 (or, if various minor details of procedure are modificated, – about 175…180 kg/m^2) would be chosen [6]. (Conversion to the measurement of acting forces gives here results about 1.75 kN/m^2.) It would be not an optimal decision, however, as some regional differences are unquestionable.

- So far as we can judge from the data at our disposal, the only part of Lithuania where the snow loads are definitely greater is the hilly region in the west [7]. Most confidently this conclusion may be applied to the upwind western side of this height (the winds from western directions bringing most of precipitation) together with the close vicinity of highest hills, – but in addition the adjacent coast of sea, while not standing out by the quantities of snow in typical winters, presumably can be joined to this region of greater loads if we turn our attention to some data of extremely severe winters and are looking for a simple subdivision of the territory wherever possible. As for the other regions, there is no firm evidence for clear distinction of any sizable territory – although an experienced meteorologist in all likelihood would be able to indicate some localities where relatively great or small values of the load are probable.

- Assuming the division of territory into regions as marked above, we obtain the following estimates of parameters characterizing the distribution of s and, consequently, the characteristic values s_k :

    (i) for the western region – $\eta_s \cong 46$ kg/m^2 and the values of m.sq.d.l.i. equal, roughly approximately, to 0.62 or even 0.66 for the «local serial» variance and 0.33 or even 0.41 for the «territorial» variance – resulting in overall estimate of m.sq.d.l.i. $\sigma_{\ln s} \cong 0.71…0.78$ and estimate of characteristic value s_k in the neighbourhood of 1.92 kN/m^2 or even considerably greater levels well above 2 kN/m^2 (unfortunately, estimates mentioned here are not exact since the variance is extremely great and the amount of information from the region is rather scant);

    (ii) for the remaining part of the country – $\eta_s \cong 49$ kg/m^2 and the values of m.sq.d.l.i. equal approximately to 0.51 for the «local serial» variance and 0.26 for the «territorial» variance – resulting in overall estimate of m.sq.d.l.i. $\sigma_{\ln s} \cong 0.58$ and estimate of characteristic value s_k in the neighbourhood of 1.6 kN/m^2.

- If we turn our attention now to the velocities of wind and, as it has been stated in our introductory remarks, confine the review only within the context of wind gusts velocities (for one year long periods), the following results are to be noted (again with dividing the country into two regions, one of them covering the western belt of territory near the sea [8]).

- For the western (seaside) region: estimates of median about 26 m/s and m.sq.d.l.i., most likely, about 0.16 for the «local serial» variance and about 0.17 for the overall estimate of variance. (The paucity of data here is even more acute than in the case of snow loads as there are just three meteorological stations in this region; however, this shortage is to some extent counterbalanced by the fact that the variance in this case is

---

[6] The multiplier 2.054 in the formula presented here is the value of standard normal variable which is exceeded with probability 0.02 predetermined for the quantile that we are estimating.

[7] It is plausible that a narrow strip of hills close by southwestern border of the country is also to be treated as a locality with greater snow loads. Nevertheless there are no observations from Lithuanian territory giving an explicit proof of such conclusion (which we can base only indirectly on some data from Poland).

[8] However, the definition of this western region as applied to wind loads must be different from the division of the territory assumed for the snow loads. (A belt much narrower than in the former case must be set off as the «seaside» region now.)

not so high.) Thus the characteristic levels of the variable under consideration (wind gust velocity) should rather be for this region about 36 or 37 m/s.

- For the remaining part of the country: estimates of median about 21 or 22 m/s and m.sq.d.l.i., most likely, about 0.1…0.14 for the «local serial» variance and about 0.14…0.15 for the «territorial» variance – which leads to the overall estimates of m.sq.d.l.i. in a rough approximation near to 0.18 – and estimates of the characteristic level of wind gust velocity near to 32 m/s.

- As a practical conclusion regarding the level of reliability requirements in local norms it may be said with reasonable confidence that snow load normatives adopted now in Lithuania are significantly below the level defined by the Eurocode methodology (at least if this methodology is applied with due consideration of the facts mentioned above). The definition of wind load parameters is more problematic and deserves more detailed further investigation; nevertheless, it is very likely that these parameters are also defined not quite appropriately. (Apart from the doubt expressed above and relating to the treatment of wind gusts, there are yet more apparent disadvantages of our Lithuanian normatives stemming from a very risky definition of partial factors of safety γ .) This matter, however, goes already clearly beyond the subject of this datasheet.

References

- 1. *EN 1991-1-3:2002. Actions on Structures – Snow Loads*. Brussels, CEN General Secretariat.
- 2. *STR 2.05.04:2003. Poveikiai ir apkrovos [statybos techninis reglamentas].*[Actions and loads – technical regulations for construction. (In Lithuanian.)] // Valstybės žinios, Nr. 59 (2003m.).

Contacts

- Dr. Regimantas Simkus
    Address: Institute of Architecture and Construction of Kaunas University of Technology street, Tunelio 60, 44405 Kaunas, Lithuania.
- e-mail: asi@asi.lt, Fax: (+370 37) 451355.
    :

COST C12 - WG2

Datasheet I.1.3.1
Prepared by: R. Vacareanu, D. Lungu, A. Aldea, C. Arion
Technical University of Civil Engineering, Bucharest

RISK ANALYSIS IN EARTHKUAKE ENGINEERING

Description

Risk represents the expectancy of damage or losses (expressed in probabilistic terms) in relation to the performance of a definite built system, as a function of duration.

The risk analysis recognizes basically the impossibility of deterministic prediction of events of interest, like future earthquakes, exposure of elements at risk, or chain effects occurring as a consequence of the earthquake-induced damage. Since the expectancy of losses represents the outcome of a more or less explicit and accurate predictive analysis, a prediction must be made somehow in probabilistic terms, by extrapolating or projecting into the future the present experience.

The general relation for the determination of the total risk can be expressed as (Whitman & Cornell, 1976):

$$P[R_i] = \sum_j P[R_i/S_j] \cdot P[S_j] \tag{1}$$

in which $P[\ ]$ signifies the probability of the event indicated within the brackets, R_i denotes the event that the state of system is i, S_j means that the seismic input experienced is level j, and $P[R_i/S_j]$ states the probability that the state of the system will be R_i *given* that the seismic input S_j takes place.

Basis of method

The probabilistic risk assessment is aiming at computing the annual probability of exceedance of various damage states for a given structural system. The consistent probabilistic approach is based on the idea of (Cornell & Krawinkler, 2000) using the total probability formula applied in a form that suits the specific needs:

$$P(\geq d_s) = \int_{PGA} \int_{Sd} \Phi(\geq d_s \mid Sd) \cdot f(Sd|PGA) \cdot f(PGA) d(Sd) d(PGA) \tag{2}$$

where:

- $P(\geq d_s)$ – annual probability of exceedance of damage state d_s
- $\Phi(\geq d_s \mid Sd)$ – standard normal cumulative distribution function of damage state d_s conditional upon spectral displacement Sd
- $f(Sd \mid PGA)$ - probability density function of spectral displacement Sd given the occurrence of peak ground acceleration PGA
- $f(PGA)$ – probability density function of peak ground acceleration PGA

One can change Eq. (2) to solve for the mean annual rate of exceedance of various damage states for a given structural system:

$$\lambda(\geq d_s) = \sum_{PGA} \sum_{Sd} P(\geq d_s|Sd) \cdot P(Sd|PGA) \cdot \lambda(PGA) \tag{3}$$

where:

– $\lambda(\geq d_s)$ – mean annual rate of exceedance of damage state d_s

- $P(\geq d_s \mid Sd)$ – probability of exceedance of damage state d_s conditional upon spectral displacement Sd

- $P(Sd \mid PGA)$ - probability of reaching spectral displacement Sd given the occurrence peak ground acceleration PGA

– $\lambda(PGA)$ – mean annual rate of occurrence of peak ground acceleration PGA.

Consequently, the probabilistic assessment of seismic risk involves the:

1. probabilistic seismic hazard assessment, $\lambda(PGA)$
2. probabilistic assessment of seismic structural response, $P(Sd \mid PGA)$
3. probabilistic assessment of seismic structural vulnerability, $P(\geq d_s \mid Sd)$

Equations (2) and (3) are disaggregating the seismic risk assessment problem into three probabilistic analysis of: hazard, structural response and vulnerability. Then it aggregates the risk

Application to structural analysis

Probabilistic seismic hazard assessment, $\lambda(PGA)$

For a given earthquake recurrence, the probability of exceeding a particular value of peak ground acceleration, PGA^*, is calculated using the total probability formula (Cornell, 1968):

$$\lambda(PGA > PGA^*) = \int\int P(PGA > PGA^* \mid m,r)\cdot f(m)\cdot f(r)dmdr \qquad (4)$$

where:

– $\lambda(PGA>PGA^*)$ – mean annual rate of exceedance of PGA^* ;

- $P(PGA>PGA^*|m,r)$ – probability of exceedance of PGA^* given the occurrence of an earthquake of magnitude m at source to site distance r;

- $f(m)$ – probability density function for magnitude;

- $f(r)$ – probability density function for source to site distance.

Probabilistic assessment of seismic structural response, $P(Sd \mid PGA)$

The seismic motions used in the analyses of structural systems consist in classes of random processes comprising at least ten samples each. The seismic motion intensity is quantified by peak ground acceleration (PGA). Elastic acceleration spectra are used to simulate samples. The input seismic motions are simulated using a stationary Gaussian model based on elastic acceleration spectra (Shingal & Kiremidjian, 1997). The time histories are generated such as to fit the given response spectrum. The probability distributions of the dynamic amplification factors are used to obtain an ensemble of response spectra corresponding to a given level of seismic motion (Vacareanu, 2000). For parametric analysis purpose, the accelerograms are simulated at predefined values of PGA.

Nonlinear dynamic analyses of structural systems are performed. The uncertainties associated with seismic demands and structural capacities need to be modeled. The Monte-Carlo technique involves the selection of samples of the random structural parameters and seismic excitations required for nonlinear analyses, the performance of nonlinear analyses and the computation of the structural response.

In brief, the Monte-Carlo simulation technique implies the following steps:

- simulation of structural parameters and seismic excitations;
- random permutations of structural parameters and of excitations;
- performing nonlinear analyses using generated samples;

- sample statistics of results of analyses.

The direct Monte-Carlo technique requires a large number of simulation cycles to achieve an acceptable level of confidence in the estimated probabilities. The Latin hypercube technique might be used to reduce the number of simulation cycles. Using the Latin hypercube technique for selecting values of the input variables, the estimators from the simulation are close to the real values of the quantities being estimated. The Latin hypercube technique uses stratified sampling of the input variables, which usually results in a significant decrease in the variance of the estimators (Rubinstein, 1981).

Spectral displacements of structural systems are calculated using nonlinear dynamic analyses and data regarding the randomness of the seismic response of the structural system are obtained. The statistic indicators of the spectral displacement obtained at each *PGA* value are used to get the parameters of a lognormal distribution function for that level of intensity of ground motion. Finally, the lognormal probability density function of spectral displacement conditional upon *PGA* is evaluated.

Probabilistic assessment of seismic structural vulnerability, $P(\geq d_s \mid Sd)$

The probabilistic assessment of seismic structural vulnerability involves the determination of the structural systems vulnerability functions. These functions describe the conditional probability of being in, or exceeding, a particular damage state, d_s, given the spectral displacement, S_d, and is defined as, *HAZUS* (1997) :

$$P[ds|S_d] = \Phi\left[\frac{1}{\beta_{ds}}\ln\left(\frac{S_d}{\bar{S}_{d,ds}}\right)\right] \tag{5}$$

where:

- $S_{d,ds}$ is the median value of spectral displacement at which the structural system reaches the threshold of the damage state, *ds*,
- β_{ds} is the standard deviation of the natural logarithm of spectral displacement for damage state *ds*, and
- Φ is the standard normal cumulative distribution function.

HAZUS (1997) includes the vulnerability function parameters, $S_{d,ds}$ and β_{ds} appropriate for each type of buildings and structural systems corresponding to *USA* practice of design and construction. In order to calibrate the vulnerability function parameters appropriate for structural systems which are different from *USA* practice, the Monte-Carlo simulation technique can be used.

Risk Analysis

Given the results of the probabilistic seismic hazard assessment, $\lambda(PGA)$, the probabilistic assessment of seismic structural response, $P(Sd \mid PGA)$, and the probabilistic assessment of seismic structural vulnerability, $P(\geq d_s \mid Sd)$ one can aggregate the risk via summation (or integration) over all levels of the variables of interest using Equation 3. The results on risk are presented as mean annual rate of exceedance of damage state d_s, $\lambda(\geq d_s)$, and as the exceedance probability of various damage states in *T* years assuming that the damage states follows a Poisson distribution:

$$P_{exc}(d_s, T) = 1 - e^{-\lambda(\geq d_s)\cdot T} \tag{6}$$

where:
- $P_{exc}(d_s, T)$ - exceedance probability of damage state d_s in time T.

Example of application

Probabilistic seismic hazard assessment, $\lambda(PGA)$
The mean annual rate of exceedance of PGA – the hazard curve - for Bucharest site and Vrancea seismic source is represented in Figure 1.

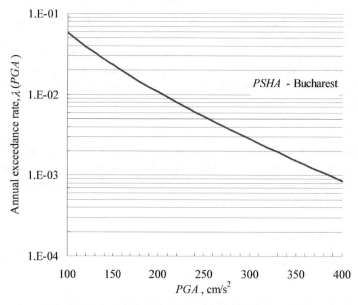

Figure 1. Hazard curve for Bucharest from Vrancea seismic source

The hazard curve can be approximated by the form $H = k_o \cdot a_g^{-k}$, where a_g is peak ground acceleration, and k_o and k are constants depending on the site (in this case k_o=1.176E-05, k=3.0865).

Probabilistic assessment of seismic structural response, $P(Sd \mid PGA)$
For parametric analysis purpose, the sample accelerograms are simulated at predefined values of PGA, as follows: 0.10g, 0.15g, 0.20g, 0.25g, 0.30g, 0.35g, 0.40g (g – acceleration of gravity).
The building analyzed has a reinforced concrete moment resisting frame structure and it was erected in early '70's. It is a thirteen-storey building, the first two storeys being of 3.60 m, all the rest of 2.75 m height. The building has two spans of 6.00 m each in the transversal direction and five spans of 6.00 m each in the longitudinal direction. The concrete is of class Bc 20 and the steel is of quality PC 52. Further details can be found elsewhere (Vacareanu, 2000).

The structural model in the transversal direction consists of six - two spans - reinforced concrete moment resisting frames acting together due to the action of horizontal diaphragms located at each storey. The computer program *IDARC 2D* (Valles et. al., 1996) is used for performing inelastic dynamic analyses. The computed mean and standard deviation of the spectral displacements are reported in Table 1.

Table 1 .Mean and standard deviation of DI

PGA, 'g	$\mu_{Sd|PGA}$	$\sigma_{Sd|PGA}$
0.1	13.1	2.2
0.15	22.8	3.3
0.2	30.0	4.0
0.25	39.5	3.8
0.3	49.8	5.3
0.35	59.0	6.4
0.4	68.7	6.9

Finally, the lognormal probability density function of spectral displacement conditional upon *PGA* is evaluated. Examples of lognormal probability density functions of *Sd* conditional upon *PGA*= 0.1g, 0.2g and 0.3g are presented in Figure 2. Using the density functions one obtains probability of reaching spectral displacement *Sd* given the occurrence of peak ground acceleration, *PGA*, *P(Sd | PGA)*.

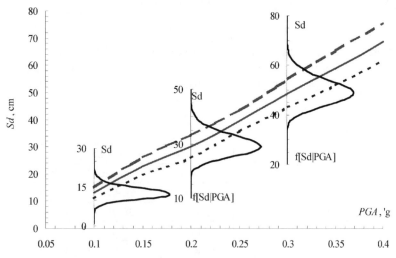

Figure 2. Mean and standard deviation of the spectral displacement

Probabilistic assessment of seismic structural vulnerability, P(≥ d$_s$ | Sd)
Once the parameters of vulnerability function, $S_{d,ds}$ and β_{ds}, are obtained – Table 2 - one can compute and plot the vulnerability functions using Eq. 5, Figure 3.

Table 2. Vulnerability function parameters (*Monte-Carlo* Simulation)

Damage state							
Slight		Moderate		Extensive		Complete	
$S_{d,ds}$, cm	β_{ds}	$S_{d,ds}$, cm	β_{ds}	$S_{d,ds}$, cm	β_{ds}	$S_{d,ds}$, cm	β_{ds}
7.82	0.66	17.88	0.66	27.94	0.76	68.16	0.91

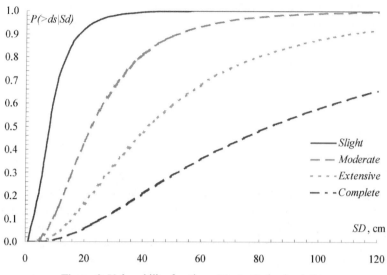

Figure 3. Vulnerability functions, Monte-Carlo simulation

Risk Analysis

The results on risk are presented as mean annual rate of exceedance of damage state d_s, $\lambda(\geq d_s)$, in Table 3. Also, in Table 3 are presented the exceedance probability of various damage states in 1 year, 50 years and 100 years assuming that the damage states follows a Poisson distribution.

Table 3. Results of seismic risk analysis

Damage state - ds	Annual exceedance rate, $\lambda(\geq d_s)$	Exceedance prob., $P_{exc}(d_s, T)$ in:		
		T=1 year	T=50 years	T=100 years
Slight	5.1E-02	5.0E-02	9.2E-01	9.9E-01
Moderate	2.6E-02	2.6E-02	7.3E-01	9.3E-01
Extensive	1.2E-02	1.2E-02	4.4E-01	6.9E-01
Complete	4.7E-03	4.7E-03	2.1E-01	3.7E-01

One can notice from Table 3 the exceedance probability of complete damage state in 1 year of $4.7 \cdot 10^{-3}$ which is much higher that the commonly accepted probabilities of failure of 10^{-4} to 10^{-5} as in the case of non-seismic loads. The main reason for this high probability comes from the design of the building which was accomplished taking into account an inferior code for earthquake resistant design (*P13-70*) combined with the low level of seismic hazard considered in the design process.

References

- Cornell, C. A. - Engineering Seismic Risk Analysis. *Bull. SSA, 56*, 1968
- Cornell C.A., Krawinkler H. 2000. Progress and Challenges in Seismic Performance Assessment. *PEER Center News* (4)1:1-3.
- *HAZUS – Technical Manual* 1997. Earthquake Loss Estimation Methodology, 3 Vol.

- C. Lomnitz, E. Rosenblueth (eds.) - *Seismic Risk and Engineering Decisions*. R. Whitman and C. A. Cornell - *Design, Chapter 9*. Elsevier, Amsterdam - Oxford - New York, 1976
- Lungu, D., Vacareanu, R., Aldea, A., Arion, C., 2000. *Advanced Structural Analysis*, Conspress, UTCB, 177p.
- Lungu, D., Aldea, A., Demetriu, S., Arion, C., (2000) Zonation of seismic hazard for the city of Bucharest – Romania *Technical Report prepared for Association Française du Génie Parasismique.*
- Rubinstein, R. Y. *Simulation and the Monte-Carlo method*. John Wiley & Sons, New York, 1981
- Singhal, A. & Kiremidjian, A. S., 1997 A method for earthquake motion-damage relationships with application to reinforced concrete frames. *Technical Report NCEER-97-0008*. National Center for Earthquake Engineering Research, State University of New York at Buffalo
- Valles R. E., Reinhorn A. M., Kunnath S. K., Li C., Madan A. – IDARC2D Version 4.0: A Computer Program for the Inelastic Damage Analysis of Buildings, *Technical Report NCEER-96-0010*, NCEER, State University of New York at Buffalo, 1996
- Văcăreanu R., 2000 – Monte-Carlo Simulation Technique for Vulnerability Analysis of RC Frames – An Example – *Scientific Bulletin on Technical University of Civil Engineering of Bucharest*, pg. 9-18

Contacts

- Dan Lungu
  Address: Reinforced Concrete Department, Technical University of Civil Engineering, 124 Lacul Tei Blvd., Bucharest, Romania
  E-mail: lungud@mail.utcb.ro; Phone/Fax: +40-21-2425804
- Radu Vacareanu
  Address: Reinforced Concrete Department, Technical University of Civil Engineering, 124 Lacul Tei Blvd., Bucharest, Romania
  E-mail: vradu@mail.utcb.ro; Phone/Fax: +40-21-2425804

STRUCTURAL SYSTEM OPTIMISATION

Description

- Presented analysis is an example of the optimization task. The objective function is minimalization of the frame members volume (mass). The restrainers are: stiffness of the joints, type of beam and column profiles and code requirements.
- Internal forces, especially bending moment distribution in frames, change due to change of joint stiffness (Brognoli, 1998), (Christopher, 1998), (Hasan, 1997).
- Design process consists not only of global analysis but also of member dimensioning according to codes regulation. Not always change in moment values in the frame involves also change in member section size, because cross-sections of standard steel members have discrete values.
- To analyze influence of joint stiffness on frame economy it is not enough to focus on moment distribution, but also on the results of frame elements dimensioning, according to limit states. When, after checking of limit states, member sizes have to be changed, also ratio of the joint to member stiffness change, what requires repeating the global analysis.

Basis of the method

- The aim of the method is optimisation of building framed structures using mixed technology and semi-rigid joints and optimal choose of joint properties.
- The analysis of six configuration of steel and composite braced and unbraced frames (Figure 1) were conducted and on the basis of the obtained results it was pointed out economical solution of such a structures. During analysis, frame structural elements (beam and columns) sections were changed according to changes of internal forced caused by joint stiffness changes.
- Efficiency of composite steel-concrete floors in the framework of buildings as well as economy confirmation of using semi-rigid joints in braced and non-slender unbraced steel and composite structures have been demonstrated.

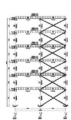
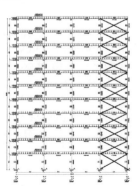

Figure 1. Two examples of frames used in the analysis.

Application to structural analysis

- Frames was analyzed in two versions: unbraced and with X-shape bracing. Bare steel frames as well as frames with composite steel-concrete floors were considered.
- Stiffness of the column bases was taken from (Wald, 1995).
- During the analysis stiffness of the joints were changed:

  - for unbraced frames initial stiffness of joints was taken as:
  - infinity (rigid joints – conventional approach),
  - $25 \dfrac{EJ_b}{L_b}$ (EC3 boundary between nominally rigid and semi-rigid joints),
  - few values from $25 \dfrac{EJ_b}{L_b}$ to $6 \dfrac{EJ_b}{L_b}$.

  - for braced frames initial stiffness of joints was taken as:
  - 0 (ideally pinned joint – conventional approach),
  - few values from $0,5 \dfrac{EJ_b}{L_b}$ to $10 \dfrac{EJ_b}{L_b}$,

  where:
  E – Young modulus,
  J_b – inertia moment of beam section,
  L_b – length of beam.

- Non-linear character of M-ϕ curves was taken into account by the method of equivalent, secant stiffness:

$$S_{wc} = \frac{S_{j.ini}}{\eta},$$

where η was taken from (Kozlowski, 1999).

- Loading of the frames were (characteristic values):
  - dead load: 30 kN/m,
  - imposed load: 12 KN/m,
  - wind loading of the value 12 kN in each frame node..
- Section of frame elements were changed according to internal forces and ULS and SLS conditions, with consideration of design practice: all beams have the same profile, section of column were changed every three storeys.
- Steel grade S235 (Fe 360) was used for each structural members. Beams were designed using IPE section and columns using HEB profiles.
- During dimensioning of elements the following assumptions were established:
  - lateral-tortional buckling of elements was neglected,
  - effective length of column was calculated with coefficients:
    - $\mu_y = 0,8$,
    - $\mu_z = 1,0$
- stiffness of the composite beams were taken as (Leon, 1996):

$$S_{r,z} = 0,6 \, S_{r,p} + 0,4 \, S_{r,w}$$

  where:
  $S_{r,p}$ – stiffness of composite beam section in midspan,
  $S_{r,w}$ – stiffness in joint section of composite beam.

- The analyses were conducted using ROBOT V6 package. "Compatibility joint" option was applied, which allows linear joint model to be implemented.
- Second order P-Δ analysis was applied.

- At each stage of the analysis, ultimate limit state (ULS) and serviceability limit state (SLS) for whole structure and each member groups were checked. In the case when any of limit state was not fulfilled, member sections and also joint stiffness related to them were changed and the global analysis was repeated. These required many recalculation of each frame.

Example of application

- Examples of the analysis results are presented graphically in Figures 2 and 3. Results of other frames analysis are included in (Kozlowski, 1999). Figures 2 and 3 show variation of the span moment M_p, moments in beam ends M_w, masses of the all steel frame m_s and composite frame m_z members, versus the secant joint stiffness S_{wc} variation. For unbraced frames, also lateral drift is shown.
- The figures are accompanied by table, which presents:
  - sections of the beam and column obtained from the ULS check,
  - saving of the steel (in %) in comparison to traditional solution, i.e. frame with pinned or rigid connections,
  - joint label, whose coordinates refer to their secant stiffness.
- In the rows concerning composite frames, additional values of steel saving are presented, taken from the comparison to steel frame with the same joint stiffness (B) and to steel frame with pinned/rigid joints (marked as *).
- The following conclusions can be drawn from braced frame analysis:
  - a) increasing joint stiffness leads to smaller beam section, due to smaller value of the maximum moment (in span), governing the beam section sizing. This is the main source of the steel saving. Simultaneously, moments in the beam ends increase and are transmitted to the columns, but it does not affect column sections because governing case for column dimensioning is out of plane buckling of internal column,
  - b) it is possible to equalize moments in the mid-span and beam ends, what can give the smallest beam section. This requires higher joint stiffness and consequently, higher cost of the joint. This can reduce profit from smaller beam section,
  - c) the higher joint stiffness the smaller sections of bracing,
  - d) in steel frames, the highest economy is achieved (steel saving up to 11,9 %), for the following joints: unstiffened flush end plate, top and seat angle, and top and seat with web angles. In these cases, beam section is reduced from IPE 400 obtained in classical design (pinned joints) to IPE 330. Even when "simple" joint is used, for example web angle, beam section is reduced to IPE 360,
  - e) in composite frames, saving of steel obtained in semi-rigid design is from 4,0 to 13,3 %, in comparison to pinned joint solution,
  - f) comparison of the steel and composite frame analysis results shows that composite frames are much more efficient. Saving of steel is from 15,5 to 25,1 % when comparison is made between steel and composite frames subjected to the same loading and of the same joint stiffness. When one compares composite semi-rigid frame with bare steel frame with pinned connections (conventional solution) this steel saving can be even 31 %.

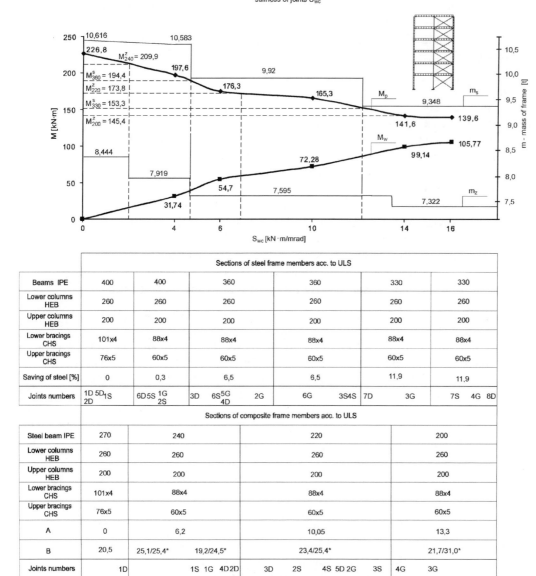

Figure 2. Results of two-bay six-storey braced frame analysis

	Sections of steel frame members acc. to ULS					
Beams IPE	400	400	360	360	330	330
Lower columns HEB	260	260	260	260	260	260
Upper columns HEB	200	200	200	200	200	200
Lower bracings CHS	101x4	88x4	88x4	88x4	88x4	88x4
Upper bracings CHS	76x5	60x5	60x5	60x5	60x5	60x5
Saving of steel [%]	0	0,3	6,5	6,5	11,9	11,9
Joints numbers	1D 5D 1S 2D	6D 5S 1G 2S	3D 6S 5G 4D 2G	6G 3S4S	7D 3G	7S 4G 8D

	Sections of composite frame members acc. to ULS			
Steel beam IPE	270	240	220	200
Lower columns HEB	260	260	260	260
Upper columns HEB	200	200	200	200
Lower bracings CHS	101x4	88x4	88x4	88x4
Upper bracings CHS	76x5	60x5	60x5	60x5
A	0	6,2	10,05	13,3
B	20,5	25,1/25,4*	19,2/24,5*	23,4/25,4*	21,7/31,0*
Joints numbers	1D	1S 1G 4D 2D	3D 2S 4S 5D 2G 3S	4G 3G

A - saving of steel in relation to composite frame with pinned joints
B - saving of steel in relation to steel frame
* - values with refer to steel frame with pinned joints
M^s_{330} - carrying capacity of steel beam IPE 330
M^z_{200} - carrying capacity of composite beam with steel beam IPE 200

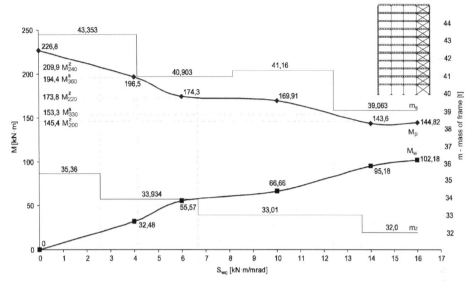

Sections of steel frame members acc. to ULS						
Beams IPE	400	400	360	360	330	330
Lower columns HEB	500	500	500	500	500	500
Internal lower columns HEB	320	320	320	320	320	320
Internal upper columns HEB	240	240	240	240	240	240
Upper columns HEB	160	160	160	180	180	180
Lower bracings CHS	101x8	101x8	101x8	101x8	101x8	101x8
Upper bracings CHS	88x5	88x5	88x5	88x4	88x4	88x4
Saving of steel [%]	0	0	5,6	5,6	9,8	9,8
Joints numbers	1D 5D 2D	1S 6D 5S	1G 2S	3D 5G 4D 6S	2G	6G	3S 4S 7D	3G 7S 4G 8D

Sections of composite frame members acc. to ULS			
Steel beam IPE	270	240	220	200
Lower columns HEB	500	500	500	500
Internal lower columns HEB	320	320	320	320
Internal upper columns HEB	240	240	240	240
Upper columns HEB	160	160	180	180
Lower bracings CHS	101x8	101x8	101x8	101x8
Upper bracings CHS	88x4	88x4	88x4	88x4
A	0	4,0	6,7	9,5
B	18,4	21,7 | 17,0/21,7*	19,8/23,8* | 15,5/23,8*	18,1/26,1*
Joints numbers	1D	1S 1G 4D 2D	3D 2S 4S 2G 5D 3S	4G 3G

A - saving of steel in relation to composite frame with pinned joints
B - saving of steel in relation to steel frame
* - values with refer to steel frame with pinned joints
M_{330}^s - carrying capacity of steel beam IPE 330
M_{200}^z - carrying capacity of composite beam with steel beam IPE 200

Figure 3. Results of four-bay eleven-storey braced frame analysis

- In this analysis only steel consumption was considered. Next step is to take into account the cost of joints and member fabrication and erection, like in (Brognoli, 1998). Other possibility of including the joint costs in economy analysis is to use the sensitivity analysis (Xu Lei, 2000).
- Presented results should encourage designers to apply the semi-rigid design concept. Theoretical background of such a design and software are now available and presented, for example, in (Chen, 1994) or (Chan, 2000).

References

- Brognoli M., Gelfi P., Zandonini R. and Zanella M. 1998. Optimal Design of Semi-Rigid Braced Frames Via Knowledge-Based Approach. *Journal of the Constructional Steel Research*, 46:1-3.
- Chan S.L. and Chui P.P.T. 2000. *Non-linear Static and Cyclic Analysis of Steel Frames with Semi-Rigid Connections,* Elsevier, Oxford, UK.
- Chen W.F. and Toma S. 1994. *Advanced Analysis of Steel Frames, Theory, Software and Application,* CRS Press, London, UK.
- Christopher J.E and Bjorhovde R. 1998. Response Characteristics of Frames with Semi--Rigid Connections. *Journal of the Constructional Steel Research*, 46:1-3.
- Hasan R., Kishi N., Chen W.F. and Komuro M. 1997. Evaluation of Rigidity of Extended End-Plate Connections. *Journal of Structural Engineering,* 123:2.
- Kozlowski A. 1999. *Shaping of the steel and composite skeletons with semi-rigid joints.* Monograph (in Polish). Rzeszow University of Technology, Rzeszow, Poland.
- Leon R.T. and Hoffman J.J. 1996. Plastic Design of Semi-Rigid Frames. *in: Connections in Steel Structures III: Behaviour, Strength and Design,* Elsevier, London, UK.
- Wald F. 1995. *Column Bases.* Česke Vysokě Učeni Technickě, Praha, Czech Rep.
- Xu Lei. 2000. Design Optimization of Semi-Rigid Steel Frames, *in: Practical Analysis for Semi-Rigid Frame Design,* World Scientific, Singapore.

Contact

- Aleksander Kozlowski

  Address: Department of Civil Engineering, Rzeszow University of Technology, 2 Poznanska street, Rzeszow, Poland.
  e-mail: kozlowsk@prz.edu.pl, Phone/Fax: +48 17 85 429 74.

COST C12 - WG2
Datasheet I.2.1.1
Prepared by: F. M. Mazzolani, G. De Matteis
University of Naples Federico II, Dept. Structural Analysis and Design

HYDROCARBON EXPLOSION LOADING

Definition

- Hydrocarbon explosion loading is a man-made action, i.e. is produced by intentional or unintentional human activities.

- Hydrocarbon explosion loading consists in a set of forces, applied dynamically to the structure.

- Hydrocarbon explosion loading may be considered as a variable action whose magnitude is not zero only for a very short period of the reference time.

Physical Features

- Hydrocarbon explosion loading consists of (Bangash, 1993):
  - a blast wave, having a steep pressure rise at the wave front and producing an overpressure load on the body;
  - a blast wind, which is the transient airflow behind the blast wave.

- The blast wind produces a drag load on the body, accompanied by an inertial load given by the time variation of blast wind velocity (Figure 1).

- Blast overpressure is likely to govern the design of containment structures (e.g., blast walls, deck platings), while structures around which air can flow are generally influenced by the effects of both blast overpressure and blast wind.

- Blast overpressure vs. time curves normally show a steep increase followed by a quick decay of the overpressure to ambient conditions and finally by a negative (suction) phase, where the load is acting towards the source of the explosion (Figure 2).

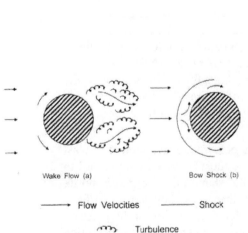

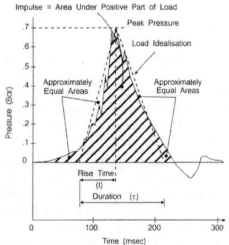

Figure 1. Different mechanisms giving rise to drag loads on bodies in: (a) low-velocity flows; (b) high-velocity flows.

Figure 2. Idealisation of the overpressure vs. time load experienced by a blast barrier as a consequence of the explosion of a hydrocarbon cloud.

Effect on constructions

- The blast overpressure can produce a premature attainment of ultimate elongation on structural material. Even if a relatively small portion of the structure due to such abnormal action collapses, this can result in the failure of a major portion of the structure, producing the so-called progressive collapse, which can be defined as an extensive structural failure initiated by local structural damage.

Modelling

- Hydrocarbon explosion loading may be considered either a scalar action (by describing the effects of blast by means of one time-varying parameter) or a vectorial action (by describing the effects of blast by means of a set of time-varying parameters). The necessity of adopting either one parameter or more depends on the nature problem under consideration.

- When both blast wave and blast wind consideration are important, more than one time-varying parameter needs be considered (The Steel Construction Institute, 1999).

- Generally, experiments and/or CFD (Computational Fluid Dynamics) simulations are necessary in order to determine the explosion loading on the structure under consideration.

- Where blast wave effects govern the design, a one-parameter description of the action may be adopted in first instance. Time-varying pressure distributions may be obtained either from experiments or from numerical simulations by means of CFD codes, such as the FLACS (FLACS, 2002).

- Blast overpressure vs. time curves may be modelled by means of:

  i) an exponential decay curve;
  ii) a linear decay curve;
  iii) a bilinear (triangular) curve;
  iv) a multilinear curve.

- Normally, the suction phase of the blast overpressure loading may be neglected.

Structural analysis method

- Structural analysis of structures subjected to hydrocarbon explosion loading should be carried out by means of finite element codes such as ABAQUS, ANSYS, LS-DYNA, NASTRAN,... preferably adopting explicit methods for the integration of the dynamic equations of motion. This is the only method of analysis able to account for both geometrical and mechanical nonlinearities, strain rate effects, non-uniform dynamic loading, complex geometries.

- Several simplified methodologies, briefly described below, are available for blast overpressure loading of SDOF systems. However, care must be taken with regard to modelling the material elasto-plastic behaviour and the shape of the pressure pulse. More precisely, the accuracy required in describing these two key features of the structural model will mainly depend on the ratio between the duration of the pressure pulse and the natural period of the structural element (The Steel Construction Institute, 1999).

- In case of structures which can be modelled as SDOFs, and blast wind effects may be neglected but any intermediate state of the structure is of interest, simplified dynamic analyses may be performed according to the "Biggs Formulae" (Biggs, 1964). These require transformation factors, available in the literature (Biggs, 1964), based on stiffness, strength and loading equivalence criteria between the actual structure and the SDOF model.

- For structures which can be modelled as SDOFs, blast wind effects may be neglected and only the final state of the structure is of interest (i.e., quasi-static and impulsive loading regimes), the so-called "Baker Methodology" (Baker, 1983) may be applied. The relevant iso-damage pressure vs. impulse charts are available in the literature (Smith and Hetherington, 1994).

Guidelines and/or codification

- The American Society of Civil Engineers (ASCE), 1964. *Design of Structures to Resist Nuclear Weapons Effects – Manual of Engineering Practice No. 42*, New York, USA.

- The Chemical Industry Association (CIA), 1976. *Guidelines for the Design of Category 1 (Blast Resistant) Control Buildings*, New York, USA.

- The American Petroleum Institute (API), 1989. *LRFD Recommended Practice for Planning, Designing and Constructing Fixed Offshore Platforms*, New York, USA.

- The United States Department of the Army, Navy, and Air Force, 1990. *Structures to Resist the Effects of Accidental Explosions – Army TM 5-1300, Navy NAVFAC P-397, AFR 88-22*, Washington D.C., USA.

- Comité Européen de Normalisation (CEN), 1998. *Draft prEN 1991-1-7 of Eurocode 1 – Actions on Structures, Part 1.7: General Actions – Accidental Actions due to Impact and Explosions*, Bruxelles, Belgium.

Example of application

- To determine the structural response of the beam under explosion loading is necessary to assume a pressure-time diagram, which can be hypothesised as a bilinear diagram, based on the following characteristics: Peak pressure (p), Rise time (t_r), Duration time (t_d).

- The response of the structure can be defined, on the natural period of the structure, as: Impulsive [$t_d/T < 0.4$], Quasi-static [$t_d/T > 2$], Dynamic [$0.4 < t_d/T < 2$]. Only in the case of the Impulsive regime the rise time (t_r) is not important and can be mislead.

- A beam having a solid rectangular cross-section, fully-clamped at both ends and subjected to a uniformly distributed blast overpressure loading is considered.

- The mid-span deflection y_m of the beam under a pressure pulse defined by a peak pressure p of 4.0 kN/m and a total duration t of 15 ms is calculated.

- The following data are assumed for the calculation:
    beam width $b = 40$ mm
    beam height $h = 10$ mm
    beam length $L = 1000$ mm
    Young's modulus $E = 200$ KN/mm^2
    Poisson's ratio $\nu = 0.30$
    mass density $\rho = 7850$ kg/m^3
    yield stress $f_y = 250$ N/mm^2

- Based on Simple Degree Of Freedom method (SDOF) the natural period of the beam can be approximated as follow:

$$T = 2\pi\sqrt{\frac{k_{lm}M}{k}} = 0.02s$$

where: $k_{lm} = 0.77$ is the load-mass factor and $k = \dfrac{384EI}{L^3}$, while M is the total mass.

- Being $\dfrac{t_d}{T} = \dfrac{0.015}{0.02} = 0.75$, the loading regime cannot be classified as quasi-static.

- <u>Quasi static asymptote</u>

   Assuming the applied pressure-time diagram as triangular with time rise $t_r = 7.5$ ms equal to the half of the duration time td = 15 ms, the maximum deflection of the beam is given at the time corresponding to a period quarter (T/4 = 5ms). Considering that $t_r \approx \dfrac{T}{4}$ the work in this range is given as half work corresponding to the peak pressure (p):

$$WK = \frac{1}{2}\left[2\int_0^{L/2} p \cdot y(x)dx \right] = \frac{pcL}{4}$$

   while strain energy is given:

$$SE = 2c^2 \frac{EI\pi^4}{16(L/2)^3}$$

   and the maximum deflection is:
   $y_m = c = 15.5$ mm

- <u>Static response</u>

   The maximum static deflection is:

$$y_m = \frac{pL^4}{384EI} = 15.6mm$$

- <u>Comments</u>

   Since the elastic response of the above beam is not quasi-static, the dynamic load factor should be determined. On the basis of pressure-impulse diagrams obtained for hydrocarbon explosion, the maximum displacement amplification in the dynamic regime could be taken equal to 2. When different beams are considered, the structural response can be also quasi-static and the dynamic load factor assumes value equal to 1. This means that the duration time of the blast pressure is longer than the natural period of the structure [$t_d/T>2$]. In such cases, the hydrocarbon explosion can be applied as a static load assuming the peak pressure of the diagram.

- <u>Linear Elastic Dynamic Analysis (via the Finite Element Method)</u>

   The obtained value of y_m is equal to 14.0 mm.

   This results does not mean that the response of SDOF system is quasi-static, in   fact the response is closer the impulsive realm than the quasi-static realm:

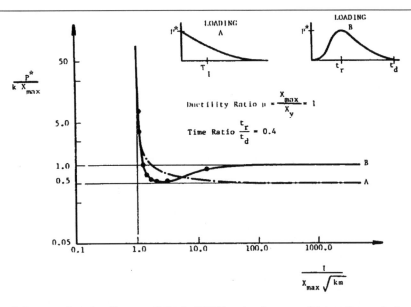

Figure 3. Pressure-Impulse diagram of elastic SDOF system in cases of detonation explosion (curve A) and hydrocarbon explosion with time ratio $t_r/t_d = 0.4$ (curve B).

Quasi-static load realm shown a DLF = 2 in case of detonation explosion [Curve A shown $\dfrac{p^*}{k \cdot x_{max}} = 0.5$], while quasi-static realm shown a DLF = 1 in case of hydrocarbon explosion [Curve B shown $\dfrac{p^*}{k \cdot x_{max}} = 1$].

As far as dynamics realm is concerned DLF, in case of hydrocarbon explosion [Curve A], first increases to the value equal to 2 starting from impulsive regime, then decreases again to the value equals to 1. The above application shows a dynamic response that is closer to the impulsive realm:

$$0.4 \prec \frac{t_d}{T} = 0.75 \prec 2 \quad \text{(Dynamic realm)}$$

$$\frac{p^*}{k \cdot x_{max}} = \frac{x_{st}}{x_{max}} = \frac{15.6mm}{14.0mm} \cong 1$$

- **Different study cases**

    In the following, different beam configurations having geometric characteristics corresponding to the usually structural applications are considered. The explosion will be considered as a triangular diagram defined by $t_d = 0.3$ s

    The following table shows the response of steel beam to the decrease of the length:

Section	I [mm⁴]	A [mm²]	L [m]	k [N/m]	k_{lm}	M [Kg]	T [s]	t_d/T	Response
HE 240 B	112590000	10600	7	26470	0,77	582	0,82	0,4	imp
	39230000	10600	7	9223	0,77	582	1,38	0,2	imp
HE 240 B	112590000	10600	6	42034	0,77	499	0,60	0,5	dyn
	39230000	10600	6	14646	0,77	499	1,02	0,3	imp
HE 240 B	112590000	10600	5	72634	0,77	416	0,42	0,7	dyn
	39230000	10600	5	25308	0,77	416	0,71	0,4	imp
HE 240 B	112590000	10600	4	141863	0,77	333	0,27	1,1	dyn
	39230000	10600	4	49430	0,77	333	0,45	0,7	dyn
HE 240 B	112590000	10600	3	336269	0,77	250	0,15	2,0	q-st
	39230000	10600	3	117167	0,77	250	0,25	1,2	dyn
HE 240 B	112590000	10600	2	1134907	0,77	166	0,07	4,5	q-st
	39230000	10600	2	395438	0,77	166	0,11	2,7	q-st

The results show different responses to the decrease of the beam length.

References

- Baker, W.E., 1983. *Explosion Hazards and Evaluation*, Elsevier.
- Bangash, M.Y.H., 1993. *Impact and Explosion – Analysis and Design*, Blackwell Scientific Publications, Oxford, UK.
- Biggs, J.M., 1964. *Introduction to Structural Dynamics*, McGraw-Hill, New York, USA.
- Bulson, P.S., 1997. *Explosive Loading of Engineering Structures: a History of Research and a Review of Recent Developments*, E&FN Spon, London, UK.
- FLACS, 2002. *The FLACS Manual (2 vols.)*, GexCon A.S., Bergen, Norway.
- Mays, G., Smith, P.D., 1995. *Blast Effects on Buildings: Design of Buildings to Optimise Resistance to Blast Loading*, Thomas Telford, London, UK.
- Smith, P.D., Hetherington, J.G., 1994, *Blast and Ballistic Loading of Structures*, Butterworth-Heinemann, Oxford, UK.
- The Steel Construction Institute, 1999. *Fire and Blast Information Group Technical Note 5 – Design Guide for Stainless Steel Blast Walls*, Silwood Park, Ascot, UK.

Contacts

- Dr. Luke A. Louca
  Address: Department of Civil and Environmental Engineering, Imperial College of Science Technology and Medicine, SW7 2BU London, United Kingdom.
  Phone: 0044.207.5946039
- Prof. Federico M. Mazzolani
  Address: Department of Structural Analysis and Design, University of Naples Federico II, P.le Tecchio, 80, I-80125 Napoli, Italy
  e-mail:fmm@unina.it - Phone: 0039.081.7682443

COST C12 - WG2
Datasheet I.2.2.1
Prepared by: S. Stamatovska
Institute of Earthquake Engineering and Engineering Seismology,
University "Ss. Cyril and Methodius", Skopje, Republic of Macedonia.

EARTHQUAKE MODELING

Definition

- An earthquake is a phenomenon during which strong vibrations occur in the ground due to release of enormous energy within a short time, causing sudden disturbance in the earth's crust or upper mantle. The energy released from the earthquake hypocenter moves through the earth towards the surface in the form of a wave motion.

- Earthquakes are random processes in respect to time, location and intensity and hence their effect upon structures is of a probabilistic nature. For the seismic design of structures, the prognosis of the intensity of the seismic effect is of a particular importance.

Physical features

- Ground motion under earthquake effects is of a probabilistic nature. The quantitative parameters of ground motion under earthquake effect are: amplitude and spectral content of seismic waves and vibrations, earthquake time duration and permanent soil deformations. These parameters depend on the energy and the proportions of the focus, the mechanism of motion in the earthquake focus, the distance between the focus and the structure, the geological composition of the medium through which the seismic waves propagate and the soil conditions under the structure. Therefore, the functional relationship between ground motion parameters and the parameters of the focus and the medium depends on a number of factors that are difficult to be mathematically considered without corresponding simplification.

- In the past, due to non-existence of instrumental records, the seismic effect was estimated based on macro-seismic data expressed by intensity degrees. The effects of strong earthquakes including geological, hydrogeological, geodetic and seismic effects upon structures as well as physical and psychological effect upon people are generalized through macroseismic data whereby lines of equal intensity, i.e., isoseismal maps are obtained. In practice, different seismic scales have been used in the world as follows: MM – Modified Mercalli scale (12 intensity degrees), which is used in USA and rarely in other countries, MSK-64 – Medvedev-Sponheuer-Karnik (12 intensity degrees), which is used in Europe, JMA – Japanese Meteorological Agency Scale(6.9 intensity degrees), which is used in Japan and new EMS-98 - European Macroseismic Scale (12 intensity degrees).

- Taken as vibration amplitude could be: the maximum (peak) amplitude; the maximum (peak to peak) amplitude; the mean square amplitude (which is equal to the square root of the squares of amplitudes in part of the record) and the spectral amplitude for selected vibration periods. The most frequently applied of all are the maximum ground acceleration–a_{max} and velocity–v_{max}. Lately, the amplitude of response spectra for damping of 5% of the critical and selected natural periods (T) of vibration have been used as parameter of ground motion. There are also amplitudes of permanent ground deformations that are connected with the size of motion in the focus and the inelastic ground motion. These amplitudes are defined based on instrumental data and occurred seismic-dislocations of the ground.

- There are different definitions of time duration of vibration. Absolute time duration is the time in the course of which the seismic signal is separated from the normal noise in the record (Bolt, 1973). Time duration of seismic energy is defined as an interval between two points of time between which almost 90% of the vibration energy is concentrated (Dobry, 1978).

- The spectral content of the vibrations is defined by the Fourier amplitude and phase spectrum and the response spectra (absolute acceleration spectrum–SA(T), relative velocity spectrum–SV(T), relative displacement spectrum– SD(T) for damping of 5%. The response spectrum represents an envelope of maximal values of the amplitude of response of a series of damped oscillators exposed to time history of acceleration. For small values of damping, the maximum relative displacement of the oscillator is the spectral value of relative displacement. The response spectra are computed by direct solution of the differential equation of motion of a damped single degree of freedom system or computation from one to another spectrum (the velocity spectrum is obtained by multiplying the displacement spectrum by ω or by dividing the acceleration spectrum by ω and is known as pseudo velocity response spectrum). The difference between the pseudo spectrum and the spectrum defined in a direct way is lesser with lesser damping ξ.

Effects on constructions

- Seismic waves propagate in all directions and cause the ground to vibrate at frequencies from about 0.1 to 30 Hz. The P and S waves mainly cause high-frequency vibrations, while the surface waves low-frequency vibrations.

- The effect on the structures of near-source and far-field ground motions is significantly different. Far-field ground motion consist mainly of the low-frequency waves, which although can be strong, have a cyclic and more predictable impact on the structure. Repeated cycles of inelastic deformations allow the energy dissipation. Near-source ground motions consist of both, strong P and strong S waves. The effect is that the horizontal pulse does not allow the structure to dissipate the energy trough the repeated cycles of inelastic behaviour. Their combination with the strong vertical accelerations can often cause a sudden failure of the structure.

Modelling

- For assessment of earthquake-resistant design it is necessary to define the design earthquake or design input motion. The methods for evaluation can be deterministic and probabilistic.

- The **deterministic approach** is based on computation of ground motion parameters by application of ground motion models for fixed distances and magnitudes of seismic sources around a certain considered site. This approach does not include the activity of the seismic source (number of earthquakes of a certain magnitude, at a certain time and within a certain area). Also, the standard deviations are used in this approach as direct multipliers of the median or mean values of the expected ground motion parameters. With this approach, the design parameter is determined as a median value plus one standard deviation, or 84,1% probability of non-exceedence of the expected values (Stamatovska, 2001; Stamatovska, 2003).

- During the last decade, the interest in application of the probability methods for evaluation of design earthquake has been significantly increased. This first of all refers to the **probabilistic seismic hazard analysis (PSHA)**. Nowadays, almost all the applied seismic hazard methodologies represent certain modification of the main approach developed by Cornell (Cornell,1968). According to this approach, the seismic hazard can be determined through involvement of defined seismic source models, models of their activity and ground motion models (McGuire, 1976). Seismic sources have been defined in respect to their spatial position, size, shape and maximum expected magnitude through mathematical modeling. The method applied for determination of the position and the geometry of seismic sources and M_{max} is based on geological criteria for the seismicity and comparison with available seismological observations. The activity of each seismic source is defined by the empirical recurrence relationship and it is expressed via cumulative number of earthquakes N with magnitude M that have occurred in the area of a considered seismic source within a certain time period (Weichert,1980). The ground motion models are defined by empirical relationships for certain ground motion parameters and earthquake parameters, such as magnitude, distance, fault type etc. (Ambraseys,1996; Sabetta and Pugliese, 1996; Sadigh,1997; Boore, 1997; Spudich, 1999, Stamatovska,2002). The results from the seismic hazard analysis are distribution functions known as hazard curves. They define the probability of occurrence of a certain value of ground motion parameter at a certain location for a determined time period (Petrovski, 2003; Stamatovska, 2001,2003; Talaganov, 2003).

- In fact, despite the evident results attained during the development of the seismic hazard methodologies, there are still some errors used to calculate the hazard curves. These are of two types: random and model errors. The model errors result from the modelling of the seismic sources, the source activity and the ground motion models. The selection of the mathematical model, i.e., whether it will represent the physical process with an adequate accuracy, as well as the data scatter around the expected average values is of a particular importance. The model errors have a significant influence on the final results from the seismic hazard analysis.

- The proper **treatment of uncertainties** is a very important subject in PSHA. The random uncertainties are considered during integration over different distributions into a simple point estimate of the hazard. The modelling uncertainties are treated by using multiple input and are displayed as a range of uncertainties associated with the point estimate. For an easier organisation and documentation, the uncertainties are treated trough a "logic tree". Each node of the logic three represents a certain PSHA input. Due to the uncertainties, it is defined by several alternatives. As a result, a family of hazard curves defining the uncertainties of the seismic hazard arising from the uncertainties in the PSHA inputs is obtained. The final result of the PSHA is the hazard curve with mean or median tendency (Petrovski, 2003).

- The **seismic design parameters** are defined on the basis of the results from the probabilistic seismic hazard analysis and applying the criterion for the acceptable hazard level (Petrovski, 2003; Stamatovska 2001,2003; Talaganov 2003).

- For the investigation of **seismic safety of structures** it is necessary to select a family of actual strong ground motion records compatible with the defined hazard curve, i.e., to account for different earthquake magnitudes, epicentral distances, duration, frequency content and other parameters. This is required because the seismic safety analysis involves determination of fragility curves of structures that are mainly based on nonlinear analyses. In this case, the selection of strong ground motion records compatible with the defined hazard curve is presented by the methodology which is based on the mathematical model for seismic hazard analysis. For each-ring-like area with a mean radius of R_i and width of ΔR_i the earthquake occurrence frequency λ_{ij} is obtained for each magnitude range of $M_j \pm \Delta M_j /2$. Using the rate of occurrences for different magnitude-distance pairs and ground motion models and performing the Monte-Carlo simulation, the participation of different magnitude-distance pairs is obtained for each peak ground acceleration level. Based on the determined participation factors, selection of strong ground motion records is made. Each record is given a certain weight in accordance with the relevant participation factor and the corresponding number of records (Petrovski, 1998).

Guidelines and/or codification

- The response spectra represent a form of presentation of spectral characteristics of ground motion that is directly applied in engineering practice. Also, they have been taken as a basis for elaboration of **seismic design regulations**.

- According to **EUROCODE 8** (Do. CEN/TC250/SC8/N317, Draft No.5, Revised final project team draft, preStage 49, May 2002), the earthquake motion at a given point on the surface is presented by an elastic ground acceleration response spectrum – "**elastic response spectrum**", for horizontal and vertical direction, for five ground types and two spectral shapes (TYPE 1 and TYPE 2), as well as for two levels of seismic action: no collapse requirement and damage limitation requirement. The horizontal seismic action is described by two orthogonal components considered as independent and represented by the same response spectrum. To avoid explicit inelastic structural analyses in design, the capacity of the structure to dissipate energy, through mainly ductile behavior of its elements and/or other mechanisms, is taken into consideration by performing an elastic analysis based on a response spectrum reduced with respect to the elastic one, the so called "**design spectrum**". This reduction is accomplished by introducing the behavior factor q. The seismic action can alternatively be presented by ground acceleration time histories and artifical accelerograms compatible with the elastic response spectrum.

- In U.S.A. (Uniform Building Code-1997, Supplement 2000), two types of earthquakes are considered as the design earthquake. The first is called "**design basis earthquake**". When the structure is subjected to this earthquake, no significant damage to the structure should occur. The other earthquake is called "**maximum capable earthquake**" which generetes the strongest ground motion at the site. When the structure is subjected to this earthquake, some demage is allowed, but the structure shoud not suffer severe damage resulting in collapse of the structure or loss of life.

- In **EUROCODE 8, Part 2, Bridges** (ENV 1998-2: 1994) the time histories for dynamic analysis of bridges in general, as well as bridges with isolation devices depend on the fundamental period of vibration of the structure. This means that the determination of time histories for dynamic analysis of bridges represents a constituent part of the design process. It makes possible that the natural period of a structure is changed in an interactive process aimed at rational design, i.e. economically justified seismically safe structure (Petrovski,1999; Hristovski,1999).

Example of application

- The example refers to the application of ENV 1998-2;1994, Clause 7.2.2 and U.B.C. 1997, Division IV- Earthquake Regulations for Seismic Isolated Structures, clause 1659.4.2. These two clauses refer to the presentation of the design earthquake with the recorded accelerogramms for base isolated structures. With this example the author would like to provides design aids with the optimisation method which is necessery to be added to these two clauses.

- A number of time histories of recorded earthquakes can satisfy the requirements pursuant to clause 7.2.2 (ENV 1998-2: 1994). However, since this clause does not anticipate a criterion for controlling the S(Ti)/Se(Ti) ratio for periods equal or close to the natural period of vibration of the bridge structures, the recommended 1.3 ratio (importance factor $\gamma I = 1.0$) could be multiply exceeded. Therefore, the author suggests that the time histories satisfying clause 7.2.2, with a scaling factor ranging between 0.5 and 2.0 and $S(T_i)/S_e(T_i)$ ratio equal or close to 1.3 be selected by an optimization method.

- Taken as an example is a bridge with a natural period of vibration of T = 1.4 sec. 10 pairs of horizontal components of 6 occurred earthquakes (Montenegro 1979, April 16; Imerial Valley , 1940 May 18; San Fernando 1971, Feb 9; Parkfield 1966, June 27; Long Beach 1933, March 10 and West Washington) have been analyzed (Table). In accordance with the stated criteria for application of the so called optimization method, the following three pairs of horizontal components of recorded accelerograms of occurred earthquakes (Montenegro 1979, April 16; Imerial Valley , 1940 May 18 and Parkfield 1966, June 27) have been selected: YUE60, El Centro and Olympia Washington. The results from the optimization method are presented graphically in Figure1.

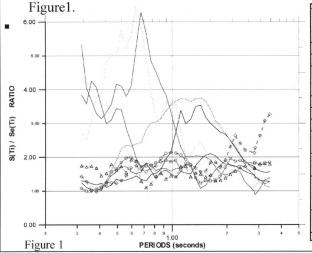

Figure 1

No.	Reccorded Accelerogram	Scaling Factor	S(Ti) / Se(Ti)
1	YUE59	1.008	3.53
2	YUE60	1.1967	1.93
3	El Centro	0.9781	1.23
4	YUE61	0.9012	3.74
5	YUE58	1.6255	2.46
6	YUE62	3.9075	2.026
7	Pacoima Dam	0.5560	2.66
8	Sholame Shandon	2.316	1.56
9	Vernon CMD Building	2.316	1.08
10	Olympia Washing	1.4666	1.50

References

- Ambraseys, N.N., K.A.Simpson and J.J.Bommer: *"Prediction of Horizontal Response Spectra in Europe"*, Earthquake Engineering and Structural Dynamics, Vol. 25, 371-400 (1996)

- Bolt,B.,*"Duration of Strong Ground Motion"*, Proc. V-th World Conf. Earth. Eng., Rome,1973,Vol.1, pp.94-102.

- Boore, D.M., W.B. Joyner & T.E.Fumal: *"Equations for Estimating Horizontal Response and Peak Acceleration from Western North American Earthquakes: A Summary of Recent Work"*, Seismological Research Letters, Volume 68, Number 1, January/February 1997

- Cornell C.A., *Engineering Seismic Risk Analysis*, Seis. Soc. Amer. Bull., V. 58, No. 58, 1968, pp. 1503-1606.

- Dobry, R., I.Idriss, E.Ng,"*Duration Characteristics of Horizontal Components of Strong Motion Earthquake Records*", Bull. Seis. Soc. Am. 68, No.5, pp.1487-1520,1978.

- Hristovski V., S. Stamatovska, V.Bickovski, D.Petrovski, PHARE Project: Detailed Design of the Upgrading of E-75 Section Demir Kapija-Udovo and Udovo-Gevgelija, Earthquake Engineering Part 1, "*Dynamic Analysis of Reinforced Concrete Viaduct 'Vodosirski'*, IZIIS, Skopje, April 1999, Report IZIIS 99-17.

- Hristovski V., S. Stamatovska, V.Bickovski, D.Petrovski, PHARE Project: Detailed Design of the Upgrading of E-75 Section Demir Kapija-Udovo and Udovo-Gevgelija, Earthquake Engineering Part 1,"*Dynamic Analysis of Reinforced Concrete Overpass 'Miletkovo'*, Skopje, May 1999, Report IZIIS 99-18.

- McGuire R. K., "EQRISK – *A Computer Program for Seismic Risk Analysis*", USGS, Open-file Report 76-67, 1976.

- Petrovski D., S.Stamatovska, A.Paskalov "*Hazard Compatible Seismic Input for Safety Analysis*", Second Japan-Turkey Workshop, February 23-25, 1998 Istanbul.

- Petrovski D., K. Talaganov, S. Stamatovska, D. Aleksovski, V. Sesov, D. Jurukovski, Phare Project: Detailed Design of the Upgrading of E-75 Section Demir Kapija-Udovo and Udovo-Gevgelija, Earthquake Engineering Part 1, "*Application of EC-8 in Design of Structures*", IZIIS, Skopje, January 1999, Report IZIIS 99-14.

- Petrovski D., S. Stamatovska, K.Talaganov, A. Paskalov, D.Hadzievski, M.Arsovski, "*Probabilistic Seizmic Hazard Analysis for NPP in Eastern Europe*", International Conference in Earthquake Engineering to Mark 40 Years from Catastrofic 1963 Skopje Earthquake and Successful City Reconstruction, Skopje-Ohrid, 26-29 August 2003.

- Sabetta F. & A. Pugliese: "Estimation of Response Spectra and Simulation of Nonstationary Earthquake Ground Motions", Bulletin of the Seismological Society of America, Vol.86.No.2, pp.337-352, April 1996.

- Sadigh, K., C.-Y.Chang, J.A.Egan, F.Makdisi & R.R.Youngs: "Attenuation Relationships for Shallow Crystal Earthquakes Based on California Strong Motion Data", Seismological Research Letters, Volume 68, Number 1, January/February, 1997.

- Spudich, P., W.B.Joyner, A.G.Lindh, D.M.Boore, B.M.Margaris & J.B.Fletcher: "SEA99: A Revised Ground Prediction Relation for Use in Extensional Tectonic Regimes", Bulletin of Seismological Society of America, 89, 5, pp.1156-1170, October 1999.

- Stamatovska S.,*" Design Response Spectra – Deterministic and Probabilistic Approach*", Albanian Journal of Natural &Technical Science (AJNTS), No.10/2001, pp. 113-123

- Stamatovska S., "*A New Azimuth Dependent Empirical Strong Motion Model for Vranchea Subduction Zone*", 12-th European Conference of Earthquake Engineering (ECEE), London, 9-13 September, 2002.

- Stamatovska S., "*Design Response Spectra for a Rock Site in the City of Skopje*", International Conference in Earthquake Engineering to Mark 40 Years from Catastrofic 1963 Skopje Earthquake and Successful City Reconstruction, Skopje-Ohrid, 26-29 August 2003.

- Talaganov K., Petrovski D., Stamatovska S., Petkovski R., Hadzievski D., "Reconciling Geophysical and Geotechnical Models for Earth Structures in the Common Political Borders – Case studies: *Seismic hazard study for lifeline systems*", First International Conference, Science and Technology for Safe Development of Lifeline Systems, Natural Risks: Developments, Tools and Techniques in the CEI Area, 4-5 November 2003, Sofia, Bulgaria.

- Weichert D.H, "Estimation of Earthquake Recurrence Parameters for Unequal Observation Periods for Different Magnitudes", Bull. Seism. Soc. Amer. Vol.70, No.4, pp1337-1347,1980

Contacts *(include at least one COST C12 member)*

- Dr. Snezana STAMATOVSKA, Associate Professor
  Address: Institute of Earthquake Engineering and Engineering Seismology University "Ss. Cyril and Methodius", Salvador Aljende 73, Skopje , Republic of Macedonia
  e-mail:snezana@pluto.iziis.ukim.edu.mk, Phone:+389-2-3176155, Fax:+389-2-3112163

LANDSLIDE MODELLING

Description

- Landslide actions are natural actions, since they are caused by the sliding of a given amount of geological material along the maximum slope direction of a hill or mountain, as a consequence of the gravity acceleration.
- Landslide actions may be considered as variable actions, acting either for very short duration and impacting the construction with a given velocity field.
- In the most general case, the sliding material consists of a cohesive or incoherent mixture of soil, water, rocks and debris.
- Within the general definition of landslides, a distinction can be made between mudflows and debris flows. On the one hand, mudflows are characterised by a cohesive mixture of soil and water, seldom including stones and debris (figure 1); on the other hand, debris flows basically consist of an incoherent amount of soil, stones and debris, or even of a single boulder rolling down the hill.

Figure 1. Effects of a mudflow (Sarno, Italy, May 1998) on a steel trestle and on several reinforced concrete buildings with masonry curtain walls.

Field of application

- During Hydrogeological Disasters numerous landslides have impacted the constructions.

- Here only mudflows giving rise to either impulsive, or dynamic, or quasi-static loading regimes on structural elements are considered.

Technical information

- The pressure of a fluid stream against a wall is given by the relevant hydrodynamic relationships. Generally speaking, it depends on the average density and velocity of the stream, on the height of the flow, as well as on the geometry of the impacted structure (i.e., its shape coefficient and the inclination angle with respect to the direction of the fluid).
- The density of the mudflow depends on the density of the dry soil, as well as on the water content of the mixture. The average density, as well as the mudflow height, may be determined either by means of post-event surveys or by means of geological investigations on the layers of the hill or mountain which are liable to slide down.
- The velocity of the mudflow depends on a number of factors, mainly: the slope of the hill or mountain, the average density and viscosity of the flowing material, the friction coefficient between the sliding material and the fixed surface underneath (e.g., the bedrock, the bed of a river or channel, the pavement of a road, etc.).
- As far as the direction of the flow is concerned, it may be assumed coincident with the maximum slope direction of the hill or mountain (in "free-field" conditions), or with the slope of the axis of the river, channel, road, in which the material is flowing.

Structural aspects

- Several types of structural failures may take place under a mudflow, depending on:
  - position of the construction with respect to the flow direction (i.e., frontal or lateral) and flow front (i.e., internal or marginal);
  - level of kinetic energy of the flow;
  - structural typology (i.e., reinforced concrete / steel framed buildings, masonry buildings, etc.) and structural geometrical parameters.
- In reinforced concrete framed buildings, the following damage typologies have been reported:
  - failure of the ground floor infill-masonry external walls, directly impacted by the flows, without significant damages to the structural parts (columns and beams);
  - collapse of isolated structural elements (mostly columns), with the formation of plastic hinges at the ends and/or at the midspan of the element, but without involving any collapse mechanism of the whole structure;
  - collapse mechanisms of the whole building (generally, ground-floor mechanisms have been detected);
  - translation of the whole building, or part of it, as a rigid body, as a consequence of the ground floor bearing structures.
- In masonry (i.e., tuff or brick) buildings, the following damage typologies have been reported:
  - failure or collapse of the ground floor bearing walls, but without involving any collapse mechanism of the whole building;
  - collapse of the whole building.

Research activity

- Theoretical and analytical analysis aiming to model the effects of the debris flows on the constructions are carrying out at the University of Salerno.

Method of analysis

- Mudflow effects on structures may be represented by means of pressure distribution or applied kinetic energy of the mudflow itself. In both cases, the applied pressure or the velocity field of the flow must be determined through the relevant hydrodynamic relationships.
- If the structure or structural element under consideration is surrounded by the mudflow, the kinetic energy is transmitted directly as impact energy. Instead, if it is located in the marginal part of the mudflow, a tangential viscous dynamic action occurs.
- Generally speaking, analysis of structures or structural elements subjected to mudflows should be carried out by means of finite element codes.
- Simplified methods, based on linear elastic-perfectly plastic or rigid-plastic material models, may be adopted in order to evaluate the collapse load of structures or structural elements, under the hypothesis of small displacements.
- In particular, rigid-plastic analysis methods may be adopted to evaluate the collapse load, under hydrodynamic actions, of:
  - masonry ground-floor panels inserted in a reinforced concrete structural framework: an arc-resisting collapse mechanism is developed, provided no appreciable damage takes place on the surrounding structural elements *(Figure 2)*;

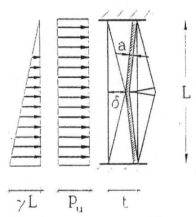

Figure 2. Collapse mechanism in a masonry panel inserted in a reinforced concrete structural framework.

  - masonry ground-floor walls, belonging to a masonry building: a bending-type vertical-resistant scheme and/or a shear-type horizontal-resistant scheme may be developed *(Figure 3)*;
  - reinforced concrete columns at ground floor: a two-plastic-hinge *(Figure 4a)* or a three-plastic-hinge *(Figure 4b)* collapse mechanism takes place, according to the fact that ground-floor mechanism or failure of the single column take place, respectively;
  - reinforced concrete ground-floor columns: a two-plastic-hinge shear-failure mechanism *(Figure 5)* may take place.

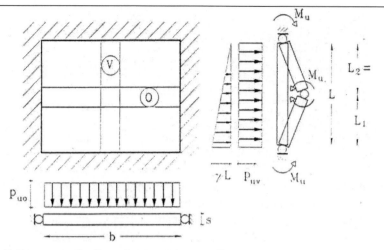

Figure 3. Collapse mechanisms in a masonry wall.

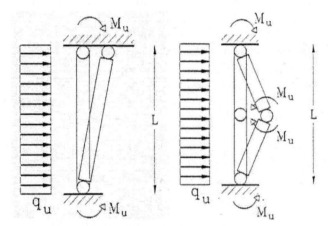

Figure 4. Two-plastic-hinge and three-plastic-hinge collapse mechanisms in reinforced concrete columns at ground floor.

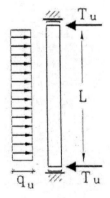

Figure 5. Two-plastic-hinge shear-failure mechanisms in reinforced concrete columns at ground floor.

Results of analysis

- With reference to partial and global possible collapse mechanisms, it is possible to evaluate the collapse resistances of structural and non-structural elements impacted by fluids in movements
- The mechanical interpretation of the damages suffered by the constructions allows to evaluate suitable ranges of the impact velocities of the debris flows on the constructions.
- The last result is the realization of guidelines devoted to allowing the rebuilding of constructions able to resist debris flows phenomena.

Further developments

- the design of direct and indirect foundations o buildings, taking into account the horizontal actions due to debris flow impact.
- the design of structures, reducing the impact surface of the single members by means of appropriate cross-section shape and planimetric position.
- the design of structures, avoiding distance between columns less than 5.0 m, in order to prevent the formation of accidental obstructions to the free flows.

References

- Faella, C., Nigro, E., 2001. Effetti delle Colate Rapide sulle Costruzioni. Parte I: Descrizione del Danno. *Meeting "Fenomeni di Colata Rapida di Fango nel Maggio '98"*, Napoli, Italy, 22nd June.
- Faella, C., Nigro, E., 2001. Effetti delle Colate Rapide sulle Costruzioni. Parte II: Valutazione della Velocità di Impatto. *Meeting "Fenomeni di Colata Rapida di Fango nel Maggio '98"*, Napoli, Italy, 22nd June.
- Faella, C., Nigro, E., 2001. Debris flow effects on constructions: damage analysis, collapse mechanisms, impact velocities, code provisions. COST C12 document No. C12-WG2-0206

Contacts

Prof. Dr. Eng. Ciro Faella
> Address:Department of Civil Engineering – Faculty of Engineering, University of Salerno, via ponte don Melillo 1, 84084 Fisciano (SA), Italy.
> E-mail: c.faella@unisa.it, Phone: 0039. 089.964043

FIRE MODELLING

Definition

- *Design fire*: a specified fire temperature development assumed for structural design purposes.

- *Fire resistance*: the ability of a structure, a part of a structure or a member to fulfil its required functions (load bearing function and/or separating function), for a specified load level, for a specified fire exposure and for a specified period of time.

- *Fire scenario*: a qualitative description of the course of a fire with time identifying key events that characterise the fire and differentiate it from other possible fires. It typically defines the ignition and fire growth process, the fully developed stage, decay stage together with the building environment and systems that will impact on the course of the fire.

- *Indirect fire actions*: internal forces and moments caused by thermal expansion.

- *Load bearing function*: the ability of a structure or a member to sustain specified actions during the relevant fire, according to defined criteria.

- *Nominal fire*: conventional design fire, adopted for classification or verification of fire resistance, e.g. the standard temperature-time curve.

- *Separating function*: the ability of a separating element to prevent fire spread (e.g. by passage of flames or hot gases) or ignition beyond the exposed surface during the relevant fire.

- *Standard temperature-time curve*: a nominal curve for representing a model of a fully developed fire in a compartment.

Physical features

- Building structures may be submitted to the effects of fire during their lifetime. The cause of the fire may be accidental or intentional (arson)

- Most casualties in building fires are caused by smoke or heat effects.

- It is nevertheless essential to maintain the integrity of the structural system during a certain amount of time (this time is called the "fire resistance" time) in order to:
  - limit the extend of the fire development (compartmentation);
  - allow evacuation of the occupants;
  - allow a safe environment for the action of the fire fighters;
  - protect neighbouring properties.

- The evaluation of the fire resistance time was historically made by experimental tests, but these tests have several severe shortcomings.

- Numerical modelling can supplement experimental testing and, in some cases, for example for structures that are too large to be tested, is actually the only way by which the fire resistance time can be evaluated.

- Modern building structures of certain sizes are usually based on a skeleton of bars (beams and columns) that can be made in various materials. The most commonly used structures are in steel, in reinforced concrete, in prestressed concrete, in composite steel-concrete combination or in timber. The load bearing function of these elements has to be ensured.

- Vertical separating walls can be present on the beams. They can be made of reinforced concrete, of masonry or of gypsum plaster boards. Horizontal floors are usually based on concrete materials if they have to provide a separating function. Timber floors are encountered in older buildings. The separating function of these elements has to be maintained.

Effects on constructions

- The elevated temperatures of the gases in the compartment as well as direct radiation from the flames heat the structure and increases its temperature.
- The strength of most manufactured building materials like steel, concrete, glass or aluminium decreases as their temperature increases. This sole effect results in the progressive decrease of the load bearing capacity of the structure.
- The stiffness of these materials also decreases as their temperature increases. The large displacements generated by this stiffness degradation can lead to additional loadings, for example via second order geometrical effects, that have also the tendency to reduce the load bearing capacity of the structure.
- If the thermal elongation that these materials exhibit is restrained, this also creates some indirect effects of action that can diminish the load bearing capacity of the structure.
- Wood behaves differently. The load bearing section of the members progressively decreases in dimensions because of the progressive extension of the charred layer from the surface to the inside of the section. The mechanical properties of the material in the residual section are not that much affected and thermal expansion is also quite inexistent. The decrease of the load bearing capacity results from the decrease of the dimensions of the load bearing elements.

Modelling

- The environment created by the fire in the compartment has to be modelled. The usual way is to represent the fire by a single gas temperature, the evolution of which is given as a function of time. More refined methods exist that allow representing the local effects of localised fires.
- Based on the fire attack, the evolution of the temperatures in the building structure has to be modelled. Uniaxial temperature fields are normally considered in flat slabs and concrete walls, whereas plane 2D temperature distributions have to be evaluated in the cross-section of beam or column type elements. The 3D effect resulting from longitudinal flux along the members is normally neglected.
- The capacity of the structure to sustain the applied loads is checked by a step by step procedure, minute after minute during the fire, up to the moment of failure. The evolution of the temperatures is also monitored in separating elements in order to check their ability to prevent transmission of the fire to adjacent compartments.

Guidelines and/or codification

- In Europe, the relevant codes for designing structures subjected to fire are the fire parts of the Eurocodes from Eurocode 1 to Eurocode 6 and Eurocode 9, see the references below.

Example of application

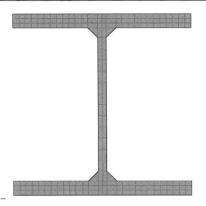

Figure 1. Discretisation of a steel cross-section

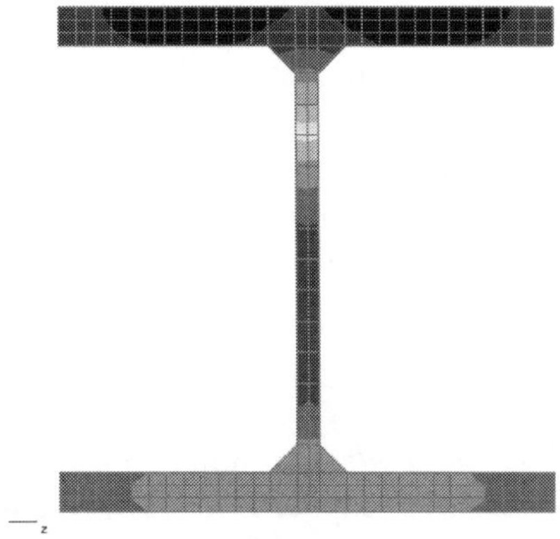

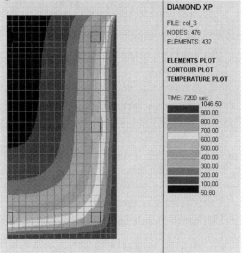

Figure 2. Isotherms in the steel section

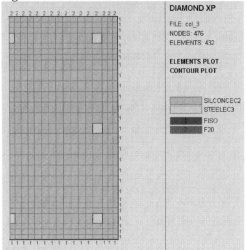

Figure 3. Discretisation of a concrete cross-section

Figure 4. Isotherms in the concrete section

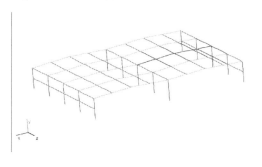

Figure 5. Discretisation of an industrial hall

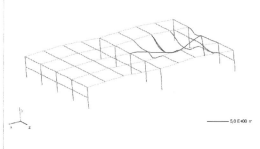

Figure 6. Deflections at failure

- Figure 1 shows the discretisation in finite elements of the cross-section of a hot rolled I profile. This profile is heated on three sides.
- Figure 2 shows the isotherms that have been calculated in this section after 1 hour of the applied fire curve.
- Figure 3 shows the discretisation in finite elements of the cross-section of a rectangular reinforced concrete section. Only ½ of the section is analysed owing to symmetry reasons. This profile is heated on three sides.
- Figure 4 shows the isotherms that have been calculated in this section after 2 hours of the applied fire curve.
- Figure 5 shows an industrial hall in which only one part of the structure, in red on the Figure, is subjected to the action of a localised fire. The columns could for example be constituted of members represented by the section and the temperatures depicted on Figure 4 and the beams by the sections and the temperatures depicted on Figure 2.
- Figure 6 shows the deflected structure at failure.

References

- *PrEN 1991-1-2, Eurocode 1 – Actions on Structures. Part 1-2 : General Actions – Actions on structures exposed to fire*, Final Draft, Stage 49 draft, European Committee for Standardization, Brussels, 10 January 2002.
- *PrEN 1992-1-2, Eurocode 2 : Design of concrete structures - Part 1.2: General rules - Structural fire design*, Draft, Stage 49, European Committee for Standardization, Brussels, October 2002.
- *PrEN 1993-1-2, Eurocode 3 : Design of steel structures. Part 1.2 : General rules. Structural fire design*, Stage 49 draft, European Committee for Standardization, Brussels, 17 April 2003.
- *PrEN 1994-1-2, Eurocode 4 : Design of composite steel and concrete structures. Part 1.2 : General rules - Structural fire design*, Final Draft, Stage 34, European Committee for Standardization, Brussels, 7 May 2003.
- *PrEN 1995-1-2, Eurocode 5 : Design of timber structures. Part 1.2: General rules - Structural fire design*, Final Draft, European Committee for Standardization, Brussels, 9 October 2002.
- *PrEN 1996-1-2, Eurocode 6 : Design of masonry structures. Part 1.2: General rules - Structural fire design*, European Committee for Standardization, Brussels
- *PrEN 1999-1-2, Eurocode 5 : Design of aluminium structures. Part 1.2: General rules - Structural fire design*, European Committee for Standardization, Brussels.
- Franssen, J.-M., 2004. *SAFIR. A Thermal/Structural Program Modelling Structures under Fire*, accepted for publication in Engineering Journal, AISC.

Contacts

- Franssen Jean-Marc
  Address: Department "Mécanique des matériaux et Structures", University of Liege, 1, Chemin des Chevreuils, 4000 Liège, Belgium.
  e-mail: jm.franssen@ulg.ac.be, Phone: +32-4-366.92.65, Fax: +32-4-366.95.34.

COST C12 - WG2

Datasheet I.2.5.1
Prepared by: Zivko Bozinovski
      Institut of Earthquake Engineering and Engineering Seismology Skopje

STRUCTURES DAMAGED DUE TO QUARRY AND MINING EFFECTS

Definition

- Loading type is man-made, dynamic, impulsive, short duration, repeated menu times

Physical features

- According to available geological documentation, usually the terrains on which the quarry is located, the villages and the ares in-between are composed of Paleozoic grannodiorites, Triassic limestone, Pliocene sediments and Quaternary deluvial, prolluvial and alluvial sediments. The bedrock is composed of Paleozoic grannodiorites, mainly compact and hard rocks, with occasional gritting material upon the surface. These sediments are most frequently mixed, badly sorted out and not layered, with different compactness and heterogeneous, with mainly weak physical - mechanical characteristics.

- The structural systems of buildings are consist of bearing walls constructed of stone or brick in both orthogonal directions, founded on strip foundation, on soil of relatively good bearing capacity. The attics and the roof structures are made of timber, with bearing capacity usually in the transverse direction of the structures. The roof covering of the structures is constructed of interlocking pantiles. The structures have been constructed for the last 40 years, by use of a material of a relatively good quality, without vertical and horizontal reinforced concrete belt courses. Only a few structures contain belt courses and partial reinforced concrete floor structures.

Effects on constructions

- Due to continuous activities in the quarry, there takes place, first of all, disturbance of the stone - mortar, i.e., brick - mortar connection. Depending on the intensity of the impact, there first of all occur micro cracks that are gradually widened and form a network of cracks.

- Stone walls with visible network of cracks between stones and mortar are frequently observed. At places with exceeded capacity of deformability of material, there occur large scale continuous cracks . ...

- The network of cracks does endanger the stability of the walls themselves, whereas the continuous cracks endanger the integrity and the stability of the bearing structural system of the structures

- The structures have visible vertical and diagonal cracks of different size in the bearing walls.

- There is also separation of transverse and longitudinal walls not only in the bearing walls but also the partition ones.

- There are visible fine cracks and cracks in the mortar of the walls and the ceilings of the structures that have not undergone interventions for the last year. However there are also such structures that had been re-plastered prior to the inspection. In these structures, only fine cracks are visible.

- Near the structures, one can see pieces of rock of different size originating from the quarry. Broken tiles that have been replaced by new ones in the meantime are also observed.
- No uneven settlement of the structures is observed.
- The maintenance of the structures is relatively good, which is proved by the fact that no continuous water leakage along the walls or the ceilings is observed.

Modelling

- Due to the occurred damage to the structures, the integrity of the bearing structural systems is considerably disturbed and it is for certain that mining in the quarry shall continue.
- It is therefore necessary to repair and strengthen these structures in order to bring them into the state prior to the mining. Such treated structures shall be able to sustain further vibrations provided that the mining be performed in accordance with strictly recommended and controlled conditions that shall be defined on the basis of detailed investigations from the aspects of intensity of mining, regime of usage, etc.
- Loading type is dynamic, impulsive, short duration.

Minimal and necessary interventions for repair and strengthening of the structures are anticipated. Repair consists of repair of elements and their connection into an integral structure that shall be capable of sustaining and transferring vertical - gravity and horizontal -vertical dynamic effects.

Vertical jackets are anticipated to be inserted at the corners and in the middle part of the bearing external walls founded on own footings while horizontal belt courses are envisaged to be incorporated at the level of the floor and roof structure.

The cost of damage depends on the extent of damage, the disposition of bearing elements in both orthogonal directions, their strength, stiffness and deformability characteristics and the necessary activities for making the structures functional. We herewith note that there is no clear relation between extent of damage and cost of damage because each concrete case needs a different level of interventions for making the structure functional.

The cost of the total damage and the maximal estimated amount to which we can decrease the damage cost due to damage from other reasons shall be defined for each individual structure.

Static analysis of the structure under vertical loads and equivalent seismic forces

During the serviceability period, the structures are exposed to the effects of vertical and horizontal, static and dynamic, permanent and occasional, normal and abnormal loads. The vertical loads, dead or live, are mainly static and depend on the purpose of the structure, the material from which they are constructed and the finishing of the elements. The mathematical model of the structure is defined as a system with concentrated masses at the level of the floor structures, connected by springs and dampers. The system is fixed on a rigid or elastic base. The stiffness of the elements and the system as a whole is defined according to standard formulae for masonry structures and corresponding fixation conditions. Stiffness is more realistically defined on the basis of the working diagrams of elements, from the beginning of loading up to failure.

Analysis of the elements up to ultimate strength and deformability states

For the structural elements with geometrical characteristics, characteristics of materials and position in the structure, analysis of elements is done and hence analysis of the structure up to ultimate states of strength, stiffness and deformability. For several types of walls as are the stone walls, brick walls, walls of stone or brick with reinforced concrete jackets, confined brick masonry with reinforced concrete vertical and horizontal belt courses and stone walls with a concrete cladding, there is a simple, but sufficiently exact way of determining the capacities of strength, stiffness and deformability, in the linear and nonlinear range of behaviour. The deformation in the same range is defined by the strength/stiffness ratio.

Nonlinear dynamic response

To predict the nonlinear dynamic response of a structure with discrete masses, the shear force - relative displacement relation should be established, i.e., the stiffness should be defined, known as hysteresis curve. The results from the experimental investigations of different types of structures point to extremely complex hysteretic behaviour of elements and the system as a whole. In practice, different authors have defined different idealized analytical hysteretic relationships and have developed several computer programmes for prediction of the nonlinear dynamic response of the structures. Recorded relationship acceleration-time is shown in Fig.1.

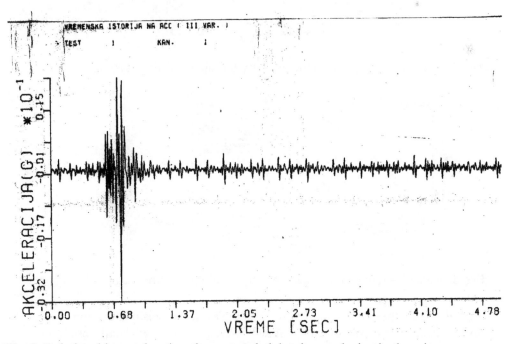

Fig. 1. Relationship acceleration-time, recorded duaring explosion in the mine.

Repair and strengthening

75

Based on the analysis of the existing state of the structure, comparison is made between the strength and deformability capacities of the bearing elements and the system as a whole in both the orthogonal directions with those required according to the regulations and those required for the analyzed response of the structure under expected earthquakes at the considered site, with intensity and frequency content. If the strength and deformability capacities are lower than the required, it is considered that the structure, it is concluded that the structure does not possess sufficient strength and deformability wherefore its repair by strengthening is necessary.

The solution for repair and/or strengthening depends on: the seismicity of the site, the local soil conditions, the type and age of the structure, the level and type of damage, time available for repair and/or strengthening, equipment and staff, restoration and architectonic conditions and requirements, economic criteria and required seismic safety. From the several analyzed possible solutions for repair and strengthening, selected is the most favourable from the aspect of stability, i.e., fulfillment of the design criteria according to the regulations, the possibility for realization of the solution, the available materials, the economic justification, the fulfillment of the sociological and aesthetic requirements. For the selected technical solution for repair and strengthening, performed is analysis of the stability of the structure (Fig. 2 and 3).

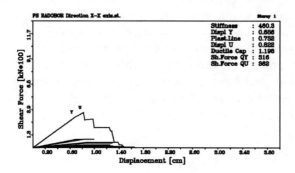

Fig. 2. Storey P - Δ diagram, stone masonry building, existing state

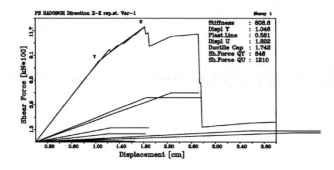

Fig. 3. Storey P - Δ diagram, stone masonry building, repaired state

For repaired and strengthened structural system, complex static, seismic, dynamic and ultimate analyses have been performed. Defined were the strength, stiffness and deformability capacities as well as the required strength, stiffness and deformability for vertical loads and seismic effects that might be expected on the considered site, with intensity and frequency content.

Analyses have shown that the strength and deformability capacities of existing structural systems are lower than the required which means that the structures do not possess sufficient strength and deformability. On the other hand, structures have suffer serious structural damage and have a considerably decreased bearing capacity. To make these structures functional, for the required seismic protection in accordance with the valid technical regulations, these have been repaired and strengthened.

Based on the strength and deformability capacities of the bearing elements and the structure as a whole and on the basis of required strength and deformability for expected seismic effects with intensity and frequency content, conclusions are drawn regarding the stability of the structure and its vulnerability level. It is of exceptional importance to bring the strength, stiffness and deformability of the structure within the frames of the requirements according to the valid technical regulations and latest knowledge on the behaviour of masonry structures exposed to gravity and seismic effects.

Conclusion on the stability of structures

Based on the achieved capacities of strength, stiffness and deformability of the bearing elements and the structures as a whole, in respect to those required for expected gravity seismic effects, with intensity and frequency content, it may be concluded that the realized solutions of repair and strengthening satisfied the requirements of the valid technical regulations and most recent knowledge on the behaviour of this type of structures.

Repair, strengthening, adaptation, reconstruction, enlargement and revitalization

The main structural system of the existing buildings usuallly consists of bearing walls constructed of stones or bricks in lime mortar in both the orthogonal directions. The foundation structure is mainly of a strip type, constructed of stone in lime mortar, with a minimal lateral extension of the walls. The floor and the roof structures are constructed of timber and the roof is covered with tile. To make these buildings functional, it is necessary to repair and strengthen them in ordar to satisfy the functioning requirements and to meet the existing technical regulations.

Based on the above mentioned data, a detail analysis of the existing principal structural system is performed by which the need of repair and strengthening of the elements of the structural system is defined in order to meet the main criteria for strength, stiffness and deformability in compliance with the valid technical regulations and the modern knowledge on behaviour of masonbry structures exposed to static and dynamic loads.

Based on the required strength and deformability characteristics of the elements and the system as a whole, several vairant solutions for strengthening of the structure are proposed and analyzed in accordance with the Book of Regulations for Construction of Buildings in Seismically Active Areas on one hand and the possibilities for adding new elements to the structure on the other.

Out of a number of variant solutions, selected is the most appropriate one from economic aspect and the aspect of satisfying the strength and deformability requirement according to the valid technical regulations.

The following is anticipated by the solution for repair and strengthening of the principal structural system:

- Guniting of all the stone walls in the basement;
- Replacement of the timber floor structures by flat reinforced concrete cross plates;
- Strengthening of the bearing walls with reinforced concrete jackets from the inside and the outside, with a thickness of minimum d = 8 cm and necessary reinforcemen. At the level of the floor structures, the jackets are connected with the horizontal belt courses.

The reinforcement concentrated in the jackets is permeated through the floor structures, while the uniformly distributed ones runs from storey to storey and is cast from place to place with the reinforcement of the floor structures or the horizontal belt courses. The jackets are appropriately connected with the stone or brick walls through dowels with an appropriate reinforcement and over metal anchors;

- Rebuilding of the existing walls by walls of solid bricks in cement lime mortar, according to the architecture of the building and the requirements for a minimal number and distribution of the bearing walls in both orthogonal directions;
- Foundation of the jackets is done via individual footings which form an entity with the existing foundation;
- Repair or construction of a new roof structure;
- Construction or revitalization of installation for the outlet of atmospheric water from the roof and in the near surrounding of the building;

The reinforced concrete elements of the structure are proportioned according to the ultimate bearing capacity theory, i.e., according to PAB 87, in compliance with the valid technical regulations.

The analysis of the strength and deformability capacity of the repaired and strengthened structure of the building provides an insight into whether the strength and deformability requirements are satisfied according to the valid regulations and recent knowledge on behaviour of such type of structures exposed to gravity and seismic effects.

Guidelines and/or codification

- Book of Regulations for Construction of Buildings in Seismically Active Areas

Example of application

- Example of damged building and their repair and strengthening well be explain.

References

- Bozinovski, Z., Stojanovski, B., and Petrusevska, R., "Repair and Strengthening of the Main Structural System by Analysis of the Stability of the Primary School Buildings in the Commune of Bitola, IZIIS Report 95-26 to 95-45, 1995.

Contacts *(include at least one COST C12 member)*

- Zivko Bozinovski

Address: Institute of Earthquake Engineering and Engineering Seismology,
              University "St. Cyril and Methodius", Skopje, Republic of Macedonia.
                e-mail: zivko@pluto.iziis.ukim.edu.mk,
                Phone:++389-2-3176 155, Fax: +389-2-3112-163

COST C12 - WG2

Datasheet I.2.5.2

Prepared by: A. Kozłowski, Z. Pisarek, B. Stankiewicz. L Ślęczka
Rzeszow University of Technology, Rzeszow, POLAND

MINING SUBSIDENCE EFFECTS ON BUILDINGS

Definition

- Underground mining activity often causes the ground movements at the surface, which can cause considerable damage of buildings and structures placed in the area above underground mining works.

- Influence of mining subsidence on building located in mining damage areas depends not only on the type of building structure, but also on forecasted parameters of mining extraction methods and kind of soil.

Physical features

- Depending on the extraction methods ("caving in" method, abandoned rooms and pillars method), underground mining works create underground voids, which may cause mining subsidence of very high accidental nature.

- Subsidence caused by underground mining activity influences building located in mining damage regions, causing (Figure 1):
    - ➢ lowering of ground level (w)
    - ➢ horizontal deformation of the ground (ε)
    - ➢ ground curvature (R)
    - ➢ slope of the ground (T)

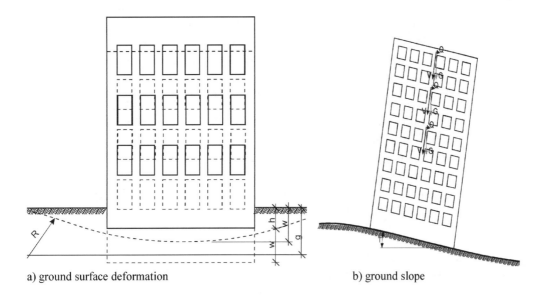

a) ground surface deformation   b) ground slope

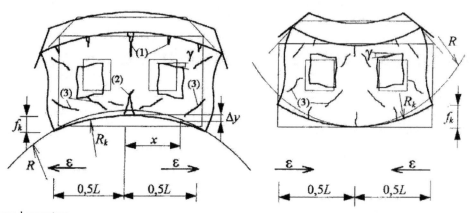

c) ground curvature

Figure 1. Effects of mining subsidence

Effects on constructions

- Building structure should be able to take following loads and forces arising as a result of continuous deformations of the ground:
  - ➢ caused by horizontal deformations of the ground, mainly taken by foundation,
  - ➢ caused by bending of the ground, which leads to arising additional internal horizontal forces on the whole height of the structure,
  - ➢ depended on slope changing of the ground, it is significant when building have more than five stores,
  - ➢ caused by shocks, they lead to arising horizontal inertial loads that have to be taken by the whole structure.
- Time depended characteristic of mining damage progress and characteristics of mining shocks should also be taken into consideration in design.

Modelling

- Horizontal forces occurring in the base of building foundation depend on: shear stresses on a bottom surface of foundations, soil pressure on vertical surface of foundations, shear stresses on vertical surface of foundations. The schema of loadings induced by horizontal soil displacements is shown in Figure 2.

  Horizontal tensile forces N can be calculated from equation:

$$N = Z + Z_b + J + H \tag{1}$$

- The value of forces Z induced by shear stresses on a bottom surface of considered continuous footing can be calculated from equation:

$$Z = b \cdot (0{,}5L - x) \cdot \Theta \tag{2}$$

where b is the width of footing or foundation plate,
L is the length of the foundation,
x is coordinate of the considered point regarding the center of the footing,

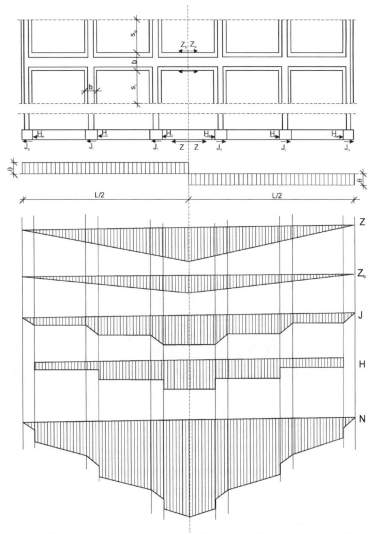

Figure 2. The schema of loadings and stresses induced by horizontal displacements of soil in the footing.

The ultimate value of shear stresses on a bottom surface of continuous footing can be predicted from equation:

$$\Theta = K \cdot (\sigma \cdot tg\Phi + c) \qquad (3)$$

where: K is the coefficient depended on soil mechanical properties and values of the normal stresses under foundations,

σ is the value of the normal stresses under foundations,

Φ is the angle of friction of the ground,

c is the cohesion of the ground.

- The value of forces Z_b in the plane parallel to soil deformations, in any points of continuous footing, can be calculated from equations:

$$Z_{bx} = \begin{cases} 0{,}75 \cdot \dfrac{h}{b} \cdot Z_x & \text{for } \dfrac{h}{b} \leq 0{,}33 \\[2mm] \dfrac{h}{b} \cdot \dfrac{D}{\Theta} \cdot Z_x & \text{for } \dfrac{h}{b} > 0{,}33 \end{cases} \qquad (4)$$

The stresses D depended on ground mechanical properties:

$$D = \gamma \cdot h_s \cdot tg\Phi \cdot tg^2\left(45^o - \frac{\Phi}{2}\right) + c \cdot \left[1 - 2 \cdot tg\Phi \cdot tg^2\left(45^o - \frac{\Phi}{2}\right)\right] \tag{5}$$

h_s is the distance between the centers of foundation cross-section and ground surface,

    γ is the unit specific weight of the ground,

- The value of tensile forces J induced by ultimate shear stresses on a bottom surface of continuous footing in its any points, can be calculated from equation:

$$J = \sum_i^n J_i \cdot s_i \tag{6}$$

where: n is number of perpendicular continuous footing, adjacent to considered continuous footing, on distance from consider point to end of foundation,

     s_i is the length of perpendicular continuous footing,

     J_i is the force induced by shear stresses on a bottom surface perpendicular to the footing i, predicted from: $J_i = b_i \cdot \Theta$, where b_i is the width of perpendicular continuous footing i.

- The tensile forces H induced by ultimate soil pressure acting on a vertical surface of continuous footing, can be calculated from equation:

$$H = \sum_i^n H_i \cdot s_i \tag{7}$$

where H_i is the force from perpendicular continuous footing i, taken as minimum value from third conditions depended on normal and shear stresses in partition between perpendicular continuous footing, and soil pressure on vertical surface of fundations.

- Building subjected to a curvature induced by ground subsidence is calculated under assumption, that shape of subsiding trough is cylindrical. A ground curvature can be convex or concave (Figure 1). Additional stresses in building depend on the soil-structure interaction phenomena.

  Two methods are usually used to predict structural behaviour of building influenced by curvature of the ground: grillage foundation method and beam analogy method.

  In grillage foundation method flexural rigidity (EJ) and shear stiffness (GA) of the beams are predicted as stiffness of all structural elements of the building (walls).

Stiffness of elastic supports for each beam of grillage foundation can be predicted from equation:

$$k_r = b_r' \cdot c_o \tag{8}$$

where b_r' is equivalent width of continuous footing, and

          c_o is comparative coefficient of ground stiffness.

Distribution of horizontal forces in structure subjected to a concave curvature is depicted in Figure 3.

The values of forces acting on separated walls can be calculated from equation:

$$N = \frac{M}{z} \tag{9}$$

where z is lever arm of internal forces, which value depends on the ratio between the height of building and its length.

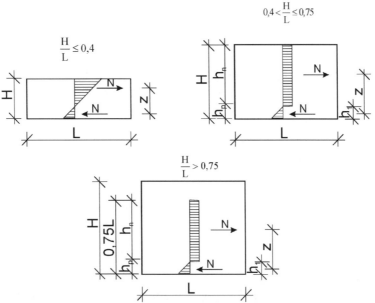

Figure 3. The distribution of horizontal forces in structure subjected to a concave curvature

In beam analogy method (Figure 4) the ground is modelled by spring elements adopting the Winkler model. Flexural rigidity (EJ) and shear stiffness (GA) of the beam are predicted as stiffness of all structural elements of the building.

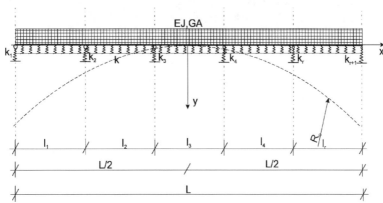

Figure 4. The ground model in beam analogy method.

Stiffness coefficient of ground is usually calculated from equation:

$$k = c_o \sum_{1}^{l} b'_{r,l} \tag{10}$$

where $b'_{r,l}$ is equivalent width of continuous footing on analyzed segment, and

c_o is comparative coefficient of ground flexibility:

$$c_o = 2 \cdot \omega \cdot \frac{E_o}{b} \tag{11}$$

where E_o is ground Young's modulus,

b is width of longitudinal continuous footing,

ω is the shape coefficient of foundation.

Guidelines and/or codification

- Buildings placed in the area above underground mining works with load-bearing walls must be design as rigid, they can not yield due to deformations compatible with deformations of subsoil.
- Each building should be checked if they are able to take horizontal loads caused by horizontal deformations of the ground. In case of the soil with original volumetric modulus of elasticity E_0 less or equal 80 MPa the analyses of influence of bending can be neglected.
- Influence of the slope of the ground should be considered when buildings have more than five stories and their height exceeds at least twice the dimension of shorter side of the floor projection of the building.
- Influence of mining shocks on buildings can be omitted if acceleration amplitude a_p does not exceed 250 mm/s^2 in buildings with five or less stories, and if acceleration amplitude a_p does not exceed 120 mm/s^2 in buildings with six to eleven stories.

References

- Ledwoń J. A., 1983. *Bauen in Bergschadengebieten*. Ernst & Sohn, Berlin, Germany.
- Deck O., Al Heib M., Homand F., 2003. Taking the soil–structure interaction into account in assessing the loading of a structure in a mining subsidence area. *Engineering Structures 25, pp. 435-448.*

Contact

- Aleksander Kozlowski
  - Address: Department of Civil Engineering, Rzeszow University of Technology, 2 Poznanska street, Rzeszow, Poland.
  - e-mail: kozlowsk@prz.edu.pl, Phone/Fax: +48 17 85 429 74.

ENERGY DISSIPATION TECHNIQUES FOR SEISMIC PROTECTION OF BUILDINGS

Description

- Use of special energy dissipation devices to absorb kinetic energy induced by earthquake actions;
- Reduction of structural damage under seismic excitation;
- Ease of application and reversibility;

Unlike the conventional approach to seismic design, based on the ductility resources of structural members and connections, the energy dissipation techniques are based on the response control concept, with the purpose of controlling and limiting the dynamic effects on the structural elements by means of special devices. The response control concept can be explained referring to the dynamic equilibrium equation of a generic structural system, expressed by the relationship:

$$m\ddot{x}(t) + c\dot{x}(t) + kx(t) + f_c(t) = F(t)$$

in which m, c and k are mass, viscous constant and stiffness of the system, respectively. $F(t)$ is the external force induced by earthquake, and f_c is the control force applied by means of additional devices. In a general form, it is given by:

$$f_c(t) = m'\ddot{x}(t) + c'\dot{x}(t) + k'x(t)$$

where m', c' and k' are mass, viscous constant and stiffness of the controlling system, respectively. Values of m', c' and k' can be chosen in such a way to modify one or more terms of the energy balance equation:

$$E_K(t) + E_\xi(t) + E_S(t) + E_H(t) = E_I(t)$$

where:

E_k is the kinetic energy of the system;

E_ξ is the energy dissipated in the special devices;

E_S is the energy due to elastic deformation in the structural system (strain energy);

E_H is the energy dissipated by cumulated plastic deformation in the structural members (hysteretic energy);

E_I is the earthquake input energy;

In energy dissipation systems special devices are used in order to reduce the amount of energy dissipated in the structures E_H, with a corresponding increase of energy dissipated by devices themselves (Figure 1). This reduces both structural response and damage in structural elements. Basically, the effect of special devices is to increase the overall damping properties of the system, thus reducing its maximum spectral acceleration (Figure 2). Possible device configurations in framed buildings are illustrated in Figure 3.

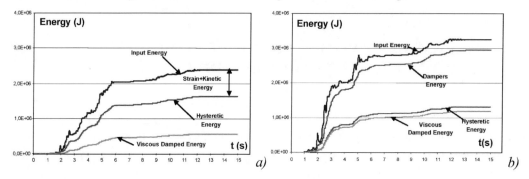

Figure 1. Diagram of dissipated energy in a structure: without dampers (a); with dampers (b)

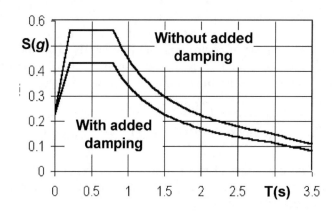

Figure 2. Spectral acceleration as a function of the structural damping

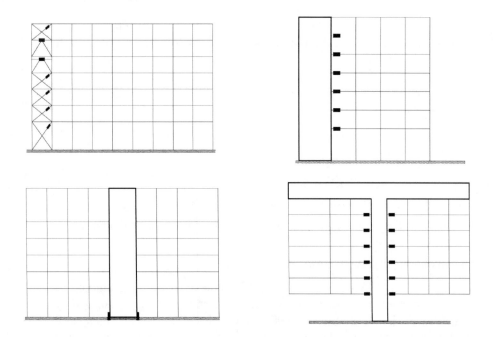

Figure 3. Possible configurations of energy dissipation devices

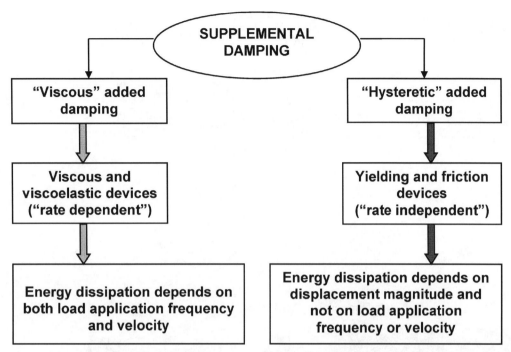

Figure 4. Basic options for structural supplemental damping

Technical Information

Energy dissipation devices can be distinguished according to the scheme illustrated in Figure 4. Accordingly, devices can be classified as follows:

- Viscous devices – They are based on the viscous properties of a fluid (Figure 5);
- Visco-elastic devices – They are based on the visco-elastic properties of a solid material (Figure 6);
- Yielding metal devices – They are based on the cycling plasticity of a metal (Figure 8);
- Friction devices – They are based on the friction between surfaces in contact (Figure 9);

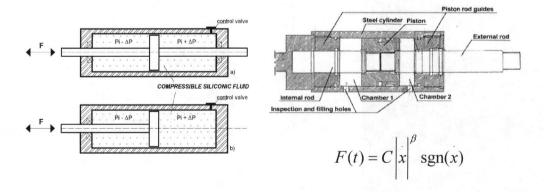

$$F(t) = C \left| \overset{.}{x} \right|^{\beta} \mathrm{sgn}(\overset{.}{x})$$

Figure 5. Viscous devices

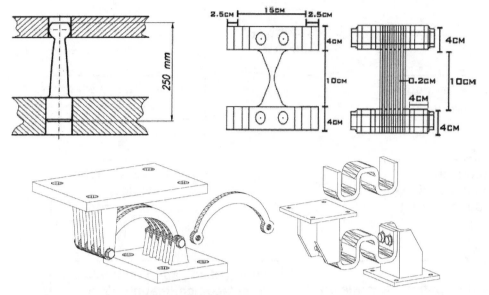

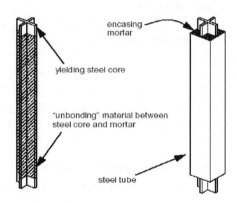

Uniform yielding steel devices

Special yielding device for bridges

Buckling restrained brace

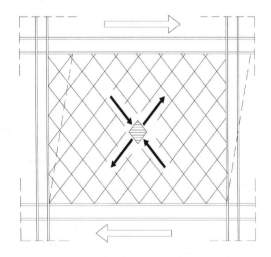

Dissipative panels

Figure 6. Yielding hysteretic devices

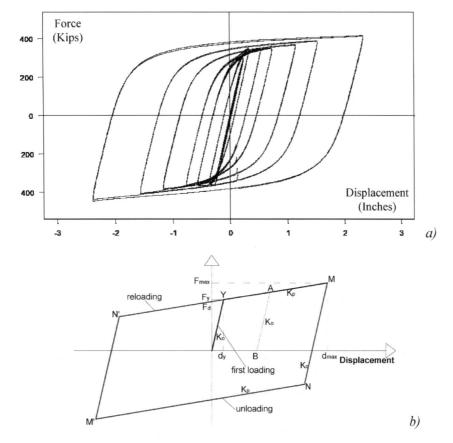

Figure 7. Hysteretic diagram for yielding devices: experimental (a), theoretical (b)

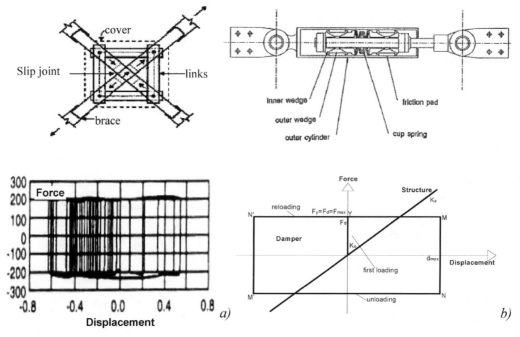

Figure 8. Friction devices with relevant experimental (a) and theoretical (b) hysteresis diagrams

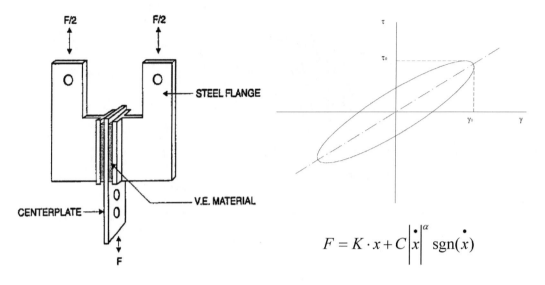

$$F = K \cdot x + C \left| \overset{\bullet}{x} \right|^{\alpha} \mathrm{sgn}(\overset{\bullet}{x})$$

Figure 9. Visco-elastic device with relevant theoretical hysteresis model

Field of Application and End Users

Such technique can be profitably used in both buildings and bridges.

- Frame buildings
    - i) Reduction of structural response under seismic actions:
    - ii) Reduction of interstorey drift;
    - iii) Reduction of damage in structural elements due to energy dissipation;
- Bridges
    - i) Reduction of top-pier displacement;
    - ii) Elimination of thermal coactive stresses (when viscous devices are used);
    - iii) Absorption of vehicle braking effect;

Structural Assessment

- Yielding hysteretic devices
    - Advantages
      Stable hysteresis loop;
      High stability and durability;
      Low cost of fabrication, installation and maintenance;
      High values of dissipated energy for relatively low values of displacement;
      Ease of substitution;
      Multi-directional behaviour;
    - Disadvantages
      Brittle behaviour in case of welding;
      Ductility properties strongly influenced by device geometry;

- Friction devices
  - Advantages

    Reduction of displacements at serviceability limit state;

    Stable hysteresis loop;

    Low cost of fabrication and installation;
  - Disadvantages

    Inaccuracy in the definition of the friction coefficient;

    High sensitivity to the compressive load between sliding surfaces;

    High imperfection sensitivity of sliding surfaces;

    Degradation of sliding surfaces after several load cycles;

- Viscous Devices
  - Advantages

    Behaviour as "shock transmitter";

    Stable hysteretic behaviour;

    Ease of connection;
  - Disadvantages

    Under some circumstances, displacements for the best performance may result too high (~10 cm);

    Possible fluid leakage from gaskets and/or rings;

    Possible aging with loss of mechanical properties of the inner fluid;

- Visco-elastic devices
  - Advantages

    High resistance to plastic fatigue;

    Good bonding with metals;

    Good hysteretic properties with comparatively low displacements;

    Low cost of fabrication, installation and maintenance;
  - Disadvantages

    High sensitivity to service temperature;

    Need for a non linear behavioural model;

- Potential (all devices)

    A great enhancement of the structural performance can be achieved without increasing bearing elements;

- Risk (all devices)

    The long-term reliability of the seismic protection system could be impaired by a poor maintenance;

Economical Aspects

- Application of such techniques showed a great convenience when applied to strategic buildings;

Robustness and Testing

- The device effectiveness has been widely tested in experimental campaigns as well as in practical applications carried out all over the world;
- Such devices are commonly used in advanced countries like USA and Japan;
- The industrialised and standardised production allows a reliable use of the devices.

Examples

Figure 10. The Central Police Station, Wellington (NZ) (1990)

Figure 11. The Fire Brigade in Naples (1985)

Figure 12. The ENEL Building in the new Management Centre of Naples (1993)

Codification

- FEMA 356. (2000). Prestandard and Commentary for the Seismic Rehabilitation of Buildings. Report by the American Society of Civil Engineering (ASCE) for the Federal Emergency Management Agency (FEMA), Washington, D.C..
- FEMA 273 (1997), NEHRP Guidelines for the seismic rehabilitation of buildings, Building Seismic Safety Council, Washington,D.C., USA (FEMA 274 Commentary).
- AISC 1999. Load and Resistance Factor Design Specification, American Institute of Steel Construction Inc., Chicago.

References

- Aiken I.D., Nims D.K. Wittaker A.S. & Kelly J.M. 1993. Testing of Passive Energy Dissipation Systems, Earthquake Engineering Research Institute California, Berkeley, California.
- Clark P., Aiken I.D., Kasai K., Ko E. & Kimura I. 1999. Design Procedures for Buildings Incorporating Hysteretic Damping Devices, Proc. of the 68th *Annual Convention*, Structural Engineers Association of California, (SEAOC), Santa Barbara, California.
- Costantinou M.C. & Symans M.D. 1993. Seismic Response of Structures with Supplemental Damping, *Structural Design Tall Buildings*, 2, 77-92.
- De Matteis G. 2002. Seismic Performance of MR Steel Frames Stiffened by Light-Weight Cladding Panels. Proc. of the 12th *European Conference on Earthquake Engineering*, London, UK, Elsevier, CD-ROM (paper n. 308).
- De Matteis G., Landolfo R. & Mazzolani F.M. 2003. Seismic Response of MR Steel Frames with low-yield Steel Shear Panels Engineering Structures. *Journal of Earthquake Wind and Ocean Engineering*, 25, 155-168.
- De Matteis G, Mazzolani F:M. & Panico S. 2004. Dissipative Metal Shear Panels for Seismic Design of Steel Frames, Prin 2001 Report.
- Holmes Consulting Group LTD. 2001. *Structure Damping and Energy Dissipation*, Wellington, New Zealand.
- Mandara A. & Mazzolani F.M. 2001. On the Design of Retrofitting by means of Energy Dissipation Devices, Proc. of 7th *International Seminar on Seismic Isolation, Passive Energy Dissipation and Active Control of Vibrations of Structures*, Assisi, Italy.

- Makris N. & Costantinou M.C. 1990. Viscous Dampers: Testing, Modelling and Application in Vibration and Seismic Isolation, Technical Evaluation Report CERF: HITEC 99-03, Civil Engineering Research Foundation, Washington D.C.
- Makris N., Costantinou M.C. 1991. Fractional Derivative Model for Viscous Dampers, *Journal Structural Engineering*, ASCE, 117, pp. 2708-2724.
- Makris N, Costantinou M.C. 1993. Models of Viscoelasticity with Complex Order Derivates, *Journal of Engineering Mechanics*, ASCE, 119(7), pp. 1463-1464.
- Makris N., Costantinou M.C. & Dargush G.F. 1993. Analytical Model of Viscoelastic Fluid Dampers, *Journal of Structural Engineering*, ASCE, 119(11), pp. 3310-3325.
- Mazzolani F.M. 1986. The Seismic Resistant Structures of the New Fire Station of Naples", Costruzioni Metalliche, 38 (6), pp. 343-362.
- Mazzolani F.M. 2001. Passive Control Technologies For Seismic-Resistant Building In Europe, Prog. Structural Engineering and Materials, 3; 277-287.
- Mazzolani F.M., Martelli A. & Forni M. 2001. Progress of Application and R&D for Seismic Isolation and Passive Energy Dissipation for Civil Buildings in the European Union", Proc. of the *7th Int. Seminar on Seismic Isolation, Passive Energy Dissipation and Active Control of Vibration of Structures*, Assisi, 2-5 Ottobre, pp.249-277, Italy.
- Mazzolani F.M. & Mandara A. 1992. Nuove Strategie di Protezione Sismica per Edifici Monumentali: il Caso della Collegiata di San Giovanni Battista in Carife (in italian), Edizioni 10/17, Salerno, Italy.
- Mazzolani F.M. & Serino G. 1997. Viscous Energy Dissipation Devices for Steel Structures: Modelling, Analysis and Application, Proc. of the *II International Conference on Behaviour of Steel Structures in Seismic Areas, STESSA '97*, Kyoto, 4-7 August, Japan.
- Skinner R.I., Heine R.G. & Kelly J.M. 1975. Hysteretic Damper for Earthquake-Resistant Structures, *Earthquake Engineering & Structural Dynamics*, Vol.3 (287-296), John Wiley & Sons.
- Soong T.T. & Dargush G.F. 1997. Passive Energy Dissipation Systems in Structural Engineering, State University of New York at Buffalo, USA, John Wiley & Sons, Chichester, England.
- Soong T.T. & Spencer B.F. Jr. 2002. Supplemental Energy Dissipation: State-of-the-Art and State-of-the-Practice, *Engineering Structures* 24(2002), pp. 243-259, Elsevier Science Ltd.
- Tehranizafeh M. 2001. Passive Energy Dissipation Device for Typical Steel Frame Building in Iran, *Engineering Structures*, 23, 643-655, Elsevier Science Ltd.
- Wittaker A.S., Bertero V.V., Alonso L.J., Thompson C.L. & School R.E. 1989a. Passive Energy Dissipation Using Steel Plate Added Damping and Stiffeness Elements, Proc. of the *International Meeting on Base Isolation and Passive Energy Dissipation Devices*, Assisi, Italia.
- Wittaker A.S., Bertero V.V., Alonso L.J., Thompson C.L. & School R.E. 1989b. Seismic Response of Steel Plate Energy Dissipator, Proc. of the *International Meeting on Base Isolation and Passive Energy Dissipation Devices*, Assisi, Italia.

Contacts

- Federico M. Mazzolani
     Address: Department of Structural Analysis and Design, University of Naples Federico II, P.le Tecchio 80, I-80125 Napoli, Italy
     E-mail: fmm@cds.unina.it
     Tel. +39-081-7682443
     Fax. +39-081-5934792

- Alberto Mandara
     Address: Department of Civil Engineering, Second University of Naples, Via Roma 29, I-81031, Aversa (CE), Italy
     E-mail: alberto.mandara@unina2.it
     Tel. +39-081-5010216
     Fax. +39-081-5037370

ISOLATION TECHNIQUES FOR
SEISMIC PROTECTION OF BUILDINGS

Description

- Structural disconnection between foundation and superstructure;
- Insertion of flexible elements at the structure-to-foundation interface in order to modify the dynamic properties of the system;
- Increase of the global system deformability, so as the magnitude of the seismic action on the structure is drastically reduced;
- Increase of energy dissipation capability of the structure under seismic loads by means of suitable devices;

Contrary to the traditional concept of seismic design, relying on the ductility resources of structural members and connections, the isolation approach is based on the response control concept, aiming at controlling and limiting the dynamic effects on the structural elements by means of special devices. The response control concept can be easily understood referring to the dynamic equilibrium equation of a generic structural system, given by:

$$m\ddot{x}(t) + c\dot{x}(t) + kx(t) + f_c(t) = F(t)$$

in which m, c and k are mass, viscous constant and stiffness of the system, respectively. $F(t)$ is the external force induced by earthquake, and f_c is the control force applied by means of additional devices. In a general form, it is given by:

$$f_c(t) = m'\ddot{x}(t) + c'\dot{x}(t) + k'x(t)$$

where m', c' and k' are mass, viscous constant and stiffness of the controlling system, respectively. Values of m', c' and k' can be chosen in such a way to modify one or more terms of the energy balance equation:

$$E_K(t) + E_\xi(t) + E_S(t) + E_H(t) = E_I(t)$$

where:

E_k is the kinetic energy of the system;

E_ξ is the energy dissipated in the special devices;

E_S is the energy due to elastic deformation in the structural system (strain energy);

E_H is the energy dissipated by cumulated plastic deformation in the structural members (hysteretic energy);

E_I is the earthquake input energy;

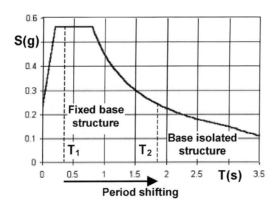

Figure 1. Basic principle of period shifting

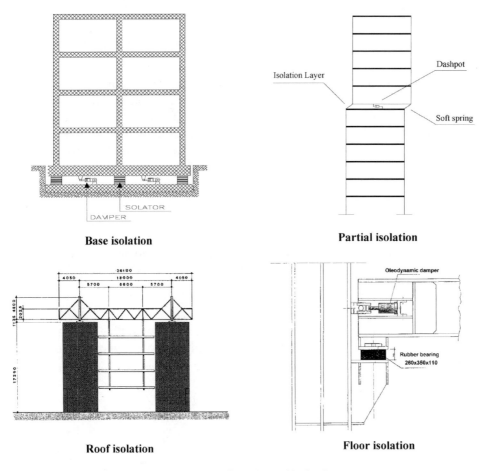

Base isolation

Partial isolation

Roof isolation

Floor isolation

Figure 2. Typical configurations of isolated structures

In isolated systems special devices (seismic isolators) are used in order to reduce the amount of the energy transmitted to the structures E_1. Basically, the effect of such devices is to shift the fundamental vibration period of the building upward, so as to reduce the value of the maximum spectral acceleration (Figure 1). A given amount of input energy can be dissipated by the devices themselves, when these devices posses special dissipative features, or can be absorbed by additional damping devices. The effect of such devices is to increase the overall damping properties of the system. This prevents the structure from an excess of displacements, with a simultaneous control of damage in structural elements. The typical structural isolation configurations are showed in Figure 2.

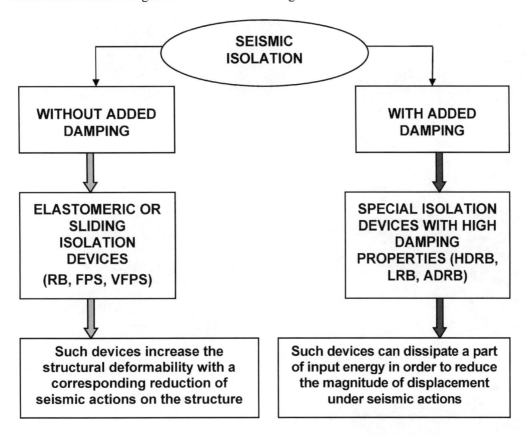

Figure 3. Basic options for base isolated structures

Technical Information

Isolation devices can be distinguished according to the scheme illustrated in Figure 3. Accordingly, most common devices can be classified as follows:

- Elastomeric devices (Figure 4)
    - High Damping Rubber Bearings (HDRB)
    - Lead Rubber Bearings (LRB)
    - Added Damping Rubber Bearings (ADRB)
    - Fibre Reinforced Rubber Bearings (FRRB)
- Sliding devices (Figure 5)
    - Flat Slider Bearings
    - Curved Slider Bearings
- Elasto-plastic Bearings (Figure 6)
- Wire-Rope Bearings (Figure 7)

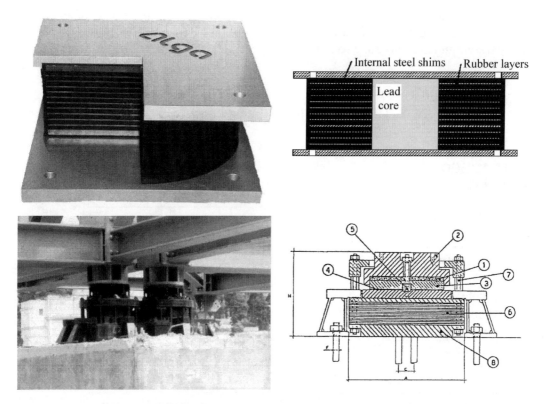

Figure 4. Elastomeric devices

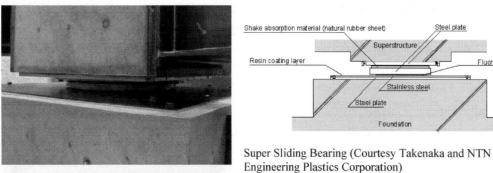

Super Sliding Bearing (Courtesy Takenaka and NTN Engineering Plastics Corporation)

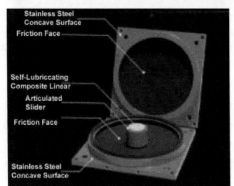

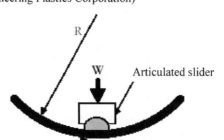

Friction Pendulum System (FPS) (Courtesy Aurotek)

Figure 5. Sliding devices

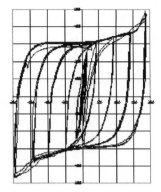

Figure 6. Elasto-plastic bearings (Courtesy TIS S.p.A.)

Figure 7. Wire-rope bearings (Courtesy Enidine Inc.)

Field of Application and End Users

- Buildings (both new and existing)
    - i)   Reduction of input energy under seismic actions;
    - ii)   Reduction of structural damage;
    - iii) Increase of global structural damping and energy dissipation;
- Bridges
    - i)   Reduction of top-pier displacement due to earthquake;
    - ii)   Reduction of vehicle braking effect;

Structural Assessment

- Elastomeric devices
    - Advantages
      High effectiveness in the reduction of both structural response and damage when properly applied, namely in case of relatively stiff structures on rigid soil;
      Horizontal deformability with high vertical load bearing capacity, in particular in case of High Damping Rubber Bearings;
      Stable hysteresis loop, in particular in case of High Damping and Lead Rubber Bearings;

High viscous damping in case of Added Damping Rubber Bearings;

Cost and weight saving with Fibre Reinforced Rubber Bearings;

- Disadvantages

Stability problems when great horizontal displacements occur;

Durability problems of some rubber materials;

Excess of deformability at the serviceability limit state;

- Sliding devices
  - Advantages

    Reduction of displacements at serviceability limit state due to friction;

    Stable hysteresis loop;

    Re-centring feature in case of curved sliders (FPS);

    Low cost of fabrication;
  - Disadvantages

    Inaccuracy in the definition of the friction coefficient due to corrosion sensitivity;

    High sensitivity to the compressive load between sliding surfaces;

    High imperfection sensitivity of sliding surfaces;

    Degradation of sliding surfaces after several load cycles;

- Elasto-plastic Bearings
  - Advantages

    Stable hysteresis loop;

    High stability and durability;

    Low cost of fabrication, installation and maintenance;

    High values of dissipated energy;

    Particularly suitable for bridges;
  - Disadvantages

    Ductility properties strongly influenced by the bearing geometry;

    Comparatively low vertical load bearing capacity;

- Wire-Rope Bearings
  - Advantages

    Good flexibility in all directions;

    Stable hysteresis loop;

    High performance in both tension and compression;

    High values of dissipated energy;

    Particularly suitable for electric lifelines;
  - Disadvantages

    Not suitable for heavily loaded applications (e.g. multi-storey buildings or bridges);

    Low vertical load bearing capacity;

- Potential

    A very high structural performance can be achieved compared with fixed base buildings;

    Optimal solution for the seismic protection of strategic buildings (hospitals, fire stations, schools, ecc.) in highly seismic areas;

- Risk

    The long-term reliability of the seismic protection system could be impaired by a poor maintenance;

Economical Aspects

- Application of such techniques showed a convenience compared with fixed base buildings in highly seismic areas;

Robustness and Testing

- The effectiveness of such technique has been widely tested in both applications and experimental campaigns carried out all over the world, including advanced countries like USA and Japan;
- The industrialised and standardised production allows a reliable use of the devices.

Examples

a) *b)*

Figure 8. Fire Centre, Naples (Italy): roof isolation (1981) (a); floor isolation (1985) (b)

Figure 9. William Clayton Building, Wellington (NZ) (1981)

Figure 10. SIP Building, Ancona (Italy) (1992)

Figure 11. Parliament House, Wellington (NZ) (seismic retrofit) (1994)

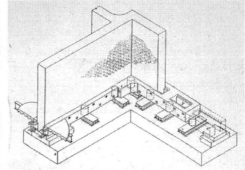

Figure 12. City and County Building, Salt Lake City, Utah (USA) (seismic retrofit) (1989)

Figure 13. Apartment building, Siracusa (Italy) (seismic retrofit) (2004)

Figure 14. Civic Centre, Naples (Italy) (seismic retrofit) (2004)

Codification

- FEMA 356. 2000. Prestandard and Commentary for the Seismic Rehabilitation of Buildings. Report by the American Society of Civil Engineering (ASCE) for the Federal Emergency Management Agency (FEMA), Washington, D.C.

References

- Aiken I.D., Clark P.W. & Kelly J.M. 1993. Design- and Ultimate-Level Earthquake Tests of a 1/2.5-Scale Base Isolated Reinforced Concrete Building. Proc. of the ATC-17-1 Seminar on Seismic Isolation, Passive Energy Dissipation and Active Control, San Francisco, California.
- Bouc R. 1971. Modele Matematique d'Histeresis, Acustica, 24(1).
- Chalhoub M.S. & Kelly J.M. 1987. Reduction of the Stiffness of Rubber Bearings due to Compressibility, Report No.UBC/SEMM-86/06, *Structural Engineering Mechanics and Materials*, University of California, Berkeley.
- Clough R.W. & Penzien J. 1975. Dynamics of Structures, McGraw Hill, London
- De Luca A., Mazzolani F.M. & Mele E. 1991. Test Results on Natural Rubber and Policloroprene Devices for Investigating the Effects of Axial Loads, Atti del Convegno Internazionale sulla Protezione Sismica degli Edifici, Ancona, 6-8 Giugno 1991
- Demetriades G.F., Costantinou M.C. & Reinhorn A.M. 1993. Study of Wire-Rope Systems for Seismic Protection of Equipment in Buildings, *Engineering Structures*, Volume 15, N° 5.
- Dolce M., Cardone D., Marnetto R., Nigro D. & Palermo G. 2003. A new Added Damping Rubber Isolator (ADRI): Experimental Tests and Numerical Simulations, Proc. of the *8th World Seminar on Seismic Isolation, Energy Dissipation and Active Vibration Control of Structures*, Yerevan, Armenia, October 6-10.
- Gent A.N. 1964. Elastic Stability of Rubber Compression Springs, *Journal of Mechanical Engineering Science*, Vol.6, No.4, pp.318-326.
- Holmes Consulting Group LTD. 2001. Base Isolation of Structures, Wellington, New Zealand.
- Hwang J.S. & Hsu T.Y. 2001.A Fractional Derivate Model to Include Effect of Ambient Temperature on HDR Bearings, *Engineering Structures* No. 23, pp.484-490.
- Hwang J.S., Wu J.D., Pan T.C. & Yang G. 2002. A Mathematical Hysteretic Model foe Elastomeric Isolation Bearings, *Earthquake Engineering and Structural Dynamics* 31; pp.771-789
- Kelly J.M. 1993. Earthquake-Resistant Design with Rubber, Springer-Verlag, London.
- Kelly J.M. 1999. Analysis of Fiber-Reinforced Elastomeric Isolators, JSSE 2(1)19-34
- Koh C.G. & Kelly J.M. 1987 Effects of Axial Load on Elastomeric Isolation Bearings, Report No.UBC/EERC-86/12, Earthquake Engineering Research Center,

University of California, Berkeley.
- Makris N. & Deoskar S. 1996. Prediction of Observed Response of Base-Isolated Structure, *Journal of Structural Engineering*, Vol.122, No.5, pp. 485-493.
- Mazzolani F.M. 1986 The Seismic Resistant Structures Of The New Fire Station Of Naples, Costruzioni Metalliche, 38 (6), pp. 343-362.
- Mazzolani F.M. 2001. Passive Control Technologies for Seismic-Resistant Building in Europe", Prog. Structural Engineering and Materials; 3; 277-287.
- Mazzolani F.M. & Serino G. 1997. Top Isolation of Suspended Steel Structures: Modelling, Analysis and Application, Proc. of the *II International Conference on Behaviour of Steel Structures in Seismic Areas, STESSA '97*, Kyoto, 4-7 August, Japan.
- Mazzolani F.M., Martelli A. & Forni M. 2001. Progress of Application and R&D for Seismic Isolation and Passive Energy Dissipation for Civil Buildings in the European Union, Proc. of the 7th *Int. Seminar on Seismic Isolation, Passive Energy Dissipation and Active Control of Vibration of Structures*, Assisi, 2-5 October, pp.249-277, Italy.
- Mazzolani F.M. & Mandara A. 1992, Nuove Strategie di Protezione Sismica per Edifici Monumentali: il Caso della Collegiata di San Giovanni Battista in Carife (in italian), Edizioni 10/17, Salerno, Italy.
- Moon B.-Y., Kang G-J, Kang B-S. & Kelly J.M. 2002 Design and Manufacturing of Fiber Reinforced Elastomeric Isolator for Seismic Isolation, Journal of Materials Processing Technology 130-131, pp. 145-150.
- Moon B-Y, Kang G-J & Kang B-S. 2003. Hole and Lead Plug Effect on Fiber Reinforced Elastomeric Isolator for Seismic Isolation, Journal of Materials Processing Technology 140, pp. 592-597.
- Naeim F. & Kelly J.M. 1999. Design of Seismic Isolated Structures – From Theory to Practice, John Wiley & Sons, Inc.
- Pranesh M. & Ravi Sinha. 2000. VFPI: An Isolator Device for Aseismic Design, *Earthquake Engineering and Structural Dynamics* 29, pp. 603-627.
- Skinner R.I, Robinson W.H & McVerry G.H. 1993. An Introduction to Seismic Isolation, John Wiley & Sons, Inc.
- Tajirian F.F., Kelly J.M. & Aiken I.D. 1990. Seismic Isolation for Advanced Nuclear Power Stations, *Earthquake Spectra*, Vol.6, No.2, pp. 371-401.
- Tsai C.S., Chiang T-C. & Chen B-J. 2003. Finite Element Formulations and Theoretical Study for Variable Curvature Friction Pendulum System, *Engineering Structures* 25, pp. 1719-1730.
- Wen Y.K. 1976 Method for Random Vibration of Hysteretic System, *Journal of the Engineering Mechanics Division*, New York, Vol.102, No.EM2, April.
- Wu Y.M. & Samali B. 2002. Shake Table Testing of a Base Isolated Model. Engineering Structures 24, pp. 1203-1215.
- Yamazaky N., Nakaucji H.& Miyazaki M. 1992. Durability of Rubber Isolators – Characterization of 100 Years Old Used Rubber Bearing, Proc. of *International Workshop on Recent Developments in Base-Isolation Techniques for Buildings*, Tokyo, April 27-30.

Contacts

- Federico M. Mazzolani
    Address: Department of Structural Analysis and Design, University of
        Naples Federico II, P.le Tecchio 80, I-80125 Napoli, Italy
        E-mail: fmm@cds.unina.it
        Tel. +39-081-7682443
        Fax. +39-081-5934792
- Alberto Mandara
    Address: Department of Civil Engineering, Second University of Naples,
        Via Roma 29, I-81031, Aversa (CE), Italy
        E-mail: alberto.mandara@unina2.it
        Tel. +39-081-5010216
        Fax. +39-081-5037370

COST C12 - WG2

Datasheet I.3.1.1.3
Prepared by: F. Dinu[1], D. Dubina[2], D. Grecea[2]

[1] Romanian Academy, Timisoara Branch/ Romania, http://acad-tim.utt.ro
[2] Politehnica" University of Timisoara, Romania, http://www.ceft.utt.ro

PARTIAL q-FACTORS FOR PERFORMANCE BASED DESIGN OF MR STEEL FRAMES

Description

- Actual seismic design codes consider explicitly only one limit state, defined as protection of occupants lives in case of a major earthquakes. Criteria for designing structures subjected to minor or moderate earthquakes are not explicitly specified

- Performance based design is a more general design philosophy, defined as a selection of design criteria and structural systems such that at the specified levels of ground motion and with defined levels of reliability, the structure will not be damaged beyond certain limit states or other useful limits (Bertero and Bertero, 2000) $\Rightarrow$ disadvantage: practical use of this new approach is rather difficult

- As an alternative, present methodology proposes a "three objectives" performance based design. Acceptance criteria for each limit state are given by partial q factors $\Rightarrow$ advantage: easy to introduce in the actual seismic codes

Field of application

- Seismic design of MR steel frames
- Methodology may also be applied to braced frames design (either concentric or eccentric)

Method of analysis

- Methodology requires the following steps:
  [1] definition of performance levels
  [2] determination of seismic hazard
  [3] selection of analysis procedure
  [4] determination of global ductility

[1] Definition of performance levels (limit states):
- Structures designed against earthquakes have to comply with specific criteria such as stiffness, strength and ductility. In the present methodology, three limit states, which are referring to conditions of drift, residual drift and rotation capacity of elements, are introduced:
- Serviceability limit state SLS (stiffness criterion): under frequent earthquakes, the building can be used without interruption and the structure remain in elastic range. Acceptance criteria - interstorey drift exceeds 0.6% of the relevant storey height
- Damageability limit state DLS (strength criterion): under rare earthquakes, the building presents important damages of non-structural elements and moderate damages of structural elements which may be repaired without high costs or technical difficulties. Structure is responding in elasto-plastic range and the determinant criterion is the resistance of member section. Acceptance criteria - permanent interstorey drift exceeds 1% of the relevant storey height

- Ultimate limit state ULS (ductility criterion) - under very rare earthquakes, the building presents major damages of both non-structural and structural elements but safety of people is guarantied. Damages are extended so that structure cannot be repaired and demolition is unavoidable. Structure is in the elasto-plastic range and the determinant criterion is the local ductility (rotation capacity of elements and connections). Acceptance criteria - plastic rotation exceeds 0,03rad

Table 1 - Acceptance criteria associated to the performance levels for MR steel frames

Performance levels (limit states)	Drift [%]	Residual drift [%]	Plastic rotation [rad]
SLS	0,6	-	-
SLD	-	1,0	-
SLU	-	-	0,03

[2] Determination of seismic hazard
- To be used in design, performance levels must be translated into seismic action, represented by magnitudes or accelerations
- If the acceleration for DLS - a_d is considered as a basic value for ground motion acceleration, the accelerations for SLS and ULS are determined with the equation (Gioncu, Mazzolani 2002):

$$a_s = 0,412a_d; \quad a_u = 1,22a_d$$

[3] Selection of analysis procedure
- Accelerations corresponding to the development of first plastic hinge and accelerations corresponding to each limit states are determined by performing an incremental non-linear dynamic analysis (Vamvatsikos and Cornell, 2002)

[4] Determination of global ductility
- The most suitable approach for seismic design based on performance is the *deformation-controlled design* while today codes are based on a *force-controlled design*, using the base shear concept. In the latter approach, the most important parameter is the behaviour q factor, based on the maximum capacity of structure to dissipate energy during the plastic deformations corresponding to ULS (see table 2)

Table 2. Design concepts, behaviour factors and structural ductility classes Eurocode 8 (1994)

Design concept		Behaviour factor q	Required ductility class
Concept b)	Low dissipative structure	1,5 - 2	L (Low)
Concept a)	Dissipative structure	1,5 - < q < 4	M (Medium)
	Dissipative structure	q ≥ 4	H (High)

- The ductility corresponding to ULS cannot be attained if higher levels of performance are required. In that case, a reduced ductility corresponding to a *partial q-factor* is attained by the structure
- The use of partial q-factor gives the possibility to implement the multiple performance design in the actual code methodology. To determine the q factor, the following equation is used:

$$q = \frac{\lambda_u}{\lambda_e}$$

Results of analysis - new buildings

- The methodology is applied on three *trial* MR frames with different geometric conditions. Two different groups of ground motion records, characterised by different corner period, are used

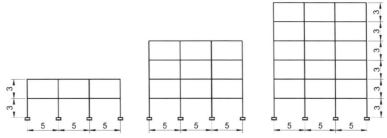

Figure 1. Frame configurations (dimensions in m)

- A series of incremental non-linear dynamic analysis were carried out using the Drain2DX computer program. The maximal accelerations for which the structures meet the specified performance criteria are determined by appropriate scaling
- q factors for rigid and semirigid MR steel frames (see table 3) are determined using equation presented above. A value of 1,0 for the q-factor is imposed for SLS, considering the structure is in elastic range

Table 3. q factor values for MR steel frames

Limit state	Rigid	Semirigid
SLS	1,0	1,0
DLS	4,2	3,4
ULS	6,0	5,3

Results of analysis - existing buildings

- Methodology may be used to verify the seismic performances of existing MR steel buildings. For demonstration, an existing steel building, completed in 2002 (see figure 2) was used (Banc Post Building in Timisoara City). The building is a medium-rise multi-storey steel structure. MR steel frames compose the main skeletal framework

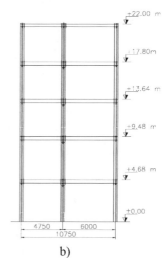

a)   b)

Figure 2. Banc Post building: a) general view; transversal frame considered in the study

- The three performance objectives and the acceptance criteria for each level are similar to those described in table 1, excepting the rotation capacity corresponding to ULS. In order to verify the rotation capacity associated to this limit state, an experimental program was developed
- The experimental program has shown plastic rotation capacities over 0,04rad. In table 3 the characteristic values associated to the three performance levels are presented

Table 4. Acceptance criteria associated to performance levels

Performance level (limit state)	Drift [%]	Residual drift [%]	Plastic rotations [rad]
SLS	0.6	-	-
SLD	-	1.0	-
SLU	-	-	0.04

- The maximal accelerations for which the structures meet the specified performance criteria are determined by appropriate scaling
- q factors are determined using equation presented above. A value of 1,0 for the q-factor is imposed for SLS, considering the structure is in elastic range
- Value of q factor used in the structure design (q_{Sd} = 5,8) is very closed to value corresponding to ULS (q = 5,7) $\Rightarrow$ the structure fulfils the requirements corresponding to ULS

Table 5. q factor values for the existing MR steel building

Limit state	q factor
SLS	1,0
SLD	4,3
SLU	5,7

Further developments

- q factor is very sensitive to the local soil conditions $\Rightarrow$ more studies are required to validate the methodology
- Application to braced frames (either concentric or eccentric)

References

- ATC 1996, *Seismic evaluation and retrofit of concrete buildings*, Vol. 1, ATC-40, Applied Technology Council, Redwood City
- Aribert, J.M., Grecea, D., 1997. *A new method to evaluate the q-factor from elastic-plastic dynamic analysis and its application to steel frames*, Proceedings of the Second International Conference STESSA 1997, Behaviour of Steel Structures in Seismic Areas, Kyoto, Japan, 3-8 August 1997, Eds. F.M. Mazzolani & H. Akiyama, Napoli
- Bertero, R.D., Bertero, V.V., 2000. *Application of a comprehensive approach for the performance-based earthquake resistant design buildings.* 12th World Conference on Earthquake Engineering, Auckland, 30 Jan. - 4 Feb. 2000, CD-ROM 0847

- Dinu, F., Grecea, D., Dubina, D., 2004. *Partial q factor values for performance based design of MR frames*, Journal of Constructional Steel Research, Elsevier, Volume 60, Issues 3-5.

- FEMA 273, 1997. *NEHRP Guidelines for Seismic Rehabilitation of Buildings*, Federal Emergency Management Agency, Washington, DC

- Dubina, D., Ciutina, A., Stratan, A., 2002. *Cyclic tests on bolted steel and composite double-sided beam-to-column joints*, Steel & Composite Structures, Volume 2, No. 2, 147-160

- Gioncu, V., Mazzolani, F.M., 2002. *Ductility of Seismic-Resistant Steel Structures. London*: SPON PRESS

- Kannan, A., Powel, G., DRAIN-2D, 1975. *A general-purpose computer program for dynamic analysis of inelastic plane structures*, EERC 73-6 and EERC 73-22 reports, Berkeley, USA

- Vayas, I., Dinu, F., 2000. *Evaluation of the seismic response of steel frames in respect to various performances*. Third International Conference on Behaviour of Steel Structures in Seismic Areas STESSA 2000, Montreal, Canada

Contacts

- Dr. Florea Dinu
  - Address: Romanian Academy, Timisoara Branch
    - Mihai Viteazul 24, Timisoara 1900 Romania
    - Tel/Fax: +40 256 403 932
    - e-mail: florin@constructii.west.ro; http://acad-tim.utt.ro
- Prof. Dan DUBINA
  - Address: "Politehnica" University of Timisoara
    - Ioan Curea 1, Timisoara 1900 Romania
    - Tel/Fax: +40 256 403 932;
    - e-mail: dubina@constructii.west.ro http://www.ceft.utt.ro

COST C12 - WG2
Datasheet I.3.1.1.4
Prepared by: A. Stratan and D. Dubina
        Politehnica University of Timisoara, Romania

ECCENTRICALLY BRACED DUAL STEEL FRAMES WITH REMOVABLE LINK

Description

- Eccentrically braced frames (EBFs) are believed to be an efficient structural system in seismic areas. They present the advantage of both good ductility (in comparison with concentrically braced frames – CBFs), and high stiffness (in comparison with moment-resisting frames – MRFs).

Field of application

- Design of multi-storey structures in high-seismicity areas is usually based on dissipative structural response, which accepts significant structural damage under the design earthquake. High economic losses following past earthquakes urged for a limitation of damage to structures in future earthquakes, leading to the development of PBD – Performance-Based Design (Hamburger, 1996). Its objectives include minimizing structural and non-structural damage under low and moderate earthquake intensities, which is equivalent to reduction of the total cost (initial and repair).

- Capacity-based design, applied in most of the current seismic design codes allows design of structures that promote plastic deformation in predefined areas only, called dissipative zones. In the case of a bolted connection between dissipative zones and the rest of the structure it is possible to replace the dissipative elements damaged as a result of a moderate to strong earthquake, reducing the repair costs.

- Different typologies of eccentrically braced frames are available (see figure 1). While all configurations in figure 1 can be applied to new constructions, the one from figure 1c can be used for seismic strengthening of existing buildings.

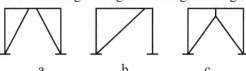

        a   b   c

Figure 1. Different typologies of eccentrically braced frames.

Technical information

- Application of removable dissipative elements this philosophy to eccentrically braced frames, where link elements serve as dissipative zones, is presented in figure 2. The connection of the link to the beams is realized by a flush end-plate and high-strength bolts. Bolted connection allows he link element to be fabricated from a lower-yield steel grade, assuring an elastic response of the elements outside link element.

- This system may be applied to both homogeneous structures (eccentrically braced frames alone) and dual ones (eccentrically braced spans combined with moment-resisting spans). The latter system has the advantage of more uniform transient and smaller permanent lateral displacements, which is beneficial for replacing damaged links, as well as for the building function (Stratan and Dubina, 2002).

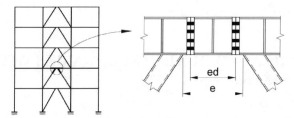

Figure 2. Bolted link concept.

Structural aspects

- Dual structural systems are formed by combining two of the basic structural systems: a MRF and a braced frame (either concentric or eccentric). Alternatively, steel shear walls may be used instead of the EBFs, a structural system popular mostly in USA and Japan.

- Little guidance on the design and seismic response of dual systems is available. Eurocode 8 (1994) states that lateral forces acting on the dual system are distributed to the subsystems according to their stiffness. Design criteria are those of the independent structural systems. The NEHRP 2000 design code requires additionally a minimum strength (25% of the total one) for the MRF subsystem in order the structure to be defined as "dual".

- Short link beams are preferred for the eccentrically braced frames, as it is easier to assure an elastic response of the rest of the structure in this case. Response of short links is characterised by a ductile behaviour, due to plastic shearing of the link web.

Research activity

- Several authors (Akiyama 1999, Iyama & Kuwamura 1999, Astaneh-Asl 2001) investigated seismic performance of dual systems. According to these studies, the potential benefits of dual structural configurations may be summarised as follows:

- Efficient earthquake resistance due to prevention of excessive development of drifts in the flexible subsystem, and dissipation of seismic energy in the rigid subsystem by plastic deformations.

- Alternative load path to seismic loading provided by the secondary subsystem (the flexible one) in the case of failure of the primary subsystem (the rigid one).

Method of analysis

- Dual structural configurations are composed of two substructures with different stiffness, strength and ductility. Their performance can be conveniently analysed using a simplified single degree of freedom system as in figure 3.

- A dual EBF – MRF structure and a homogeneous EBF (see figure 4) were designed according to Eurocode 3 (1993) and Eurocode 8 (1994). A series of pushover and time-history analyses were carried out using the Drain-3dx computer program.

- Two different sets of ground motion records were used, each comprised of seven historical and semi/artificial accelerograms. Ground motion sets have different corner period T_C, denoting different local soil conditions.

- Seismic performance was evaluated for three limits states: serviceability (SLS), ultimate (ULS) and collapse prevention (CP) limit states. The performance level associated to the limit states were: interstorey drift limitation to 0.006h for SLS, attainment of the available capacity of elements (0.03 rad in beams and columns, and 0.1 rad in links) for ULS, and dynamic instability for CP. The corresponding acceleration multipliers were $\lambda=0.5$, $\lambda=1.0$ and $\lambda=1.5$, respectively.

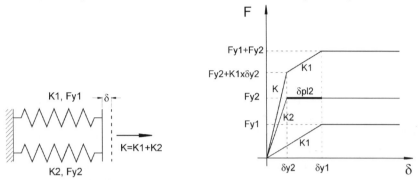

Figure 3. Simplified model of a generalized dual system.

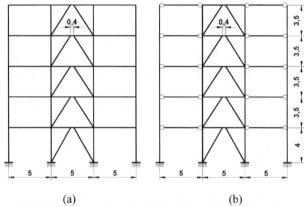

(a)   (b)

Figure 4. Dual EBF-MRF (a) and standard EBF (b) configurations (dimensions in m).

- An experimental program was carried out to determine cyclic performance of bolted links and to check the feasibility of the suggested solution (see figure 5a). The complete ECCS (1985) loading procedure was applied, consisting of one monotonic and two cyclic tests.

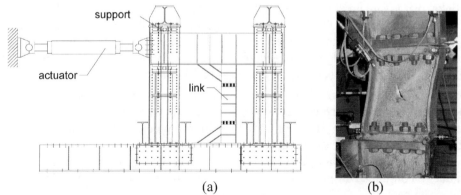

(a)   (b)

Figure 5. Experimental set-up for removable bolted links (a), and failure of the LL4-c1 specimen (b).

- Phenomena that control seismic response of dual structural configurations may be divided into two groups: (1) effects of global force-displacement relationship and ground motion characteristics, which can be modelled using single degree of freedom systems, and (2) distribution of ductility demands and degradations within the structure, which require a multi-degree of freedom models.

- At the level of global force-displacement relationship, dual structural configurations are characterised by: (1) similar strength and stiffness in comparison with homogeneous configurations and (2) higher maximum strength and hardening. For structures with the fundamental period of vibration higher than the ground motion control period T_C, inelastic displacement demands are close to the elastic displacement demands. Therefore, hardening and strength reserve do not affect ductility demand of dual configurations, their performance being similar to the one of homogeneous structures. Different global force-displacement relationship becomes important when the structure fundamental period of vibration is less than the ground motion control period T_C. Inelastic displacement demands are higher in this case than the elastic ones, and the hardening behaviour of dual configurations become efficient in reducing ductility demands.

- Dual structural configurations are efficient in reducing permanent deformations as long as the flexible substructure remains in the elastic range. From the practical point of view, this objective may be accomplished by a higher steel grade in the moment-resisting spans, increasing in this way the strength, but not the stiffness of the flexible subsystem. Reduction of permanent displacements is maximum for structures with fundamental period of vibration in the constant velocity spectral region.

- Superior performance of dual configurations in comparison with homogeneous ones becomes important for structures with low ductility, dual configuration reducing ductility demands and risk of dynamic instability.

- At the level of multi-storey dual configurations, the flexible substructure improves seismic performance by promoting more uniform displacement demands over the height and reducing the risk of partial plastic mechanism.

- Experimental investigations on removable bolted links demonstrated the technological feasibility of the solution. Performance of short removable links and possibility to be easily replaced makes them attractive for dual eccentrically braced frames.

- Very short links, that assure an elastic behaviour of the connection are preferred, due to much easier replacement of damaged links. Concentration of damage in the removable link (performing like passive energy dissipation devices) may be accomplished by the capacity design principles, including fabrication of the link from a steel with lower yield strength in comparison with rest of the structure.

- Longer links and closer stiffener spacing imposed higher demands on the connection. Cyclic response elements for which connection represented the weaker element were characterised by: (1) a reduction of maximum force in comparison with elements dominate by web shear; (2) a pinching behaviour with stiffness and strength degradation in cycles of constant amplitude; (3) failure by gradual strength degradation due to bolt thread stripping. Post elastic connection response was ductile, due to thread stripping.

- Response of short links was governed by web shear (see figure 5b), stiffener spacing being important for their performance. In the case of rare spacing of stiffeners, inelastic response of short links was determined by plastic web buckling, which lead to strength degradation by alternative buckling in the direction of the two diagonals. Closer stiffener spacing limited plastic web buckling, leading to: (1) attainment of the maximum possible shear strength; (2) a stable hysteretic response; (3) a larger rotation capacity, but also (4) a more rapid failure by web tearing on the panel edges.

- With the exception of very short links, connections were partial-strength. On the basis of the experimental program, in order to prevent excessive connection damage, it is recommended to limit link length ed to 0.8My/Vy. Design strength of short removable links limited to this length may be computed as for classical short links. Full bolt preloading resulted in higher initial stiffness, a more stable hysteretic response and a larger deformation capacity, and therefore is recommended for removable short links. Semi-rigid connections with flush end plate reduce substantially initial stiffness of removable short links in comparison with classical solution.

Further developments

- Future research should address analysis of more structural configurations, finite element analysis of link behaviour and development of a simplified model of the bolted link to be incorporated into global structural analysis programs.

References

- Akiyama H., 1999. Behaviour of connections under seismic loads. *Control of semi-rigid behaviour of civil engineering structural connections. COST C1. Proceedings of the international conference, Liege, 17-19 September 1998.*

- Astaneh-Asl A., 2001. Seismic Behavior and Design of Steel Shear Walls, *Steel TIPS, USA.*

- Hamburger, R.O. (1996). Implementing performance-based seismic design in structural engineering practice. In: *Proceedings of 11th World Conference on Earthquake Engineering, Acapulco, Mexico.* Paper no. 2121. Oxford: Pergamon.

- Iyama, J., Kuwamura, H., 1999. Probabilistic advantage of vibrational redundancy in earthquake-resistant steel frames. *Journal of Constructional Steel Research, 52: 33-46.*

- NEHRP 2000. Building Seismic Safety Council, BSSC (2001). *NEHRP Recommended Provisions for Seismic Regulations for New Buildings and Other Structures, Part 1 — Provisions and Part 2 — Commentary.* Federal Emergency Management Agency, Washington D.C.

- Stratan, A. and Dubina, D. (2002). Control of performance of dual frames with eccentric bracing, *Proc. Stability and Ductility of Steel Structures – SDSS 2002, Budapest, Hungary, 26-28 september 2002.*

- Stratan, A. and Dubina, D. (2004). Bolted links for eccentrically braced steel frames. *Proc. Connections in Steel Structures V: Innovative Steel Connections, June 3-5, 2004, Amsterdam, Holland* (in print).

Contacts

- Aurel Stratan

    Address: Department of Steel Structures and Structural Mechanics, Politehnica University of Timisoara, Str. Ioan Curea nr. 1, 300224 Timisoara, Romania.
    e-mail: astratan@ceft.utt.ro, Phone: +40.256.403932, Fax: +40.256.403932

- Dan Dubina

    Address: Department of Steel Structures and Structural Mechanics, Politehnica University of Timisoara, Str. Ioan Curea nr. 1, 300224 Timisoara, Romania.
    e-mail: dubina@constructii.west.ro, Phone: +40.256.403932, Fax: +40.256.403932

COST C12 - WG2

Datasheet I.3.1.1.5

Prepared by: G. De Matteis, S. Panico

Univ. Naples Federico II – Dept. Structural Analysis and Design

USE OF SHEAR PANELS FOR SEISMIC PROTECTION OF STEEL STRUCTURES

Description

- Metal shear panels may be adopted to resist lateral forces and to control the dynamic response of steel framed buildings according to passive seismic protection strategy.

- Due to their considerable shear stiffness and strength, shear panels can be favourably used as a seismic resistance system under moderate earthquake loading.

- They can be also adopted as dissipative elements for the seismic protection of the primary structure under strong loading conditions, due to the large energy dissipation capacity related to the large portion where plastic deformations take place.

- The dissipative function of shear panels is activated by means of interstorey drifts occurring in the primary structure.

Field of application

- Owing to their high lateral stiffness, shear panels may be profitable used as a sort of upgrading system for frames designed according to strength only and therefore unable to meet also the serviceability limit state requirements prescribed by current structural codes.

Technical information

- Shear panels may be basically used into a lateral forces resisting system of steel buildings according to two types of structural schemes:

    - standard scheme, where beam-to-column joints are considered to be simply pinned and shear walls are assumed to be the only lateral-force-resisting system in the structure (Figure 1a).

    - dual scheme, where steel shear walls and moment frames behave as a combined system to resist external lateral forces. The moment resisting frame is considered as the primary structure acting as back up system to shear walls, which however absorb the majority of the lateral forces at least in the early deformation stages (Figure 1b).

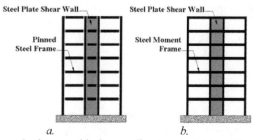

Figure 1. Typical structural schemes with shear walls: Standard (a) and Dual (b) structural system

- Several structural geometric configuration of panel distribution throughout the structure could be adopted, the most common being: single wall (Figure 2a), coupled wall (Figure 2b), out-rigger wall (Figure 2c) and mega-truss (or mega-frame) wall configuration (Figure 2d).

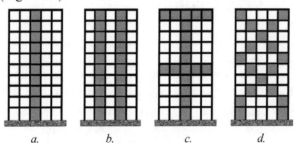

a. b. c. d.

Figure 2. Main structural configurations: a) Single wall, b) Coupled walls, c) Out-rigger walls, d) Mega-truss (or mega-frame) walls

- Shear panels can be inserted into the single mesh of the frame as a large panel rigidly and continuously connected along columns and beams, serving also as cladding panel (Figure 3a).
- Smaller element installed at the nearly middle height of the storey could be used as well, according to different schemes, namely partial bay type (Figure 4b), bracing type (Figure 3c) and pillar type (Figure 3d).

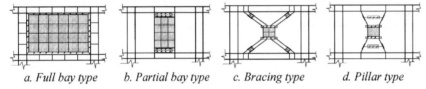

a. Full bay type   b. Partial bay type   c. Bracing type   d. Pillar type

Figure 3. Typical arrangements for shear panels.

- Shear panels can be made of stiffened and un-stiffened plates (Figure 4). The use of stiffeners allows to delay shear buckling phenomena in the plastic field so to obtain a stable inelastic cyclic behaviour under shear action.

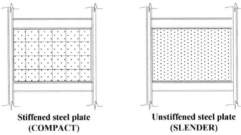

Stiffened steel plate
(COMPACT)

Unstiffened steel plate
(SLENDER)

Figure 4. Un-stiffened (right side) and stiffened (left side) shear panels.

Structural aspects

- Introduction of shear panels into steel framed structures allows the improvement of structural performance levels under lateral loads due to the increasing of stiffness, strength and ductility.
- A steel frame equipped with shear walls has to be intended as a composite (dual)

118

system, where the primary structure exhibits elastic deformations only under moderate earthquakes, while it becomes a useful supplementary energy dissipation system for medium and high intensity earthquakes, developing plastic hinges in beams and columns. On the other hand, shear panels have to be intended as the main energy dissipation system, supplying also additional lateral stiffness and strength to the whole structure (Figure 5).

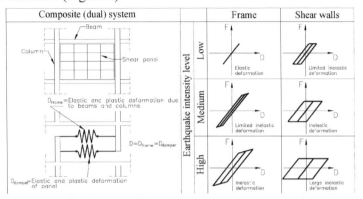

Figure 5. Schematised modelling for frame-shear wall combined systems

Research activity

- Ongoing research activities focuses on the local and global seismic behaviour of shear panel systems made of both steel thin plates (Figure 6) and low yield steel plates (Figure 7).

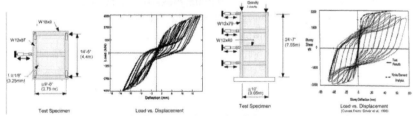

Tests at the Univ. of British Columbia, (Kulak, 1991; Rezai et al., 1998)

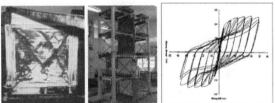

Tests at the Univ. of Alberta (Timler and Kulak, 1983; Driver et al, 1996)

Figure 6. *Research activities on steel thin plate shear walls*

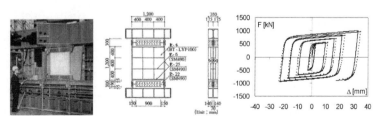

Figure 7. *Research activities on low yield steel shear panels.*

- The low yield strength metals with high E/f_y ratio allow to build shear panels both with larger plate width-to-thickness ratio and able to ensure the energy dissipation yet for smaller deformation levels, as in the case of wind and moderate earthquakes, working as dampers also for the serviceability limit state.

- An experimental research project has been recently undertaken at the Department of Structural Analysis and Design of the University of Naples Federico II (De Matteis et al. 2003), aiming at investigating on the possibility to use aluminium with high degree of purity to build shear panels as an alternative metal to low yield strength steel (Figure 8).

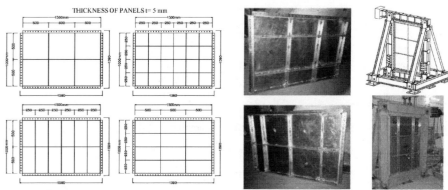

Figure 8. Ongoing experimental tests campaign on pure aluminium shear panels

Method of analysis

- Resisting and dissipative actions of shear panels take place due to shear mechanism, by means of either pure shear stress action or tension field action (Figure 9).

- According to AISC specification for plate girders (AISC 1999), shear panels behaving with a pure shear dissipative mechanism can be identified as "compact shear panels". Whereas steel shear panels dissipating energy by means of the tension field action can be denoted as "slender shear panels". An intermediate class of shear walls could be also identified as "non-compact shear panels", when plate shear yielding is reached before buckling occurrence, but with available plastic deformation limited in relation to the required one.

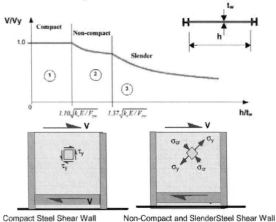

120

Figure 9. Shear energy dissipation mechanism

- For reason due to numerical processing time, it is necessary to set up simplified models reproducing the effect of shear panels in terms of strength, stiffness and dissipative behaviour.

- With reference to slender shear panels, several studies have been already carried out demonstrating the effectiveness of the so-called strip model to predict the results of experimental tests (Thorburn et al. 1983, Driver et al. 1998, Rezai et al. 2000) (Figure 10). The strip model consists in replacing shear plate by a series of diagonal struts, capable of transmitting tensile forces only, positioned and oriented in the same direction as the principal tensile stresses in the panel.

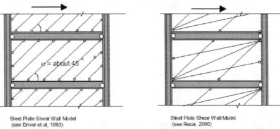

Figure 10. Two models proposed for replacing shear panel with truss members

- For reason due to numerical processing time, it is necessary to set up simplified models reproducing the effect of shear panels in terms of strength, stiffness and dissipative behaviour.

- The strip model could be extended and applied for compact and non-compact shear panels as well. In such a case it has to be considered that compression principal stresses are not negligible and must be taken into account.

- When the effect of flexural interaction between shear panels and the boundary members is negligible, a further simplification of the model can be based on the adoption of two trusses only, which are placed through the two diagonal directions, connecting the opposite corners of the frame mesh according to the X-bracing model (Figure 11).

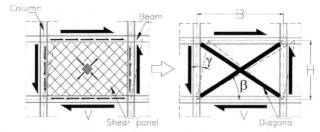

Figure 12. X-bracing model for compact and non-compact shear panels

Further developments

- Next developments should be devoted to set up adequate design criteria for steel frame-shear wall dual structural systems.

- Besides, criteria for the selection of geometrical characteristics of shear panels in order to ensure an adequate dissipative behaviour should be provided.

References

- De Matteis, G. 2002. Seismic performance of MR steel frames stiffened by light-weight cladding panels. *Twelfth European Conference on Earthquake Engineering*, London, UK, Elsevier, CD-ROM (paper n. 308).

- De Matteis G., Mazzolani F.M., Panico S. 2003. Pure aluminium shear panels as passive control system for seismic protection of steel moment resisting frames. *Proceedings of the 4th STESSA Conference* (Behaviour of Steel Structures in Seismic Areas), Naples, Italy.

- De Matteis G., Mazzolani F.M., Panico S., 2004. Seismic protection of steel buildings by pure aluminium shear panels, *13th World Conference on Earthquake Engineering*, Vancouver, B.C., Canada, August 1-6, (Paper no. 2704) CD-ROM, 2004.

- Driver, R.G., Kulak, Elwi, A.E., Kennedy, D.J.L. 1998. FE and simplified models of steel plate shear wall. *Journal of Structural Engineer*. ASCE Vol. 124, No. 2, Feb., pp. 112-120.

- Kulak, G.L. 1991. Unstiffened Steel Plate Shear Walls. In Structures subjected to repeated loading – *stability and strength*, eds. R. Narayanan and T. M. Roberts, Elsevier, London (UK), 237-276.

- Miller, C.J. & Serag, A.E. 1978. Dynamic Response of Infilled Multi-Storey Steel Frames. *Proc. 4th Int. Spec. Conf. Cold-formed Steel Structures*, Missouri-Rolla, 557-586.

- Nakagawa, S., Kihara, H., Torii, S., Nakata, Y., Matsuoka, Y.,Fyjisawa, K., et al. 1996. Hysteretic behaviour of low yield strength steel panel shear walls: experimental investigation. *Proc. Of the 11th WCEE*, Elsevier, CD-ROM, Paper No. 171.

- Panico, S., De Matteis, G., Mazzolani, F. M., 2003. Numerical investigation on pure aluminium shear panels. *XIX Convegno C.T.A.*, Vol. 2, pp. 459-470, Genova.

- Rezai, M., Ventura, C.E., Prion, H.G.L. 2000. Numerical investigation of thin unstiffened steel plate shear walls. *Proceedings 12th World Conference on Earthquake Engineering*.

- Tanaka, K., Torii, T., Sasaki, Y., Miyama, T., Kawai, H., Iwata, M., and Wada, A. 1998. Practical application of Damage Tolerant Structures with seismic control panel using low yield point steel to a high-rise steel building. *Proc. Structural Engineering World Wide*, Elsevier, Paper T190-4- CD-ROM.

- Thorburn, L.J., Kulak, G.L., Montgomery, C.J. 1983. Analysis of steel plate shear walls. *Structural Engineering* Report No. 107, University of Alberta, Canada.

Contacts *(include at least one COST C12 member)*

- Prof. F. M. Mazzolani
  Address: Department of Structural Analysis and Design, University of Naples 'Federico II', P.le Tecchio, 80, Naples, Italy.
  e-mail: fmm @unina.it
  Phone: (+39) 0817682443
  Fax: (+39) 0815934792

Prepared by: L. Fiorino, G. Della Corte

Department of Structural Analysis and Design

University of Naples "Federico II"

SEISMIC BEHAVIOUR OF COLD-FORMED STEEL LIGHTWEIGHT LOW-RISE BUILDINGS

Description

- In cold-formed steel lightweight low-rise buildings the main structure are generally made of galvanized steel profiles and wood-based and/or gypsum-based sheathings assembled together by means of mechanical fasteners.

- If sheathing elements have adequate strength and stiffness, and if there are adequate connections to the profiles, then roofs, floors and walls can perform as diaphragms under seismic actions ("sheathing braced design" approach).

Field of application

- Cold-formed steel lightweight profiles generally used as secondary structural elements are employed more and more as main structural members for residential, industrial and commercial constructions.

- The use of cold-formed steel lightweight low-rise buildings has increased significantly in country located in seismic zones in recent years.

Technical information

- There are three basic systems of constructions for low-rise houses, which are named (*i*) "stick-built" constructions, (*ii*) "panelised" constructions, and (*iii*) "modular" constructions.

- The stick-built constructions (Figure 1) are directly derived from the wood construction systems and are, undoubtedly, the most common type of cold-formed structural system. They are obtained by the assembling on the job-site of cold-formed members with C or Z section. Namely, the vertical-load bearing system is obtained by means of single or coupled back-to-back C-section (studs), interconnected at each end by 'wall tracks', in such a way to realize 'stud walls'. The horizontal floor system is usually made of 'floor joists' interconnected on the upper side by a 'floor deck' made of wood, steel or composite materials. Steel trusses, whose members have C and/or L sections, often realize the roof structure.

- The level of prefabrication increases going from the stick-built constructions to the panelised constructions (Figure 2). In the latter, wall panels are realized in the workshop and then assembled together by means of mechanical fasteners.

- The higher prefabrication level is reached in the modular construction systems (Figure 3), where wall panels are assembled together in the workshop, in order to realize entire modules to be later assembled on site by means of mechanical fasteners.

- The seismic performance of house structures built with cold-formed steel members is conditioned by the horizontal (roof and floors) and vertical (stud walls) Lateral Force Resisting System (LFRS) behaviour. Generally, both horizontal and vertical lateral support systems are cold-formed steel frames sheathed with structural panels. Consequently, the same basic parameters influence their shear response.

Figure 1. Stick-built construction.

Figure 2. Panelised construction.

Figure 3. Modular construction.

Structural aspects

- Two main design methodologies for the evaluation of the lateral load resistance of a Cold-Formed Steel Stud Shear Wall (CFSSSW) exist, following methods similar to that used for wood-framed walls: (*i*) "segment" method and (*ii*) "perforated shear wall" method.
- The segment method is a traditional shear wall design methodology. The contributions from various connections and wall portions that are not part of the "designed" LFRS are neglected, as shown in Figure 4.
- The perforated shear wall method is relatively very simple, empirical method. It requires that a fully-sheathed wall line with perforations for windows and doors be

restrained at the ends with a hold-down bracket or adequate corner framing in lower capacity shear walls, as shown in Figure 4.

- For a CFSSSW laterally braced by structural sheathings, the possible failure modes are: sheathing failure, sheathing-to-frame connections failure, frame-to-foundation failure. For each of these failure modes, the evaluation of the unit shear strength values (v) may be obtained through two main ways: (*i*) semi-empirical methods and (*ii*) experimental methods.

- Reference of relevant related codes:

AISI (1996) Cold-Formed Steel Design manual. AISI (American Iron and Steel Institute). Washington DC.

AISI (2001) North American Specification for the design of Cold-Formed Steel structural members (November 9, 2001 draft edition). AISI (American Iron and Steel Institute), Washington DC.

AS/NZS 4600 (1996) Cold-formed steel structures. AS/NZS (Australian Standards/New Zealand Standards). Sydney.

CSSBI (1991) Canadian Steel framing design manual, CSSBI (Canadian Sheet Steel Building Institute). Cambridge, Ontario.

ENV 1993-1-3 (1996) Eurocode 3: Design of steel structures – Part 1-3: General rules - Supplementary rules for cold formed thin gauge members and sheeting. CEN (European Committee for Standardization). Bruxelles.

IBC (2000) International Building Code: 2000. International Code Council, Inc. Falls Church, VA, USA.

IRC (2000) International Residential Code for One- and Two-Family Dwelling: 2000. International Code Council, Inc. Falls Church, VA, USA.

NASFA (2000) Prescriptive Method For Residential Cold-Formed Steel Framing (Year 2000 Edition). NASFA (North American Steel Framing Alliance). Lexington, KY, USA.

prEN 1998-1 (2001) Eurocode 8: Design of structures for earthquake resistance – Part 1: General rules – Seismic actions and rules for buildings. CEN (European Committee for Standardization). Bruxelles.

UBC (1997) Uniform Building Code: Volume 2. International Conference of Building Officials. Whittier, CA. USA.

Research activity

- To investigate on the seismic behaviour of CFSSSW sub-assemblages, a rather large number of experimental programs have been conducted in USA, Canada, Australia, Romania and Italy.

- It is possible to individuate the following basic factors that influence the lateral behaviour of CFSSSWs:

- From the examination of these experimental data, it is possible to individuate the following basic factors that influence the lateral behaviour of CFSSSWs: (*i*) sheathing (type, thickness, orientation); (*ii*) framing (stud size, thickness and spacing, presence of X-bracing); (*iii*) fastener (type, size, spacing); (*iv*) geometry (wall aspect ratio, openings); (*v*) type of loading (monotonic or cyclic); (*vi*) construction techniques and anchorage details.

- The specific research activities are focused on the evaluation of seismic performance of sheathed cold-formed steel stud shear walls.

- This specific research activities carried out by the University of Naples "Federico II" is focused on the evaluation of seismic performance of sheathed cold-formed steel stud shear walls as part of a more comprehensive research program, devoted to study innovative steel structures for the seismic protection of buildings.

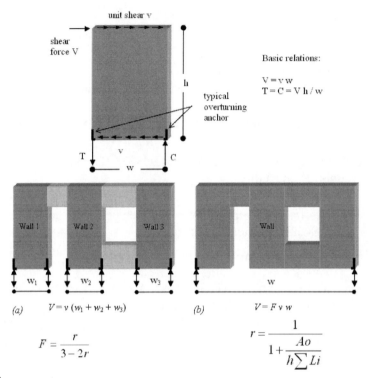

Basic relations:

$$V = v\,w$$
$$T = C = V\,h\,/\,w$$

$$(a) \quad V = v\,(w_1 + w_2 + w_3)$$

$$F = \frac{r}{3 - 2r}$$

$$(b) \quad V = F\,v\,w$$

$$r = \frac{1}{1 + \dfrac{Ao}{h\sum Li}}$$

where:
r: is the sheathing area ratio;
Ao: is the total area of openings,
h: is the height of the wall;
Li: is the length of the full height wall segment

Figure 4. Two alternative design methods: the segment method (a) and the PSW method (b)

Method of analysis

- The main part of the specific research activity is based on an experimental program on two identical stud shear wall sub-assemblies. One sub-assemblage has been tested under monotonic loading, the other has been instead subjected to a purposely developed cyclic loading history.

- The generic stud shear wall sub-assembly specimen represents a good model of typical lateral load resisting systems of stick-built house structures. It is made-up by a floor supported by two walls, as shown in Figure 5.

- Two types of load are uniformly applied to the floor: gravity and racking loads. In particular, the racking loads are applied following both monotonic and cyclic loading regime.

- This testing apparatus allowed the capacity of the horizontal floor panels to transmit loads to the vertical wall panels to be checked, up to failure of the vertical stud-to-panel connections (Figure 6).

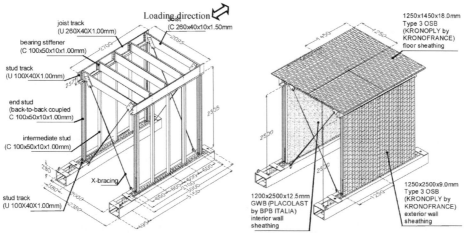

(a) Specimen without sheathings. (b) Specimen with wall and floor sheathings.
Figure 5. Global 3D view of the generic sub-assemblage.

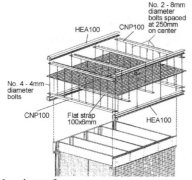 

(a) Load transfer system. (b) Global 3D view.
Figure 6. Test setup.

Results of analysis

- All the components of this structural system can be designed according to capacity design principles, imposing collapse in the vertical shear walls' connections (most ductile collapse mechanism), without significant increase of the cost.

- In the monotonic test the collapse mechanism was invariant during the increasing lateral displacement, whilst in the cyclic test some modifications (more brittle collapse mechanism) occurred after that the peak lateral load was achieved. These modifications produced significant strength degradation in the cyclic test, after the achievement of the peak load, which was stronger than the one observed during the monotonic test.

- The lateral-load response for displacements lesser than the ones corresponding to the peak strength was very similar for all tested walls, whilst significant differences were observed for larger values of the displacement. This indicates a relatively more unreliable response in the unstable branch of the behaviour.

- The horizontal diaphragm can adequately transfer the horizontal loads to the vertical shear walls, without any appreciable damage.

Further developments

- Further analysis should be performed for a deeper interpretation of experimental results and the development of theoretical/numerical models for both the prediction and the simulation of the seismic response of this constructive system.

References

- Della Corte, G., Fiorino, L., Landolfo, R., Di Lorenzo, G. 2003. Seismic performance of steel stud shear walls: planning of a testing program. In proceedings of the *4th International Conference on Behaviour of Steel Structures in Seismic Areas (STESSA 2003)*. Napoli. Mazzolani F.M. (ed.). A.A. Balkema Publishers.

- Fiorino, L. 2003. *Seismic Behavior of Sheathed Cold-Formed Steel Stud Shear Walls: An Experimental Investigation*. Ph.D. Dissertation, Department of Structural Analysis and Design, University of Naples "Federico II".

- Lawson, R.M. & Ogden, R.G. 2001. Recent developments in the light steel housing in the UK. In proceedings of the *9th North Steel Cconference of Construction Institute*. Helsinki.

- Schuster, R.M. 1996. Residential applications of cold-formed steel members in North America. In proceedings of the *5th International Structural Stability Research Council – SSRC*. Chicago.

Contacts

- Raffaele Landolfo
  - Address: Department of Constructions and Mathematical Methods in Architecture, University of Naples "Federico II",
    Via Monteoliveto, 80134 Napoli, Italia.
    e-mail: landolfo@unina.it
    Phone: ++39 081 7682447, Fax: ++39 081 5934792

COST C12 - WG2

Datasheet I.3.1.1.7

Prepared by: V. Bosiljkov, M. Tomaževič
               Section for Earthquake engineering, Slovenian National Building and
               Civil Engineering Institute

CONTEMPORARY MASONRY SUBJECTED TO SEISMIC LOADING

Description

- Evolution of masonry construction has been in the past a casual process rather than one preceded by scientific or engineering analysis and some of the innovative construction techniques can be in the long term incomparably more harmful to the masonry construction than (in the short term) they can provide benefits by the means of increasing the performance and economy of masonry.

- This problem is particularly pronounced and can be closely correlated to the development and optimisation of the vertically perforated lightweight clay blocks with very thin shells and webs, with different types of the execution of the perpend joint and different pattern of perforations and their application in the earthquake prone areas.

- All these improvements of the blocks, though not significantly influencing the collapse mechanisms of the masonry when subjected to compressive gravity loads, can govern the overall brittle behaviour of masonry under seismic conditions as they reduce robustness of masonry units and thus overall behaviour of masonry walls as structural elements.

Field of application

- Although execution of the perpend joints (tested ones are presented in Figure 1) and its influence on the shear and compressive strength of the masonry is partly incorporated in the new EC 6, there is still lack of knowledge of the overall resistance of such masonry under seismic loading.

-

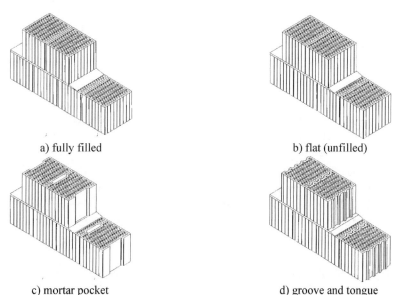

a) fully filled   b) flat (unfilled)

c) mortar pocket   d) groove and tongue

Figure 1. Characteristic types of the perpend joints

- Traditionally, only the construction of masonry walls with fully grouted (filled) perpend joints has been allowed in seismic zones.
- Perpend joints (according to EC 6) can be considered to be filled if mortar is provided to the full height of the joint over a min. of 40% of the width of the unit.

Technical information

- In order to investigate the possibility of optimization of masonry construction with masonry units and technologies not being in complete conformity with the requirements of Eurocode 8 for the construction of buildings in seismic zones, it has been agreed that the walls made of hollow clay units with maximum 50 % of volume of holes (ENV 1998-1-2 requirement 50 %) and minimum thickness of shells of 12 mm (ENV 1998-1-2 requirement 15 mm) as typical representative of units available on the market, shall be tested. The units can be classified into Group 2 units according to Table 3.1 of prEN 1996-1.
- Especially shaped hollow masonry units with dimensions 245 x 300 x 240 mm (length x width x height), which had the same general hole pattern but head faces adjusted to different types of head joints, have been manufactured for this research project. General shape of hole patterns of the units is presented in Fig. 2.
- General purpose class M5 mortar shall be used for the construction of test specimens.

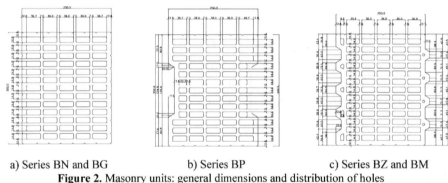

a) Series BN and BG   b) Series BP   c) Series BZ and BM

Figure 2. Masonry units: general dimensions and distribution of holes

Structural aspects

- The existing model for evaluation of the shear strength of masonry presented in EC 6 makes a distinction between masonry with filled and unfilled head joints and gives two equations for the determination of the characteristic shear strength.
- However, there are no comments on the influence of the execution of the head joints on the overall stiffness of the masonry elements and its dissipation capacity. When calculating the earthquake resistance of the masonry structure according to the allowable stresses method, the absence of those data's are tolerable, but when calculating the design resistance according to the limit state method, those data are essential.
- Up to recently requirements for the masonry units (Group 2) which were stated in EC 6 (1995) and EC 8 (1994) were as follows:

  min. compressive strength normal to the bed face 2.5 MPa (EC 6), min. compressive strength parallel to the bed face in the plane of the wall 2.0 MPa (EC 6),

max. percent of voids 50% (EC 6),

min. thickness of outer shell 15 mm (EC 8) and

fulfilment of the vertical joints 100% (EC 8).

- Being pushed by the demands of industry, most of these requirements have been redrawn from the codes or left to the national authorities for their consideration. Present situation according to EC 6 (2002) is that the selection of suitable units according to grouping specified in EC 6 is up to National Annexes.

- According to the newest version of EC 8 (2002), all old demands have been redrawn and a new demand has been stated so that *units should have sufficient robustness in order to avoid local brittle failure when subjected to seismic loads*. Unfortunately in the draft version of EC 8 (2002) and accompanying standards for the units, neither definition nor criteria for robustness have been defined up to now.

- Not many data are available in the literature concerning the overall resistance of the masonry with and without filled head joints. One of the fist experimental approaches on that subject was made by Barth and Marti (1997). Through mono and biaxial compressive tests of the masonry specimens with different inclination of the bed joints towards the main compressive stresses, they have tested the masonry specimens made from smaller blocks with filled and unfilled head joints. Their conclusions were that depending on the bed joint inclination towards the main compressive stresses within the masonry, the strength of masonry with unfilled head joints was considerably reduced in comparison with the masonry with fully filled head joints. The reduction of stiffness was even more pronounced than the reduction of strength. The latter was attributed either to the slip mechanisms of failure of the whole row of the units within the specimen or due to the early crushing of the units where (because of the slightly concave shape over the height of the unit) concentration of the stresses have early occurred.

- Recently Capozucca et al. (2000) through compressive and diagonal shear tests (monotonic loading) have tested masonry made from two types of hollow clay blocks (45% of voids) with filled, unfilled and with groove and tongue system of execution of the head joint. His results on the shear strength of the masonry were in accordance with the results of Meyer (2000), who also has found that the shear strength of the masonry with unfilled joints are lesser than the masonry with the filled head joints. They both had concluded that the coefficient of the reduction for the shear strength of the masonry with unfilled head joints according to the EC 6 is too high. Although their test methods for verification of the shear resistance of the masonry due to seismic loading can be also disputable, both of them did not present any results on the deformation characteristics or on the ductility of that type of the masonry.

Research activity

- Attention is focused on the behaviour of masonry made from brittle units with different types of perpend joints.
- Four different types of perpend joints were investigated (Figure 1).
- Masonry was subjected to monoaxial compressive loading and to seismic (biaxial) loading.

Method of analysis

- For each of the series of different execution of perpend joints, 6 masonry specimens having dimensions 100/150/30 cm (l/h/t) were constructed. Three specimens were tested under compressive loading (compression tests), whereas the other three were tested by subjecting them to constant vertical and cyclic horizontal loading (shear tests).

- Compression tests: mechanical properties of masonry under compression were determined in a 5000 kN hydraulic testing machine, by subjecting the specimens to compressive loading with constant velocity of application and step-wise increasing magnitude of compressive force and unloading at each step. Compressive strength, modulus of elasticity as well as vertical stress–strain relationships of the masonry were determined.

- There is no harmonized approach in testing resistance of the masonry due to the seismic loading. For this research, the walls have been tested as vertical cantilevers fixed to the testing floor and subjected to constant vertical load. Programmed lateral displacements have been imposed at the level of the bond-beam at the top of the walls. The reinforced concrete foundation blocks on which the walls have been constructed, have been fixed to the strong floor by means of bolts, prestressed in order to prevent any lateral motion or rocking of wall specimens.

- Horizontal load were applied in the form of programmed displacements, cyclically imposed in both directions, with programmed, step-wise increased amplitudes up until the collapse of the specimens. At each displacement amplitude, the loading was repeated three times. During the tests, forces, displacements, and rotations of the specimens were measured, and possible displacements and rotations of the foundations were monitored.

- The test set-up consisted of a steel testing frame and two programmable hydraulic actuators, fixed to the frame in order to simulate constant gravity loads and lateral in-plane seismic loads. Test set-up is schematically presented in Figure 3, whereas a typical specimen during the lateral resistance test is shown in Figure 4.

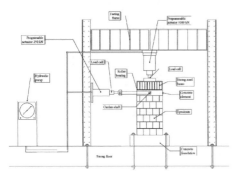

Figure 3. Lateral resistance test set-up   **Figure 4.** Typical wall during lateral resistance test

Results of analysis

- Compressive tests: No significant difference in the values of compressive strength of masonry has been observed as regards the different types of head joints. The differences are in good correlation with the actual strength of the units and mortar.

- Compressive tests: During the tests, vertical cracks have been first observed in the shells of the units. Then, the shells started buckling and falling out and, finally, crushing of units took place, randomly distributed over the surface of the specimens.
- Shear tests: All walls failed in shear, as expected. However, brittle local failure of units, i.e. buckling and crushing of thin shells and webs, was the predominant phenomenon which determined the failure mode. The walls failed in a non-ductile, brittle mode, soon after the maximum resistance has been attained.
- Shear tests: The displacement capacity of the tested walls, expressed by displacement capacity indicators with regard to rotation of the walls at the attained maximum resistance and ultimate rotation just before collapse θ_u/θ_{Hmax}, is small, below the expected values for unreinforced masonry walls.
- Masonry made from brittle units is not capable to overtake large in-plane deformation that may arise within the masonry under seismic loading.
- No significant difference in energy dissipation capacity of the different tested types of the walls, expressed in terms of energy dissipation indicators $I_{E,dis} = E_{hys}/E_{inp}$ (ratio between the cumulative dissipated hysteretic energy and cumulative input energy) and coefficient of equivalent viscous damping ξ, has been observed, which could have been attributed to the influence of different types of the head joints.
- The values of mechanical properties of masonry did not depend on the type of the head joints

Further developments

- Robustness of masonry units and masonry bond are important parameters, which define the behaviour of masonry when subjected to seismic loads, thus further work should be focused on definition of robustness of masonry units and methodology for its evaluation.
- Harmonized (codified) testing procedure for seismic resistance of masonry should be determined as well.
- Two models for the numerical analysis of the geometry of the blocks are in the stage of preparation. First one for simulating the behaviour of masonry blocks due to the monotonic compressive loading and other for modelling the behaviour of the block due to biaxial loading (Figure 5).

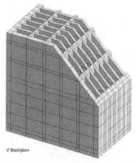

Figure 5. FEM model for biaxial loading of blocks

References

- European Committee for Standardisation, 1995. **EUROCODE 6.**, Design of Masonry Structures. Part 1-1: general rules for buildings – rules for reinforced and unreinforced masonry, ENV 1996 1-1: Brussels: CEN, June.

- CEN/TC 250, 2002. **EUROCODE 6**, Design of Masonry Structures – Part 1-1: Common rules for reinforced and unreinforced masonry structures, prEN 1996-1-1

- **EUROCODE 8**, 1994. 'Design provisions for earthquake resistance of structures - Part 1-3: General rules - Specific rules for various materials and elements', November.

- PrEN 1998-1:200X, **EUROCODE 8,** 2002. Revised Final PT Draft (preStage 49) Design provisions for earthquake resistance of structures - Part 1-3: General rules - Specific rules for various materials and elements, Draft May,

- Barth, M. and Marti, P., 1997. Versuche an knirsch vermauertem Backsteinmauerwerk, *Institut für Baustatik und Konstruktion ETH Zürich*, August.

- Capozucca, R., Cerri, M.N. and Zanarini, G., 2000. Shear strength of brickwork masonry with different types of mortar joints, *12th IBMAC, Madrid, Spain, 25-28 June*.

- Hauck, E. and Jung, E., 1991. Improvement of the coefficient of thermal conductivity of lightweight clay bricks and blocks, *Jahrbuch für die Ziegel-, Baukeramik- und Steinzeugröhren-Industrie*, Beuverlag GMBH; Wiesbaden und Berlin.

- Meyer, U., 2000. In plane shear strength of clay unit masonry – recent German test results, *12th IBMAC, Madrid, Spain, 25-28 June*.

Contacts

- Vlatko Bosiljkov, Ph.D.
    Address: Section for Earthquake engineering, Slovenian National Building and Civil Engineering Institute, Dimiceva 12, 1000 Ljubljana, Slovenia
    e-mail: vlatko.bosiljkov@zag.si,
    Phone: ++386 1 2804280,
    Fax:++386 1 2804484
- Prof. dr. Miha Tomaževič
    Address: Section for Earthquake engineering, Slovenian National Building and Civil Engineering Institute, Dimiceva 12, 1000 Ljubljana, Slovenia
    e-mail: miha.tomazevic@zag.si,
    Phone: ++386 1 2804250,
    Fax:++386 1 2804484

EXPERIMENTAL APPROACH IN DETERMING THE STRUCTURAL BEHAVIOUR FACTOR FOR URM BUILDINGS

Definition

- The capacity of the structural systems to resist seismic actions in the nonlinear range permits generally their design for forces smaller than those corresponding to a linear elastic analysis.

- Behaviour factor q is an approximation of the ratio of the seismic forces that the structure would experience if its response was completely elastic, to the minimum seismic forces that may be used in the design with a conventional elastic model, still ensuring a satisfactory response of the structure.

- It may depends on the ductility of the structure, the strength reserves that physicaly exist in a structure (structural redundancy) and effective damping of the construction.

Physical Features

- The safety of a structure against earthquakes is a probabilistic function, which depends on the expected seismic action and resistance capacity of a structure.

- For all structural members, their design resistance capacity R_d, calculated by taking into account characteristic strength values and partial safety factors γ_M of members materials, shall be greater than the design value of combined action effect E_d, incuding seismic action effect $E_d \leq R_d$.

- Considering the regularity of masonry buildings, whose response is nog significantly affected by contributions from higher modes of vibration (Tomaževič, 1999), lateral force method of analysis will provide adequate results. Following this method, the seismic base shear force F_b for each horizontal direction in which the building is analysed, is determined as: $F_b = S_d(T_1)m\lambda$, where $S_d(T_1)$ is the ordinate of the design spectrum at period T_1, T_1 is fundamental period of vibration of the building for lateral motion in the direction considered, m is total mass of the building above the foundation and λ is correction factor.

- Since masonry buildings are rigid structures with natural periods of vibration T_1 ranging between T_B and T_C, where the response specturm is flat, the ordinate of design spectrum can be determined by $S_d = a_g S \eta \dfrac{2,5}{q}$, where a_g is design graound acceleration on type A ground, S – soil factor, η – damping correction factor and q – structural behaviour factor.

- As specified in EC8, the capacity of structural system to resist seismic actions in the non-linear range generally permits the design for forces smaller than those corresponding to a linear elastic response.

- The definition of behaviour factor q is explained in following figure, where the seismic response curve of an actual structure, idealized as a linear elastic-perfectly plastic envelope, is compared with the reponse of a perfectly elastic structure having the same initial elastic stiffness characteristics.

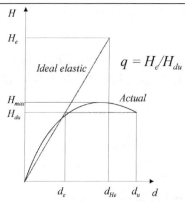

Figure 1. Definition of structural behaviour factor q

- As a result of the energy dissipation capacity of the actual structure, expressed by the global ductility factor $\mu_u = d_u/d_e$, there is no need for the structure to be designed for strength, i.e. for the expected elastic load H_e. The structure is designed for the ultimate design load H_{du} and the ratio between the two is called bhe behaviour factor q.
- Following the definition, for behaviour factor, it can be also expressed in the temrs of global ductility factor μ_u as $q = \sqrt{(2\mu_u - 1)}$.

Effect on constructions

- For better understanding of the behaviour of masonry structures under seismic actions, the experimental results are essential. The results of these tests consistently revealed that masonry, when properly proportioned, detailed and constructed with good workmanship, provides adequate resistance against seismic forces.
- When subjected to seismic actions the structures can dissipate energy. It becomes apparent either through large deformations in the case of ductile materials (e.g. steel) or cracking in the case of more brittle materials (Unreinforced Masonry – URM).

Modelling

- In order to experimentally evaluate behaviour factor for masonry structures, six models representing buildings with two different structural configurations and constructed with two different types of masonry materials have been tested on a simple uni-directional seismic simulator, a two-story terraced house with main structural walls orthogonal to seismic motion (models M1 - Figure 2) and a three-story apartment house with uniformly distributed structural walls in both directions (models M2 - Figure 3). Four models of the first and two models of the second type have been tested. In the case of the terraced house, two models (Figures 4 and 5) have been built as either partly or completely confined masonry structures.
- Models have been built at a 1:5 scale with special model materials designed to fulfill the requirements of complete model similitude. The correlation between the model and prototype characteristics of masonry has been verified by testing a series of model and prototype-size walls under compression and a combination of compression and cyclic lateral loading (Figure 6).

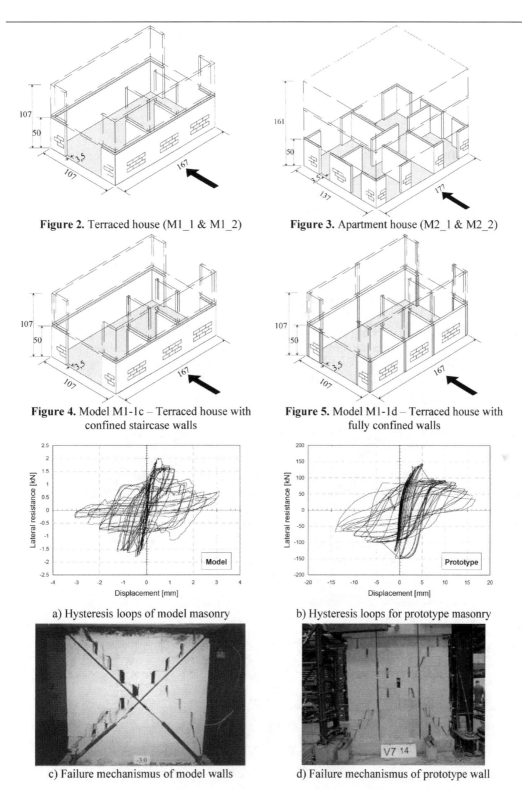

Figure 2. Terraced house (M1_1 & M1_2)

Figure 3. Apartment house (M2_1 & M2_2)

Figure 4. Model M1-1c – Terraced house with confined staircase walls

Figure 5. Model M1-1d – Terraced house with fully confined walls

a) Hysteresis loops of model masonry

b) Hysteresis loops for prototype masonry

c) Failure mechanismus of model walls

d) Failure mechanismus of prototype wall

Figure 6. Correlation between the results of typical model (a&c) and prototype wall (b&d) lateral resistance tests

- The north-south component of the earthquake acceleration record obtained at Petrovac

137

during the April 15, 1979, earthquake in Montenegro, with peak ground acceleration of 0.43 g has been used to drive the shaking-table. The intensity of shaking was controlled by adjusting the maximum amplitude of the shaking-table displacement, obtained by numerical integration of the accelerogram used as the input in each successive test run, scaled according to the laws of model similitude (Figure 7).

a)   b)

Figure 7. Earthquake acceleration record (a) and (b) typical scaled accelerogram

■ All models failed in shear, as expected. Regardless to the structural type and configuration, shear cracks developed in structural walls in the direction of seismic motion, subsequently leading to stiffness and strength degradation and final collapse of the models. The unreinforced terraced house model M1-1 and apartment house model M2-1, made of materials simulating calcium silicate masonry units collapsed immediately after the first damage occurred, whereas respective models M1-2 and M2-2, made of materials simulating hollow clay units, though damaged, withstood additional shaking before collapse. The behavior of partly and fully confined terraced house models M1-1c and M1-1d was significantly improved. Typical damage to the models just before collapse is shown in Figures 8.

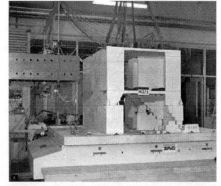

a) Apartment house just before the collapse   b) Terraced house just before the collapse

c) Confined terraced house model M1-1c just   d) Fully confined terraced house model M1-1d
before collapse

Figure 8. Typical damage to the models just before collapse

- On the basis of the recorded displacement and acceleration response time histories and taking into account the masses of the models, concentrated at each floor level, the maximum values of the base shear developed in the models during the individual phases of testing, have been calculated. The values have been expressed in a non-dimensional form in terms of the base shear coefficient (BSC), which is the ratio between the base shear resisted and the weight of the model. The values are plotted against the first story rotation angle (the ratio between the relative story displacement and story height), hence obtaining the lateral resistance - displacement envelopes of a critical story in a non-dimensional form (Figure 10).

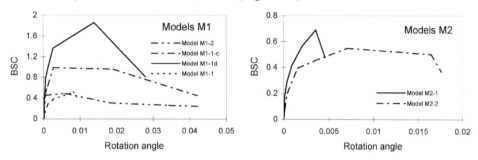

Figure 10. Experimentally obtained base shear coefficient - story rotation angle relationships

- For each of the tested models, the theoretical elastic response in terms of the calculated maximum elastic base shear has been compared with the maximum base shear value BSC_{max}, evaluated on the basis of the known masses of the models and measured acceleration response, as well as with the value BSC_u obtained by the idealization of the experimental resistance envelopes shown in Figure 10. In order to retain the character of simplicity, experimentally obtained base shear coefficient - story rotation angle relationships have been idealized as bilinear elastic-plastic relationships. The values of behavior factors q, resulting from experimental envelopes, are presented in Table 1.

Table 1. Values of structural behavior factor q evaluated from experiments

Model	$q = BSC_e/BSC_{max}$	$q = BSC_e/BSC_u$	$q = (2 \mu_u - 1)^{1/2}$	EC8 provisions
M1-1	1.34	1.53	2.09	1.5-2.5
M1-2	1.84	2.06	2.20	1.5-2.5
M1-1c	2.44	2.56	2.99	2.0-3.0
M1-1d	1.63	1.91	2.88	2.0-3.0
M2-1	1.55	1.91	2.61	1.5-2.5
M2-2	1.74	1.85	2.84	1.5-2.5

Structural analysis method

- In general, the seismic analysis of masonry structures can be done either through usage of lumped parameter models (LPM), structural element model (SEM) or finite element models (FEM). Since for ordinary masonry buildings, the first vibration mode shape is the predominant one, there is usually no need for sophisticated non-linear dynamic and the LPM models are quite rare.

- SEM approximates the actual structural geometry more accurately by describing individual structural elements such as piers and walls. In the case of single or multistory buildings due to their regularity and simplicity an equivalent static analysis in two orthogonal directions by using SEM can provide reliable information regarding the seismic safety under expected seismic loads. Nevertheless, since seismic loading can exercise the structural system to and beyond its maximum resistance capacity, the SEM models usually have to be used with static non-linear analysis. In that case a step-by-step procedure is followed, using decreased stiffness values under increasing lateral loads. Nonlinear element behavior is prescribed in the form of nonlinear lateral deformation-resistance relationships, depending on the boundary conditions and failure mode of masonry elements. Usually the bi-linear or tri-linear behavior of SE is considered. The storey resistance envelope is calculated by stepwise drifting of the storey for small values. The SE's are deformed equally (due to the rigidity of floor structure) and internal forces are induced according to the assumed shape of resistance envelope of each SE. In the case of torsional effects (due to relatively displacement of the mass centre to the centre of stiffness of the storey) the displacements of individual SE are modified.

- FEM can be adopted for the masonry depending from the type of the problem as simplified micro-modelling technique, homogenization approach or macro-modelling approach.

Guidelines and/or codification

- EUROCODE 8, 'Design provisions for earthquake resistance of structures - Part 1-3: General rules - Specific rules for various materials and elements', November 1994.

Example of application

- Following the different methods of evaluation of the seismic resistance of masonry buildings and their comparison to the actual resistance, the benefits of consideration of structural behaviour factor q is apparent from the following figure.

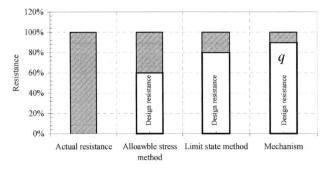

Figure 11. Typical correlation between actual and design resistance of structure (Tomaževič, 1999)

References

- Paulay, T. & Priestley, M.J.N., 1992. *Seismic design of reinforced concrete and masonry buildings*, Jonh Wiley & Sons, INC.
- Tomaževič, M., 1999. *Earthquake-resistant design of masonry buildings*, Imperial

College Press, London,

- Tomaževič, M., Weiss, P., & Bosiljkov, V., 2004. Structural behaviour factor for masonry structures, *13th World Conference on Earthquake Engineering*, Vancouver, B.C., Canada, August 1-6.

Contacts

- Vlatko Bosiljkov, Ph.D.
  Address: Section for Earthquake engineering, National Building and Civil
  Engineering Institute, Dimiceva 12, 1000 Ljubljana, Slovenia
  Phone: ++386 1 2804280
  Fax:++386 1 2804484
  E-mail: vlatko.bosiljkov@zag.si
- Prof. dr. Miha Tomaževič
  Address: Section for Earthquake engineering, National Building and Civil
  Engineering Institute, Dimiceva 12, 1000 Ljubljana, Slovenia
  Phone: ++386 1 2804250
  Fax:++386 1 2804484
  E-mail: miha.tomazevic@zag.si

COST C12 – WG2

Datasheet I.3.1.2.1

Prepared by: _F.M. Mazzolani[1], A. Mandara[2], G. Laezza[2]_

[1] University Federico II of Naples
[2] Second University of Naples

PASSIVE CONTROL OF EXISTING BUILDINGS

Description

- Dissipation of kinetic energy induced by earthquake actions;
- Reduction of structural response under seismic excitation;
- Use of both viscous and plastic-threshold devices;
- Ease of application and reversibility;
- Liable to be profitably used in the field of historical and monumental constructions;

Technical Information

- The technique is based on the use of special devices purposely conceived for dissipating a share of the energy entering the structure under seismic actions; this energy dissipation prevents the construction to be severely damaged as a consequence of earthquake actions;

- This technique is recommended for the seismic upgrading of existing constructions when it is possible to find at least two sub-systems (m_1 - k_1, m_2 - k_2, ... m_i - k_i) with different dynamic properties ($m_1/k_1 \neq m_2/k_2 \neq ... \neq m_i/k_i$), typically a heavy but rather deformable system, connected to a much stiffer sub-assemblage, used as a retaining element for the weaker one (Figure 1). In all cases, devices or dissipative ties are placed in such a way to exploit the dissimilar behaviour between connected parts, hence producing energy dissipation and response damping.

- Devices liable to be used are typically of two types:
  i) Viscous devices (Figure 2);
  ii) Plastic-threshold devices (Figure 3);

- Viscous devices can be sized according to a optimisation procedure, in which the value of the viscous constant c of the linear device behaviour law $F = c \cdot |\dot{x}|^\beta \cdot sign(\dot{x})$ is evaluated that minimises the structural response. Optimal design charts, obtained from the SDOF model and relevant parameters showed in Figure 4, are given in Figure 5.

- In case of non linear devices ($\beta \neq 1$), which is the case when siliconic fluid is used, an equivalent linear damper with $\beta = 1$ can be considered according to the procedure illustrated in Figure 6.

- Plastic-threshold devices additional to viscous dampers can be sized according to the maximum capability of the system, in order to limit the magnitude of transmitted forces. As a rule it should result $F_{2u} \geq 0.5F_{1u}$, F_{iu} being the ultimate load of the i-th sub-system.

- Both viscous constant and plastic threshold values can be also related to the desired performance level, as defined in Table1. The corresponding design value of device properties can be chosen according to curves of Figure 7 for a given performance level.

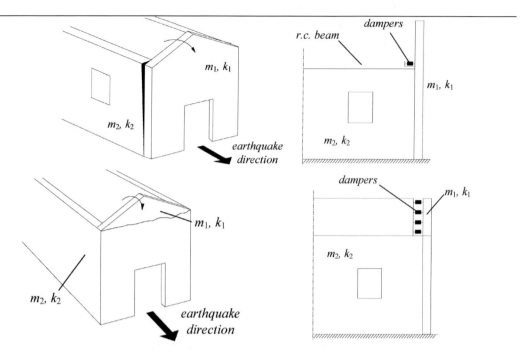

Energy dissipation devices for retaining a façade or a tympanum.

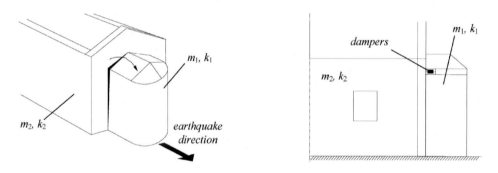

Energy dissipation devices for retaining an abside

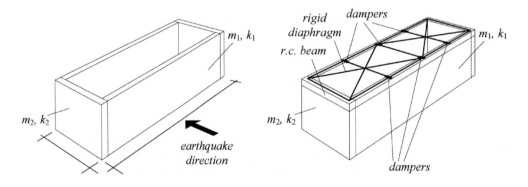

Energy dissipation devices for the protection of a single-storey masonry building.

Figure 1. Possible location of energy dissipation devices

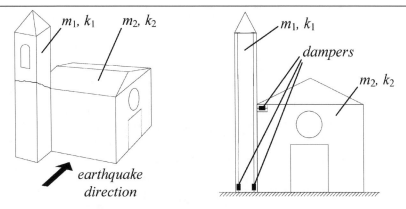

Energy dissipation devices for the protection of a bell tower.

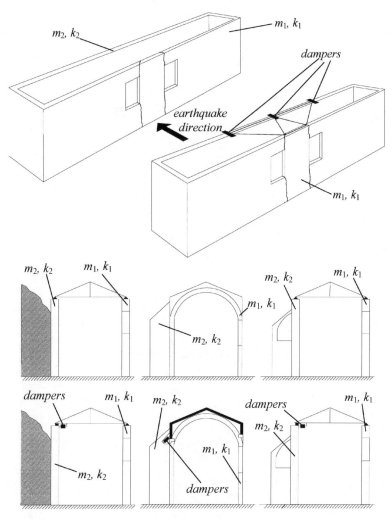

Possible location of energy dissipation devices in case of buildings having walls with strongly different properties.

Figure 1. Possible location of energy dissipation devices (ctd)

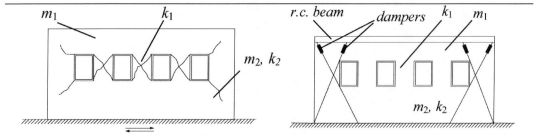

Bracing of a masonry wall by means of dissipative ties.

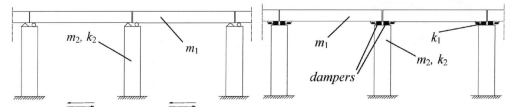

Energy dissipation devices for the protection of bridges.

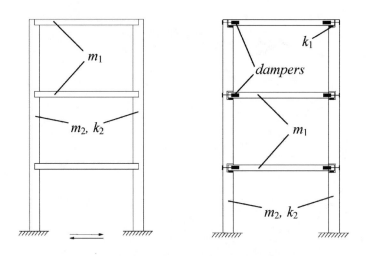

Energy dissipation devices for the protection of a multi storey building.

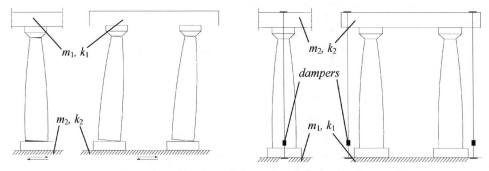

Protection of a colonnade by means of dissipative ties.

Figure 1. Possible location of energy dissipation devices (ctd)

146

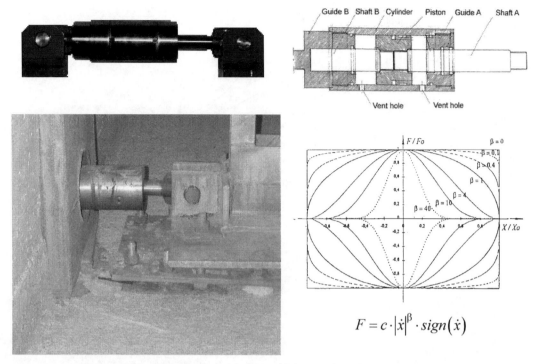

$$F = c \cdot |\dot{x}|^{\beta} \cdot sign(\dot{x})$$

Figure 2. Oleodynamic viscous devices with hysteretic cycle for several β values

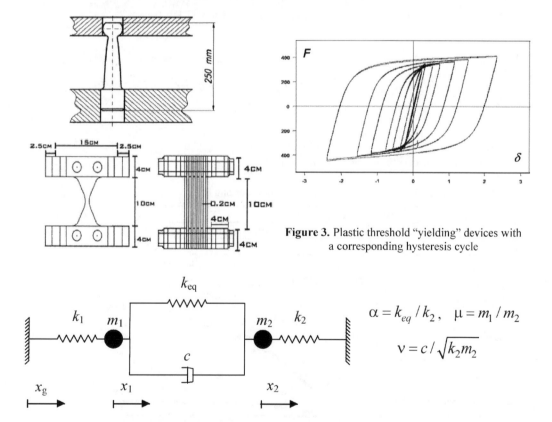

Figure 3. Plastic threshold "yielding" devices with a corresponding hysteresis cycle

$$\alpha = k_{eq} / k_2, \quad \mu = m_1 / m_2$$

$$\nu = c / \sqrt{k_2 m_2}$$

Figure 4. Simplified SDOF model for the evaluation of optimal viscous constant and relevant nondimensional design data.

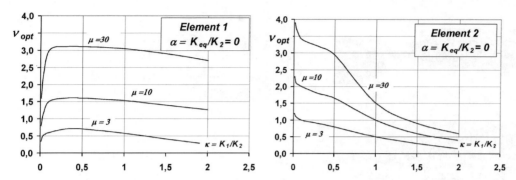

Figure 5. Optimal viscous values for the simplified system of Figure 4.

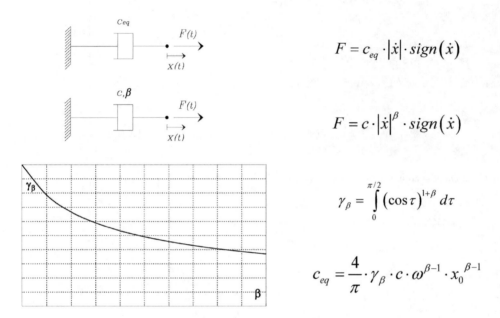

$$F = c_{eq} \cdot |\dot{x}| \cdot sign(\dot{x})$$

$$F = c \cdot |\dot{x}|^{\beta} \cdot sign(\dot{x})$$

$$\gamma_{\beta} = \int_{0}^{\pi/2} (\cos\tau)^{1+\beta} \, d\tau$$

$$c_{eq} = \frac{4}{\pi} \cdot \gamma_{\beta} \cdot c \cdot \omega^{\beta-1} \cdot x_0^{\beta-1}$$

Figure 6. Equivalent linear-non linear damping for SDOF system

Table 1. Performance levels for primary masonry structures according to FEMA 356.

Performance level		
Immediate occupancy (**IO**)	Life safety (**LS**)	Collapse prevention (**CP**)

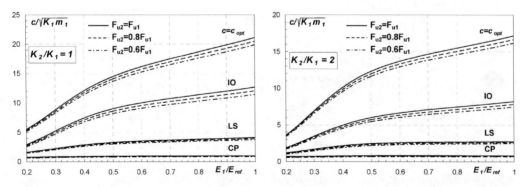

Figure 7. Nondimensional viscous constant for a given performance level versus E_1/E_{ref} ($E_{ref} = 660\text{N/mm}^2$).

Field of Application and End Users

- Masonry buildings
    - i) Reduction of structural response under seismic actions;
    - ii) Increase of in-structure damping and energy dissipation;
- Bridges
    - i) Reduction of top-pier displacement;

Structural Assessment

- Advantages
    - High effectiveness in the reduction of both structural response and damage when properly applied, namely when relative displacement between structural sub-systems can be exploited;
    - Ease of application on existing buildings and bridges, including historical or monumental cases;
    - Full application reversibility, which makes this technique particularly suitable to seismic retrofit of historical and monumental buildings;

- Disadvantages
    - Depending on the structural configuration, sometimes the system effectiveness can not be fully exploited;
    - As a rule, a careful evaluation of dynamic response of the structure is required in design in order to get the best performance from the system; this may sometimes require an accurate non linear modelling of the whole structure;

- Potential
    - A greatly improved structural performance can be achieved without the use of intrusive strengthening operations;

- Risk
    - The long-term reliability of the seismic protection system could be impaired by a poor maintenance; for this reason it is advised to place devices in a easily accessible position.

Economical Aspects

- Application of such techniques showed a convenience compared with most of traditional strengthening options; in particular it proved competitive compared with intrusive strengthening techniques for masonry based on the use of steel bars and injected mortar.

- The device effectiveness has been widely tested in experimental campaigns, theoretical research and practical applications carried out all over the world in many fields of engineering;
- Applications based on the use of such devices are very common in advanced countries like USA and Japan;
- The industrialised and standardised production allows a reliable use of the devices;
- An advanced state-of-the-art has been reached in the field of both research and codification, which provides the designer with all tools and knowledge necessary for a correct implementation of such techniques.

Examples

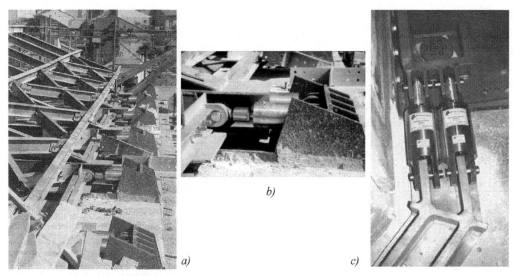

Figure 8. Viscous devices installed on the church of S. Giovanni Battista, Carife (a,b) (1990) and on the Basilica of S. Francesco, Assisi (c – Courtesy FIP Industriale S.p.A.) (1999), Italy

Figure 9. Viscous devices in a building of the University Federico II of Naples (1996)

Figure 11. Special oleodynamic (a) and yielding (b) devices in an industrial shed (c) in Sarno (SA) and the corresponding scheme of roof diaphragm with indication of device location (d) (2004)

Codification

- FEMA 356. 2000. *"Prestandard and Commentary for the Seismic Rehabilitation of Buildings"*. Report by the American Society of Civil Engineering (ASCE) for the Federal Emergency Management Agency (FEMA), Washington, D.C.

References

- Holmes Consulting Group LTD, 2001. Structure Damping and Energy Dissipation, Wellington, New Zealand.
- Mandara A. & Mazzolani F.M. 2001a. On the Design of Retrofitting by means of Energy Dissipation Devices, Proc. of *7th International Seminar on Seismic Isolation, Passive Energy Dissipation and Active Control of Vibrations of Structures*, Assisi, Italy.
- Mandara A. & Mazzolani F.M. 2001b. Energy Dissipation Devices in Seismic Up-Grading of Monumental Buildings. Proc. of *III Seminar on Historical Constructions, Guimaraes*, Portugal.
- Mazzolani F.M. 2001. Passive Control Technologies for Seismic-resistant Building in Europe", Prog. Structural Engineering and Materials 3, 277-287.
- Mazzolani F.M. & Mandara A. 1992., Nuove Strategie di Protezione Sismica per Edifici Monumentali: il Caso della Collegiata di San Giovanni Battista in Carife (in italian), Edizioni 10/17, Salerno, Italy.
- Mazzolani F.M. & Mandara A. 1994. Seismic Upgrading of Churches by Means of Dissipative Devices. In F.M. Mazzolani & V. Gioncu (eds), Behaviour of Steel Structures in Seismic Areas STESSA '94. E & FN SPON, London, p. 747-758.
- Mazzolani F.M. & Mandara A. 2002a. Modern Trends in the Use of Special Metals for the Improvement of Historical and Monumental Structures, *Engineering Structures* 24, 843-856, Elsevier.
- Mazzolani F.M. & Mandara A. 2002b. Seismic Protection of Historical Constructions by Means of Passive Control Techniques, Proc. of *12th European Conference on Earthquake Engineering*, London.
- Mazzolani F.M., Martelli A. & Forni M. 2001. Progress of Application and R&D for Seismic Isolation and Passive Energy Dissipation for Civil Buildings in the European Union", Proc. of *7th Int. Seminar on Seismic Isolation, Passive Energy Dissipation and Active Control of Vibration of Structures*, Assisi, 2-5 Ottobre, pp.249-277, Italy.
- Mazzolani F.M., Mandara A. & Froncillo S. 2003. Dissipative Steel Structures for Seismic Up-grading of Long-Bay Masonry Buildings. In F.M. Mazzolani (ed.), Behaviour of Steel Structures in Seismic Areas STESSA 2003, Balkema, Amsterdam.
- Mazzolani F.M., Mandara A. & Laezza G. 2003. Seismic Up-grading of Masonry Structures by Steel Braces and Dissipative Devices. Atti della III Settimana delle Costruzioni in Acciaio, Genova.
- Mazzolani F.M., Mandara A., Froncillo S. & Laezza G. 2003. Design Criteria for the Use of Special Devices in the Seismic Protection of Masonry Structures. Proc. of the *8th World Seminar on Seismic Isolation, Energy Dissipation and Active Vibration Control of Structures*, Yerevan, Armenia.
- Naeim F. & Kelly J.M. 1999. Design of Seismic Isolated Structures – From Theory to Practice, John Wiley & Sons, Inc.
- Skinner, R. I, Robinson, W. H & McVerry G. H. 1993. An Introduction to Seismic Isolation, John Wiley & Sons, Inc.
- Soong T.T. & Dargush G.F. 1997. Passive Energy Dissipation Systems in Structural Engineering, State University of New York at Buffalo, USA, John Wiley & Sons, Chichester, England.

Contacts

- Federico M. Mazzolani
  Address: Department of Structural Analysis and Design, University of Naples Federico II, P.le Tecchio 80, I-80125 Napoli, Italy
  E-mail: fmm@cds.unina.it
  Tel. +39-081-7682443
  Fax. +39-081-5934792

- Alberto Mandara
  Address: Department of Civil Engineering, Second University of Naples, Via Roma 29, I-81031, Aversa (CE), Italy
  E-mail: alberto.mandara@unina2.it
  Tel. +39-081-5010216
  Fax. +39-081-5037370

COST C12 - WG2

Datasheet I.3.1.2.2

Prepared by: Beatrice Faggiano

Department of Structural Analysis and Design

University of Naples "Federico II", Italy

DOG-BONE TECNIQUE FOR THE SEISMIC IMPROVEMENT OF STEEL MOMENT RESISTING FRAMES

Description

- The "dog-bone connection", namely Reduced Beam Section (RBS), is a structural detail aimed at enhancing the seismic behaviour of moment resisting steel frames. It allows to shift the plastic hinge from the beams' ends by means of a controlled weakening of the beam itself in a relevant position, hence ensuring a better structural performance under cyclic loads at both local and global levels.

- The dog-bone idea was proposed by Plumier since 1990 and first applied by Georgescu in 1996. However the type and the technology of the RBS beam-to-column connection, as well as its behaviour under cyclic load conditions, were comprehensively investigated, in U.S.A., after the Northridge earthquake (Engelhardt et al., 1996).

Field of application

- The Reduced Beam Section (RBS) connection is proposed as an upgrading or a retrofitting system for moment resisting steel frames in seismic areas.

Structural aspects

The dog-bone concept

The RBS protects the node from plastic deformations, producing a ductility enhancement:

- at local level
  - the inelastic deformations in the beam do not involve connection welds;
  - the cross section reduction is applied to the beam flange width, hence the flange slenderness ratio b/t reduces and the local instability phenomena are delayed;
- at global level
  - global ductility benefits from the increased rotational capacity;
  - dog-bones encourage the attainment of more dissipative collapse mechanisms.

The dog-bone geometry

The RBS profile should be curve shaped:

- the tapered element shall not present any geometrical discontinuities, in order to avoid the trigger of cracks during inelastic excursions;
- the curved RBS behaves with the largest rotational capacity as respect to polyline shaped solutions;
- the curve tapering is the most economic technology.

The dog-bone design parameters

- the amount of cross-section resistance reduction ($m_{db}=M_{pl,db}/M_{pl}$),

- the reduced width of the beam flanges (B_r),
- the location of the RBS mid-section as respect to the beam-end (e).

M_{pl} = beam plastic moment $M_{pl,db}$ =dog-bone plastic moment M_{Sd} =bending moment due to design loads
Figure 1. The Dog-Bone Conception

Research activity

- Definition of a design strategy of steel MRF modified by means of dog-bones.
- Assessment of the reliability of the RBS connection system for the seismic retrofitting or upgrading of steel MR frames. The influence of dog-bones on the seismic response of MR frames is evaluated by means of static inelastic pushover analyses and expressed in terms of the main factors characterising the structural behaviour, such as maximum bearing capacity, global ductility, collapse mechanism.

Method of analysis

Definition of the design procedure for MRF with RBS (Faggiano et al. 2002)

1) Design of steel MR frames according to traditional methods;
2) Insertion in the beams of appropriately designed dog-bones.
   - Macro-design of the dog-bone: estimation of the plastic moment reduction ratio m_{db}, associated to the dog-bone position along the beam, e;

$$e_{FEMA} = a + (l_{db} / 2) \text{ (FEMA 350, 2000)}$$

$$e_{MP} > \sqrt{2 \cdot (1 + m_{db}) \cdot \frac{M_{pl}}{q}} + 2\sqrt{\frac{M_{pl}}{q}} \text{ (Montuori and Piluso, 2000)}$$

$$e_{MP} = e_{FEMA} \cdot \zeta = 1.5 \times 0.75 \cdot d \cdot \zeta \text{ (Faggiano et al., 2003)}$$

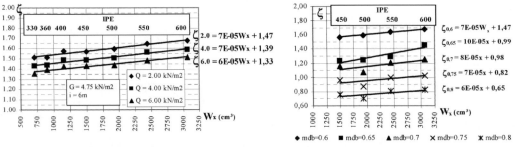

Figure 2. The coefficient ζ as a function of IPE profiles (W_x), vertical loads and m_{db}.

- Micro-design of the dog-bone: determination of the dog-bone mid-section flange width, B_r, and design of the constructional detail.

$$B_r = \frac{[k_1 + 2 \cdot B \cdot k_2] \cdot m_{db} - k_1}{2 \cdot k_2} \quad k_1 = t_w (h - 2t_f)^3 \quad k_2 = 3t_f \cdot h^2 - 6t_f^2 \cdot h + 4t_f^3$$

$B_r^* = m_{db} \, B$ (no web contribution to the bending resistance, which induce a scatter of about 15÷20% in excess)

Study cases

- Multi-storey steel MRF configurations representative of the common practice in steel buildings have been analysed:
  - First numerical investigation campaign:

    5 storeys (S), 2 and 4 bays (B) MRFs, three different bay spans (L=6.0, 9.0, 12.0m), inter-story height equal to 3.5m for current levels, 4.5m for ground level, a fix amount of the RBS plastic moment reduction ratio $m_{db} = M_{pl,db}/M_{pl}$ equal to 0.6, it appearing the most significant and commonly used value;
  - Second numerical investigation campaign:

    in order to estimate the impact of both number of stories and m_{db} ratios on the global behaviour of MR frames, 2 bays, 5 stories and 8 stories MR steel frames have been studied, considering different amount of plastic moment reduction ratios (m_{db} equal to 0.6, 0.65, 0.7, 0.75 and 0.8), bay spans equal to L=9.0m.

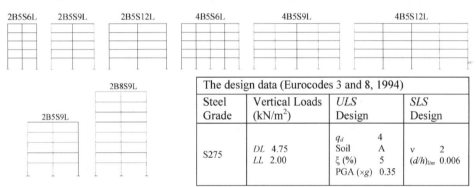

The design data (Eurocodes 3 and 8, 1994)			
Steel Grade	Vertical Loads (kN/m²)	*ULS* Design	*SLS* Design
S275	DL 4.75 LL 2.00	q_d 4 Soil A ξ (%) 5 PGA (×g) 0.35	ν 2 $(d/h)_{lim}$ 0.006

Figure 3. Study cases

- Static inelastic pushover analyses have been performed using the computer program SAP2000 Non-linear 7.12 (1999).

- P-Δ effects have been considered and the inelastic behaviour of dissipative zones has been defined by concentrated rigid-plastic hinges at the structural members end.

- Dog-bones have been modelled as "non prismatic section frame" elements; for the sake of simplicity, the curve profile has been linearized.

Figure 4. The dog-bone modelling by means of SAP2000 Non-linear 7.12

Results of analysis

A performance-based comparison between the different RBS design criteria can be developed on the basis of the pushover behavioural curves, i.e. top sway displacements Δ versus base shear V (the structures designed by means of traditional methods are referred as BASE, the structures modified by means of the FEMA dog-bones, referred as FEMA, and the structures modified by means of the MP dog-bones, referred as MP):

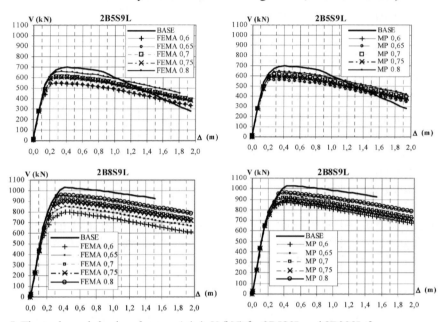

Figure 5. The pushover behavioural curves Δ (m), V (kN) for 2B5S9L and 2B8S9L frames

- The RBS elements are able to protect nodes from plastic deformations.
- Inelastic deformations in the beam do not involve connection welds, the RBS undergoing inelastic deformations instead of the beam-ends.
- The number and the length of bays does not affect the RBS efficacy.
- The presence of dog-bones within the beams of a frame has a negligible influence on both the structural weight and the overall lateral flexibility.
- The overall bearing capacity of frames modified by means of dog-bones is smaller than the one of the BASE frame.
- The ductility coefficient of frames modified by RBS, $\mu=\Delta_{0.03}/\Delta_y$ (where Δ_y is the top sway displacement at the activation of the first plastic hinge, this situation being assumed as the first yielding condition, $\Delta_{0.03}$ is the top sway displacement at the attainment of the inelastic rotation equal to 0.03rad in the most engaged plastic hinge, this situation being assumed as a conventional collapse condition), matches the design requirements, it being always larger than the design q-factor ($q_d=4$).
- With regard to the collapse mechanism, it can be observed that:
  - beam plastic hinges only form in dog-bone elements;
  - RBSs induces a decrease of both the plastic hinge number in column and the extent of plastic rotations;
  - RBSs allow the attainment of failure modes, which involve a higher number of storeys, it resulting in higher dissipation capacity of structures.

- The RBSs position along the beams has no influence on failure modes.
- Small amounts of cross-sectional resistance reduction, such as $m_{db} = 0,75 - 0,8$, are unsuitable, because with such values the RBS should be located very close to the beam in order to develop the plastic hinge in itself, in this way plastic deformation could anyway involve the beam-to-column connection.
- In order to establish the most convenient method for the RBS frame design, the following simple performance coefficient can be evaluated, which takes into account the ductility coefficient (μ) and the adimensional overall bearing capacity (V_{max}) of the structure:

$$\pi' = \mu \times \frac{V_{max}}{V_d}$$

As far as such product is larger, the best structural performance is exhibited by the corresponding frame. The best performance is offered by the m_{db}=0.6-FEMA case.

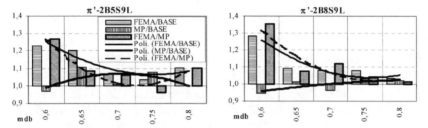

Figure 6. The performance coefficient π' for 2B5S9L and 2B8S9L frames

Further developments

- The MR-frame performance improvement induced by RBS connection type has been appreciated by means of a wide program of both numerical and experimental investigations. Nevertheless, traditional RBS design evidently applies with some difficulties to enhance the ductility properties of existing structures, because these are completed, so that the primary structure is partially or fully inaccessible to further interventions. Actually, in case of retrofitting, the current section reduction should be obtained at the expenses of the lower beam flange alone, realizing a non-symmetric constructional detail, which has not yet beneficed of full attention:
  - Basic aspects concerning the use of non-symmetric RBS connections as a moment-resisting retrofitting systems should be examined, such as the geometric characteristics of the weakening detail and the RBS design methodology;
  - The actual ductility category of the reduced beam cross-section should be considered, in order to calibrate the conventional collapse situation to the actual enhanced rotational capacity.
- The RBSs connection efficiency as a seismic upgrading or retrofitting system should be finally validated by means of dynamic time-history seismic analyses.

References

- Chen, S.J., Tsao, Y.C. and Chao, Y.C. 2001. Enhancement of Ductility of Existing Seismic Steel Moment Connections. *Journal of Structural Engineering*, May, 538-545.
- Engelhardt, M.D. and Sabol, T.A. 1997. Seismic-resistant steel moment connection:

developments since the 1994 Northridge Earthquake. *In:* Zandonini R. Elnashai A. and Dexter R. (eds). *Progress in structural engineering and materials*, 1(1): 68-77.

- Engelhardt, M.D., Winneberger, T., Zekany, A.J., Potyraj, T.J. 1996. The Dogbone Connection: Part II, *Modern Steel Construction*, August: 46-55.
- Faggiano, B. and Landolfo, R. 2002. Seismic Analysis and Design of Steel MR frames with "dog-bone" connections. *In Proceedings of the 12th European Conference on Earthquake Engineering,* London, U.K., Elsevier Ltd,: paper 309.
- Faggiano, B. and Mazzolani, F.M., 1999. Proposal for improving the steel frame ductility by weakening. *In: Proceedings of the XVII C.T.A. Congress*, Napoli, Italia, 1, 269-280.
- Faggiano, B., de Cesbron de la Grennelais, E. and Landolfo, R. 2003. Design criteria for RBS in MR frame retrofitting. *In Proceedings of the 4th International Conference STESSA "Behaviour of steel structures in seismic areas", 9-12 June,* Napoli, Italia, F.M. Mazzolani ed., A.A. Balkema Publisher, The Netherlands, pp. 683-690.
- FEMA 350 2000. *Recommended Seismic Design Criteria for New Steel Moment-frame Buildings.* July.
- Georgescu, D. 1996. Recent developments in theoretical and experimental results on steel structures. Seismic resistant braced frames. *Costruzioni Metalliche*; 1: 39-52.
- Montuori, R and Piluso, V. 2000. Plastic design of steel frames with dog-bone beam-to-column joints. In: F.M. Mazzolani and R. Tremblay (eds). *Proceedings of the STESSA 2000 Conference*, Montreal, Canada. Rotterdam: A.A. Balkema: 627-634.
- Plumier, A. 1990. New idea for safe structure in seismic zone. In: *IABSE Symposium*, Brussels, Belgium.
- SAC 96-03 1997,. *Interim guidelines.* FEMA 267/A, SAC joint Venture, California U.S.A.

Contacts

- Raffaele Landolfo
    Address: Dept. of Constructions and Mathematical Methods in Architecture, Faculty of Architecture, University of Naples"Federico II", Via Monteoliveto 3, 80134, Naples, Italy
    e-mail: landolfo@unina.it, Ph: +39-081-768-2447 Fax: +39-081-593-4792
- Federico M. Mazzolani
    Address. Dept. of Structural Analysis and Design, Faculty of Engineering, University of Naples"Federico II", P.le Tecchio 80, 80125, Naples, Italy
    e-mail: fmm@unina.it, Ph: +39-081-768-2443 Fax: +39-081-593-4792

COST C12 - WG2

Datasheet I.3.1.2.3

Prepared by: G. De Matteis, S. Panico

Univ. Naples Federico II – Dept. Structural Analysis and Design

USE OF STEEL BRACINGS FOR SEISMIC UP-GRADING OF EXISTING STRUCTURES

Description

- Steel bracings are lateral resisting systems usually used for seismic protection of both new and existing steel and r.c. framed buildings.

- Steel bracings are traditionally based on the use of concentric steel members arranged into a frame mesh (Concentrically Braced Frame – CBF), according to single bracing, cross bracing, chevron bracing and any other concentric bracing scheme (Figure 1).

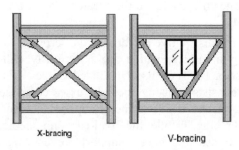

Figure 1. Some typical steel bracing arrangements

- Concentrically steel bracings posses high stiffness and strength for moderate intensity earthquakes. On the other hand, they may have an unfavorable hysteretic behaviour under severe earthquake, due to buckling of compression diagonal member.

- To improve the global seismic performance of the structure, also in terms of dissipative capacities, special damping devices are placed in the conventional bracing systems in order to modify an ordinary bracing system in a dissipative bracing system (Figure 2).

- Many hysteretic dampers are based on metal yielding technology.

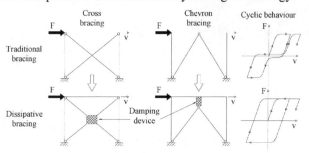

Figure 2. Traditional and dissipative bracing

Field of application

- Steel bracing systems may be profitable used as both an upgrading system existing frames and as primary lateral load resisting system for new building framed structures.

- Their use is particularly advantageous also in case of seismic retrofitting of existing r.c., since they increase both damping and stiffness of the whole structure, avoiding to enhance the ductility characteristics of beam-column joints.

Technical information

- A number of dissipative steel bracing systems have been proposed and used worldwide.

- Cross bracings can behave as dissipative members if braces are designed in such a way plastic mechanisms due to material yielding are exploited before the buckling occurrence in compression member.

- To this purpose, Low Yield Strength Point (LYSP) steel could be used. It allows the adoption of stiff and low-resistance braces due to the greater E/fy ratio with respect to ordinary steel (Izumi et al. 1992).

- Another possibility to improve the dissipative performance of cross bracing system is using special types of bracing members, which are notoriously called Buckling Inhibited Brace (BIB) (Chen & Lu 1990) or also Unbonded Brace (UB) (Clark et al. 2000) (Figure 3). The system consists in using special trusses composed by a steel core, as load-carrying element, placed inside a lateral support element, in order to obtain a buckling restrained bracing. While the load-carrying element takes the tensile and compressive axial forces, the lateral support prevents buckling of the central core when the member is compressed, owing to appropriate lateral restraining mechanisms.

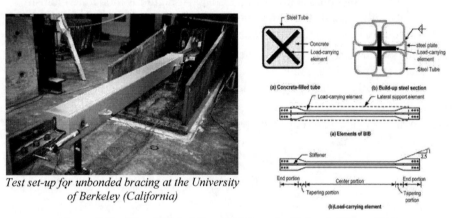

Test set-up for unbonded bracing at the University of Berkeley (California)

Figure 3. Buckling Inhibited Brace (BIB) or also Unbonded Brace (UB)

- Special devices can be adopted as bracing system. A first example of special dissipative device were based on the use of a rectangular frame inserted into a frame mesh and connected to its joints by tension-only braces (Ciampi 1993). It is made of thick steel plate adequately shaped in order to have a uniform flexural resistance for a linearly variable asymmetrical flexural moment, which is the one occurring for lateral deformation of the external frame (Vulcano 1991) (Figure 4a).

- Other dissipative devices can be obtained by assembling a number of thick elements arranged as a sandwich and pin-connected at the ends of the two parts of a diagonal member *(Figure 4b)*.

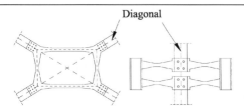

Figure 4. Typical dissipative systems for braced frames

- As far as the chevron bracing scheme is concerned, the transformation from traditional bracing to a dissipative scheme can be obtained by inserting special dissipative devices between the joint of the diagonal members and the beam (Figure 5a). The basic design principle of the system is that, while plastic deformations occur in the dissipative device, the diagonals have to remain elastic both in tension and in compression.

- The simplest scheme is based on the transformation of a conventional concentric bracing into an eccentric bracing system by means of a steel link, which is fixed to the beam and pin-joined to the diagonals (Ghobarah & Abou Elfath 2001) (*Figure* 5b). Cyclic behaviour of the link element allows a large amount of the input energy to be dissipated without any damage of the external framed structure.

- The recent technological evolution of the link is represented by so-called Added Damping And Stiffness (ADAS) elements (Aiken & Whittaker 1993). It is an energy dissipation device mounted atop chevron bracing and differentiated for shape, dissipative mechanism and utilized material. From the shape point of view, several types of devices have been proposed, namely X-shaped (fig. 5c), E-shaped (fig. 5d), U-shaped (fig. 5e), W-shaped (fig. 5f), honeycomb-shaped (fig. 5g) and so on. As far as the dissipative mechanism is concerned, devices that can dissipate energy for flexural, shear, axial (fig. 5a,b,c,d,e,f,g) and torsion (fig. 9h) deformations, or also by their combination, have been proposed. Finally, ADAS elements can be differentiated in relation to the basic material, the common adopted materials being mild steel, low-yield strength steel, lead, shape-memory alloys and aluminium.

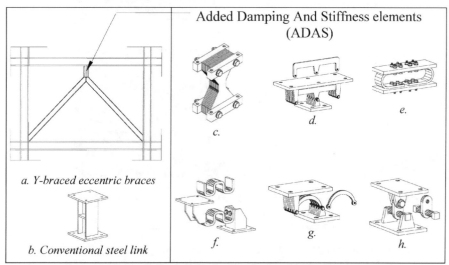

Figure 5. Typical dissipative chevron bracing systems

Structural aspects

- The introduction of bracing systems into steel framed structures allows the improvement of structural performance levels under lateral loads due to the increasing of stiffness, strength and ductility.

- Framed structures equipped with bracing systems have to be intended as composite (dual) systems, where under moderate earthquakes the primary structure exhibits elastic deformations only, while it could be a useful supplementary energy dissipation system for medium and high intensity earthquakes.

Research activity

- A wide experimental investigation has been recently undertaken at the University of Naples Federico II. The experimental activity concerns the investigation of different metal-based upgrading technologies applied to an existing full-scale r.c. building, designed in the Seventies according to the gravity loads only. Among different metal-based seismic upgrading techniques, the Y-shaped eccentrics bracing system (Figure 6a,b) and the dissipative concentric bracing system (Figure 6c,d) have been considered. In the former case, the link has been vertically placed and directly attached to the slab of the existing structure, while in the latter case buckling inhibited braces made of common mild carbon steel have been applied (Mazzolani et al. 2003).

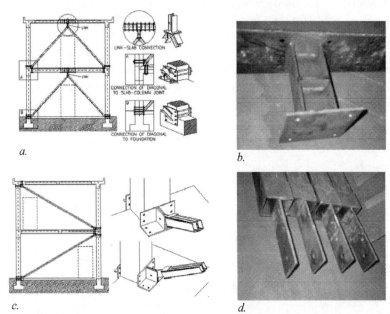

Figure 6. Some metal-based seismic upgrading techniques of existing r.c. frame:
a,b   Y-shaped eccentric bracing system with detail of the link;
c,d   Buckling inhibited braces with detail of the braces

- A practical application of dissipative devices in the seismic retrofitting of existing r.c. frames is represented by Domiziano Viola and La Vista school buildings located in Potenza (Italy). This retrofit system was built in 1999 by using traditional steel bracings connected to the structure by means of cover plates that behave as dissipative elements by shear plastic deformation of mild steel (Figure 7). These devices were patented by Dolce and Marnetto and tested at the Department of Structures of the University of Basilicata (Martelli & Forni 2000).

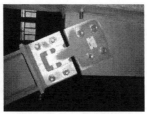

Figure 7. A seismic retrofitting system based on dissipative cover plates

- Dissipative devices based on Shape Memory Alloys (SMA) represent a new trend in the field of passive seismic protection of new and existing buildings (Dolce & Marnetto 2000, Dolce et al. 2000). These devices allow to eliminate some limitations of the current passive protection technologies, such as problems due to ageing and durability, maintenance, installation complexity or replacement and geometry restoration after strong earthquakes, as well as variable performances depending on temperature (Figure 8).

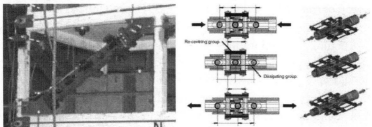

Figure 8. Bracing device with SMA and functioning scheme of the functional groups.

Method of analysis

- A structure supplied with energy dissipation devices could be represented as a composite systems with 'parallel components". The whole structure can be modeled as two independent systems: "the primary structure" and "the energy dissipation system" (Wada et al. 1992). The former denotes the system composed by beams and columns of the main frame. The latter system is the one composed by the dissipative devices, which are mainly designed to resist lateral seismic loads, but provide also a contribution to the stiffness, strength and damping of the whole structure.

- Displacement-based approach may be used as design method, in accordance to the performance-based design method. The main concept of displacement-based design consists in designing the structure in such a way under an assigned design earthquake (ground motion level), corresponding to a predefined recurrence interval, the maximum interstorey drift does not exceed the target value (performance level).

- This method is applied according to the so-called 'substitute structure approach' (Priestley 1992), where the structure is substituted by an equivalent linear elastic scheme having both stiffness and damping equal to the actual ones (Mendhekar & Kennedy 2000), so to employ the elastic displacement spectra as design tool. The 'substitute structure approach' can be applied according to the iterative procedure given by Lin et al. (2003).

- The first guidelines and commentary for the seismic rehabilitation of buildings supplied with dissipative devices are presented in the FEMA 273 (1997) and were developed by Whittaker et al. (1997).

- A simplified procedure for seismic retrofitting of framed buildings by means of low yield shear walls has been recently proposed by De Matteis and Mistakidis (2003).

References

- Aiken, I.D. and Whittaker, A.S. 1993. Development and application of passive energy dissipation techniques in the U.S.A. *Proc. of the International Post-SmiRT Conference Seminar on Isolation, Energy Dissipation and Control of Vibration of Structures*, Capri, (Italy).

- Chen, C.C. and Lu, L.W. 1990. Development and experimental investigation of a ductile CBF system. *Proceedings of the 4th National Conference on Earthquake Engineering*, Palm Springs, Calif., Vol.2, p.p. 575-584.

- Ciampi, V. 1993. Development of passive energy dissipation techniques for buildings. *Proc. of the International Post-SmiRT Conference Seminar on Isolation, Energy Dissipation and Control of Vibration of Structures*, Capri, (Italy).

- Clark, W., Kasai, K., Aiken, I.D., Kimura, I. 2000. Evaluation of design methodologies for structures incorporating steel unbonded braces for energy dissipation. *Proceedings of the 12th World Conference on Earthquake Engineering*, Auckland, New Zealand, Paper No. 2240, CD-ROM.

- De Matteis, G., Mistakidis, E.S. 2003. Seismic retrofitting of moment resisting frames using low yield steel panels as shear walls. *Proc. of the Fourth International Conference on the seismic behaviour of steel structures in seismic areas* (STESSA 2003), A.A. Balkema, Rotterdam, The Netherlands, 677-682.

- Dolce, M. and Marnetto, R. 2000. Seismic devices based on shape memory alloy. *Proceedings of the 12th World Conference on Earthquake Engineering*, Auckland, New Zealand, CD-ROM.

- Dolce, M., Cardone, D., Nigro, D. 2000. Experimental tests on seismic devices based on shape memory alloys. *Proc. of the 12th World Conference on Earthquake Engineering*, Auckland, New Zealand, CD-ROM.

- FEMA, NEHRP 1997. Guidelines for the seismic rehabilitation of buildings, Washington, DC: *Federal Emergency Management Agency*.

- Ghobarah, A. and Abou Elfath, H. 2001. Rehabilitation of a reinforced concrete frame using eccentric steel bracing . *Engineering Structures* , 23, 745-755.

- Izumi, M., et al. 1992. Low cycle fatigue tests on shear yielding type low-yield stress steel hysteretic damper for response control, part 1 and 2. *Proc. Annu. Meeting of the Arch. Inst. Of Japan*, Tokyo, Japan, 1333-1336.

- Lin, Y.Y., Tsai, M.H., Hwang, J.S., Chang, K.C. 2003. Direct displacement-based design for building with passive energy dissipation systems. *Engineering Structures*, 25: 25-37.

- Martelli, A. and Forni, M. 2000. The most recent applications of seismic isolation and passive energy dissipation. *3rd International Workshop on Structural Control*,

Paris, France.

- Mazzolani, F.M., Calderoni, B., Della Corte, G., De Matteis G., Faggiano B., Landolfo, R., Panico, S. 2003. Full-scale testing of different seismic upgrading techniques on an existing RC building. *Proc. of the Fourth International Conference on the behaviour of steel structures in seismic areas* (STESSA 2003), Mazzolani F.M. Editor, A.A. Balkema, Rotterdam, The Netherlands, 719-726.

- Mendhekar, M.S., Kennedy, D.J.L. 2000. Displacement-based design of buildings – theory. *Engineering Structures*, 22: 201-209.

- Priestley, M.J.N. 1992. Displacement-based seismic assessment of reinforced concrete buildings. *Journal of Earthquake Engineering*, 1(1):157-92.

- Vulcano, A., 1991. Use of steel braces with damping devices for earthquake retrofit of existing framed structure, *Costruzioni Metalliche*, No. 4 .

- Wada, A., Connor, J.J., Kawai, H., Iwata, M., Watanabe, A. 1992. Damage Tolerant Structures. *Fifth US-Japan Workshop on the Improvement of building structural design and construction practices*, ATC-15, California, USA.

- Whittaker, A., Constantinou, M., Kircher, C. 1997. Development of analysis procedures and design guidelines for supplemental damping. *Proc. of the International Post-SmiRT Conference Seminar on Isolation, Energy Dissipation and Control of Vibration of Structures*, Taormina, (Italy), August 25 to 27.

Contacts *(include at least one COST C12 member)*

- Prof. F. M. Mazzolani

  Address: Department of Structural Analysis and Design, University of Naples 'Federico II', P.le Tecchio, 80, Naples, Italy.
  e-mail: demattei@unina.it
  Phone: (+39) 0817682443
  Fax: (+39) 0815934792
  e-mail fmm@unina.it

Datasheet I.3.1.2.4
Prepared by: G. De Matteis[1], A.S. Mistakidis[2]

[1] Dept. Struct. Analysis and Design, Univ. Naples Federico II, Italy
[2] Department of Civil Engineering, University of Thessaly, Volos, Greece

USE OF SHEAR PANELS FOR SEISMIC UP-GRADING OF EXISTING STRUCTURES

Description

- Steel plate shear walls are more and more used for building protection against strong earthquake and wind forces (De Matteis et al 2003).

- They can be profitable used as up-grading system for existing reinforced concrete structure in seismic areas.

Field of application

- The application of steel shear walls is mainly related to new high-rise steel building in case of high seismic hazard. They provide a competitive method also for the seismic up-grading of existing steel or reinforced concrete structures.

- The last years, interest is emerging in rehabilitating older reinforced concrete frame buildings with steel plate infill panels (Robinson and Ames 2000).

- Their use is particularly advantageous for seismic retrofitting of existing r.c., since they increase both damping and stiffness of the whole structure, avoiding to enhance the ductility characteristics of beam-column joints.

- Other direct benefits of their use are improvement of structural energy dissipation capability and increase of building lateral strength and stiffness. They provide also several other advantages, as e.g. speed of erection, reduced foundation cost and increased usable space in buildings.

Technical information

- Shear panels can be made of stiffened and un-stiffened plates. The use of stiffeners allows to delay shear buckling phenomena in the plastic field so to obtain a stable inelastic cyclic behaviour under shear action.

- For the construction of un-stiffened steel shear panels, the use of conventional steel is more appropriate and convenient. Due to buckling phenomena occurring in the slender plate, tension field develops within steel panel in shear.

- The use of stiffened plates made of special low yield strength steel has been recently proposed in Japan (Tanaka et al 1998). The combination of low-yield steel and shear panels is particularly convenient. Due to the adopted chemical composition (very small amounts of carbon and alloying elements), low-yield point (LYP) steel has a nominal yield stress of about 90-120 MPa, the same Young's modulus as conventional steel and a nominal elongation over 50% (Nakagawa et al 1996).

- The above properties give to the steel panel the following advantages:
    - The shear panel will undergo large inelastic deformations at the first stages of the loading process, thus enhancing the energy dissipation capability of the whole system in a wide range of deformation demand.
    - The use of a plate subjected to uniform in-plane shear force allows the

yielding of the material to be spread over the entire panel, ensuring a very large global energy dissipation capability.

- Low-yield strength shear panels are characterised by a very stable hysteretic response up to large deformations, with a conspicuous strain-hardening under load-reversals and with limited strength and stiffness degradation arising after plate buckling.

Structural aspects

- The shear wall panels can be installed in the existing system in various different ways (see Figure 1).

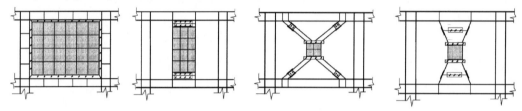

Figure 1a. Full bay type   **Figure 1b.** Partial bay type   **Figure 1c.** Bracing type   **Figure 1d.** Pillar type

- Concerning the connections with the structural members of the existing structure, difference between steel and reinforced concrete structure has to be made .

- For steel structures, a typical connection detail of steel plate shear wall to boundary beams and columns is shown in Figure 2a. If the connection is a welded connection, it should be designed such that the connection plates (fin plates) and welds develop the "expected shear yield" strength of the wall. If the connection is a field-bolted connection, the bolts should be slip critical and should develop the "expected shear strength" of the wall. Even if bolts are slip critical, it is expected that during the cyclic loading of the wall, the bolts slip before the tension field yields. However, such slippage will occur at a load level considerably above the service load level and not only is not harmful but can be useful in improving the seismic behaviour.

- For concrete structures, a common method to connect the shear panel with the existing concrete structure is the use of drilled or adhesive anchors. Recently, a connection scheme has been devised by a research project currently being conducted at the University of Alberta (Driver and Grondin 2001) which is not only able to transfer the interfacial forces but also to provide confinement and shear reinforcement to the concrete columns, thereby enhancing their ductility (Figure 2b). The steel infill plate is connected to the concrete columns using a series of tube collars.

Research activity

- A wide research activity is in progress aiming at establishing adequate design method for seismic retrofitting of existing reinforced concrete structures by using either slender or compact steel shear panels.

- The research is developed in cooperation between University of Naples Federico II and University of Thessaly.

- Analytical, numerical, and experimental analysis are carried out.

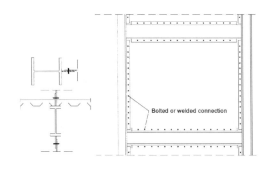

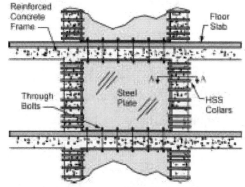

Figure 2a. Typical connection with steel members

Figure 2b. Typical connection with concrete members

Method of analysis

- The first part of the research is addressed at setting up simplified rules for retrofitting design, which should provide effective methods for the selection of appropriate panel geometry.
- The applied method is framed within the performance based design approach.
- The method is based on the selection of the target spectral displacement desired for the retrofitted structure. Then the characteristics of steel shear panels allowing the shifting of the performance point to this spectral displacement is determined.
- Figure 3 illustrates the way this information may be obtained for an example structure. In this figure, the capacity spectrum for the unretrofitted structure is shown with the solid line. Moreover, it is assumed that the design engineer has determined after the examination of the damage state that exists at each of the various points on the capacity curve, the spectral displacement levels that correspond to each performance level (Immediate Occupancy - IO, Life Safety – LS, Structural Stability - SS). Let us assume that the design engineer sets as a performance objective for the building the Life Safety one. Therefore, the target displacement for the retrofitted building is set to 3.7 cm. This is illustrated in the figure by the vertical dashed line drawn at a spectral displacement of 3.7 cm.
- The next step is to determine an appropriate initial stiffness for the retrofitted structure.
- As an approximation (based on the "equal displacements" simplifying assumption), an estimate of the initial period required for the retrofitted structure can be obtained by extending the vertical line that corresponds to the desired target displacement until it intersects with the elastic response spectrum (demand spectrum for 5% viscous damping).
- A radial line drawn from the origin of the demand/capacity spectrum plot through this intersection defines the minimum initial stiffness for the retrofitted structure T_{ret} expressed as a period with units of seconds.
- Period of retrofitted structure can be calculated from the equation

$$T_{ret} = 2\pi \sqrt{\frac{S_d}{S_{ae}}}$$

where S_d is the target displacement and S_{ae} is the spectral acceleration corresponding to the intersection of the target displacement line with the elastic response spectrum.

- The target stiffness for the retrofitted structure can be calculated from the equation

$$K_{ret} = K_{ini} \left(\frac{T_{ini}}{T_{ret}} \right)^2$$

where K_{ini} and T_{ini} are the initial stiffness and period respectively of the unretrofitted structure and K_{ret} is the stiffness required for the retrofitted structure.

- Once the required stiffness has been determined, the stiffness K_p of the shear steel panel can be determined from the equation

$$K_{ret} = K_{ini} + K_p$$

- To determine the strength of the retrofitted structure, an assumption about its damping properties is necessary. As a simplifying assumption it is reasonable to consider that the retrofitted structure will be able to provide at least the same level of damping as that of the initial structure.

- The approximate solution for the performance point of the retrofitted structure is obtained as the intersection of the vertical line at the desired target spectral displacement with the demand spectrum that corresponds to the damping level of the initial structure. This point is annotated in the figure as the "desired performance point". A horizontal line extending from the desired performance point to the y axis indicates the minimum spectral acceleration capacity required for the retrofitted structure. With this information known, the required ultimate base shear capacity for the retrofitted structure can be obtained form the following equation

$$V_{ret} = \frac{S_{a_{ret}}}{S_{a_{ini}}} V_{ini}$$

where V_{ini} is the ultimate base shear capacity of the initial structure, V_{ret} is the required ultimate shear capacity of the retrofitted structure, $S_{a_{ini}}$ and $S_{a_{ret}}$ is the ultimate spectral acceleration for the initial and retrofitted structures respectively.

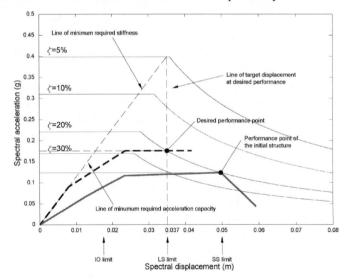

Figure 3. Preliminary design of retrofitting with steel shear walls

Results of analysis

- In the following, a numerical application is provided, according to the above design method. The original building is a real concrete structure which was built in Greece in 1968. The structural scheme is given in Figure 4.

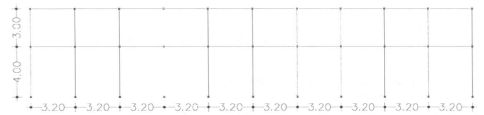

Figure 4. The analysed structure

- The non linear static analysis of the structure is given in Figure 5. In order to achieve a "Life safety" performance level, a maximum displacement of about 3 cm is required. The required period for the retrofitted structure is therefore T_{ret}=0.55 sec. Assuming that the retrofitted structure will be able to provide the same level of damping as the one of the original structure (about 25%), the following characteristics for the structure are obtained: K_{ret} = 2.62 · K_{ini} = 40.610 kN/m and V_{ret}=1.93 · V_{ini} = 497.9 kN. Therefore, the minimum required stiffness of the shear panel is K_p = 25110 N/mm and the required strength is V_p = 239.9 kN.

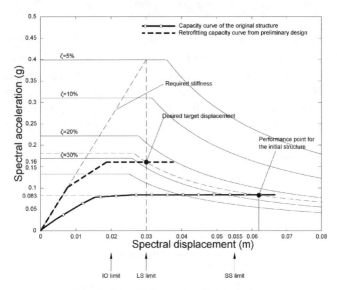

Figure 5. Preliminary retrofitting design

- The configuration of the retrofitted structure is depicted in Figure 6, where it is shown that the steel shear panels have been installed in the central bay of the existing frame, acting as a bracing system throughout the longitudinal direction. The material used for the LYSP is supposed to have: $f_{0.2}$ = 85.8 MPa and f_u= 236.20 MPa.

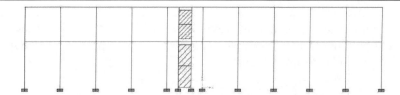

Figure 6. Proposed arrangement of the retrofitting system

- The actual panel configuration has been determined by a trial and error procedure, on by applying three different numerical models according to the following steps.
  1. Determination of appropriate panel geometry for retrofitting purposes by means of simplified analysis (strip model analysis);
  2. Check of the mechanical behaviour of the selected shear panels under monotonic loading by means of sophisticated FEM analysis (using the ABAQUS F.E. code);
  3. Check of the hysteretic behaviour of the selected shear panels submitting the aforementioned F.E. model in cyclic loading.
- Finally, the panel geometric characteristics presented in Table 1 have been adopted.

Table 1. Geometric characteristics of the adopted panel configurations

	Width (m)	Height (m)	Thickness (mm)
1st floor	1.20	3.75	4.00
2nd floor	1.20	2.75	1.70

- The diagram of Figure 7 provides the Sa-Sd behaviour of the retrofitted structure. The required spectral displacement for the retrofitted structure has been calculated equal to 29 mm using the "Displacement Coefficient" method. It has been also verified that this displacement is compatible with the strength and ductility of the existing reinforced concrete members. Also, it is worth noting that the equivalent damping of the final retrofitting design resulted to be ζ=22%, therefore very close to the one assumed in the preliminary design (ζ=25%).

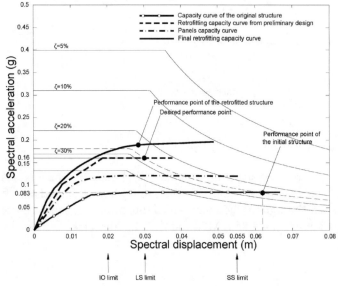

Figure 7. Final retrofitting design

- The behaviour of the preliminarily selected panel has been checked by means of a sophisticated FEM analysis (Figure 8). Stiffeners having a depth of 55 mm have been considered.

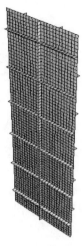

Figure 8. The finite element model used for shear panels (ABAQUS)

- The obtained results are plotted in the load-displacement diagram of Figure 9. The comparison between the curve obtained by using simplified strip model theory and the one obtained by ABAQUS adopting the elastic-perfectly plastic material having a yield stress equal to 1.5 f0.2 is very satisfactory. It is worth mentioning that the difference between the simplified elastic-perfectly material model and the actual one is quite limited for the displacement range which is of interest in the specific analysis performed here (29-35 mm). Moreover, it is important to observe that the adopted stiffener configuration for the selected panel is appropriate, as it prohibits the development of buckling phenomena at the panel even for deformations larger than the required one.

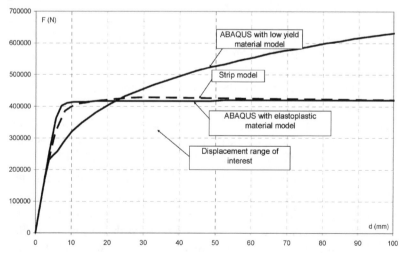

Figure 9. Comparison of the shear panel behaviour obtained by ABAQUS model and SAP2000 model

- In order to verify the design of select panel geometry and in particular of the adopted stiffeners, a cyclic analysis of the panel has been carried out as well. The analysis has been refereed to constant displacement amplitude equal to ± 35 mm, with a number of cycles equal to 20.

- The results are shown in Figure 10. It can be observed that the hysteretic behaviour of the system is excellent, confirming the possibility to adopt selected shear panels as dissipative devices for improving the seismic behaviour of the existing structure.

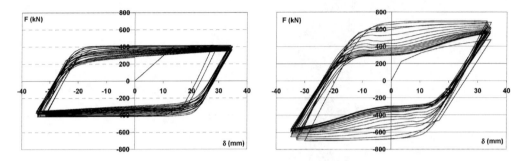

Figure 10. Cyclic behaviour of the panel by assuming a) the simplified elastic-plastic material model and b) the actual low yield steel material model

Further developments

- The next developments of the research will be devoted to the application of the proposed design procedure to different study cases.
- Then, appropriate criteria for the correct selection of the geometrical characteristics of shear panels in order to ensure an adequate dissipative behaviour will be provided.
- Also, the seismic behaviour of the retrofitted structure should be analysed through sophisticated FEM analysis in order to identify the actual contribution provided by shear panels on the global response of the structure.
- Finally, the developments of experimental investigation would be strongly advisable to corroborate the effectiveness of the proposed innovative technique.

References

- De Matteis G., Landolfo, R. and Mazzolani F.M. 2003. Seismic response of MR steel frames with low-yield steel shear panels, *Journal of Structural Engineering*, (25): 155-168.
- Driver R.G. & Grondin G.Y. 2001. Shear walls now performing on the main stage, *Modern Steel Construc., AISC,* Sept.
- Nakagawa, S., Kihara, H., Torii, S., Nakata, Y., Matsuoka, Y., Fujisawa K., and Fukuda, K. 1996. Hysteretic Behavior of Low Yield Strength Steel Panel Shear Walls: Experimental Investigation. *Proc., 11th WCEE*, Elsevier, Paper No. 171.
- Robinson, K. & Ames, D. 2000. Steel plate shear walls: Library seismic upgrade. *Modern Steel Construction, AISC*, Jan.: 56-60.
- Tanaka, K., Torii, T., Sasaki, Y., Miyama, T., Kawai, H., Iwata, M, and Wada, A. 1998. Practical Application of Damage Tolerant Structures with Seismic Control Panel Using Low Yield Point Steel to a High-Rise Steel Building. *Proc., Struct. Eng. World Wide*, Elsevier, Paper T190-4.

Contacts

- Prof. G. . De Matteis

    Address: Department of Structural Analysis and Design, University of
    Naples 'Federico II', P.le Tecchio, 80, Naples, Italy.
    e-mail: demattei@unina.it
    Phone: (+39) 0817682444
    Fax: (+39) 0815934792

- Prof. E. S.. Mistakidis

    Address: Structural Analysis and Design Laboratory, Department of Civil
    Engineering, University of Thessaly, 38334 Volos, Greece
    e-mail: emistaki@uth.gr
    Phone: (+30) 24210/74171
    Fax: (+30) 24210/74124

Prepared by: G. Della Corte, E. Barecchia

Department of Structural Analysis and Design, Faculty of Engineering
"Federico II", Naples.

SEISMIC UPGRADING OF RC STRUCTURES BY MEANS OF COMPOSITE MATERIALS

Description

- This datasheet provides discussion about the use of composite materials for strengthening existing reinforced concrete structures. First, general information on composite materials and techniques is given. Second, a full-scale testing of a seismic repairing/up-grading intervention is described.

Field of application

- Information found in this datasheet apply to strengthening existing reinforced concrete structures.

Technical information

The following definitions are basically taken from *fib* (2001).

- FRP is the acronym of Fibre Reinforced Polymers.
- EBR is the acronym of Externally Bonded Reinforcement.
- Fibre is a general term used to refer to filamentary materials. They can be manufactured in continuous or discontinuous form, as unidirectional or bi-directional reinforcement.
- Matrix is a binder material in which reinforcing fibres are embedded. The function of the matrix is to protect the fibres against abrasion or environmental corrosion, to bind the fibres together and to distribute the load.
- Adhesive is a substance applied to mating surfaces to bond them together. The purpose of adhesive is to provide a shear load path between the concrete surface and the composite material.
- FRP materials consist of a large number of small, continuous, directionalized, non-metallic fibres, bundled in a resin matrix. Depending on the type of fibre they are referred to as AFRP (aramid FRP), CFRP (carbon FRP) or GFRP (glass FRP).
- Composite materials for strengthening of civil engineering structures are available today mainly in the form of: 1) thin unidirectional strips made by pultrusion; 2) flexible sheets or fabrics, made of fibres in one or at least two different directions, respectively. However, other forms are also available for special purpose applications.

Structural Aspects

- Basic mechanical properties of FRP materials may be estimated if the properties of the constituent materials (fibres, matrix) and their volume fraction are known. This may be accomplished by applying the "role of mixtures" simplification as follows:

$$E_f = E_{fib} V_{fib} + E_m V_m$$

$$f_f \approx f_{fib}\, V_{fib} + f_m\, V_m$$

where E_f = Young's modulus of FRP in fibre direction, E_{fib} = Young's modulus of fibres, E_m = Young's modulus of matrix, V_{fib} = volume fraction of fibres, V_m = volume fraction of matrix, f_f = tensile strength of FRP in fibre direction, f_{fib} = tensile strength of fibres and f_m = tensile strength of matrix.

- Typical stress-strain diagrams for unidirectional composites, under short-term monotonic loading, are given in the following figure, in comparison with a typical mild steel.

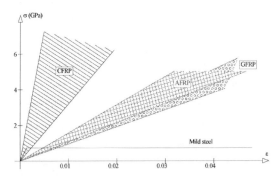

Figure 1. Comparison between typical FRP and steel materials

- Failure modes of a reinforced concrete element strengthened with externally bonded FRP reinforcement may be divided in two classes: *a*) those where full composite action of concrete and FRP is maintained until the concrete reaches crushing in compression or the FRP fails in tension; *b*) those where composite action is lost prior to class *a*) failure (debonding or peeling-off).

- Bond failure in the case of EBR implies the complete loss of composite action between the concrete and the FRP reinforcement. Bond failure may occur at different interfaces between the concrete and the FRP reinforcement, as described below: a) debonding in the concrete near the surface or along a weakened layer; b) debonding in the adhesive; c) debonding at the interfaces between adhesive and concrete or adhesive and FRP; d) debonding inside FRP.

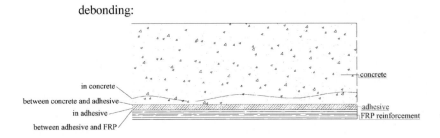

Figure 2. Different possible locations for bond failures (debonding).

- In ultimate limit state (ULS) verification, a parabolic-rectangular stress-strain relationship can be assumed for concrete. Analogously, for the steel reinforcement, the assumption of a standard elastic perfectly-plastic stress-strain relationship can be adopted.

- The tensile stress-strain behavior of the FRP can instead be idealized by means of a linear response, defined as:

$$\sigma_f = E_{fu} \, \varepsilon_f \leq f_{fd}$$

where $E_{fu} = f_{fk}/\varepsilon_{fuk}$ is the modulus of elasticity at rupture, based on the characteristic values of the FRP tensile strength and ultimate strain.

Flexural strengthening:
- Reinforced concrete elements, such as beams and columns, may be strengthened in flexure through the use of FRP composites epoxy-bonded to their tension zones, with the direction of fibres parallel to that of high tensile stresses.
- The most desirable failure mode of the critical cross section is by yielding of the tensile steel reinforcement followed by crushing of concrete, while the FRP is intact.
- The design bending strength of the strengthened cross section is calculated based on principles of RC design.
- The analysis should take into account that the RC element may not be fully unloaded when strengthening takes place.

Shear strengthening:
- Shear strengthening of RC members using FRP may be provided by bonding the external reinforcement with the principal fibre direction as parallel as practically possible to that of maximum principal tensile stresses, so that the effectiveness of the FRP is maximized.
- The nominal shear capacity can be computed by adding the strength contribution provided by FRP to that provided by concrete and steel stirrups.

Torsion strengthening:
- Strengthening for increasing torsional capacity may be required in conventional beams and columns as well as in bridge box girders.
- The principles applied to strengthening in shear are also valid in the case of torsion, with a few minor differences.
- An externally bonded FRP jacket will provide contribution to torsional capacity only if full wrapping around the element's cross section is applied.

Column confinement:
- Column wrapping by externally bonding FRP can enhance their strength and ductility. The beneficial effects of a confinement action of fibres are multiple: a) the improvement of the concrete compressive behaviour (increase of strength and ductility; prevention of concrete cover spalling); b) the lateral support to longitudinal reinforcement under compression.
- Predictive equations of FRP-confined concrete properties are available in the technical literature (see *fib*, 2001). One main difference between the FRP and steel confinement actions lies in the fact that fibres exhibit a linear stress-strain relationship up to fracture. Therefore, the lateral confining pressure is steadily increasing with increasing the lateral concrete section dilation.
- Based on the confined concrete properties the strength increase of the column cross section can be evaluated.
- Column wrapping also improves the flexural ductility. This is a relatively important aspect in the case of RC structures located in seismic regions. The improvement of the curvature ductility obviously implies an improvement of the global structural ductility.

Research activity

- This part of the datasheet presents an experimental investigation on the application of composite materials for seismic upgrading/repairing of existing gravity-load designed (GLD) reinforced concrete (RC) buildings. The testing activity is carried out on a full-scale substructure obtained starting from an existing building. A brief summary of this testing activity is given in the following of this datasheet. More detailed information can be found in Mazzolani et al. (2004).
- The building is located in the Bagnoli district of Naples, in the area of the steel mill named ILVA (former Italsider). After the political decision to dismiss the steel mill and convert the industrial area into a tourist and residential one, the building was destined to be demolished. But, the research team coordinated by prof. Mazzolani, suggested to carry out an 'intelligent' demolition, by using the building as a sample specimen for an experimental investigation.
- In order to increase the potential number of specimens and to experiment different upgrading solution, it has been decided to cut the slabs at the first and second floor, in such a way to divide the building into six separate sub-structures to be tested.

Figure 3. The building before the testing activity

Figure 4. The building after removal of partitions and subdivision into six separate sub-structures

- The geometry of each sub-structure has been measured and data about the structural member size, the slab arrangement, the steel reinforcement details, the mechanical properties of in-sito materials, have been acquired. The following figure illustrates the geometry of sub-structure n. 3. For the latter, the following testing activity has been planned:
  1. Static pushover test up-to-collapse of the original structure.
  2. Repairing of the damaged structure, by using carbon fiber reinforced polymers (CFRP).
  3. Repetition of a static pushover test of the repaired structure.

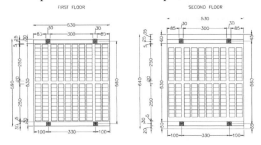

Figure 5. Geometry of a typical sub-structure

Results of analysis

- Results of the first lateral loading test on the original structure are shown in the following Figures. A monotonically increasing lateral load has been applied at the top slab of the structure, by means of two hydraulic actuators applying a resultant lateral force through the center of stiffness of the structure. A top-story sway collapse mechanism, with two plastic hinges forming at the base and the top of each of the second story columns, characterized the damage pattern.

Figure 6. The pushover test on the existing structure (without strengthening)

- Numerical models of the tested structure have been implemented, using both plastic-hinge and fibre beam-column finite elements. The following Figures shows two 3D views of the geometry of the numerical model and the results in terms of base-shear versus top-story displacement. The comparison with the experimental data shows a good agreement between the numerical and experimental results.

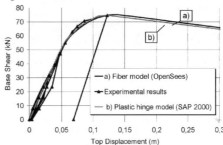

Figure 7. Comparison of model and experimental results

- The repairing of the damaged structure is based on the application of CFRP, used both in the form of pre-manufactured cured straight strips, for reinforcing columns in flexure, and in the form of wet lay-up unidirectional fibre sheets, for improving shear resistance and flexural ductility of columns.
- Namely, the seismic repairing of the structure has been addressed to the rehabilitation of the damaged columns.
- Simple calculations easily lead to the conclusion that if the initial strength of the columns has to be rehabilitated then their ductility is inevitably impaired (failure of columns' sections takes place in the form of fibres' failure), owing to the very small quantity of fibres needed.
- In order to obtain a more ductility-oriented hierarchy of failures within the section, a larger quantitative of fibres is required, producing a significant increase of the column flexural strength. Namely, the bending strength of the repaired columns is about three times the bending strength of the initial ones.
- This situation is particularly favourable, because the required increase in the bending strength is approximately the minimum increment of plastic strength required in the columns in order to shift plastic hinges from columns to beams.
- Then, the design of the repairing system has been guided by the aim to change the collapse mechanism, from a story sway to a global type one. Figure 8 illustrates the design drawings of the repairing system.

- Transverse CFRP sheets were applied in order to increase the columns' shear strength (as required by the increase in flexural strength). In any case, transverse CFRP sheets also improve the columns' ductility, by means of the confinement action, which could be useful in case of an undesired plastic hinging in columns.

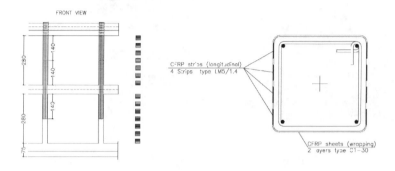

Figure 8. Design drawings of the seismic repairing intervention based on CFRP

Further developments

- The physical pushover test up to collapse of the repaired structure will be next carried out, in order to assess the efficiency of the designed CFRP-based seismic repairing intervention.

References

- Concrete Society 2002. Design guidance for strengthening concrete structures using fibre composite materials. Technical Report No. 55 of a Concrete Society Committee.
- Cosenza E. & Nanni A. (editors) 2001. *Composites in Construction: A Reality*, Department of Structural Analysis and Design, University of Naples Federico II, Naples, Italy.
- *fib* (CEB-FIP) 2001. *Externally bonded FRP reinforcement for RC structures.* Technical report on the 'Design and use of externally bonded fibre reinforcement polymer reinforcement (FRP RBR) for reinforced concrete structures', Task Group 9.3 FRP (fiber reinforced polymer) reinforcement for concrete structures, July.
- Mazzolani F.M., Dolce M., Landolfo R., Nicoletti M. 2004. Seismic upgrading of RC buildings by means of innovative techniques: the ILVA-IDEM project. *Proceedings of the XI Italian Congress of Earthquake Engineering*, 25-29 January, Genova, Italy.
- Teng J.G., Chen J.F., Smith S.T., Lam L. 2001. *FRP-Strengthening of RC structures*, J. Wiley & Sons.

Contacts

- Prof.. Federico M. Mazzolani
  Address: Department of Structural Analysis and Design – University Federico II of Naples, P.le Tecchio 80, 80125 Naples, Italy.
  e-mail:fmm@unina.it, Phone: 0039 081 7682443, Fax: 0039 081 5934792

COST C12 Final Conference
Datasheet I.3.1.2.6
Prepared by: A. Kudzys[1], R. Simkus[2]

[1] Vilnius Gediminas Technical University, Lithuania
[2] Institute of Architecture and Construction of Kaunas Technological
University, Lithuania

INJECTION RENOVATION OF CRACKED RC JOINTS

Definition

- Under earthquake motions the joints of cast-in-situ RC moment-resisting frames from beams and columns and frame-type systems from plain (girdles) floors and walls receive tremendous stresses from adjacent members causing inelastic strains, bending and shear cracks.

- Moment-resisting RC frames and frame-type systems subjected to low-cycle seismic actions belong to hysteretic structures which must be able to dissipate energy by means of ductile hysteretic behaviour of joints and horizontal members (beams or slabs) but their vertical members (columns or walls) must remain elastic at all cases.

- Non-linear structural behaviour and resistance of cracked beam-column joints of RC frames subjected to low-cycle actions are rather well investigated [ACI-ASCE Committee 352, 1985; AIJ, 1988; NZS 4203, 1992; CEB, 1996], but do not exist information's and data on floor-wall joints of frame-type systems.

- The primary design criteria for structural performance of rigid joints of RC frames and systems under reversal loading in an inelastic stress-strain state are their strength and ductility which must be in force after renovation procedures.

- Resin grout injections and compound covers are progressive measures for the renovation of RC beams and slabs [Kamaitis Z., Kudzys A., 1982], therefore they may be successfully applied for restoration of repairable members and joints of cracked RC frames and frame-type systems.

- Injection reparability of cracked members and joints, their structural performance (strength and ductility) after the renovation may be checked and estimated objectively only according to experimental trial data and maintenance practice.

Renovation quality and methodology

- Resin grout injections and compound covers belong to rational repair practice for the renovation of RC frames, construction and civil engineering works cracked and failed due to abnormal construction or use loads, storm wind gusts, snow pressures, earthquake motions and other actions.

- Good penetration, assured adhesive ability and low viscosity of resins or other binding agents allow diminishing the pressure of grout injections and at the same to simplify the renovation technology.

- Resin injections not only help to renovate fractured, disturbed and cracked zones of frame members and joints but they can also allow to increase effectively a bond between concrete and reinforcement bars, a rigidity, a stiffness and a strength in shear of frame structures exposed to lateral low-cycle actions.

- Means, materials and technology peculiarities for beam-column or floor-wall joint renovations depend on concrete cracking and cover damage intensities, therefore injection renovation methods and their practical application must be adapted to structural reconstruction conditions.

- Injection renovations can not be used for strengthening of joints of RC frames or frame-type systems which beams or floor slabs lost their initial positions and changed in shape due to inelastic strains of reinforcement bars.

Floor-wall joints and their trial investigation

- Experimental research were carried out in order to investigate and ascertain the structural behaviour, failure mechanism and performance of primary and injection renovated interior and exterior joints of multi-storey RC buildings subjected to lateral out-plane seismic actions [Kudzys A., 2002].

- The test specimens (Fig. 1) presented full-scale fragments of multi-storey RC frame-type systems consisted from cast-in-situ walls and girdles floors, and imitated model fragments of multi-storey uniplanar RC frames.

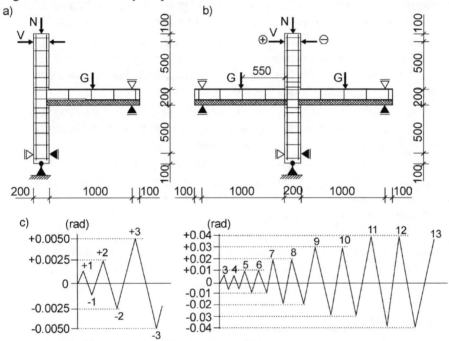

Figure 1. Test fragments of exterior (a) and interior (b) joints, their loading program (c)

- Cast-in-situ walls and continuous multihollow floors with stay-in-place precast thin mould slabs corresponded to requirements of advantage construction technology in RC house- and office-building.

- Special test equipment with an actuator which simulated reiterated reversal lateral forces V was constructed, erected and used measuring at all loading steps strains of horizontal and vertical components, shear deformations and horizontal displacements of joint concrete cores, cracking development and strains of reinforcement bars.

- After the first stage testing of primary fragments the specimens were purposely returned to the initial position, repaired by injections of epoxy resin grouts into concrete cracks and replacing damaged concrete covers with epoxy compounds and tested under the same loading program.

- Non-linear concrete and reinforcement properties, inelastic bond slip versus shear stress relationship, joint core concrete cracking and horizontal force-displacement envelope curves of joints for computer simulation sufficiently concurred with experimental trial results.

Performance of primary and renovated joints

- The expedience to renovate cracked joints of existing RC frames or frame-type systems by resin grout injections is corroborated by experimental trial data on structural behaviour of primary and renovated joint fragments.

- Cracking processes and failure natures (Fig. 2), shapes of hysteresis loops of displacement storey drift angles (Fig. 3) and energy dissipation ratio λ_{ed} (Fig. 4) for primary and repaired joints were identical and may be assessed using the same analysis models.

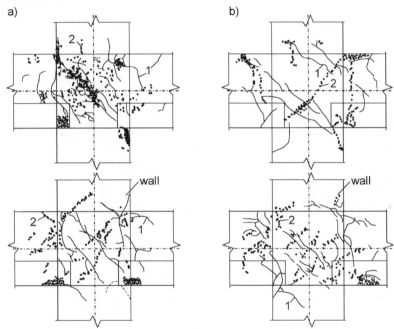

Figure 2. Carked joints of primary (a) and repaired (b) specimens when cracks are caused by forward (1) and backward (2) lateral (shear) forces

- According to the test data, resisting mechanism of primary and renovated joints of RC frame-type systems exposed to low-cycle seismic actions might be presented and assessed by the main concrete strut model (Fig. 5) which is acceptable both for primary and renovated structures (Fig. 2).

- Predicted resistance R_{pr} of frame joints may be calculated by the formula:

$$R_{pr} = 0.85 \ f_c \ b_j \ \chi \sqrt{z_h^2 + z_v^2}$$

where f_c is concrete cylinder compressive strength; b_j is the effective joint width; χ is the cracking factor which values are 0.35, 0.4 and 0.45 when the ratio h_h/h_v is 1.5, 1.25 and 1.0, respectively.

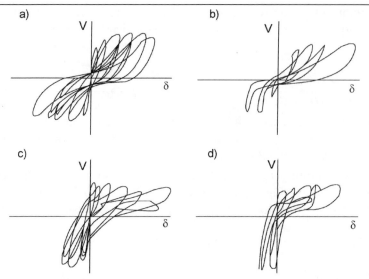

Figure 3. Hysteresis loops of angle δ versus shear force V of primary (a, c) and repaired (b, d) internal (a, b) and external (c, d) joints

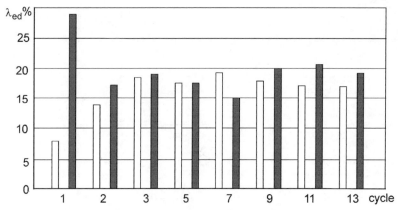

Figure 4. Energy dissipation ratio λ_{ed} for primary (1) and repaired (2) specimen of internal frame joint

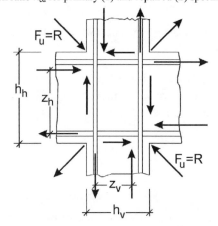

Figure 5. Design mechanical model and inner forces for interior joints of RC frames and frame-type systems

- Due to strengthening of the bond between concrete and reinforcement bars of joints repaired by resin injections their load-carrying capacity is not less than resistance of primary structures (Fig. 6).

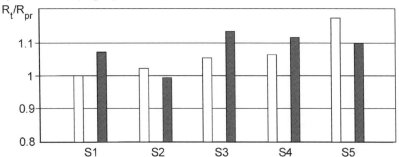

Figure 6. Ratio of test (R_t) and predictive (R_{pr}) strengths of primary (1) and repaired (2) specimens of internal frame joints

- Injection of bending and shear cracks by resin grouts and covering of fractured and destroyed concrete surfaces by resin compounds are simple, rational and effective means for the renovation of RC frames and frame-type systems being able to withstand unfavourable and dangerous seismic actions.

References

- ACI – ASCE Committee 352, 1985. Recommendations for design of beam-column joints in monolithic reinforced concrete structures. *Journal of the ACJ*, May-June.

- AIJ – Architectural Institute of Japan, 1988. Design guidelines for earthquake resistant reinforced concrete buildings based on ultimate strength concept, October.

- New Zealand Standard NZS 4203, 1992. Code of practice for design of concrete structures. Standard Association of New Zealand, Wellington.

- CEB, 1996. RC frames under earthquake loading. State of art report. Thomas Teilford, London.

- Kamaitis Z., Kudzys A., 1982. Rèparation du bèton prècontraint fissure par injections des polymers. *The Nonth International Congress of the FIP,* Stockholm.

- Kudzys A., 2002. Structural behaviour of joints of multi-storey concrete buildings, Vilnius, Technika.

Contacts

- Dr. Regimantas Simkus
  Address: Institute of Architecture and Construction of Kaunas University of Technology street, Tunelio 60, 44405 Kaunas, Lithuania.
  e-mail: asi@asi.lt, Fax: (+370 37) 451355.

COST C12 - WG2

Datasheet I.3.2.1
Prepared by: F.Wald[1],L. Simões da Silva[2]
> [1] Czech Technical University in Prague
> [2] Coimbra University

FIRE DESIGN OF STEEL JOINTS

Definition

- Steel structures under fire expand when heated and contract on cooling. Subject to fire steel looses both its strength and stiffness. The effect of restrained to thermal movement introduce high strains in both the steel member and the associated connections.

- Fire tests on steel structures have shown that the temperature within the joints is lower compare to connecting steel members. This is due to the additional material around a joints (column, end-plate, concrete slab etc.) which significantly reduces the temperatures within the connections compared to those at the centre of supported beam.

- Traditionally the fire protection is applied to the member and its connections. The level of protection is based on that applied to the connected members taking into account the different level of utilisation that may exist in the connection compare to the connected members. A more detailed approach is the component one together with a method for calculation the behaviour of welds and bolts at elevated temperature. By using this approach the connection moment, shear and axial capacity can be evaluated at elevated temperature taking into account the changes in internal forces during fire event.

Field of application

- Prediction of behaviour of the steel and steel and concrete composite structures under exceptional loading conditions.

Technical information

- The procedure consists of two parts: calculating the temperature distribution in the joints and calculating the joint componets resistance at high temperature (including connectors welds, bolts).

- The strength and stiffness of a bolt reduces with increasing temperature [Sakumoto et al, 1992], [Kirby, 1995]. The results of this work has been included in EN 1993-1-2.

- The design strength of a full penetration butt weld, for temperatures up to 700°C, should be taken as equal to the strength of the weaker part of the joint using the appropriate reduction factors for structural steel. For temperatures higher than 700°C the reduction factors given in EN 1993-1-2 for fillet welds can be applied to butt welds.

- The thermal conductivity of steel is high. Nevertheless, because of the concentration of material within the joint area, a differential temperature distribution should be considered within the joint. Various temperature distributions have been proposed or used in experimental tests by several authors. According to prEN 1993-1-2, the temperature of a joint may be assessed using the local massivity value (A/V) of the joint components. As a simplification, a uniform distributed temperature may be assumed within the joint; this temperature may be calculated using the maximum value of the ratios A/V of the adjacent steel members. For beam-to-column and beam-to-beam joints, where the beams are supporting any type of concrete floor, the temperature may be obtained from the temperature of the bottom flange at mid span.

- The component method, that consists of the assembly of extensional springs and rigid links, may be adapted and applied to the evaluation of the behaviour of steel joints under elevated temperatures. Depending on the objective of the analysis, a simple evaluation of resistance or initial stiffness may be pursued or, alternatively, a full non-linear analysis of the joint may be performed [Simões da Silva et al, 2001], taking into account the non-linear load deformation characteristics of all the joint components, thus being able to predict the moment-rotation response. To evaluate the non-linear response of steel joints in fire, knowledge of the mechanical properties of steel with increasing temperature is required, see Fig. 1. In the context of the component method, this is implemented at the component level. The elastic stiffness K_e, is directly proportional to the Young's modulus of steel and the resistance of each component depends on the yield stress of steel, see Fig. 2.

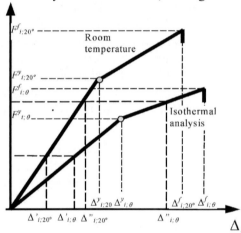

Figure 1. Isothermal force-deformation response of component [Simões da Silvaet al, 2001].

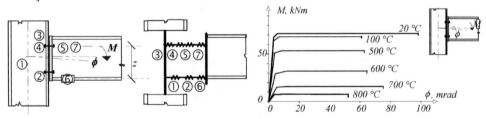

Figure 2. Example of the prediction of the joint behaviour by component method [Simões da Silvaet al, 2001].

Structural aspects

- Significant developments have been made in the analysis of the behaviour of steel framed structures under fire conditions over the last ten years. Due to the high cost of full-scale fire tests and size limitations of existing furnaces, these studies are based on the observation of real fires and on tests performed on isolated elements subjected to standard fire regimes, which serve as reference heating, but do not model the natural fire. However, the failure of the World Trade Centre on 11th September 2001 and, in particular, of building WTC 7, alerted the engineering profession to the possibility of connection failure under fire conditions. Many aspects of behaviour occur due to the interaction between members and cannot be predicted or observed from isolated tests, such as global or local failure of the structure, stresses and deformations due to the restraint to thermal expansion by the adjacent structure, redistribution of internal forces, etc. Otherwise, it is known that even nominally 'simple' connections can resist significant moments at large rotation. At the severe deformation of a structure in fire, moments are transferred to the connection and to the adjacent members, and hence they may have a beneficial effect on the survival time of structure.

Research activity

- The experimental results on the response of steel connections under fire conditions are relatively recent and limited, partly because of the high cost of the fire tests and the limitations on the size of furnace used. Only a limited connections tests have been performed and they have concentrated on obtaining the moment-rotation relationships of isolated connections. It is doubtful whether these results will be useful when dealing with the behaviour of frame connection.

- The development of the of the Building Research Establishment (BRE) has provided the opportunity to carry out several research projects that included full-scale fire tests, , see Fig. 3 to 6.

- The test and numerical studies of subassembly, based on the Cardington Laboratory results are under development.

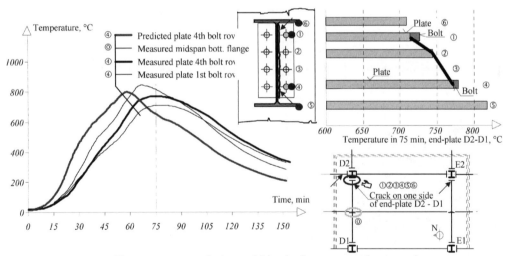

Figure 3. Temperature variation within the beam-to-column major axes end plate connection [Wald et al, 2004].

191

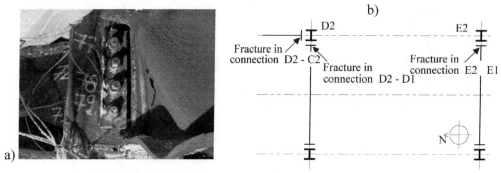

Figure 4. Fracture of the end-plate in the heated affected zone of the welds, a) connection E2-E1, b) cracked connections [Wald et al, 2004].

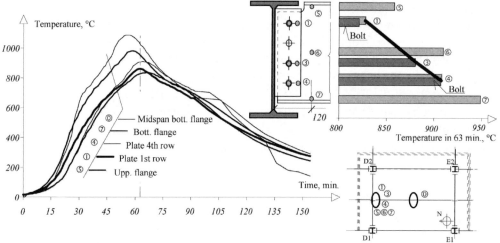

Figure 5. Temperature variation within the beam-to-beam fin plate connection [Wald et al, 2004].

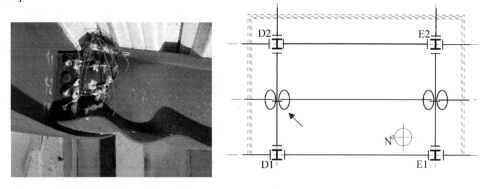

Figure 6. Elongation of holes in the beam web in fin plate connection [Wald et al, 2004].

Method of analysis

- Temperature distribution models based on experiments.
- Component method evaluated by experiments on components and on assemblies.

Results of analysis

- Good prediction of the temperature in case of multi-storey building.
- Very accurate prediction of connection behaviour by component method.

Further developments

- The detailed description of the component behaviour.
- Precision of the prediction models of the temperature distribution on joints.
- The internal forces into joints under the exceptional loading by fire.

References

- Al-Jabri, K.S., Burgess, I.W. and Plank, R.J., 1997. "Behaviour of steel and composite beam-to-column connections in fire - volume 1". Research Report DCSE/97/F/7, Department of Civil and Structural Engineering, University of Sheffield.

- Al-Jabri, K.S., Lennon, T., Burgess, I.W. and Plank, R.J., 1997. "Behaviour of steel and composite beam-column connections in fire, *Journal of Constructional Steel Research*, 46, pp. 1-3.

- Beneš, M., 2004. Equivalent T stub in tension under fire, PhD, thesis, CTU in Prague, , p. 124.

- Bailey, C.G., Lennon, T., Moore, D.B., 1999. The behaviour of full-scale steel-framed building subject to compartment fires, The Structural Engineer, Vol.77/No.8, p. 15-21.

- El-Rimawi J.A., Burgess I.W., Plank R.J., 1997. The Influence of Connections Stiffness and Behaviour of Steel Beams in Fire, Journal of Constructional Steel Research, Vol. 43 (1-3), pp. 1-15.

- Franssen J. M., 2002. Numerical determination of 3D temperature fields in steel joints. Second International Workshop "Structures in Fire", Christchurch.

- Kruppa, J., 1979. "Resistance au feu des assemblages par boulons haute resistance", St. Remy-les-Chevreuse, Centre Technique Industriel de la Construction Metalique, France,.

- Kirby B. R., 2001. The behaviour of high-strength grade 8.8 bolts in fire, Journal of Constructional Steel Research, Vol. 33 (1-2), pp. 3-37.

- Lawson, R.M. , 1990. "Behaviour of steel beam-to-column connections in fire", The Structural Engineer, 68(**14**), London, pp. 263-271.

- Liu T.C.H. , 1996. Finite element modeling of behaviour of steel beam and connections in fire, Journal of Constructional Steel Research, 36 (2), pp. 181-199.

- Leston-Jones, L.C., Burgess, I.W., Lennon, T. and Plank, R.J., 1997. "Elevated temperature moment-rotation tests on steelwork connections". *Proceedings of Institution of Civil Engineers. Structures & Buildings*; 122(**4**): pp. 410-419.

- Moore D.B., 1997. Full-scale fire tests on complete buildings, in *Fire, static and dynamics tests of building structures*, Proceedings of the second Cardington Conference Cardington (eds G.S.M. Armer and T. O'Dell), E&FN SPON.

- prEN 1993-1-2, 2003. Eurocode 3: Design of Steel Structures, Part 1.2: General

Rules, Structural Fire Design, European Norm, CEN, Brussels.

- prEN 1993-1-8, 2003. Eurocode 3: Design of Steel Structures, Part 1.8: Design of Joints, European Norm, CEN, Brussels.

- Sakumoto Y., Keira K., Furumura F., Ave T., 1993. Tests of fire-resistance bolts and joints, Journal of Structural Engineering, 119 (11), pp. 3131-3150.

- Simões da Silva L., Santiago A., Vila Real P, 2001. A component model for the behaviour of steel joints at elevated temperatures, Journal of Constructional Steel Research, Vol. 57 (11), pp. 1169-95.

- Spyrou S, Davison B., Burgess I., Plank R., 2002. Component-based studies on the behaviour of steel joints at elevated temperatures. in Lamas, A. and Simões da Silva, L. (eds.), *Proceedings of Eurosteel 2002 – 3rd European Conference on Steel Structures*, pp. 1469-1478, Coimbra, Portugal.

- Spyrou, S. and Davidson, J.B., "Displacement measurement in studies of steel T-stub connections", *Journal of Constructional Steel Research*, 57, pp. 647-659, 2001.

- "The performance of beam/column/beam connections in the BS476: art 8 fire test", British Steel (Swinden Labs), *Reports* T/RS/1380/33/82D and T/RS/1380/34/82D.

- Wald, F., Simões da Silva, L., Moore, D., Lennon, T., Chladná, M., Santiago, A., Beneš M. and Borges, L., 2004. Experimental behaviour of steel structure under natural fire, The Structural Engineer (submitted for publication).

- Wald F., Luís Simões da Silva L., Santiago, Moore D., 2004. Experimental behaviour of steel joints under natural fire, AISC-ECCS meeting, Amsterdam (submitted for publication).

Contacts *(include at least one COST C12 member)*

- František Wald

  Address: Department of Steel structures, Czech Technical University in Prague, Thakurova street 7, Prague, Czech Republic.
  e-mail: wald@fsv.cvut.cz, Phone: +42024354757, Fax: +420233334766

COST C12 - WG2

Datasheet I.3.2.2

Prepared by: P. Vila Real[1], L. Simões da Silva[2], N. Lopes[1]

[1] Department of Civil Engineering, University of Aveiro, Portugal
[2] Department of Civil Engineering, University of Coimbra, Portugal

DESIGN OF STEEL BEAM-COLUMNS UNDER FIRE CONDITIONS

Description

- The behaviour of beams submitted to lateral-torsional buckling is strongly related with the shape of its bending moment diagrams. The part 1-1 of the Eurocode 3 has already recognised the effect of the moment distribution but its fire part not. Recent work from the authors has shown that in case of fire, as at room temperature, the type of load acting on the beam must be taken into account. The behaviour of beam-columns submitted to lateral torsional buckling is also affected by the shape of the moment distribution as it will be shown in this datasheet.

Field of application

- Recently part 1-1 of Eurocode 3 has suffered significant changes in the design of beam-columns and unrestrained beams with lateral-torsional buckling.

- In order to study the possibility of having, in part 1-1 and part 1-2 of Eurocode 3, the same approach for the design of beam-columns, a numerical research was carried out by the authors, which concluded that the new approach (part 1-1) could be used in case of fire.

- An improvement of the analysis of the lateral-torsional buckling of beams under fire conditions has been made by the authors using an approach similar to the one implemented in part 1-1 of Eurocode 3.

- In the present datasheet this new approach for the lateral-torsional buckling has been used with the formulae for the design of beam-columns at elevated temperature based on prEN 1993-1-1 combined with the formulae from prEN1993-1-2.

Technical information

- A simply supported beam-column with fork supports has been chosen to explore the validity of the beam safety conditions, as shown in figure 1. With respect to the bending moment variation along the member length, two values, (-1, 0), of the ψ ratio (see fig. 1) have been investigated.

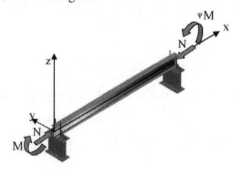

Figure 1. Simply supported beam-column with non-uniform bending.

- The case $\psi = 1$ was not studied here because this case is not affected by the new procedure for lateral-torsional buckling. In this case the proposed formulae for the evaluation of the lateral-torsional buckling resistance of steel beams remains the same as those proposed in the prEN 1993-1-2. The parametric study of beam-columns for $\psi = 1$ has already been made by Vila Real et al (2003b).

- An IPE 220 steel section of grade S 235 has been used.

- Uniform temperature in the cross-section has been used so that comparison between the numerical results Vila Real et al (2004) and the eurocode could be made. In this datasheet the temperature used was 600 °C, deemed to adequately represent the majority of practical situations.

- A lateral geometric imperfection given by the following expression was considered:

$$y(x) = \frac{l}{1000} \sin\left(\frac{\pi x}{l}\right)$$

- An initial rotation around the longitudinal axis with a maximum value of $l/1000$ rad at mid span was also introduced.

- Finally, the residual stresses adopted are constant across the thickness of the web and flanges. A triangular distribution as shown in figure 2, with a maximum value of 0.3×235 MPa has been used (ECCS, 1984).

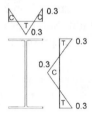

Figure 2. Residual stresses: C – compression; T – tension

Structural aspects

- According to the proposal of prEN 1993-1-2, the design buckling resistance moment of a laterally unrestrained beam with Class 1 or 2 cross-section type, can be determine by the following reduction factor:

$$\chi_{LT,fi} = \frac{1}{\phi_{LT,\theta,com} + \sqrt{[\phi_{LT,\theta,com}]^2 - [\lambda_{LT,\theta,com}]^2}} \qquad (1)$$

- Vila Real et al (2003a) has made a new proposal that adopts a modified reduction factor for lateral-torsional buckling, $\chi_{LT,fi,mod}$, given by

$$\chi_{LT,fi,mod} = \frac{\chi_{LT,fi}}{f} \quad \text{but} \quad \chi_{LT,fi,mod} \leq 1 \qquad (2),$$

where f depends on the loading type and is given by the following equation

$$f = 1 - 0.5(1 - k_c) \qquad (3)$$

where k_c is a correction factor according to table 1.

Table 1. Correction factors k_c for the new proposal

Moment distribution	Class 1, 2, 3 sections
	k_c
  M ψM $-1 \leq \psi \leq 1$	$0.6 + 0.3\psi + 0.15\psi^2$ but $k_c \leq 1$

- For fire loading, according to the new version of part 1-2 of Eurocode 3, the interaction equations for beam-columns are:

$$\frac{N_{fi,Ed}}{\chi_{z,fi} A k_{y,\theta} \dfrac{f_y}{\gamma_{M,fi}}} + \frac{K_{LT} M_{y,fi,Ed}}{\chi_{LT,fi} W_{pl,y} k_{y,\theta} \dfrac{f_y}{\gamma_{M,fi}}} \leq 1 \tag{4}$$

where

$$K_{LT} = 1 - \frac{\mu_{LT} N_{fi,Ed}}{\chi_{z,fi} A k_{y,\theta} \dfrac{f_y}{\gamma_{M,fi}}} \quad \text{but} \quad K_{LT} \leq 1.0 \tag{5}$$

and

$$\mu_{LT} = 0.15 \overline{\lambda}_{z,\theta} \beta_{M,Lt} - 0.15 \quad \text{but} \quad \mu \leq 0.9 \tag{6}$$

Here $\chi_{z,fi}$ is the reduction factor for flexural bucking around zz axis, and $\chi_{LT,fi}$ is the reduction factor for lateral-torsional buckling, given by (1).

The results obtained with these formulae considering or not the new proposal for the lateral-torsional buckling are shown in figures 3 and 4.

- Vila Real et al (2003b) have proposed the following interaction formulae for beam-columns in case of fire:

$$\frac{N_{fi,Ed}}{\chi_{y,fi} \dfrac{N_{fi,Rk}}{\gamma_{M,fi}}} + k_{yy,fi} \frac{M_{y,fi,Ed} + \Delta M_{y,fi,Ed}}{\chi_{LT,fi} \dfrac{M_{y,fi,Rk}}{\gamma_{M,fi}}} \leq 1 \tag{7a}$$

$$\frac{N_{fi,Ed}}{\chi_{z,fi} \dfrac{N_{fi,Rk}}{\gamma_{M,fi}}} + k_{zy,fi} \frac{M_{y,fi,Ed} + \Delta M_{y,fi,Ed}}{\chi_{LT,fi} \dfrac{M_{y,fi,Rk}}{\gamma_{M,fi}}} \leq 1 \tag{7b},$$

where χ_{fi} are the reduction factors for flexural bucking around yy and zz axis, and $\chi_{LT,fi}$ is the reduction factor for lateral-torsional buckling, calculated according to (1). The factors $k_{yy,fi}$ and $k_{zy,fi}$ are the interaction factors in case of fire that can be determined by two methods (denoted "EC3 Method 1, fi" and "EC3 Method 2, fi" in figures 5 to 8) described in Vila Real et al (2003b). Also, if the new proposal of Vila Real et al (2003b) for lateral-torsional buckling is used, two other methods are obtained (denoted "EC3 Method 1, fi / f" and "EC3 Method 2, fi / f" in figures 5 to 8).

- The procedure for the evaluation of the interaction factors for "EC3 Method 1, fi" is based on method 1 at room temperature, that is reported in Annex A of part 1.1 of EC3 and was developed by a French-Belgian team (Boissonnade, 2002) combining theoretical rules and numerical calibration to account for all the differences between the real model and the theoretical one.

- "EC3 Method 2, fi" is related to method 2 at room temperature, which is described in Annex B of part 1.1 of EC3 and results from an Austrian-German proposal (Greiner, 2001) that attempted to simplify the verification of the stability of beam-columns, all interaction factors being obtained by means of numerical calibration. These factors are not clearly understandable from a physical point of view, but this simple formulation reduces the possibility of mistakes.

- Figures 5 and 6 show the influence of considering or not the new proposal for LTB with method 1 and method 2 adapted to elevated temperature, for $\psi = 0$ and $\psi = -1$ respectively.

- Finally in figures 7 and 8, all the three methods studied here, Lopes et al (2004), considering the new proposal for the lateral-torsional buckling of beams are plotted together, showing good agreement with the numerical results.

Research activity

- Two new formulae for the design of beam-columns at room temperature have been proposed in prEN 1993-1-1 (2003) as the result of extensive work by two working groups that followed different approaches, namely, a French-Belgian team and an Austrian-German one.

- Under fire conditions, in prEN 1993-1-2 (2002), the proposed formulae for the design of beam-columns in case of fire have not changed and are still based on ENV 1993-1-1 (1992).

- In order to study the possibility of having, in part 1-1 and part 1-2 of the Eurocode 3, the same approach for beam-columns, a numerical research has been made concluding that it is possible to use the formulae from the part 1-1 (2003) provided that some factors are modified to consider high temperatures (Vila Real et al, 2003b).

- Significant changes, proposed in prEN 1993-1-1 (2003), have been introduced in the evaluation of the lateral-torsional buckling resistance of unrestrained beams at room temperature leading to results which are safe but less conservative than those obtained by the approach utilised in ENV 1993-1-1 (1992) in the case of non-uniform bending.

- In accordance with the safety format of the lateral-torsional buckling code provisions for cold design, an alternative proposal for rolled sections or equivalent welded sections subjected to fire was presented by Vila Real et al (2003a), that addressed the issue of the influence of the loading type on the resistance of the beam, leading to better agreement with the real behaviour while maintaining safety.

- The objective of the present datasheet is to evaluate the proposals made by Vila Real et al (2003b) in terms of a consistent safety check for the stability of beam-columns subjected to lateral-torsional buckling under fire loading, but using the new proposal for lateral-torsional buckling of unrestrained beams in case of fire Vila Real et al

(2003a). This new proposal has been also used with the design formulae for beam-columns from the prEN 1993-1-2 (2002).

Method of analysis

- Using the specialised finite element code SAFIR (Franssen, 2003), results of second-order analysis, including imperfections, for a range of lengths, levels of axial force and loading cases, are compared with the codified interaction formulae from Part 1-2 of Eurocode 3 (denoted in figures 3 and 4 "prEN 1993-1-2" when the new proposal for lateral-torsional buckling (Vila Real et al, 2003a) is not considered and "prEN 1993-1-2 / f" when this new proposal is included) and with the proposed adaptation (Vila Real et al, 2003b) to fire loading of method 1 and method 2 in prEN 1993-1-1 (2003), henceforth denoted "EC3 Method 1, fi / f" and "EC3 Method 2, fi / f" or "EC3 Method 1, fi" and "EC3 Method 2, fi", again if the new proposal for LTB is considered or not. Finally, the safety of these proposals is discussed and established.

Results of analysis

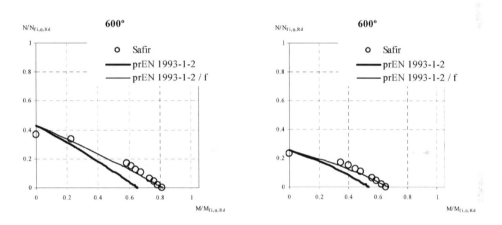

c) L=2000mm; $\bar{\lambda}_{LT,fi}$ =0.62; $\bar{\lambda}_{y,fi}$ =0.29; $\bar{\lambda}_{z,fi}$ =1.06   e) L=3000mm; $\bar{\lambda}_{LT,fi}$ =0.84; $\bar{\lambda}_{y,fi}$ =0.43; $\bar{\lambda}_{z,fi}$ =1.59

Figure 3. Interaction diagrams for combined moment and axial load at 600 °C, prEN 1993-1-2, considering or not the new proposal for lateral-torsional buckling, for ψ= 0

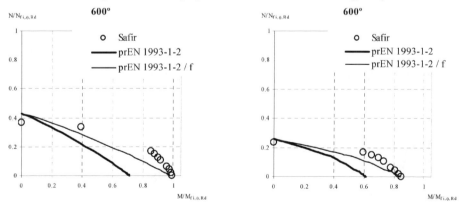

c) L=2000mm; $\bar{\lambda}_{LT,fi}$ =0.51; $\bar{\lambda}_{y,fi}$ =0.29; $\bar{\lambda}_{z,fi}$ =1.06   e) L=3000mm; $\bar{\lambda}_{LT,fi}$ =0.69; $\bar{\lambda}_{y,fi}$ =0.43; $\bar{\lambda}_{z,fi}$ =1.59

Figure 4. Interaction diagrams for combined moment and axial load at 600 °C, prEN 1993-1-2 considering or not the new proposal for lateral-torsional buckling, for ψ= -1

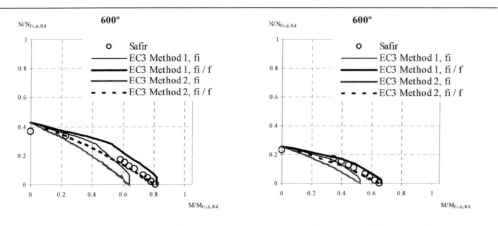

c) L=2000mm; $\bar{\lambda}_{LT,fi}$ =0.62; $\bar{\lambda}_{y,fi}$ =0.29; $\bar{\lambda}_{z,fi}$ =1.06　　　*e) L=3000mm; $\bar{\lambda}_{LT,fi}$ =0.84; $\bar{\lambda}_{y,fi}$ =0.43; $\bar{\lambda}_{z,fi}$ =1.59*

Figure 5. Interaction diagrams for combined moment and axial load at 600 ℃, adapted prEN 1993-1-1 considering or not the new proposal for lateral-torsional buckling, for ψ= 0

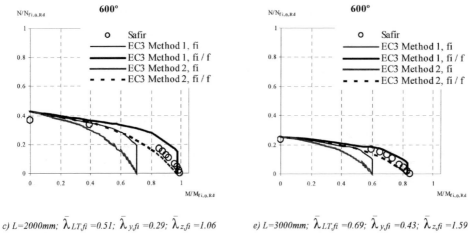

c) L=2000mm; $\bar{\lambda}_{LT,fi}$ =0.51; $\bar{\lambda}_{y,fi}$ =0.29; $\bar{\lambda}_{z,fi}$ =1.06　　　*e) L=3000mm; $\bar{\lambda}_{LT,fi}$ =0.69; $\bar{\lambda}_{y,fi}$ =0.43; $\bar{\lambda}_{z,fi}$ =1.59*

Figure 6. Interaction diagrams for combined moment and axial load at 600 ℃, adapted prEN 1993-1-1 considering or not the new proposal for lateral-torsional buckling, for ψ= -1

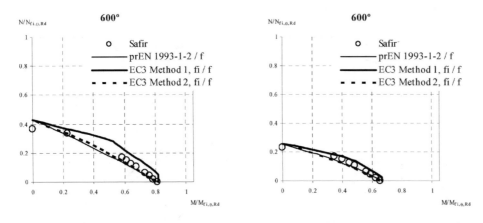

c) - L=2000mm; $\lambda_{LT,fi}$ =0.62; $\lambda_{y,fi}$ =0.29; $\lambda_{z,fi}$ =1.06　　　*e) - L=3000mm; $\lambda_{LT,fi}$ =0.84; $\lambda_{y,fi}$ =0.43; $\lambda_{z,fi}$ =1.59*

Figure 7. Interaction diagrams for combined moment and axial load at 600 ℃, adapted prEN 1993-1-1 considering or not the new proposal for lateral-torsional buckling, for ψ= 0

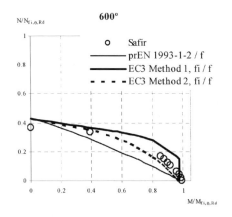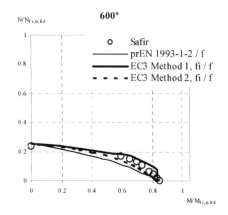

c) - L=2000mm; $\bar{\lambda}_{LT,fi}$ *=0.51;* $\bar{\lambda}_{y,fi}$ *=0.29;* $\bar{\lambda}_{z,fi}$ *=1.06*   *e) - L=3000mm;* $\bar{\lambda}_{LT,fi}$ *=0.69;* $\bar{\lambda}_{y,fi}$ *=0.43;* $\bar{\lambda}_{z,fi}$ *=1.59*

Figure 8. Interaction diagrams for combined moment and axial load at 600 °C, adapted prEN 1993-1-1 considering or not the new proposal for lateral-torsional buckling, for ψ= -1

- It has been shown that the proposed method for the lateral-torsional buckling of unrestrained steel beams at high temperatures, introduce great improvements in the design curves of beam-columns under fire conditions.

- Although the new proposal based on the method 1, in some cases, is not in the safe side when compared with the numerical results all the proposals presented here are in general in very good agreement with this results.

- In addition, if method 1 and method 2 are adopted, this has the advantage of using the same formulae at room temperature and at elevated temperature, being in line with the procedure which used to be in usage in the Eurocodes. These aspects should be considered when the new proposal for the resistance of beam-columns has to be chosen in the next revision of the Eurocode 3.

- The study presented in this datasheet strongly recommends the use of one of these proposals in future versions of part 1-2 of the Eurocode 3.

Further developments

- More numerical and experimental tests are needed, so that the validity of these proposals can be thoroughly assessed, namely the use of different steel grades and cross sectional shapes.

References

- Boissonnade, N.; Jaspart, J.-P.; Muzeau, J.-P.and Villette, 2002, *New Interaction Formulae for Beam-Columns in Eurocode 3: The French-Belgian Approach*, Proceedings of Third European Conference on Steel Structures – Eurosteel 2002, António Lamas and Luís Silva (eds), Coimbra, Portugal.

- ECCS – EUROPEAN CONVENTION FOR CONSTRUCTIONAL STEELWORK, 1984, Technical Committee 8 – Structural Stability, Technical Working Group 8.2 – System, *Ultimate Limit State Calculation of Sway Frames With Rigid Joints*, first edition.

- Eurocode 3, 2003, *Design of Steel Structures – part 1-1. General rules and rules for buildings.* prEN 1993-1-1:2003, Stage 34 Draft, European Committee for Standardisation, Brussels, Belgium.

- Eurocode 3, 2002, *Design of Steel Structures – part 1-2. General rules – Structural fire design.* Draft prEN 1993-1-2:20xx, Stage 34, European Committee for Standardisation, Brussels, Belgium.

- Eurocode 3, 1992, *Design of Steel Structures – part 1-1. General rules and rules for buildings.* ENV 1993-1-1, Commission of the European Communities, Brussels, Belgium.

- Franssen, J.-M., 2003, *SAFIR. A Thermal/Structural Program Modeling Structures under Fire*, Proc. NASCC conference, American Inst. for Steel Constr., Baltimore.

- Franssen, J. M., 1990, *The unloading of building materials submitted to fire*, Fire Safety Journal, Vol. 16, pp. 213-227.

- Franssen, J.-M. 1989, *Modelling of the residual stresses influence in the behaviour of hot-rolled profiles under fire conditions* (in French), Construction Métallique, Vol. 3, pp. 35-42.

- Greiner, R., 2001, *Background information on the beam-column interaction formulae at level 1*, ECCS TC 8 Ad-hoc working group on beam-columns paper No.TC8-2001, Technical University Graz.

- Lopes, N., Simões da Silva, L., Vila Real, P., Piloto, P., 2003 *"New Proposals for the Design of Steel Beam-Columns under Fire Conditions, Including a New Approach for the Lateral-Torsional Buckling of Beams"*, accepted for publication, Computer & Structures, PERGAMON.

- Souza, V., Franssen, J.-M., 2002, *Lateral Buckling of Steel I Beams at Elevated Temperature – Comparison between the Modelling with Beam and Shell Elements*, Proc. 3rd European Conf. on Steel Structures, ISBN: 972-98376-3-5, Coimbra, Univ. de Coimbra, A. Lamas & L. Simões da Silva ed., pp. 1479-1488.

- Vila Real, P.M.M., Lopes, N., Simões da Silva, L., Franssen, J.-M., 2003a, *Lateral-torsional buckling of unrestrained steel beams under fire conditions: improvement of EC3 proposal*, submitted to Computers & Structures.

- Vila Real, P.M.M., Lopes, N., Simões da Silva, L., Piloto, P., Franssen, J.-M., 2003b, *Towards a consistent safety format of steel beam-columns: application of the new interaction formulae for ambient temperature to elevated temperatures*, Steel & Composite Structures, an International Journal, TECHNO-PRESS vol. 3, n. 6, p.p. 383-401.

- Vila Real, P. M. M.; Lopes, N.; Simões da Silva, L.; Piloto, P.; Franssen, J.-M., 2004, *Numerical Modelling of Steel Beam-Columns in Case of Fire – Comparisons with Eurocode 3*, Fire Safety Journal, ELSEVIER, 39/1, pp. 23-39.

Contacts *(include at least one COST C12 member)*

- Paulo M. M. Vila Real
  Address: Department of Civil Engineering, University of Aveiro, 3810 Aveiro, Portugal
  e-mail: pvreal@civil.ua.pt, Phone: +351-234-370049,
  Fax: +351-234-370094

COST C12 - WG2

Datasheet I.3.2.3

Prepared by: B. Faggiano, G. Della Corte

Department of Structural Analysis and Design

University of Naples "Federico II", Italy

POST-EARTHQUAKE FIRE RESISTANCE OF MOMENT RESISTING STEEL FRAMES

Description

- A fire coming soon after an earthquake finds a different, more vulnerable, structure as respect to the initial, undamaged one. Depending on the extent of damage, the fire resistance of the structure could be significantly reduced.

Field of application

- Risks coming from fires following strong earthquakes are very high. In fact, ensuing losses may be comparable to those caused by the shaking:
  - the increase in the time needed to firemen for reaching the place of the fire (due to traffic congestion, collapsed constructions, rubble in the streets, concomitance of multiple fires), the possible difficulties in water supply and the decrease of the collapse time of the structure may result in high risk.
- In view of the development of a comprehensive methodology of performance-based design of buildings, the fire-safety codes should distinguish between structures located in seismic and non-seismic areas, by requiring more stringent fire resistance provisions for those buildings potentially subjected to seismic actions.

Structural aspects

Seismic damage modelling

The following schematisation has been adopted:

1) 'geometrical damage', the change of the initial structure geometry owing to the residual deformation produced by plastic excursions during the earthquake:
   - for an ideal Elastic Perfectly Plastic (EPP) structure, undergone a seism, the residual values of the inter-story drift angles is the main damage index (Figure 1);
   - the residual inter-story drift is the origin of residual P-Δ effects, which produce an increase in the stress-state of the structure under loads acting after the earthquake.

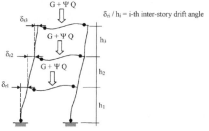

Figure 1. Residual deformed configuration of the structure after the earthquake

2) 'mechanical damage', the degradation of mechanical properties of structural components engaged in the plastic range of deformation during the earthquake:

- for steel structures the main causes of degradation are: a) both local and overall buckling of members; b) repetition of plastic deformations during the earthquake;
- in case of well-engineered steel structures, the assumption of non-degrading structural components is realistic in the range of plastic deformation induced by earthquakes at the design performance level;
- for structures not adequately designed against earthquakes and/or in case of very rare earthquakes, the effect of strength and stiffness degradation should be taken into account;
- for MR steel frames designed according to Eurocode 8, strength degradation becomes significant only for very large values of the peak ground acceleration.

Fire modelling (Figure 2)

- The principal factors that influence a fire are: building division into compartments, fire load and rooms' ventilation.
- The more significant effect in the structural field is the increase of temperature, hence the usual representation of a fire event is the time-temperature curve (t-T).
- Methods for fire modelling are: the method of the nominal curves, which are expressed by analytical relationships, simple and identical; the method of the parametrical curves, which are expressed by more complex relationships.

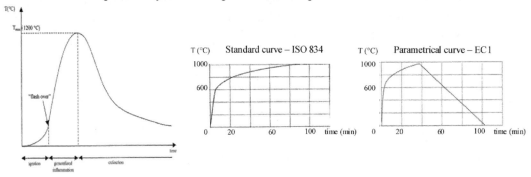

Figure 2. Typical time-temperature curves of fire

Material modelling at high temperatures

The steel behaviour at high temperatures, such as the steel stress-strain-temperature relationship and the thermal properties (thermal elongation, thermal conductivity and specific heat vs. temperature relationships) can be modelled according to Eurocode 3 (CEN 1994).

Research activity

With reference to steel moment resisting frames:

- Definition of the post-earthquake structure physical state.
- Evaluation of the fire resistance ratings of structures damaged by the earthquake, in

terms of time up to collapse, according to the chosen thermal program corresponding to the fire event.

- Evaluation of the behaviour of structures under fire conditions, at growing temperatures, according to the chosen thermal program corresponding to the fire event, in terms of stresses, deformations and collapse mechanisms.

Method of analysis

The study has been subdivided in two phases:

1) parametrical analysis of simple portal steel frames (Figure 3)

   for focusing on the parameters potentially affecting the problem and for individuating the range of residual inter-story drift angles for which the problem is significant.

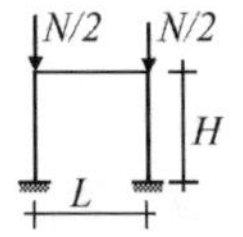

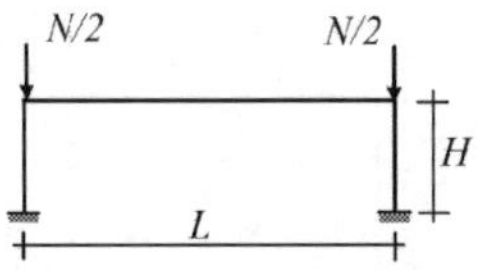

	L/H	I_b/I_c	M_{pb}/M_{pc}	N/N_{cr}	δ/H
		0.63	0.54		0.0-0.385
	1			0.05-0.30	
		4.50	2.22		0.0-0.355
		0.53	0.47	0.05-0.30	0.0-0.365
	3				
		4.72	2.27	0.025-0.30	0.0-0.620

L/H	ratio between the beam span and the story height
I_b/I_c	ratio between the beam and column moment of inertia
M_{pb}/M_{pc}	ratio between the beam and column section flexural plastic strength
N/N_{cr}	ratio between the vertical load and its elastic critical value
δ/H	level of geometrical damage, measured by residual story drift angle

Figure 3. Simple portal steel frames under study

2) evaluation of the fire resistance rating reduction of frames extracted from civil ordinary buildings (Figures 4-5-6):

   - 5 and 10 story frames, extracted from a perimeter moment-resisting frame system, and a 10 story frame, extracted from a spatial frame system;
   - two design strategies: ultimate limit state (ULS) only considered; both serviceability and ultimate limit state (SLS+ULS) considered (Eurocode 8, CEN 1998).

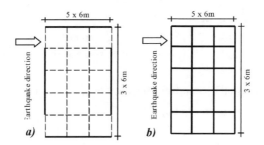

Figure 4. Plan layout of the perimeter (*a*) and the spatial (*b*) frame structural systems.

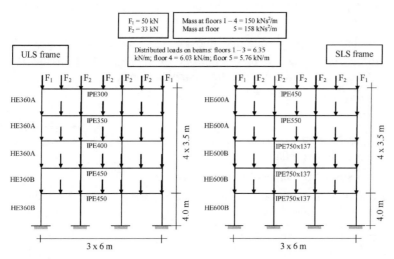

Figure 5. The 5 story frames

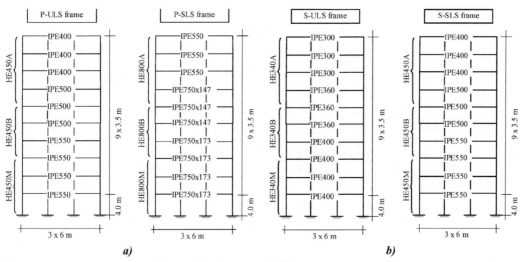

Figure 6. (a) 10 story perimeter frames; (b) 10 story spatial frames.

Model assumptions:

- The fire scenario is assumed as the most severe for the earthquake-damaged structure, obtaining the minimisation of the failure time of the structure in fire:
  - fire acting at the base of the frame in all the bays;
  - temperatures uniformly distributed in the compartment and monotonically increasing according to the standard ISO-834 time-temperature curve (Eurocode 1, CEN 1997).
- Strain-hardening of steel and thermal elongation of members has been considered.
- Structures have been supposed non-protected against fire.

Structural analysis under fire

1) Seismic analysis
   - The numerical code SAP 2000 vers. 7.12 (Wilson, 1998) has been used.

206

- Determination of the post-earthquake geometry.
  - Single-story frames: residual inter-story drifts were generated artificially for conducting a parametric analysis.
  - Multi-story frames: residual inter-story drifts were generated using real ground acceleration records, for directly assessing the relationship between the level of seismic intensity and the level of fire resistance rating reduction.

2) Fire analysis
   - The finite element program for structural calculation, SAFIR 2002, developed by J. M. Franssen at the University of Liége, has been used.
   - The failure time of the structure in fire has the meaning of the limiting time at which a structural instability phenomenon occurs.

Results of analysis

Portal frames

- The effect of the geometrical damage is quite insensitive to the level of vertical loads.
- Simple abaci could be developed for computing fire resistance rating reduction at increasing levels of residual story drifts.

Multi-storey frames

- The effect of the seismic design option:
  - for ULS-designed structures, the fire resistance rating reduction is always appreciable.
  - for (SLS+ULS)-designed structures, the fire resistance rating reduction is relatively smaller: at the design seismic intensity level, it is < 10%.
  - for all the frames, the fire resistance rating reduction is non-negligible for very rare earthquakes, i.e. earthquakes with a mean return period > 475 years.
- The effect of the structural system layout:
  the reduction is smaller in case of spatial frame systems, as respect to the case of perimeter frames.
- The type of collapse mechanism:
  - during the pre-earthquake fires, the undamaged frames collapsed on themselves, without appreciable lateral displacement;
  - the earthquake-induced damage produces a lateral stability type of collapse mechanism.

Post-earthquake fire resistance ratings have been normalized by the pre-earthquake values ($t_f/t_{f,0}$) and represented as a function of spectral accelerations, which have been normalized by their design values ($S_{a,e}/S_{a,e,d}$) (Figures 7-8)

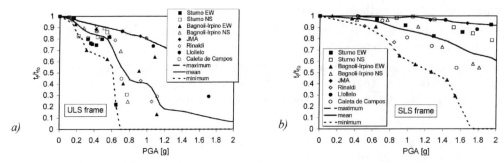

Figure 7. 5 story MRF: Normalised fire resistance rating vs. site seismic intensity: a) ULS frame; b) SLS+ULS frame.

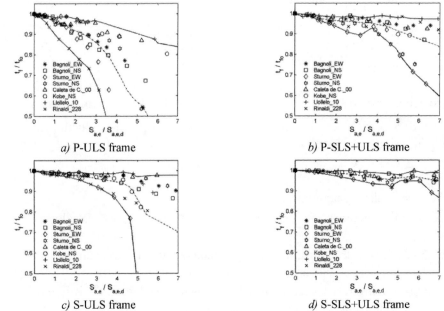

a) P-ULS frame   *b)* P-SLS+ULS frame

c) S-ULS frame   *d)* S-SLS+ULS frame

Figure 8. 10 story MRF: Normalized fire resistance rating vs. normalized seismic intensity

Further developments

The development of a quantitative proposal for fire-safety codes is necessary. It requires both to refine modelling and to increase the number of simulated cases.

Further investigations will be developed by using the finite element program for structural calculation, ABAQUS, in order to perform as a first step the seismic analysis of the examined structures, as a second step the fire analysis of the same structures damaged by the earthquake, with the actual damage distribution within the frames.

References

- Anderberg, Y. 1988. Modelling Steel Behaviour. *Fire Safety Journal, 13(1), 17-26.*
- Buchanan, A.H. 1994. Fire Engineering for a Performance Based Code. *Fire Safety Journal, 23(1), 1-16.*
- Buchanan, A.H. 2001. *Structural Design for Fire Safety.* John Wiley & Sons,

Chichester, England.

- CEN 1994. ENV 1998. *Eurocode 8: Design provisions for earthquake resistance of structures.*
- CEN 1995. ENV 1991-2-2. *Eurocode 1: Basis of design and actions on structures – Part 2-2: Actions on structures – Actions on structures exposed to fire.*
- CEN 1995. ENV 1993-1-2. *Eurocode 3: Design of steel structures. – Part 1.2: General rules – Structural fire design.*
- Della Corte, G., De Matteis, G., Landolfo, R., Mazzolani, F.M. 2001. Seismic Analysis of MR Steel Frames Based on Refined Hysteretic Models of Connections. *Journal of Constructional Steel Research*, 58(10), 1331-1345.
- Della Corte, G., Landolfo, R. 2001. Post-Earthquake Fire Resistance of Steel Structures. *In E. Zio, Demichela M., Piccinini N. (eds), Safety and Reliability, Towards a Safer World (Proceedings of the European Conference on Safety and Reliability – ESREL 2001),* Torino: Politecnico di Torino, 3: 1739-1746.
- Della Corte, G., Mazzolani, F.M. 2002. Seismic Stability of Steel Frames. *In Ivanyi M. (ed.), Proceedings of the International Colloqium on Stability and Ductility of Steel Structures, September*, Budapest.
- Della Corte, G., Landolfo, R., Mazzolani, F.M. 2003. Post-earthquake fire resistance of moment-resisting steel frames. *Fire Safety Journal*, 38 (2003), 593-612, Elsevier Ltd.
- EQE International 1995. *The January 17, 1995 Kobe Earthquake – An EQE Summary Report. April.*
- Fajfar, P. 2002. Structural Analysis for Earthquake Engineering – A Breakthrough of Simplified Non-Linear Methods. *In Proceedings of the 12th European Conference on Earthquake Engineering, Keynote Lecture, London, Elsevier Science Ltd,* Oxford, UK, CD-ROM.
- Feasey, R. & Buchanan, A. 2002. Post-Flashover Fires for Structural Design. *Fire Safety Journal*, 37, 83-105.
- FEMA 350, 2001. *Seismic Design Criteria for New Moment-Resisting Steel Frame Construction.* Federal Emergency Management Agency.
- Franssen, J.M., Kodur, V.K.R., Mason, J. 2002. *User's manual for SAFIR 2001 free – A computer program for analysis of structures submitted to the fire.* University of Liege, Department: Structures du Génie Civil, Service: Ponts et Charpentes.
- Hamburger, R.O., Foutch, D.A., Cornell, C.A. 2000. Performance Basis of Guidelines for Evaluation, Upgrade and Design of Moment-Resisting Steel Frames *In: Proceedings of the 12th World Conference on Earthquake Engineering*, Auckland, New Zealand, Paper No. 2543, CD-ROM.
- Krawinkler, H. 2000. System Performance of Steel Moment Resisting Frame Structures. *In Proceedings of the 12th World Conference on Earthquake Engineering*, Auckland, New Zealand, Paper No. 2545, CD-ROM.
- Robinson, J. 1998. Fire – A Technical Challenge and a Market Opportunity. *Journal of Constructional Steel Research*, 46: 1-3, Paper no. 179.
- Trifunac, M.D., Brady, A.G. 1975. A Study on the Duration of Earthquake Ground Motion. *Bull. seism. soc. Am.*, 65, 581-626.
- Wang, Y.C. & Kodur, V.K.R. 2000. Research Toward Use of Unprotected Steel Structures. *ASCE, Journal of Structural Engineering*, Vol. 126, No.12, December,

1442-1450.

- Wilson, E.L. 1998. Three Dimensional Static and Dynamic Analysis of Structures. Computers & Structures Inc., Berkeley, California, USA.

Contacts

- Federico M. Mazzolani

    Address: Dept. of Structural Analysis and Design, Faculty of Engineering, University of Naples"Federico II", P.le Tecchio 80, 80125, Naples, Italy

    e-mail: fmm@unina.it, Ph: +39-081-768-2443, Fax: +39-081-593-4792

- Raffaele Landolfo

    Address: Dept. of Constructions and Mathematical Methods in Architecture, Faculty of Architecture, University of Naples"Federico II", Via Monteoliveto 3, 80134, Naples, Italy

    e-mail: landolfo@unina.it, Ph: +39-081-768-2447, Fax: +39-081-593-4792

COST C12 - WG2

Datasheet I.3.2.4

Prepared by: G. De Matteis[1], Beatrice Faggiano[1], F. Wald[2]

[1] University of Naples "Federico II", Italy
[2] Czech Technical University in Prague, Czech Republic

BEHAVIOUR OF ALUMINIUM STRUCTURES IN FIRE

Description

- The methods of structural analysis in fire condition have been assessed for both steel and reinforced concrete structures. Their extension to the aluminium alloy structures requires that the specific mechanical properties of the material as a function of the temperature must be considered. In particular the influence of the shape of the constitutive laws and thus of the kinematic strain hardening on the global behaviour of the structure in the post elastic field have to be held in due account.

- For the development of complex analysis methods for aluminium alloys structures under fire, accurate material models have to be implemented in finite element programs for structural analysis.

Field of application

- The widespread employment of aluminium alloys in the field of the construction of bridges, footbridges, aerials, offshore platforms, thanks to lightness, versatility and resistance to the atmospheric agents, requires exhaustive provisions for the design under fire of the structures in aluminium alloys, consisting in sound methodologies of calculus, at different levels of refinement.

Codification

European scale

- Eurocode 1-Part 2-2, for the applied loads in fire conditions;
- Eurocode 3- Part 1-2, for steel structure under fire;
- Eurocode 9- Part 1-2, for aluminium alloys' structure under fire.

Structural aspects

In the design under fire situations may be distinguished tree basic steps. The modelling of the temperature development in the fire compartment, the transfer of the heat into the structure and the structural behaviour affected by high temperatures and elongation and shortening of heated members.

Fire modelling (Figure 1)

- The principal factors that influence a fire are: building division into compartments, fire load, rooms' ventilation, and the quality of compartment surfaces.

- The more significant effect in structural field is the increase of temperature, so the usual representation of a fire event is the time-temperature curve (t-T).

- Methods for the modelling of a fire are: the nominal curves, which coming from the fire tests of structural elements and are expressing the time temperature relation only, the parametric curves, which are modelling the analytical relationships between the basic fire parameters; and fluid dynamics, which are modelling the fire using the discrete and step by step procedures.

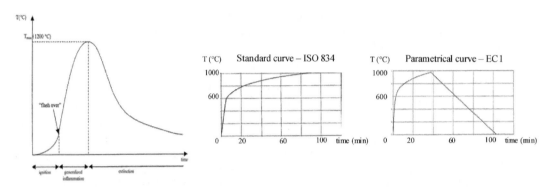

Figure 1. Typical time-temperature curves of fire

Material modelling

- Modelling the constitutive law of the aluminium alloys is difficult already at ambient temperature, because: a) the material presents different mechanical characteristics within the same alloy type, in relation to the production state and to the undergone treatments; b) the material is lacking of yielding and characterized by a not negligible non linear hardening.

- For structural analysis under high temperatures, Eurocode 9 implicitly adopts a simplified elastic-perfectly plastic relationship, supplying the variation of the elastic modulus and of the conventional yielding stress with the temperature (Figure 2a).

- A better interpretation of the evolution of the mechanical characteristics of the material with the temperature, can be achieved by applying the model of Ramberg and Osgood, diffusely used for the modelling in non linear field of the aluminium alloy at ambient temperature (Figure 2b)

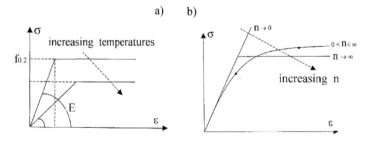

Figure 2. Typical constitutive laws for the aluminium alloys: (a) elastic-perfectly plastic law (b) Ramberg-Osgood relationship.

The critical temperature of aluminium alloys may be expressed by

$$\theta_{a,cr} = C \, ln\left(\frac{1}{A \, \mu_0^D} - 1\right) + B \,,$$

where the degree of utilisation $\mu_0 = E_{fi,d} / R_{fi,d,0}$ may not be taken lass than 0,015, where $E_{fi,d}$ is the design effect of actions for the fire design situation according to EN 1991-1-2 and $R_{fi,d,0}$ is the corresponding design resistance of the steel member for fire design situation at time t. The invariable are summarized in table Tab. 1. The accuracy of the prediction varies. The maximal non-conservative deviations from the correct values are included.

Tab. 1 Constants for calculation of critical temperature of aluminium alloys, the alloys with a limited accuracy are marked by Italic (Strejček and Wald, 2004)

Alloy	Temper	Invariable				Deviation
		A	B	C	D	
EN AW-5052	*O*	*0,9905*	*428*	*74,88*	*0,2063*	*14,7 %*
EN AW-5052	*H34*	*0,9797*	*420*	*90,06*	*0,1273*	*27,0 %*
EN AW-5083	O	0,9942	430	62,53	0,1485	10,5 %
EN AW-5083	H113	0,9843	424	89,97	0,2711	0,9 %
EN AW-5454	*O*	*0,9885*	*424*	*74,01*	*0,1519*	*15,7 %*
EN AW-5454	*H32*	*0,9806*	*422*	*85,83*	*0,1427*	*15,6 %*
EN AW-6061	T6	0,9957	427	65,38	0,1169	1,8 %
EN AW-6063	T6	0,9902	422	74,06	0,1048	8,7 %
EN AW-6082	T6	0,9826	420	89,37	0,1377	4,6 %
EN AW-3003	O	0,9806	424	95,59	0,3199	4,5 %
EN AW-3003	H14	0,9753	412	95,87	0,1263	9,4 %
EN AW-5086	O	0,9843	424	89,97	0,2711	0,8 %
EN AW-5086	*H112*	*0,9826*	*428*	*78,80*	*0,2438*	*19,6 %*
EN AW-7075	*T6*	*0,9763*	*412*	*94,12*	*0,1143*	*12,0 %*

Research activity

Prediction of the mechanical response of aluminium alloy structures exposed to fire by achieving:

1) adequate knowledge of the material behaviour under high temperatures;

2) developing accurate structural analyses in ultra elastic field, which takes correctly into account the behavioural characteristic of the base material, such as the strain-hardening and the limited deformation capability.

Method of analysis

1) Evaluation of the principal mechanical parameters as a function of temperatures for the examined aluminium alloy for structural uses (Figure 3):
   - the elastic modulus (E_T);
   - the stress at the conventional elastic limit ($f_{0,2,T}$), it corresponding to a residual deformation equal to 0.2%;
   - the ultimate strength $f_{t,T}$;
   - the ratio $f_{t,T}/f_{0,2,T}$, it representing the material strain hardening;
   - the elongation at collapse $\varepsilon_{u,T}$.

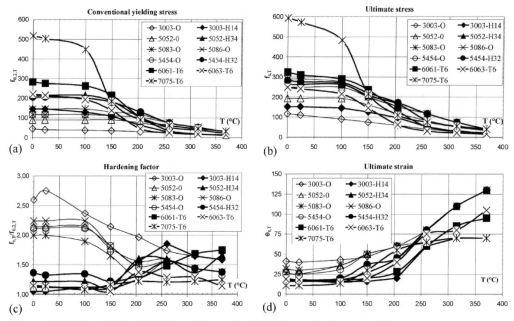

Figure 3. (a) $f_{0.2,T}$, (b) $f_{t,T}$, (c) $f_{t,T}/f_{0.2,T}$, (d) $\varepsilon_{u,T}$, for 10000 hours of exposure to fire.

2) Extension to high temperatures of the Ramberg-Osgood relationship, by introducing the variation law with the temperature of the relevant mechanical parameters (Figure 4).

$$\varepsilon_T = \frac{\sigma}{E_T} + 0.002 \left(\frac{\sigma}{f_{0.2,T}} \right)^{n_T} \text{ with } n_T = \frac{\log \dfrac{0.002}{\varepsilon_{u,T}}}{\log \dfrac{f_{0.2,T}}{f_{t,T}}} \text{ (hardening factor)}$$

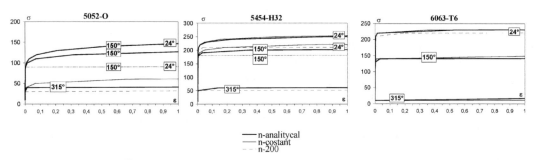

Figure 4. The σ(N/mm²)-ε laws for the aluminium alloys at different temperatures.

3) Structural analysis under fire:

- Aims

evaluation of the fire resistance of structures, in terms of time up to collapse, according to the chosen thermal program corresponding to the fire event: the design of structures under fire must assure that in case of fire, the static safety is guaranteed for a pre-fixed period of time, to be associated to the class of fire resistance imposed by the standard rules.

- Implementation

The finite element program for structural calculation, SAFIR 2002, developed by J. M. Franssen of the University of Liége, has been used. The software has been implemented introducing in the library of materials also the aluminium alloys indicated in the Eurocode 9, according to the modified Ramberg-Osgood relationship.

- Study case

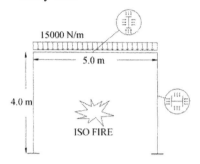

ALLOY	f_d (N/mm²)	Column HEB	Beam IPE
3003-O	45	160	450
3003-H14	145	160	300
5052-O	90	160	330
5052-H34	215	160	240
5454-O	117	160	300
5454-H32	207	160	240
5083-O	145	160	300
5086-O	117	160	300
6061-T6	283	160	240
6063-T6	220	160	240
7075-T6	517	160	240

Figure 5. Definition of the study case

Results of analysis

- The simplified elastic-perfectly plastic mechanical model of the material is not able to well characterize the behaviour at high temperatures (T< 200°C) of the aluminium alloys: it is necessary considering an opportune modelling of the strain hardening.

- The adoption of simplified mechanical model of the elastic-perfectly plastic type entails a fire resistance reduction in some cases not negligible, therefore resulting too much precautionary.

- The structural analysis in fire conditions of a simple study case has pointed out the urgency of deepening the study of the fire resistance of the structures in aluminium alloys, since times up to collapse result extremely low (Figure 6).

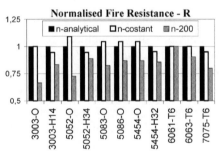

Figure 6. Fire resistance for the study cases

Further developments

Presented results are related to a simple study case and they refer to a specific structural scheme, a standard fire model and the worse exposure condition of the frame members (on 4 sides of the profile). Considerable margins of improvement of the performances to fire of the structure exist:

- it is possible to operate on the type of structural members, playing on the factor S/V, which represents the ratio between the surface (S) struck by the thermal flux and the volume (V) of the structural element: profiles with double T cross section are the most unfavourable, while the hollow profiles are the more suitable;

- it is opportune designing in adequate manner the condition of division into compartments of the locals, for bounding the heaviness of the fire effects on the structural frame;

- the fire event should be modelled by means of the parametrical curves;
- it is necessary to identify the possible application fields of such materials for structural use in fire risk situations.

References

- ASM Specialty Handbook 1993. *Aluminium and aluminium alloys*. Edited by J.R. Davis Davis &Associates.
- CEN (European Communities for Standardisation), ENV 1991-2-2 1997. *Eurocode 1: Basis of design and actions on structures - Part 2-2: Actions on structures - Actions on structures exposed to fire.*
- CEN (European Communities for Standardisation), ENV 1992-1-2 1998. *Eurocode 2: Design of reinforced concrete structures - Part 1-2: General rules-Structural fire design.*
- CEN (European Communities for Standardisation), ENV 1993-1-2 1998. *Eurocode 3: Design of steel structures - Part 1-2: General rules-Structural fire design.*
- CEN (European Communities for Standardisation), ENV 1999-1-2 1998. *Eurocode 9: Design of aluminium structures - Part 1-2: General rules-Structural fire design.*
- Conserva, M., Donzelli, G., Trippodo, R. 1992. *Aluminium and its application*. EDIMET.
- Faggiano, B., De Matteis, G., Landolfo, R., Mazzolani, F.M. 2003. On the behaviour of alluminium alloy structures exposed to fire. *In Proceedings of the XIX National Congress C.T.A. "III week of steel Constructions", 28-30 September,* Genova, 393-406 pp., Italia.
- Faggiano, B., De Matteis, G., Landolfo, R., Mazzolani, F.M. 2004. Effects of high temperatures on the resistance of aluminium alloy structures. *In Proceedings of the 7th International Conference on Modern Building Materials, Structures and Techniques, Vilnius, Lithuania, 19-21 May,* in press.
- Franssen, J.M.. 1998. SAFIR98a - User's Manual. University of Liege, Belgium.
- Lundberg, S. 1995. Design for fire resistance. *In Training in Aluminium Application Technology (TALAT) EU-COMETT Program,* F. Ostermann Ed., Aluminim Training Partnership, Brussels: Section 2500.
- Mazzolani, F.M. 1994. *Aluminium alloy structures*. 2nd edn., E&FN Spon, London, UK.
- Strejček M., Wald F. 2003. Critical temperature of aluminium alloys, Internal report, CTU in Prague.
- Wald, F., Bosiljkov, V., De Matteis, G., Haller, P., Vila Real, P. 2002. Structural integrity of buildings under exceptional fire. *In Proceeding of the 1st Cost C12 Seminar, Lisbon 18-19 April.*

Contents

- Federico M. Mazzolani

  Address. Dept. of Structural Analysis and Design, Faculty of Engineering, University of Naples"Federico II", P.le Tecchio 80, 80125, Naples, Italy
  e-mail: fmm@unina.it, Ph: +39-081-768-2443 Fax: +39-081-593-4792

- Raffaele Landolfo

  Address: Dept. of Constructions and Mathematical Methods in Architecture, Faculty of Architecture, University of Naples"Federico II", Via Monteoliveto 3, 80134, Naples, Italy
  e-mail: landolfo@unina.it, Ph: +39-081-768-2447 Fax: +39-081-593-4792

BEHAVIOUR OF RC ELEMENTS AND RC FRAMES EXPOSED TO FIRE

Description

- The structural response of centrically and eccentrically loaded reinforced concrete columns and plane frame structures exposed to standard fire is considered in this analysis.

Field of application

- The fast development of the construction industry and the increasing prevalence of multi-storey structures have imposed the need that the skeleton (the frame structure) "survives" the fire without collapsing over the surrounding structures and inducing even greater losses. The columns as structural elements have an important role in preventing loss of global stability of structures. If these elements do not suffer failure, damages shall be of a local character.

Technical information

- Columns analused in this study are fixed at A and free to move vertically at B, allowing a free expansion in longitudinal direction (Figure 1a).

- Figure 1b shows the cross section of the centrically loaded columns (M=0) exposed to fire from all sides. Because of the symmetry of the cross section and the symmetry of the fire load, only one quarter of the columns is analyzed.

- The influence of: dimensions of the cross section; concrete cover thickness; type of aggregate; compression strength of concrete; steel ratio and intensity of the axial force are analyzed and the results are presented. In this case the support conditions and column length have a negligible influence on the fire resistance (Lin, 1992), so they are not varied.

- The most usual case in practice is when eccentrically loaded columns are part of a wall that separates the fire compartment, so they are exposed to fire only from the inside of the compartment (Figure 1c).

- The influence of: dimensions of the cross section, concrete cover thickness, steel ratio, intensity of the axial force and bending moment are analyzed as significant factors affecting the fire resistance of eccentrically loaded RC columns.

- Three-bay, two-story reinforced concrete frame exposed to different fire scenarios is analyzed too. The frame is assumed to be fixed at the base, the uniform load is q=45KN/m' (the same for the two stories), all columns and beams are designed with the same cross section (40×40cm), columns are symmetrically reinforced with 8ϕ16mm. The concrete cover thickness for all elements is a=2.5cm. The yield strength of reinforcement at ambient temperature is $f_y = 400Mpa$, and the compressive strength of concrete is $f_c = 30Mpa$.

- ISO 834 standard fire model is used. Temperature dependent thermal and mechanical properties of concrete and steel are taken as recommended by Eurocode 2, part 1.2.

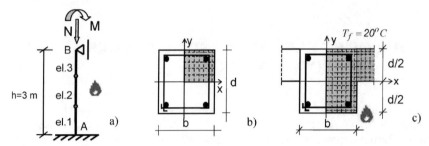

Figure 1. a) Geometry, support conditions. Discretization of cross-section when columns are: b) exposed to fire from all sides; c) part of a wall that separates the fire compartment

Structural aspects

- To define the fire resistance of frame structures, experimental investigations of models are almost impossible. The time dimension of spreading of the temperature field is practically impossible to be simulated on a model of small proportions. Hence, model investigations can hardly be accepted due to the high cost.

- For the last twenty years, particular importance has therefore been given to analytical definition of the problem. Generally, analytical computation of the frame structure for the case of fire means proof that the structure, or its elements, loaded by a defined load and exposed to thermal effect, satisfies certain functional requirements, expressed through the ultimate state of bearing capacity and usability.

Research activity

- Determining the fire response of reinforced concrete structure is a complex nonlinear analysis problem in which the strength and stiffness of a structure as well as internal forces continually change due to restraints imposed by the structural system on free thermal expansion, shrinkage, or creep to maintain compatibility among all structural elements.

- Due to the complexity of the problem and the incremental nature of behaviour of the elements composing the structure related to the time-dependent temperature gradient, the problem can be solved numerically by integration in time domain. At each given time step, it is necessary to use an iterative approach in order to define the deformed configuration of the structure under satisfied conditions as to the equilibrium of the forces from the external loads and the inner stresses. To solve this problem the computer program FIRE is used.

Method of analysis

- The program FIRE carries out the nonlinear transient heat flow analysis (modulus FIRE-T) and nonlinear stress-strain response associated with fire (modulus FIRE-S).

- A nonlinear heat flow analysis is used to account for the temperature dependence of material properties and the thermal boundary conditions. This nonlinearity requires the use of an iterative procedure within any given time step.

- While cracks appear, or same parts of the element crush, the heat penetrates in the cross section easier, but in this study it is neglected.

- In modelling the fire response of reinforced concrete structures, it has been assumed that the heat flow is separable from the structural analysis.

- The modulus FIRE-S accounts for: dimensional changes caused by temperature differentials, changes in mechanical properties of materials with changes in temperature, degradation of sections by cracking and/or crushing and acceleration of shrinkage and creep with an increase of temperature.

- Because frames are modelled as an assemblage of members connected to joints, the basic analytical problem is to find the deformation history of the joints when external loading at the joints and temperature history within the members are specified. A direct stiffness method approach of nonlinear structural analysis is implemented in this program.

- An iterative procedure within a given time step is required to solve the problem because the structural stiffness depends on the deformed shape, on the current temperature distribution and the previous response history of the structure.

- The used analysis procedure does not account for the effects of large displacements on equilibrium equations.

Results of analysis

Centrically loaded columns exposed to fire from all sides:

- Dimensions of the cross section, intensity of the axial force (load ratio $\alpha = \sigma_c / f_c(20^\circ C)$) and the type of the aggregate (siliceous or carbonate) are significant factors affecting fire resistance of these columns. If columns are designed according to current design standards, the fire resistance can be directly defined from the diagrams on Figure 2.

- When dimensions of the cross section are small, temperature penetrates deeper after a short time period, mechanical characteristics of the concrete are reduced and the fire resistance is low.

- With the increase of the strength of concrete, the fire resistance increases in proportion to the assigned load capacity.

- When the load ratio is increased, the fire resistance is decreased, and opposite. If load ratio $\alpha = 0.3$ is decreased to $\alpha = 0.2$, the fire resistance is increased for 30%.

- If carbonate aggregate concrete is used instead of siliceous aggregate concrete, the fire resistance is higher for 30% in average. Experimental results indicate even higher effect, which leads to a conclusion that recommendations for the temperature dependant physical characteristics of carbonate aggregate concrete, given in Eurocode 2, part 1.2 are conservative, but are on the side of safety.

- The steel ratio has a negligible effect on the fire resistance of these columns. If the column 30×30cm has concrete cover thickness $a = 4$cm and the steel ratio is increased from 1% to 1.5%, but in same time the axial force is increased (the load ratio α remains constant), the reinforcement remains cold for a long time period and the effect is positive, but not more than 5% in average. If the dimensions of the cross section are larger the concrete core remains cold for a long time period and has dominant influence on the bearing capacity of the column, so the positive effect of the steel ratio increase is less.

- From the same reasons, the concrete cover thickness has a negligible effect, too.

- Because of the symmetry of the axial load and fire exposure, support conditions and height of the column have no influence on the fire resistance. The effect of axial restraint on the fire resistance is negligible. Experimental results indicated that fully restrained column did not significantly decrease its fire resistance and only columns with high slenderness ratio failed in buckling.

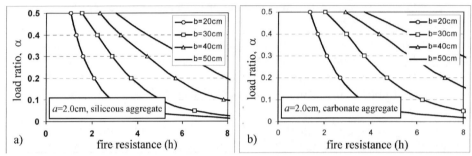

Figure 2: Effect of cross section dimensions and load ratio upon the fire resistance of centrically loaded columns exposed to fire from all sides: a)siliceous, b) carbonate aggregate concrete

Eccentrically loaded columns that are part of a wall:

- These columns have the lowest fire resistance. Dimensions of the cross section; concrete cover thickness; steel ratio; support conditions; fire scenario and intensity of the axial force and bending moment are significant factors affecting fire resistance of these columns. If the columns are designed according to current design standards, the fire resistance can be directly defined from the diagrams on Figure 3.

$$\eta = \frac{N}{N_{u,max}} \quad , \quad \beta = \frac{M}{M_u}$$

where: η-load coefficient for axial force, N-axial force before action of fire, $N_{u,max}$-ultimate axial force when the bending moment is zero, β-load coefficient for the bending moment, M-bending moment before action of fire, M_u-ultimate bending moment corresponding to N_u

- The effect of increasing the steel ratio is positive. It was not a case when the column was centrically loaded ($\beta = 0$). When the bending moment is increased and the axial force is decreased, the positive effect is more expressive.

- For optimally loaded columns ($\eta \le 0.3$) the concrete cover thickness has a positive effect in increasing the fire resistance, but not more than 5%, for columns with small dimensions, and up to 10%, for larger columns (for columns 30×30cm correction has to be made according to the values on Figure 4).

RC frame structures:

- If fire occurs in one compartment, the heated elements will tend to expand and push against the surrounding part of the construction. In turn, the unheated part of the construction exerts compressive forces on the heated elements. The compressive force acts near the exposed surface of the element and can be quite large. This force acts in a manner similar to an external prestressing force, which increases the positive moment capacity. The compressive force is generally great enough to increase the fire resistance significantly.

- The differential heating in the cross section of the elements exposed to fire results in a redistribution of moments, i.e., the negative moment increases, while the positive moments decrease (Figure 5). Generally, the redistribution that occurs is sufficient to cause yielding of the negative moment reinforcement, while the concrete at the bottom of the cross section crushes (Figure 7). The resulting decrease in positive moment means that the positive moment reinforcement can be heated to a higher temperature before failure will occur (Figure 6).

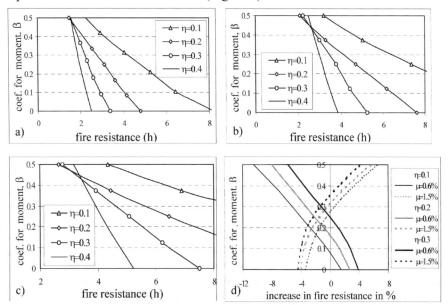

Figure 3: Effect of load intensity upon the fire resistance of ecc. loaded column:
a)30×30cm, b) 40×40cm, c) 50×50cm (concrete cover *a*=2cm),
d) effect of steel ratio upon the fire resist. of 40×40cm column (refer. column μ=1%)

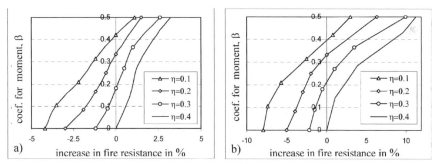

Figure 4: Increase in fire resistance of ecc. loaded columns 30×30cm, when the concrete cover thickness is increased: a) a=3cm, b) a=4cm, (referent column a=2cm)

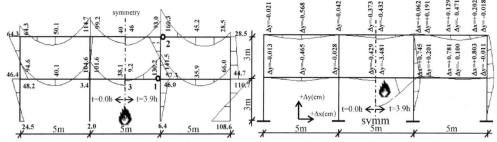

Figure 5: Bending moment redistribution when fire is in the first story middle bay

Figure 6: Deformation history of the frame when fire is in the first story middle bay

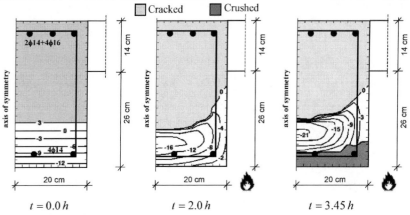

Figure 7: Time degradation profiles for beam at the interior support, when fire is in the first story exterior bay

Further developments

- Parametric studies, helping to identify important design considerations, could be easily achieved throughout implementation of this program. The time response capability of FIRE can also be used to assess potential modes of failure more realistically and to define the residual capacity of structure after attack of fire.

References

- Bazant, Z.P. & Kaplan, M.F. 1996. *Concrete at high temperature: Material properties and mathematical models*. London: Longman Group Limited, UK
- Cvetkovska, M. 2002. Nonlinear stress strain behaviour of RC elements and RC frames exposed to fire. *Phd. thesis*. Skopje: University St.Cyril and Methodius
- Cvetkovska, M & Lazarov. L. 2004. Nonlinear stress strain behavior of RC columns exposed to fire. *2nd International Conference, Lifetime-oriented Design Concepts*. Ruhr University-Bochum, Germany
- Iding, R., Bresler, B. & Nizamuddin, Z. 1977. FIRES-RC II A computer program for the fire response of structures-reinforced concrete frames. *Report No. UCG FRG 77-8*. Berkeley: University of California, USA
- Lin, T.D. Zwiers, R.I. Burg, R.G. Lie, T.T. & McGrath R.J. 1992. Fire resistance of reinforced concrete columns. *PCA Research and Development Bulletin RD101B*. Skokie, Illinois: Portland Cement Association

Contacts *(include at least one COST C12 member)*

- Assist. Prof. Meri Cvetkovska
    Address: Department of Structural Mechanics, Civil Engineering Faculty, University "St. Cyril and Methodius", Partizanski Odredi, 24, 1000 Skopje, R. Macedonia
    e-mail:cvetkovska@gf.ukim.edu.mk, Phone: +389 2 3116 066, Fax: +389 2 3117 367

COST C12 - WG2

Datasheet I.3.3.1

Prepared by: G. De Matteis, I. Langone, F.M. Mazzolani

Univ. Naples Federico II - Dept. Structural Analysis and Design

EFFECTS OF GAS EXPLOSIONS IN PRECAST R.C. STRUCTURES

Description

- A number of precast load bearing walls were developed soon after the II World War, due to the necessity to provide housing to the homeless from the war.

- Progressive collapse can be defined as a collapse of a relatively small portion of a structure due to an abnormal action, which results in the failure of a major portion of the structure. So the progressive collapse is an extensive structural failure initiated by local structural damage.

- Gas-explosion is defined as: "Fast chemical reaction of gas in air; it happens at high temperatures and high pressure and having as a result the propagation of a pressure wave" (see Figure 1).

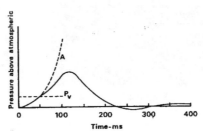

Figure 1. Pressure wave due to the gas-explosion

- The approaches for reducing the risk of progressive collapse may be categorised as in the following.

- *Event control* refers to avoiding or protecting against an incident that might lead to progressive collapse.

- *Indirect design* is used to develop resistance against progressive collapse by specifying a minimum level of strength, stiffness and ductility.

- *Direct design* considers explicitly the resistance of a structure against progressive collapse (*specific local resistance method*) and its ability to absorb possible damages (*alternate path method*).

- When *Specific local resistance method* is applied, precast bearing walls are designed to absorb the energy due to the explosion by ductile connections.

- When *alternate path method* is applied, floor slabs are designed to absorb the debris of above floor slab due to wall failure.

Field of application

- Numerous cases of progressive collapse due to the gas-explosions happened in recent years. In particular, in Great Britain, in the period 1968-1970, 122 cases of progressive collapse due to the gas-explosion developed. The collapse of Ronan Point building (London, 1968) is a dramatic example of progressive collapse. The explosion occurred on the 18th storey, with a pressure peak of 80 kN/m^2 about; as a result, an entire corner of building collapsed.

- Damages in terms of human lives, economics and socials aspects due to progressive collapse under gas-explosion could be compared to the ones caused by earthquake.
- Connection system of precast elements is very important to provide stability and robustness to the structures; several cases of "Progressive collapse" are due to the absence of well-designed connections.
- Wall-to-wall connections could be adopted to reduce the risk of progressive collapse under an unforeseeable action in both new and old buildings (see Figure 2).
- Adequate wall-to-wall connections should be adopted for buildings with precast load bearing walls, having high risk of gas explosion.
- Wall-to-wall connections could be adopted in a limited compartment of the storey only.

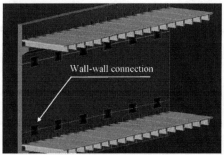

Figure 2. Typical precast R.C. panel

Technical information

- Wall-to-wall connections can be realised by using steel plates which are welded on the steel elements placed on the panel wall.
- Steel plates may be welded *in-situ* by both lateral and front welding.
- Wall-to-wall connections are on both the internal and external side of the precast panel.

Structural aspects

- Bearing capacity of the connection system depends essentially on the tensile strength of steel plate *(Figure 3)*.

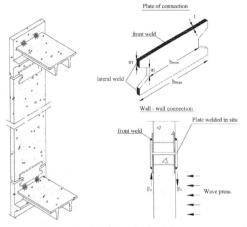

Figure 3. Wall-wall connection

- The shear forces may produce an important effects only near the wall-to-wall gap.
- Lateral welding ensure that bending deformation of steel plates are not important in comparison the axial deformation.
- Connection ductility is related to the length of plastic zone developing in the plate. As a first indication, such a length could be assumed equals to three times the wall-to-wall gap.

Research activity

- At the current state of the research, only theoretical methods and numerical investigation have been employed.
- Experimental investigations could be developed carrying out both static and dynamic tests.

Method of analysis

- In the applied theoretical model, the energy balance method is used to estimate the maximum response of precast panel under gas-explosion.
- The behaviour of steel connection has been represented by means of an elastic-plastic model.
- The behaviour of precast concrete wall has been represented by means of an elastic model.
- History pressure-time has been represented by means of a triangular model with time rise to the peak pressure equals to zero.
- Impact on the lower storeys is studied considering that the concrete slabs have a rigid-plastic behaviour. Even when a storey collapse is accepted, then the collapse of this storey must not be extended to either the whole building or a part of it.
- To avoid storey by storey collapse, the floor slab should have a sufficient robustness to absorb the kinetic energy transmitted by the upper collapsed part.
- In the applied numerical model, the response of the precast panel is represented in a Pressure-Impulse diagram considering different ductility levels.
- The combination of pressure and impulse that fall in the left part and below of the obtained curve are representative of safe conditions, while those belonging to right part and above the obtained curve will produce the failure of the panel (see Figure 4).
- The safety coefficient to be adopted on the design process to ensure that the floor slab does not collapse under debris impact is related to partial safety factors applied adopted for normal and abnormal actions.

Results of analysis

- Precast walls can be designed as "key element" under a gas explosion by applying adequate wall-to-wall connections.
- Otherwise, if an higher damage level (one storey collapse) is accepted, floor slabs have to possess sufficient "robustness" so to avoid progressive collapse in the lower part of the building.

- When considering a three support beam scheme for the floor slab under an impact due to the upper storey, the minimum robustness of the slab necessary to avoid progressive collapse is obtained by adopting a safety coefficient factor for the slab equal to about 4.

Pressure-Impulse diagram for damage to one way slab *(ductility connection $\mu = \varphi_{pl} / \varphi_{el} = 1.5$)*

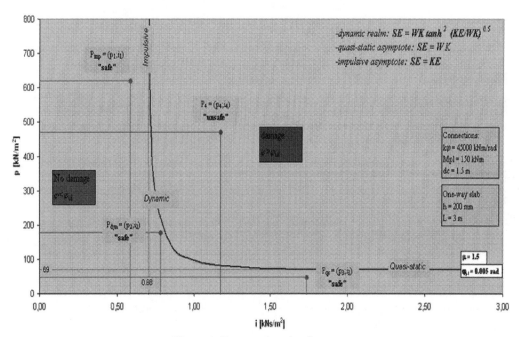

Figure 4. *Pressure-Impulse diagram*

Further developments

- Experimental analysis based on dynamic and static loads applied should be carried out.

- The Pressure-Impulse diagram that have been proposed represent practical tools for designing a precast panel under a gas explosion.

- Obtained results should be checked and extended by means FEM models, which could be applied for a precise prediction of precast wall response under gas explosion. The development of such model is now under progress.

References

- AD131 *Structural Integrity – Tying to BS 5950: Part 1, New Steel Construction*, Feb. 1983, p. 29-30.

- AD063 *Accidental damage - tying, Steel Construction Today*, March 1992, p. 58-59.

- AD104 BS 5950 – *Tying forces: Part 1, Steel Construction Today*, March 1992, p. 88.

- Armer G.S.T.: *A general review of recent BRE research into behaviour of panel structures*, NR B 485/77, BRE 1977, Garston, p. 12.

- Armer G.S.T.: *Abnormal dynamic loading of a multi-storey panel structure*, NR B 484/77, BRE 1977, Garston, p. 15.

- Baker W. E., Spivey K. H. and Baker Q.A., *Energy-Balance Methods for Structural Response and Damage Analysis Under Transient Loads*, Structural Design for Hazardous Loads 229-237, E and FN Spon, London, 1992.

- Corley W. G. and Oesterle R. G., *Dynamic Analysis to Determine Source of Blast Damage*, Abnormal Loading on Structures, E and FN Spon, London 2000.

- Craig R. R., *Structural Dynamics*, John Wiles and Sons, New York, 1982.

- Elliot C. L., Mays G. C. and Smith P. D., *The Protection of Buildings Against Terrorism and Disorder*, Engrs Structs and Bldgs 287-350, 1992.

- Elliot K. S., *Precast Concrete Structures*, Butterworth – Heinemann, Oxford, 2002.

- Esper P. and Hadden D., *Dynamic Response of Buildings and Floor Slabs to Blast Loading*, Abnormal Loading on Structures, E and FN Spon, London 2000.

- Eurocode 3, Part 1-8, Design of steel structures – *Design of joints*, 30 April 2002.

- Eurocode 1, Part 1-1, General Actions, *Imposed loads for buildings*, April 2002.

- Eurocode 1, Part 1-7, General Actions - *Accidental actions due to impact and explosions*, Stage 34 Draft, 2002.

- Kumar S.: *Lateral loading tests on an 18-storey large panel model structure*, NR B 486/77, BRE 1977, Garston, p. 12.

- Langone I., *Thesis: "Structural Integrity of Buildings with precast load bearing walls in case of explosion"*, University Federico II of Naples, Naples, 2003.

- Mainstone R. J., *The Hazard of Internal Blast in Buildings*, Building Research Establishment, Garston, Watford, 1972.

- Mainstone R. J., *The Hazard of Explosion, Impact, and Other Random Loadings on Tall Buildings*, Building Research Establishment, Garston, Watford, 1974.

- Mainstone R. J., *The Response if Buildings to Accidental Explosions*, Building Research Establishment, Garston, Watford, 1976.

- Moore D.B.: *The UK and European Regulations for Accidental Actions*, BRE, Garston, 2003.

- Owens G.W., Moore D.B.: *The robustness of simple connections, The Structural Engineer*, Vol. 70., No.3, Feb. 1992.

- Smith P. D. and Hetherington J. G., *Blast and Ballistic Loading of Structures*, Butterworth – Heinemann, Oxford, 1994.

- Waddell J. J., *Precast Concrete: Handling and Erection*, American Concrete Institute, Detroit, 1974.

Contacts
- Prof. Federico. M. Mazzolani
    Address: Department of Structural Analysis and Design, University of Naples 'Federico II', P.le Tecchio, 80, Naples, Italy.
        e-mail:fmm@unina.it; Ph.: (+39) 0817682443; Fax: (+39) 0815934792

Datasheet I.3.4.1

Prepared by: A.C.W.M. Vrouwenvelder[1], A.M. Gresnigt[2]

[1]Delft University of Technology / TNO Building and Construction Research.
[2]Delft University of Technology.

ACCIDENTAL ACTIONS IN BUILDINGS

Description

- Accidental actions such as explosion and collision due to traffic may cause collapse of a part of a building or even complete collapse of a building.
- Buildings should be designed such that accidental actions to a reasonable degree do not lead to structural catastrophes.

Field of application

- Buildings to be designed to limit the effect of accidental actions such as explosions and impact by road or rail traffic or fork lift trucks.
- Often relatively simple measures may provide sufficiently robustness to prevent disproportionate damage. In traditional design standards usually little attention is paid to accidental actions and the need for robustness.

Codification

- General principles for classification of actions on structures, including accidental actions and their modelling in verification of structural reliability, are introduced in EN 1990 Basis of Design. In particular EN 1990 defines the various design values and combination rules to be used in the design calculations. A detailed description of individual actions is then given in various parts of Eurocode 1, EN 1991.
- Part 1-7 of EN 1991 (stage 34) covers accidental actions and gives rules and values for impact loads due to road, train and ship traffic and loads due to internal explosions. The impact loads are mainly concerned with bridges, but also impact on buildings is covered. Explosions concern explosions in buildings and in tunnels.
- It should be kept in mind that the loads in the main text are rather conventional. More advanced models are presented in the annex C: "Dynamic design for impact". Apart from design values and other detailed information for the loads mentioned above, the document also gives guidelines how to handle accidental loads in general. In many cases structural measures alone cannot be considered as very efficient.
- In annex A: "Design for consequences of localised failure in a building structure from an unspecified cause" deemed to satisfy rules are given to guarantee some minimum robustness against identified as well as unidentified causes of structural damage.

Basis of applications

- Design for accidental situations in the Eurocodes is implemented to avoid structural catastrophes. As a consequence, design for accidental design situations needs to be included mainly for structures for which a collapse may cause particularly large consequences in terms of injury to humans, damage to the environment or economic losses for the society. In practice this means that Eurocode 1, Part 1-7 accepts the principle of safety differentiation. In fact three different safety categories, called consequences classes, are distinguished:

- o CC1 – consequences class 1 : Low consequences of failure
- o CC2 – consequences class 2 : Medium consequences of failure
- o CC3 – consequences class 3 : High consequences of failure

For these categories different methods of analysis and different levels of reliability are accepted.

- Another part of the philosophy is that local failure due to accidental actions is accepted provided that it does not lead to overall failure. This distinguishing between component failure and system failure is mandatory to allow a systematic discrimination between design for variable actions and accidental actions.

- A result of the acceptance of local failure (which in most cases may be identified as a component failure) provided that it does not lead to a system failure, is that redundancy and non-linear effects both regarding material behaviour and geometry play a much larger role in design to mitigate accidental actions than variable actions. The same is true for a design, which allows large energy absorption.

- In order to reduce the risk involved in accidental types of load one might, as basic strategy, consider probability reducing as well as consequence reducing measures, including contingency plans in the event of an accident. Risk reducing measures should be given high priority in design for accidental actions, and also be taken into account in design. Design with respect to accidental actions may therefore pursue one or more as appropriate of the following strategies, which may be mixed in the same design:

1. *preventing the action* from occurring (e.g. in the case of bridges, by providing adequate clearances between the vehicles and the structure) or reducing the probability and/or magnitude of the action to an acceptable level through the structural design process (e.g. providing sacrificial venting components with a low mass and strength to reduce the effect of explosions). Note: The effect of preventing actions may be limited; it is dependent upon factors which, over the life span of the structure, are commonly outside the control of the structural design process. Preventive measures often involve periodic inspection and maintenance during the life of the structure.

2. *protecting the structure* against the effects of an accidental action by reducing the actual loads on the structure (e.g. protective bollards or safety barriers).

3. *ensuring that the structure has sufficiently robustne*ss by adopting one or more of the following approaches :
   - o *designing key elements* of the structure on which its stability depends, to be of enhanced strength so as to raise the probability of their survival following an accidental action.
   - o *designing structural members and connections to have sufficient ductility* capable of absorbing significant strain energy without rupture.
   - o *incorporating sufficient redundancy* in the structure so as to facilitate the transfer of actions to alternative load paths following an accidental event. It means designing in such a way that neither the whole structure nor an important part thereof will collapse if a local failure (single element failure) should occur.

4. *applying prescriptive design/detailing rules* which provide in normal circumstances an acceptably robust structure (e.g. tri-orthogonal tying for resistance to explosions, or minimum level of ductility of structural elements subject to impact). For prescriptive rules Part 1-7 refers to relevant EN 1992 to EN 1999.

- The design philosophy necessitates that accidental actions are treated in a special manner with respect to load factors and load combinations. Partial load factors to be applied in analysis according to the strategy to design no. 3 are defined in EN 1990, Eurocode: Basis of Structural Design, to be 1,0 for all loads (permanent, variable and accidental) with the following qualification in: "Combinations for accidental design situations either involve an explicit accidental action A (e.g. fire or impact) or refer to a situation after an accidental event ($A = 0$)".

- After an accidental event the structure will normally not have the required strength in persistent and transient design situations and will have to be strengthened for a possible continued application. In temporary phases there may be reasons for a relaxation of the requirements e.g. by allowing wind or wave loads for shorter return periods to be applied in the analysis after an accidental event.

Impact actions from vehicles

- In the case of hard impact, design values for the horizontal actions due to impact on vertical structural elements (e.g. columns, walls) in the vicinity of various types of internal or external roads may be obtained from Table 1.

- The forces F_{dx} and F_{dy} denote respectively the forces in the driving direction and perpendicular to it. There is no need to consider them simultaneously.

- For impact from lorries on the supporting sub-structure the resulting collision force F should be applied at any height between 0,5 - 1,5 m above the level of the carriageway or greater where a protective barrier is provided. The recommended application area is 0,5 m (height) by 1,50 m (width) or the member width, whichever is the smaller.

- In addition to the values in Table 1, Eurocode 1, Part 1-7 specifies more advanced models for non-linear and dynamic analysis in an informative annex.

Table 1: Recommended minimum equivalent static design forces due to vehicular impact on members supporting structures over or adjacent to roadways.

Category	Minimum Force $F_{d,x}$ [kN]	Minimum Force $F_{d,y}$ [kN]
Motorways and country national roads	1000	500
Country roads in rural area	750	375
Roads in urban area	500	250
Courtyards and parking garages with access - to cars - to lorries (vehicle weight >3,5 ton)	50 150	25 75

Accidental action due to natural gas explosion

- Key elements of a structure should be designed to withstand the effects of an internal natural gas explosion, using a nominal equivalent static pressure is given by:

$$p_{Ed} = 3 + p_{stat} \qquad (1)$$

or

$$p_{Ed} = 3 + 0,5 p_{stat} + 0,04/(A_v/V)^2 \qquad (2)$$

whichever is the greater, where p_{stat} is the uniformly distributed static pressure at which venting components will fail [kN/m^2], A_v is the area of venting components [m^2] and V is the volume of the room [m^3].

- The explosive pressure acts effectively simultaneously on all of the bounding surfaces of the room. The expressions are valid for rooms up to a volume of 1000 m^3 and venting areas over volume ratios of $0,05\,\mathrm{m}^{-1} \leq A_v/V \leq 0,15\,\mathrm{m}^{-1}$

- An important issue is further raised in Part 1-7, Section 5.3 Principles for design, Clause (4). It states that the estimated peak pressures may be higher than the values given in Annex D but these can be considered in the context of a maximum load duration of 0,2 s and plastic ductile material behaviour (assuming appropriate detailing of connections to ensure ductile behaviour). The point is that in reality the peak will generally be larger, but the duration is shorter. So combining the loads from the above equations with a duration of 0,2 s seems to be a reasonable approximation.

Design example for an explosion in a living compartment

- Consider a living compartment in a multi-storey flat building (Figure 1). Let the floor dimensions of the compartment be 8 x 14 m and let the height be 3 m. The two small walls (the facades) are made of glass and other light materials and can be considered as venting area. These walls have no load bearing function in the structure. The two long walls are concrete walls; these walls are responsible for carrying down the vertical loads as well as the lateral stability of the structure. This means that the volume V and the area of venting components A_v for this case are given by:

$$A_v = 2 \cdot 8 \cdot 3 = 48\,\mathrm{m}^2$$

$$V = 3 \cdot 8 \cdot 14 = 336\,\mathrm{m}^3$$

So the parameter A_v/V can be calculated as:

$$A_v/V = 48/336 = 0,143\,\mathrm{m}^{-1}$$

- As V is less then 1000 m^3 and A_v/V is well within the limits of 0,05 m^{-1} and 0,15 m^{-1} it is allowed to use the loads given in the code. The collapse pressure of the venting panels p_{stat} is estimated as 3 kN/m^2. Note that these panels normally can resist the design wind load of 1,5 kN/m^2. The equivalent static pressure for the internal natural gas explosion is given by:

$$p_{Ed} = 3 + p_{stat} = 3 + 3 = 6\ \mathrm{kN/m}^2$$

or

$$p_{Ed} = 3 + p_{stat}/2 + 0,04/(A_v/V)^2 = 3 + 1,5 + 0,04/0,143^2 = 3 + 1,5 + 2,0 = 6,5\,\mathrm{kN/m}^2$$

This means that we have to deal with the latter.

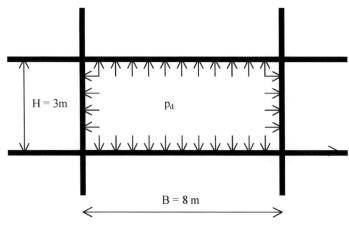

Figure 1: Load arrangement for the explosion load.

- The load arrangement for the explosion pressures is presented in Figure 1. According to Eurocode EN 1990, Basis of Design, these pressures have to be combined with the self-weight of the structure and the quasi-permanent values of the variable loads. Let us consider the design consequences for the various structural elements.

Bottom floor

- Let us start with the bottom floor of the compartment. Let the self-weight be 3 kN/m² and the live load 2 kN/m². This means that the design load for the explosion is given by:

$$p_{da} = p_{SW} + p_{Ed} + \psi_{1LL} \; p_{LL} = 3,0 + 6,5 + 0,5 \cdot 2,0 = 10,5 \text{ kN/m}^2$$

The design for normal conditions is given by:

$$p_d = \gamma_G \, \xi \, p_{SW} + \gamma_Q p_{LL} = 0,85 \cdot 1,35 \cdot 3,0 + 1,5 \cdot 2,0 = 6,4 \text{ kN/m}^2$$

- We should keep in mind that for accidental actions there is no need to use a partial factor on the resistance side. So for comparison we could increase the design load for normal conditions by a factor of 1,2. The result can be conceived as the resistance of the structure against accidental loads, if it designed for normal loads only:

$$p_{Rd} = 1,2 \cdot 6,4 = 7,7 \text{ kN/m}^2$$

So a floor designed for normal conditions only should be about 30 percent too light.

- It is now time to remember the clause in Part 1-7, Section 5.3, mentioned before. If we take into account the increase in short duration of the load we may increase the load bearing capacity by a factor φ_d given by:

$$\varphi_d = 1 + \sqrt{\frac{p_{SW}}{p_{Rd}}} \sqrt{\frac{2u_{max}}{g(\Delta t)^2}} \qquad (3)$$

where $\Delta t = 0,2$ s is the load duration, $g = 10$ m/s² is the acceleration of the gravity field and u_{max} is the design value for the mid-span deflection at collapse. The background of equation (3) is explained at the end of this section.

- The value of factor φ_d of course depends on the ductility properties of the floor slab and in particular of the connections with the rest of the structure. It is beyond the scope of this datasheet to discuss the details of that assessment, but assume that u_{max} = 0,20 m is considered as being a defendable design value. In that case the resistance against explosion loading can be assessed as:

$$p_{REd} = \varphi_d p_{Rd} = \left(1 + \sqrt{\frac{3}{7,7}}\sqrt{\frac{2 \cdot 0,20}{10 \ (0,2)^2}}\right) \cdot 7,7 = 12,5 \ \text{kN/m}^2$$

Therefore, the bottom floor system is safe in this case.

Upper floor

▪ Let us next consider the upper floor. Note that the upper floor for one explosion could be the bottom floor for the next one. The design load for the explosion in that case is given by (upward value positive!):

$$p_{da} = p_{SW} + p_E + \gamma_Q \psi \ p_{LL} = -3,0 + 6,5 + 0 = 3,5 \ \text{kN/m}^2$$

▪ So the load is only half the load on the bottom floor, but will give larger problems anyway. The point is that the load is in the opposite direction of the normal dead and live load. This means that the normal resistance may simply be close to zero. What we need is top reinforcement in the field and bottom reinforcement above the supports. The required resistance can be found by solving p_{Rd} from:

$$\varphi_d p_{Rd} = \left(1 + \sqrt{\frac{p_{SW}}{p_{Rd}}}\sqrt{\frac{2u_{max}}{g(\Delta t)^2}}\right) \cdot p_{Rd} = 3,5 \ \text{kN/m}^2$$

Using again $p_{SW} = 3 \ \text{kN/m}^2$, $\Delta t = 0,2\text{s}$, $g = 10 \ \text{m/s}^2$ we arrive at $p_{Rd} = 1,5 \ \text{kN/m}^2$. This would require about 25 percent of the reinforcement for normal conditions on the opposite side.

▪ An important additional point to consider is the reaction force at the support. Note that the floor could be lifted from its supports, especially in the upper two stories of the building where the normal forces in the walls are small. In this respect edge walls are even more vulnerable. The uplifting may change the static system for one thing and lead to different load effects, but it may also lead to freestanding walls. We will come back to that in the next paragraph. If the floor to wall connection can resist the lift force, one should make sure that the also the wall itself is designed for it.

Walls

▪ Finally we have to consider the walls. Assume the wall to be clamped in on both sides. The bending moment in the wall is then given by:

$$m = 1/16 pH^2 = 1/16 \cdot 6,5 \cdot 3^2 = 3,7 \ \text{kNm/m}$$

▪ If there is no normal force acting in the wall this would require a central reinforcement of about 0,1 percent With the reinforcement assumed to have a yield strength $f_y = 300 \ \text{N/mm}^2 = 300000 \ \text{kN/m}^2$, the corresponding bending capacity can be estimated as:

$$m_p = \omega d \cdot 0,4d \cdot f_y = 0,001 \cdot 0,2 \cdot 0,4 \cdot 0,2 \cdot 300000 = 4,8 \ \text{kNm/m}$$

▪ Normally, of course normal forces are present. Leaving detailed calculations as being out of the scope of this document, the following scheme looks realistic. If the explosion is on a top floor apartment and there is an adequate connection between roof slab and top wall, we will have a tensile force in the wall, requiring some additional reinforcement. In our example the tensile force would be for a middle wall:

$$(p_E - 2p_{SW})B/2 = (6,5 - 2 \cdot 3) \cdot 8/2 = 2 \ \text{kN/m}$$

and for an edge wall:

$$(p_E - p_{SW})B/2 = (6.5-3) \cdot 8/2 = 14 \text{ kN/m}$$

- If the explosion is on the one but top story, we usually have no resulting axial force and the above mentioned reinforcement will do. Going further down, there will probably be a resulting axial compression force and the reinforcement could be diminished ore even left out completely.

Background of equation (3)

- The background of equation (3) is as follows: Consider a spring mass system with a mass m and a rigid plastic spring with yield value F_y. Let the system be loaded by a load $F > F_y$ during a period of time Δt. The velocity of the mass achieved during this time interval is equal to:

$$v = (F - F_y) \Delta t / m$$

The corresponding kinetic energy of the mass is then equal to:

$$E = 0.5 \, m v^2 = 0.5 (F - F_y)^2 \, \Delta t^2 / m$$

By equating this energy to the plastic energy dissipation, that is we put $E = F_y \, \Delta u$, we may find the increase in plastic deformation Δu.

$$\Delta u = 0.5 (F - F_y)^2 \, \Delta t^2 / m F_y$$

As $mg = F_{SW}$ we may also write:

$$\Delta u = 0.5 (F - F_y)^2 \, g \, \Delta t^2 / F_{SW} \, F_y$$

Finally we may rewrite this formula in the following way:

$$F = F_y \left(1 + \sqrt{\frac{F_{SW}}{F_{Rd}}} \sqrt{\frac{2u_{max}}{g(\Delta t)^2}} \right) = F_y \, \varphi_d$$

For the floor slab the forces F may be replaced by the distributed loads p.

Robustness of buildings

- The rules drafted for Annex A of EN1991-1-7 were developed from the UK Codes of Practice and regulatory requirements introduced in the early 70s following the partial collapse of a block of flats in east London caused by a gas explosion. The rules have changed little over the intervening years. They aim to provide a minimum level of building robustness as a means of safeguarding buildings against a disproportionate extent of collapse following local damage being sustained from an accidental event.

- The rules have proved satisfactory over the past 3 decades. Their efficacy was dramatically demonstrated during the IRA bomb attacks that occurred in the City of London in 1992 and 1993. Although the rules were not intended to safeguard buildings against terrorist attack, the damage sustained by those buildings close to the seat of the explosions that were designed to meet the regulatory requirement relating to disproportionate collapse was found to be far less compared with other buildings that were subjected to a similar level of abuse.

- Annex A also provides an example of how the consequences of building failure may be classified into "Consequences classes" corresponding to reliability levels of robustness. These classes are summarized in Table 2.

Member states are invited to provide rules for a national classification system.

Table 2: Summary of Categorization of Consequences Classes.

Consequences Class	Example structures
Class 1	low rise buildings where only few people are present
Class 2, lower risk group	most buildings up to 4 stories
Class 3, upper risk group	most buildings up to 15 stories
Class 4	high rise building, grand stands etc.

Design rules for robustness

Recommended strategies

- Adoption of the recommended strategies should ensure that the building will have an acceptable level of robustness to sustain localised failure without a disproportionate level of collapse.
- For buildings in Consequences Class 1, provided the building has been designed and constructed in accordance with the rules given in EN1992 to EN1999 for satisfying stability in normal use, no further specific consideration is necessary with regard to accidental actions from unidentified causes.
- For buildings in Consequences Class 2 (Lower Group) it is recommended to provide effective horizontal ties, or effective anchorage of suspended floors to walls for framed and load-bearing wall construction.
- For buildings in Consequences Class 2 (Upper Group) it is recommended:
  - to provide effective horizontal ties respectively for framed and load-bearing wall construction, together with effective vertical ties in all supporting columns and walls, or alternatively,
  - to ensure that upon the notional removal of each supporting column and each beam supporting a column, or any nominal section of load-bearing wall (one at a time in each storey of the building) that the building remains stable and that any local damage does not exceed a certain limit.

    The limit of admissible local damage may be different for each type of building. The recommended value is 15 % of the floor, or 200 m^2, whichever is greater, in each of two adjacent storeys. Where the notional removal of such columns and sections of walls would result in an extent of damage in excess of the above limit, or other such limit specified in the National Annex, then such elements should be designed as a "key element" for an accidental design action $A_d = 34$ kN/m^2.

    In the case of buildings of load-bearing wall construction, the notional removal of a section of wall, one at a time, is likely to be the most practical strategy to adopt.

- For buildings in Consequences Class 3 a systematic risk assessment of the building should be undertaken taking into account all the normal hazards that may reasonably be foreseen, together with any abnormal hazards. The depth of the risk analysis is up to the local authorities.
- Below the rules are summarised. A distinction is made between framed structures and load-bearing wall construction.

Class 2, Lower Group, Framed Structures:

▪ Effective horizontal ties should be provided around the perimeter of each floor (and roof) and internally in two right angle directions to tie the columns to the structure of the building (Figure 2).

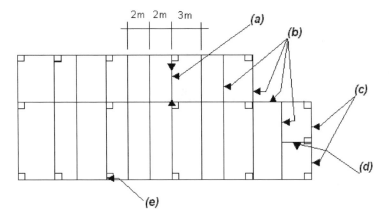

(a) 6 m span beam as internal tie

(b) All beams designed to act as ties

(c) Perimeter ties

(d) Tie anchoring column

(e) Edge column

Figure 2: Example of effective horizontal tying of a 6 storey framed office building.

▪ The ties should be continuous and be arranged as closely as practicable to the edges of floors and lines of columns and walls. At least 30 % of the ties should be located within the close vicinity of the lines of columns and walls.

▪ Each continuous tie, including its end connections, should be capable of sustaining a design tensile load of T_i for the accidental limit state in the case of internal ties, and T_p, in the case of perimeter ties, equal to the following values:

  o For internal ties: $T_i = 0.8(g_k + \psi q_k)s L$ or 75kN, whichever the greater.

  o For perimeter ties: $T_p = 0.4(g_k + \psi q_k)s L$ or 75kN, whichever the greater.

where g_k and q_k are the characteristic values in [kN/m^2] of the self weight and imposed load respectively; ψ is the combination factor, s [m] is the spacing of ties and L [m] is the span in the direction of the tie.

Class 2, Lower Group, Load-bearing wall construction:

▪ A cellular form of construction should be adopted to facilitate the interaction of all components including an appropriate means of anchoring the floor to the walls.

Class 2, Upper Group, Framed Structures:

▪ Effective horizontal ties should be provided as for Class 2 Lower Group above.

▪ Effective vertical ties should be provided. Each column and wall should be tied continuously from the foundations to the roof level. In the case of load-bearing wall construction the vertical ties may be considered effective if:

  o In the case of masonry walls their thickness is at least 150 mm thick and if they have a minimum compressive strength of 5 N/mm^2 in accordance with EN1996-1-1.

  o The clear height of the wall, H, measured in metres between faces of floors or roof does not exceed 20t, where t is the thickness of the wall in metres.

  o The vertical tie force T is :

$$T = \frac{34\,A}{8000}\left(\frac{H}{t}\right)^2 \; \text{N, or 100 kN/m of wall, whichever is the greater.}$$

where A is the cross-sectional area in mm^2 of the wall measured on the plan, excluding the non- load bearing leaf of a cavity wall.

o The vertical ties are grouped at 5 m maximum centres along the wall and occur no greater than 2,5 m from an unrestrained end of wall.

Class 2, Upper Group, Load-bearing wall construction:

▪ Provide continuous effective horizontal ties in the floors. These should be internal ties distributed throughout the floors in both orthogonal directions and peripheral ties extending around the perimeter of the floor slabs within a 1,2 m width of slab.

▪ Rules for horizontal ties similar to those for framed buildings except that the design tensile load in the ties are as follows:

o For internal ties: $\quad T_i = \dfrac{F_t(g_k + \psi\, q_k)}{7,5}\dfrac{z}{5}$ kN/m but $\geq F_t$

o For perimeter ties: $\quad T_p = F_t$

where $F_t = (20 + 4n_s)$ kN/m with a maximum of 60 kN/m; n_s represents the number of storeys; g, q and ψ have the same meaning as before, and z is the lesser of the greatest distance in metres in the direction of the tie, between the centres of the columns or other vertical load-bearing members whether this distance is spanned by a single slab or by a system of beams and slabs, or 5 times the clear storey height H.

Examples

Class 2, Upper Group, Framed Structure

▪ Consider a 5 storey building with story height $H = 3,6$ m. Let the span be $L = 7,2$ m and the spacing of the ties $s = 5,5$ m. The loads are $q_k = g_k = 4$ kN/m^2 and $\psi = 1,0$. In that case the required internal tie force may be calculated as:

$$T_i = 0,8 \cdot (4 + 4) \cdot 5,5 \cdot 7,2 = 253 \; \text{kN} > 75 \text{ kN}$$

For steel grade FeB 500 this force corresponds to a steel area $A = 506$ mm^2 or 2 ø18 mm. The perimeter tie is simply half the value. Note that in continuous beams this amount of reinforcement usually is already present as upper reinforcement anyway.

▪ For the vertical tying force we find:

$$T_v = (4 + 4) \cdot 5,5 \cdot 7,2 = 288 \; \text{kN/column}$$

With FeB 500 this correspondents to $A = 576$ mm^2 or 3 ø18 mm.

Class 2, Upper Group, Framed Structure

▪ Consider a 5 storey building with story height $H = 3,6$ m. Let the span be $L = 7,2$ m and the spacing of the ties $s = 5,5$ m. The loads are $q_k = g_k = 4$ kN/m^2 and $\psi = 1,0$. In that case the required internal tie force may be calculated as:

$$T_i = 0,8 \cdot (4 + 4) \cdot 5,5 \cdot 7,2 = 253 \; \text{kN} > 75 \text{ kN}$$

For steel grade FeB 500 this force corresponds to a steel area A = 506 mm2 or 2 ø18 mm. The perimeter tie is simply half the value. Note that in continuous beams this amount of reinforcement usually is already present as upper reinforcement anyway.

- For the vertical tying force we find:

$$T_v = (4+4) \cdot 5{,}5 \cdot 7{,}2 = 288 \quad \text{kN/column}$$

With FeB 500 this correspondents to $A = 576$ mm^2 or 3 ø18 mm.

Class 2, Upper Group, Load Bearing Wall

- For the same starting points we get $F_b = \min(60, 40) = 40$ and $z = 5H = 18$m and from there for the internal and perimeter tie forces:

$$T_i = 60 \, \frac{4+4}{7{,}5} \, \frac{18}{5} = 230 \quad \text{kN/m}$$

$$T_p = 40 \quad \text{kN/m}$$

- The vertical tying force is given by:

$$T_v = \frac{34 \cdot 0{,}2}{8} \left(\frac{3{,}6}{0{,}2} \right)^2 = 300 \quad \text{kN/m}$$

For many countries this may lead to more reinforcement then usual for this type of structural elements.

Acknowledgement

- This datasheet is based on work carried out in the Leonardo da Vinci Pilot Project CZ/02/B/F/PP-134007.

References

- CIB, 1992. *Actions on structures impact*, CIB Report, Publication 167, Rotterdam.

- Damgaard Larsen, Ole, 1993. *Ship Collision with Bridges, The Interaction between Vessel Traffic and Bridge Structures*, Structural Engineering Documents, IABSE, Zurich.

- ENV 1991-2-7, 1998, Eurocode 1: Basis of design and actions on structures – Part 2-7: Actions on structures - *Accidental actions due to impact and explosions*, CEN, Brussels. Will be withdrawn after completion of prEN 1991-1-7.

- Project Team prEN 1991-2-7, 1997, *Background document to ENV 1991-2-7: Accidental actions due to impact and explosions*, TNO, Delft.

- prEN 1991-1-7, Draft stage 34, May 2004, Eurocode 1 - Actions on structures - *Accidental actions due to impact and explosions*, CEN, Brussels.

- Stufib Studiecel, 2004. *Incasseringsvermogen van bouwconstructies* (*Robustness of building structures*, in Dutch), STUFIB, Delft.

- Vrouwenvelder, A.C.W.M., 2000. *Stochastic modelling of extreme action events in structural engineering*, Probabilistic Engineering Mechanics 15, pp. 109-117.

- Vrouwenvelder, A.C.W.M., 2004. *Accidental actions on Buildings*, Leonardo da Vinci Pilot Project CZ/02/B/F/PP-134007, Delft.

- Vrouwenvelder, A.C.W.M., 2004. *Accidental actions on Bridges*, Leonardo da Vinci Pilot Project CZ/02/B/F/PP-134007, Delft.

- Institution of Structural Engineers. *Safety in tall buildings and other buildings with large occupancy*, London.

Contacts

- Prof. ir. A.C.W.M. (Ton) Vrouwenvelder

    Address: TNO Building and Construction Research, P.O. Box 49, 2600 AA Delft, The Netherlands.

    e-mail: A.Vrouwenvelder@bouw.tno.nl

    Phone: +31 15 276 3230 - Fax: +31 15 276 3018

- Associate Prof. ir. A.M. (Nol) Gresnigt

    Address: Delft University of Technology, Faculty of Civil Engineering and Geosciences, P.O. Box 5048, 2600 GA Delft, The Netherlands.

    e-mail: A.M.Gresnigt@citg.tudelft.nl

    Phone: +31 15 278 3382 - Fax: +31 15 278 3173

COST C12 - WG2

Datasheet I.4.1.1
Prepared by: L. Ślęczka
Rzeszów University of Technology.

ANALYTICAL ASSESSMENT OF LOW CYCLE FATIGUE

Description

- Fatigue is the loss of the strength (or other important property of structural element) caused by repeated stress fluctuations.

- Low cycle fatigue (LCF) is one of fatigue mode of steel elements or connections, which are repeatedly stressed into the plastic region, when plastic strains control and determine their behaviour (Figure 1).

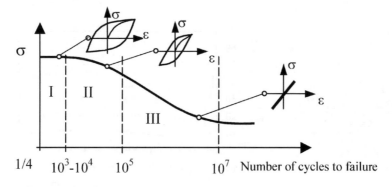

Figure 1. Complete Wöhler's curve with quasi-static (I), low cycle fatigue (II) and high cycle fatigue (III) areas (Łubiński et al, 1995).

Field of application

- Up till now LCF problems in civil engineering structures are connected with seismic actions on structures, pressure vessels and sometimes with thermal cycling.

- As a consequence of continuous progress in field of materials and in methods of analysis, modern structures become much more lightweight and have longer spans than in the past. Because the major part of actions on structures are variable actions, so problems with fatigue, fracture and dynamic behaviour are still arising and start to be important in other types of steel structures.

- Although in steel structures designed in elastic range the nominal strain value does not exceed the yield strain, existence of macro-geometry effects, structural discontinuities and local notches produce plastic areas (Figure 2).

- Nowadays it is also observed general tendency to application of semi-rigid joints and more and more frequent utilization of plastic design. In case of loading different from once acting monotonic loading, cyclic deformation phenomena can occur in areas with plastic strain. Such phenomena can lead to initiation of fatigue cracks or to deterioration in stiffness or strength even after small number of load cycles.

Technical information

- Connections in steel structures are particularly exposed to fatigue cracking. Significant geometric effects as rapid change of cross-sectional area, disturbances of internal forces caused by arrangement of bolts or welds (Figure 2), existence of local notches and secondary bending from eccentricities) produce stress concentrations and elastic–plastic behaviour of connections.

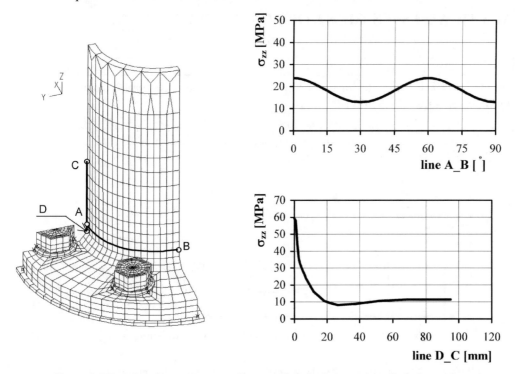

Figure 2. Examples of stress concentration zones in bolted connection of tubular section

- Similar stress concentrations occur in steel elements and they are caused by structural discontinuities as openings, curvature of elements, shear lag effects and bending stresses in shells.

Structural aspects

- There are different definitions of a failure criterion in term of LCF. According to (Calado et al, 1998) failure criterion can be based on deterioration of strength, stiffness and energy dissipation capacity, but failure criterion in local strain approach is based on initiation of first crack.
- In general, existence of fatigue cracks in steel structures is not allowed. LCF procedures give relationship between strain range (or amplitude) and number of cycles to crack initiation.
- LCF procedures based on Manson-Coffin's or Langer's equations have the range of validity from 0.5 to even 10^6 cycles, so they are perfectly suitable for assessing fatigue life of elements and connections in the range of quasi-static, LCF and part of high cycle fatigue (Figure 1).

Research activity

- Performed analysis and tests show that initiation of fatigue cracks can occur after small number of load cycles, both in elements and connections of steel structures, e.g. (Seeger, Zacher, 1994), (Zandonini, Bursi, 2002), (Ślęczka, 2004).

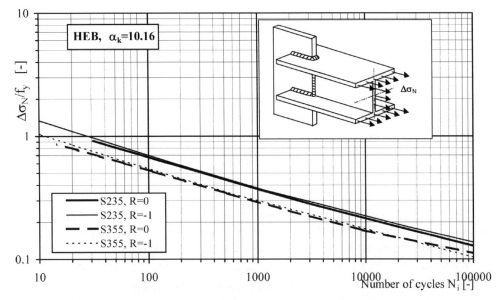

Figure 3. Fatigue life of HEB connections with gusset plate

- Analysis show that the mechanism of low cycle fatigue is connected not only with fluctuating loading but also with the influence of stress concentrations on that mechanism.

Method of analysis

- One of the low cycle fatigue life assessment method is so called "Local Strain Approach", based on local strain state in notch root. This method enables to represent the most strained region of structure by small specimen made of the same material, and submitted to the same strain history that occurs at analysed region, usually notch root (Figure 4) (Kocańda, Kocańda, 1989).

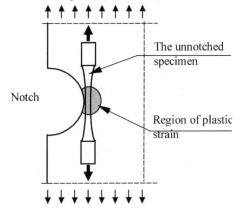

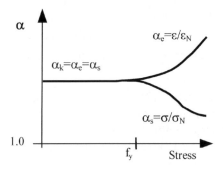

Figure 4. Smoothed specimen for simulating the cyclic behaviour of material in concentrations zone

Figure 5. Elastic stress concentrations factor α_k and elastic-plastic stress α_s and strain α_e concentrations factor

- Equations giving the relationships between the local strain amplitude ε_a and number of cycles to crack initiation N were proposed by many authors. The most popular is Morrow equation:

$$\varepsilon_{ac} = \varepsilon_{ae} + \varepsilon_{ap} = \frac{\sigma'_f}{E}(2N)^b + \varepsilon'_f(2N)^c \qquad (1)$$

where σ'_f and ε'_f are the fatigue strength and the fatigue ductility coefficients; b and c are the fatigue strength and the fatigue ductility coefficients. They can be determined by cyclic testing or from handbooks (Boller and Seeger, 1987).

- The structural analysis by means of finite element method enables to obtain the notch strain amplitude ε_a according to external loading. Finite element analysis with elastic-plastic material model gives the values of ε_a direct, but is computational expensive.

- The simplest and the quickest way to evaluate the strain field in stress concentration zones is to use linear elastic models. However, due to material yielding, the real values of stresses and strains are different, so "elastic" stress/strain must be transformed into elastic-plastic values. The relationship between nominal stress σ_N (far from place of concentrations) and notch elastic-plastic stress-strain is given by Neuber's rule $\alpha_k^2 = \alpha_\sigma \cdot \alpha_\varepsilon$ (Figure 5) and stabilized cyclic stress-strain curve, often modelled by Ramberg-Osgood equation:

$$\varepsilon = \left(\frac{\sigma}{K'}\right)^{1/n'} + \frac{\sigma}{E} \qquad (2)$$

where the cyclic hardening coefficient K' and the cyclic hardening exponent n' can also be determined by cyclic testing or from handbooks (Boller and Seeger, 1987).

- From computational point of view, there is need to use relatively very small size of finite elements in the high stress regions, so calculation of α_k using one model is often impractical, even using elastic analysis. So, the analysis can often be divided into global finite element analysis at the geometrical stress level, coupled with a local finite analysis of the notch area. Stress concentration factors for geometrical stresses (or hot spot stress) α_{kG} can be established separately using 3D mesh and the influence of the local notch (as local geometry of the weld toe) can be obtained from FE analysis of a small region in the vicinity of the weld. As a result, additional stress concentration factor α_{kL} can be established, as a correction factor for different types of welds. The effect of total discontinuity is obtained by multiplying partial factors $\alpha_k = \alpha_{kG} \cdot \alpha_{kL}$.

- Although it is well known from years that fatigue cracks initiate at places with maximum strain concentration, rapid development of the Local Strain Approach has started nowadays with common using of computer technology (Finite Element Method). The stages of local strain approach are schematically shown in Figure 6.

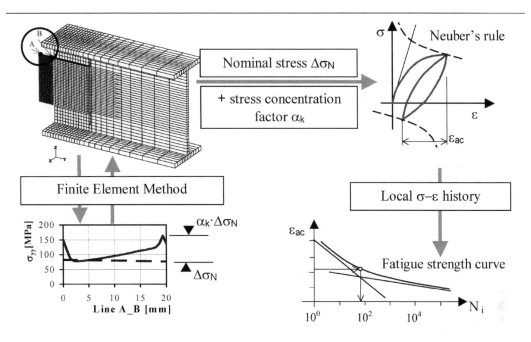

Figure 6. The determination of Low Cycle Fatigue life based on local strain approach.

Results of analysis

- Tendency to use semi-rigid joints, more and more frequent utilization of plastic design and tendency to simplification of connections increases the number of structural elements and situations, where the fatigue or low cycle fatigue can be detrimental.

- Fatigue or low cycle fatigue may happen in many structural elements for which the live load is a large percentage of the total load.

- Although the applied nominal load in steel structures do not vary in fully cyclic type, existence of stress concentrators can lead to cyclic deformation phenomena in elements and fatigue cracking after small number of load cycles.

- Some codes have recommendation, in which the number of load cycles corresponding to predominantly static loading is predicted (e.g. $N \leq 10^4$), so the effect of variability of loading should be taken into account when the number of cycles is greater than this value. In some cases this statement can be insufficient.

Further developments

- The local strain approach is promising tools for predicting fatigue life for elements and connections in steel structures.

- The local strain approach can be used for damage analysis for individual components of joints. It can help to extend component approach, adopted in current codes for static loading, to other loading situations.

References

- Boller Ch., Seeger T., 1987. *Materials data for cyclic loading.* Volume 1-5, Amsterdam, Elsevier.

- Calado L., Castiglioni C. A., Bernuzzi C., 1998. *Behaviour of steel beam-to-column joints under cyclic reversal loading: an experimental study.* In T. Usami and Y. Itoh (Eds.), Stability and Ductility of Steel Structures, Elsevier.

- Kocańda S., Kocańda A., 1989. *Low cycle fatigue of metals* (in Polish), PWN, Warszawa.

- Łubiński M., Goss Cz., Leśniewski J., 1995. *Durability of steel structures in respect to low-cycle fatigue* (in Polish). In J. Murzewski (Ed.) Proc. IX Int. Conf. on Metal Structures, Kraków, Poland, 26-30 June 1995, Vol. 3, pp.157-166.

- Seeger T., Zacher P, 1994. *Lebensdauervorhersage zwischen Traglast und Dauerfestigkeit am Beispiel ausgeklinkter Träger,* Bauingenieur 69 pp.13-23.

- Ślęczka L., 2004. *Low cycle fatigue strength assessment of butt and fillet weld connections.* J. Construct. Steel Res., 60: 701-12.

- Zandonini R., Bursi O.S., 2002. Monotonic and hysteretic behaviour of bolted endplate beam-to-column joint. In Chan, Teng and Chung (Eds.) Advances in Steel Structures, Vol. 1, Elsevier.

Contacts

- Dr Lucjan Ślęczka
  Address: Department of Building Structures,
  Rzeszów University of Technology,
  W. Pola 2, 35-959 Rzeszów, Poland
  E-mail: sleczka@prz.rzeszow.pl, Phone/fax: +48-17-8542974

SHAKEDOWN OF BEAMS WITH SEMI-RIGID JOINTS

Description

- Building structures are subjected to various permanent and variable loading (wind, snow, imposed loads), which change independently. Instead of real values of loading, we should tell about intervals of loading, given by loading codes.

- In case of repeated variable loads exceeding elastic range, there may occur in the structure such phenomena as a brittle fracture due to low-cycle fatigue (alternating plasticity) or progressive growth of plastic deformation during consecutive cycles of loading (incremental collapse).

- Plastic deformation in the structure may also leads to creating of the residual stresses. It may happened that internal residual stresses created by plastic deformation during first few loading stages cause that response of the structure in the following loading cycles is elastic (**shakedown** of the structure).

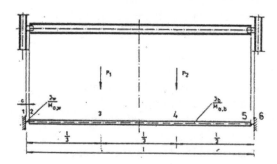

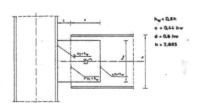

Figure 1. Analysed beam with semi-rigid joints   **Figure 2.** Semi-rigid joint used in the analysis

Field of application

- Structures subjected to repeated variable loading, when plastic methods for global analysis are used, should be analyzed using shakedown theory.

- This is especially important for the semi-continuous structures like frames and beams, with semi-rigid and/or partial-strength joint, where the first plastification occurs in the joint. Even in case of elastically designed frames subjected to exceptional loading (e.g. earthquake), alternating plasticity may occur.

Technical information

- The steel beam with semi-rigid connection is analysed (Figure 1) to show differences in results obtained by limit and shakedown theories.

- Span of the beam was 6,0 m. Beam was loaded by two concentrated loads which values can be changed independently.

- Beam was of IPE 240 section of S235 steel grade.

- Flexible, welded joint shown in Figure 2 was chosen for this example (Kozlowski, 1988). Joint consists of gusset plate welded around to the beam web by filled welds.

Structural aspects

- Eurocode 3 and other design recommendations permit to use plastic global analysis. In the old EC3 version (ENV, 1992) in the point 5.2.1.4(11) it was stated: "in the case of building structures, it is not normally necessary to consider the effects of alternating plasticity". New version (prEN, 2003) in annex AB allows in plastic analysis to increase permanet and variable action proportionaly. That means, that EC3 recommend the theory of limit analysis.

Research activity

- Ultimate limit theory enables to obtain the magnitude of the loading which, applied to a given structure, turn it into mechanism due to the formation of a necessary number of plastic hinges.

- Complete solution in theory of limit analysis, which is the solution of boundary problem for differential equations of plastic yield theory, is as a rule difficult to obtain. In practical analysis, kinematic simplified method is used (Heyman, 1971), (Hodge, 1959).

- Theory of limit analysis can be applied only in case of one parameter loading, i.e. when all loads increase proportionally until the structure collapse.

- Most structures are subjected to repeated loadings which can change independently. In the plastic design, in case of variable loads shakedown theory should be used (Konig, 1974).

- Three possible cases may occur:
  1. during consecutive cycles of loading, a progressive growth of plastic strains leads to excessive deformations (incremental collapse),
  2. during consecutive cycles, plastic strains have opposite values, which lead after some cycles to brittle fracture (Iow-cycle fatigue),
  3. after some cycles, internal residual force field develops, which makes structure response elastically in successive cycles (shakedown of structure).

- Two basic statements are used in shakedown theory for framed structures: static (Bleich's) and kinematic (Neal, 1956).

- In practical cases, loading related to incremental collapse is determined using Neal's kinematic state, expressed as:

$$\sum_{i=1}^{n} M_{0i} \cdot |\varphi_i| \leq \sum_{i=1}^{n} N_i \cdot \varphi_i, \tag{1}$$

where:

M_{0i}, φ_i – plastic moment capacity, rotation of i section,

N_i – envelope of bending moments obtained in elastic global analysis of the structure for assumed range of variable loads:

$$N_i = \begin{array}{l} \max M_i^e \text{ when } \varphi_i > 0 \\ \\ \min M_i^e \text{ when } \varphi_i < 0 \end{array}$$

max, min M_i^e – extreme values of bending moments in i section, taken from elastic envelope.

- Loading related to low-cycle fatigue is determined using the third Bleich's equation:

$$\max M_i^e - \min M_i^e < 2\, M_{ei}\,, \tag{2}$$

where:

M_{ei} – elastic moment resistance of i beam section.

Method of analysis and results

- Possible failure mechanisms used in kinematic limit theory method are shown in Figure 3 (Kozlowski, 1988).

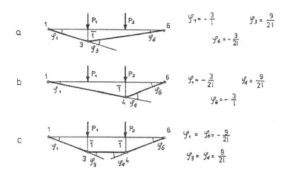

Figure 3. Possible failure mechanisms

- Virtual work principle can be written as:

$$W_E = W_I \tag{3}$$

where:
W_E – external loads work,
W_I – internal loads work.
For a and b mechanism:

$$W_E = \sum P_i \cdot w_i = P \cdot 1 + P \cdot 0{,}5 = 1{,}5 \cdot P\,, \tag{4}$$

$$W_I = \sum M_{oi} \cdot |\varphi_i| = M_{0w} \cdot \varphi_1 + M_{0b} \cdot \varphi_3 + M_{0w} \cdot \varphi_6\,, \tag{5}$$

ratio of plastic moment resistances of beam section to the joint: $k = \dfrac{M_{0b}}{M_{0w}}$,

$$W_I = \frac{M_{0b}}{k} \cdot \frac{3}{L} + M_{0b} \cdot \frac{9}{2 \cdot L} + \frac{M_{0b}}{k} \cdot \frac{3}{2 \cdot L} = 4{,}5 \frac{M_{0b}}{L}\left(1 + \frac{1}{k}\right). \tag{6}$$

Substituting (4) and (6) to (3) we obtain:

$$P_0 = 3 \frac{M_{0b}}{L}\left(1 + \frac{1}{k}\right). \tag{7}$$

For semi-rigid joint shown in Figure 2, $k = 2{,}865$, so $P_0 = 4{,}05 \dfrac{M_{0b}}{L}$.

For c mechanism:

$$W_E = 2 \cdot P \cdot 1 = 2 \cdot P\,, \tag{8}$$

$$W_I = M_{0w} \cdot \varphi_1 + M_{0b} \cdot \varphi_3 + M_{0b} \cdot \varphi_4 + M_{0w} \cdot \varphi_6\,, \tag{9}$$

$$W_I = 2 \cdot \frac{M_{0b}}{k} \cdot \frac{9}{2 \cdot L} + 2 \cdot M_{0b} \cdot \frac{9}{2 \cdot L} = 9 \frac{M_{0b}}{L}\left(1 + \frac{1}{k}\right), \tag{10}$$

substituting to (3) and taking into account that $k = 2{,}865$:

251

$$P_0 = 6{,}07 \, \frac{M_{0b}}{L} \, . \tag{11}$$

Finally, the carrying capacity of the beam is the minimum value of P_{0i}:

$$P_0 = \min P_{0i} = \mathbf{4{,}05} \, \frac{M_{0b}}{L} \, . \tag{12}$$

- Load multiplayer calculated using shakedown kinematic method:

$$\lambda_{(i)} = \min \frac{\sum\limits_{1}^{n} M_{0i} \cdot |\varphi_i|}{\sum\limits_{1}^{n} N_i \cdot \varphi_i} \, . \tag{13}$$

For mechanism *a* and *b*:

$$\lambda_i^{(a)} = \lambda_i^{(b)} = \frac{\dfrac{M_{0b}}{k} \cdot \dfrac{3}{L} + M_{0b} \cdot \dfrac{9}{2 \cdot L} + \dfrac{M_{0b}}{k} \cdot \dfrac{3}{2 \cdot L}}{0{,}2048 \cdot P \cdot L \cdot \dfrac{3}{L} + 0{,}1285 \cdot P \cdot L \cdot \dfrac{9}{2 \cdot L} + 0{,}2048 \cdot P \cdot L \cdot \dfrac{3}{2 \cdot L}} =$$

$$= 3{,}0 \frac{M_{0b}}{L} \left(1 + \frac{1}{k} \right) = 4{,}05 \frac{M_{0b}}{L} \tag{14}$$

For mechanism *c*:

$$\lambda_i^{(c)} = \frac{2 \dfrac{M_{0b}}{k} \cdot \dfrac{9}{2 \cdot L} + 2 \cdot M_{0b} \dfrac{9}{2 \cdot L}}{2 \cdot 0{,}2048 \cdot P \cdot L \cdot \dfrac{9}{2 \cdot L} + 2 \cdot 0{,}1285 \cdot P \cdot L \cdot \dfrac{9}{2 \cdot L}} = 3{,}0 \frac{M_{0b}}{L} \left(1 + \frac{1}{k} \right) = 4{,}05 \frac{M_{0b}}{L} \, . \tag{15}$$

- Load multiplayer calculated according to static method:

$$\lambda_{(n)} = \min \frac{2 \cdot a_i \cdot M_{0i}}{\max M_i^{\,e} - \min M_i^{\,e}} \, , \tag{16}$$

where:

$a_i = \dfrac{W_e}{W_{pl}}$ - ratio of elastic to plastic section modulus,

for rectangular sections: $a = 0{,}67$
for IPE sections: $a = 0{,}88$.

For sections *1* and *6*:

$$\lambda_1 = \lambda_6 = \frac{2 \cdot 0{,}67 \cdot \dfrac{M_{0b}}{k}}{0 - (-0{,}2048 \cdot P \cdot L)} = 2{,}28 \frac{M_{0b}}{P \cdot L} \ . \tag{17}$$

For sections *3* and *4*:

$$\lambda_3 = \lambda_4 = \frac{2 \cdot 0{,}88 \cdot M_{0b}}{0{,}1285 \cdot P \cdot L - 0} = 13{,}7 \frac{M_{0b}}{P \cdot L}. \tag{18}$$

Finally, shakedown capacity P_s is the lowest value:

$$P_s = \min P_{si} = \mathbf{2{,}28} \ \frac{M_{0b}}{L} \ . \tag{19}$$

- In the presented example, carrying capacity calculated according to shakedown theory ($P_s = 2{,}28 \ \dfrac{M_{0b}}{L}$) is much smaller than obtained from limit theory ($P_0 = 4{,}05 \ \dfrac{M_{0b}}{L}$) what means that ultimate limit theory may overestimate structure capacity.

Further developments

- Real structures, which are subjected to repeated variable loading should be analyzed using shakedown theory. Also for structures, where exceptional loading may occur, some parts of structure may be overstrengthed and plastic deformation may take place. In such case, shakedown theory should also be applied.
- The same procedure as shown for presented example should be applied to framed and other structures subjected to variable loading.

References

- ENV 1993-1-1. 1992. Eurocode 3. Design of steel structures. Part 1-1. Genaral rules and rules for buildings. CEN, Brussels.
- Heyman J., 1971. *Plastic Design of Frames*. Cambridge University Press. UK.
- Hodge P.G., 1959. *Plastic Analysis of Structures*. Mc Graw-Hill. New York.
- Konig J.A., Sawczuk A., Paprocka-Grabczyńska W., 1974. *Shakedown Analysis of Frames and Beams. (In Polish)*. Research and Design Center for Metal Structures "Mostostal" Procc. No 5. Warsaw, Poland.
- Kozłowski A., 1988. *Carrying Capacity of I-beam with Flexible Connections.* (in Polish). Ph.D. Thesis. Institute of Building Technology. Warsaw.
- Neal B. G., 1956. *The Plastic Methods of Structural Analysis*. Chapman and Hall.
- prEN 1993-1-1. 2003. Eurocode 3. Design of steel structures. Part 1-1. General rules and rules for buildings. CEN. Brussels.

Contact

- Aleksander Kozlowski
  Address: Department of Civil Engineering, Rzeszow University of Technology, 2 Poznanska street, Rzeszow, Poland.
  e-mail: kozlowsk@prz.edu.pl, Phone/Fax: +48 17 85 429 74.

LOW CYCLE FATIGUE OF STEEL COLD-FORMED MEMBERS

Description

- The behaviour of light gauge steel members is characterised by local buckling phenomena which determine a reduction of load bearing capacity .

- According to EC3, this kind of profiles have to be classified as slender sections (class 4).

- The use of cold-formed thin-walled steel members in seismic structures is nowaday strongly limited by code prescriptions (EC8), which require for this structures a q-factor equal to 1, so that the design for the action of a severe earthquake must be carried out practically in the elastic field.

Field of application

- Cold-formed steel structural elements are very suitable for low cost and fast realization buildings.

- The light weight of this kind of members determines also a reduction of the seismic actions, in comparison with "traditional" members.

Technical information

- Experimental tests have been carried out on five identical cold-formed thin walled beams made by a double back-to-back coupled unlipped channel profiles.

- The specimen was about 2.10 m long and were linked together by a 20 mm thick vertical inner plate placed at mid-length. Two 15 mm thick end plates are welded to the members at both edges of beam to allow the connection with external restraints. Moreover, the profiles are stiffened by means of 4 mm thick vertical plates parallel to the web and welded to the flanges at mid-length, corresponding to the load application zone, and four further horizontal 4 mm thick batten plates welded to the top and the bottom flanges (Figure 1) (Formisano, 2003).

- With reference to the material, the average yield and ultimate stresses of the used steel were f_y=364MPa and f_u=425MPa, respectively.

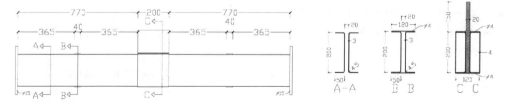

Figure 1. Longitudinal and transversal sections of the tested specimen

Structural aspects

- The monotonic behaviour of this kind of member (in term i.e. Force-Displacement relationship) is characterised by an initial elastic branch up to the attainment of the local buckling load. Afterwards there is a significant reduction of the load bearing capacity when increasing the displacement. The slope of the softening branch is initially hard, while after becomes softer up to the residual bearing capacity, which is a little percentage of the maximum one.

- The cyclic behaviour is characterised by a significant degradation of both resistance and stiffness, when local buckling load is overpassed in the first cycle. The reduction of the load bearing capacity, the available ductility (rotation capacity) and the energy dissipation are important aspects to be analysed in order to define the possibility of using these members in seismic zones.

Research activity

- A building subjected to seismic actions gives different response according to its ductility, which represents the capacity to undergo plastic deformations and energy dissipation before brittle failure. Ductility of the structure is linked together with the ductility of the members, the connections and the material. In case of steel structures, ductility is influenced by fatigue and buckling phenomena, which can lead to premature collapse. In particular, it is important to keep into account the low-cycle fatigue in evaluating the behaviour of steel buildings in seismic areas.

- In this situation, because of the ductility can play only a secondary role, fatigue could determine the collapse of the structure. In general, fatigue life depend on two factors:
  1) amplitude loading history: increasing amplitude of loops lead to a reduction of the useful life of the building;
  2) fatigue resistance of structural elements, which substantially depends from their geometrical and mechanical properties, represented by means of a numerical constant called K.

- For a structural detail, the fatigue failure prediction function, used by the S-N curve approach, can be expressed by the following equation:

$$N \cdot \Delta\sigma^m = K \qquad (1)$$

where N is the number of cycles to failure and $\Delta\sigma$ the stress (strain) range, while non-dimensional constant m and the dimensional parameter K depend on both the typology and the mechanical properties of the steel components (Castiglioni et al., 1995).

- In the Log-Log domain equation (1) can be re-written as:

$$Log(N) = Log(K) - mLog(\Delta\sigma) \qquad (2)$$

This relation represents a straight line with a slope equal to $-1/m$ called fatigue resistance line, which identify the safe and unsafe regions.

- This law has been proposed by Woehler in case of high cycle fatigue and adopted in the current design phases of the constructive details.

- Ballio and Castiglioni showed the possibility to extend this approach also in low cycle fatigue design.

- In low cycle fatigue design, according to Calado and Castiglioni criterion (Calado, 2000), the number of cycles corresponding to the collapse, when constant amplitude loading history is applied, is that corresponding to the first cycle satisfying the following condition:

$$\frac{E_{cf}}{E_{c0}} \leq \alpha_f \qquad (3)$$

where E_{cf} represents the ratio between the absorbed energy of the considered component at the last cycle before collapse above the energy that might be absorbed in the same cycle in the case of elastic-perfectly plastic behaviour, while E_{c0} represents the same ratio but with reference to the first cycle in the plastic range. The value of α_f, which depends on several factors (such as type of joint and the steel grade of the component), should be determined by fitting experimental results but it is possible to define it a priori: Calado and Castiglioni proposed a value of $\alpha_f = 0.5$.

Method of analysis

- Three point bending tests have been carried out in the elastic and post-elastic field on simply supported beams, aiming to determine the structural response of the member to progressively increasing external loads.

- The main purpose of the experimental program is the evaluation of the whole load-displacement curve, thus allowing the determination of stiffness and bearing capacity degradation, as well as the failure mechanism of the tested members in cyclic tests.

- The tests were carried out using a testing equipment consisting of a reaction steel frame and a hydraulic jack.

- The jack is controlled by means of an electronic device, which is also employed for data acquisition. The beam was restrained by a hinge on the left side and a simple support on the other side (Figure 2).

- Lateral bracing system was adopting in the cyclic test to avoid flexural-torsional buckling in the specimen (Figure 3).

- The tests were carried out under displacement control, using a constant displacement rate equal to 1.0 mm/s, for the sake of obtaining quasi-static loading procedure. Besides, every increasing displacement step was followed by a hold-step of 20 s.

- Deflections and deformations were measured by using strain gauges (SG) and displacement transducers (LVDT), with an accuracy of 0.0001 mm, located in appropriate positions in the central and lateral zones of the member. Particularly mid-span deflection and edge rotation have been monitored during testing (Figure 4).

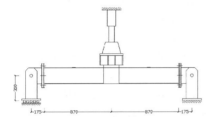

Figure 2. Experimental set-up for three-point bending loading test conducted on the beam restrained by means of a hinge (left side) and a simple support (right side).

Figure 3. Lateral bracing system adopted in the experimental program.

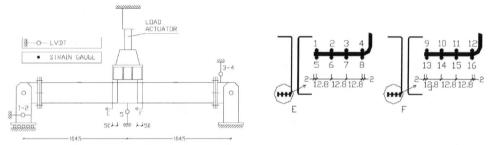

Figure 4. LVDTs and strain gauges location along the beam in the cyclic test.

Results of analysis

- The cyclic tests (T03, T04 and T05) presented a similar behaviour, characterized by a progressive deterioration of strength and stiffness, typical of steel members when affected by local buckling (Figure 5).

- In all performed cyclic tests, local buckling started in right bottom flanges of both profiles. During the test, as the number of cycles (N) increased, cracks always arise in the flanges, in the buckled zone at the flange-to-web connection. These cracks gradually developed toward the free edge of the flange. When the whole length of the flange was involved by the crack, the beam lost quite at all its load bearing capacity and the test was ended. The phases corresponding to the crack beginning and to its full development are shown in Figures 6a and 6b, respectively.

- With reference to beam T03 (displacement amplitude $\Delta = 15$ mm – see Figure 5a), a maximum force (F_{max}) equal to 72.3 kN were recorded for a deflection of 13.1 mm in the first cycle, when local buckling was already developed. The strength and stiffness degradation was very significant in the first four cycles. First cracks arise at the 15th cycle and involved the whole flanges at the 47th cycle.

- For beam T04 ($\Delta = 30$ mm – see Figure 5b), the maximum force reached was equal to 76.7 kN, at a deflection of 14.5 mm. Also in this test the first three-four cycles were the most important for beam degradation. The cracks started at the 7th cycles and reached the free edges of flanges during the 21st cycle.

- For beam T5 ($\Delta = 45$ mm – see Figure 5c) the maximum load bearing capacity was 72.9 kN when the deflection was equal to 13.5 mm. The strength and stiffness deterioration was very important in the first two cycles. The cracks started in the 3th cycle and was fully developed in the 6th cycle.

- The ratio between the energy dissipated during the generic cycle (E_i) and the energy dissipated during the first cycle (E_1), drawn as function of the number of cycles (N), is presented for the three cyclic tests in Figure 5d. It can be clearly observed that the increasing of the displacement amplitude of the test causes a more significant reduction of structural dissipation as the number of cycles grows.

- In the field of low cycle fatigue design, on the basis of the previously described research activities, the evaluation of the low cycle fatigue is based on the behaviour of the cold-formed steel specimens under consideration, characterised by loading cycles with displacement amplitudes of $\pm$ 15, 30 and 45 mm.

- Differently by the traditional approach to fatigue, according to the Woehler's law, it is more convenient to adopt global quantities (e.g. displacements or internal forces) instead of local ones (like stress or strain) as parameter S, because they are connected to the global seismic behaviour of the structure and more easily predictable in the design phase. In this paper we refer to mid-span displacement as key parameter S to be used in the S-N relationship.

■ The number of cycles to failure (N_f) can be derived by adopting a specific failure criterion. In this paper three different criteria have been considered for determining N_f:

    a) N_f is the number of the cycle in which the cracks have developed in the whole length of flanges and started to interest the web too;

    b) N_f is the number of the cycle in which the reaction force became less than 20% of the maximum reached one (residual bearing capacity – RBC = 20%);

    c) N_f is the number of the cycle in which the dissipated energy E_i (area enclosed below the curve in the plane F-δ) is equal to the 50% of that dissipated in the first cycle E_1, as demonstrated by Calado and other authors.

■ According to these criteria, the experimental tests have given the results showed in Table 1. The values of N_f corresponding to the (c) criterion has been obtained as shown in Figure 5d, where the relationship α_i versus N is depicted. In the Table it can be observed that the criteria (b) and (c) practically give the same results.

■ In Figure 7, in the Log(S)-Log(N) diagram, the full dots represents the fatigue failure results obtained from tests applying both the criterion (a) (Fig. 7a) and the criterion (c) (or (b)) (Fig. 7b). On the same Figures the corresponding fatigue resistance lines, obtained as the best linear fitting of the experimental dots, are depicted.

■ It is worth to note that the slope of the lines in both the cases is significantly higher than the ones given by Eurocode 3 (m = 3 or 5) (CEN,1995) for fatigue design of steel structural details, being quicker the degradation of the structure when cycles increases.

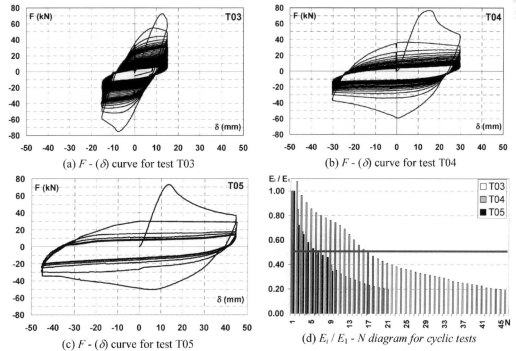

(a) F - (δ) curve for test T03

(b) F - (δ) curve for test T04

(c) F - (δ) curve for test T05

(d) E_i / E_1 - N diagram for cyclic tests

Figure 5. Results of cyclic tests.

Figure 6. Typical propagation of a crack: start (a) and full development (b).

TEST	S (mm)	Nf		
		(a)	(b)	(c)
T03	15	47	18	17
T04	30	21	6	5
T05	45	6	3	3

Table 1. Number of cycles to failure according to different failure criteria.

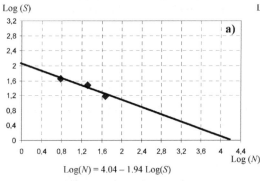

$Log(N) = 4.04 - 1.94\ Log(S)$

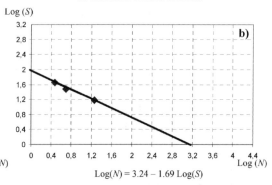

$Log(N) = 3.24 - 1.69\ Log(S)$

Figure 7. Evaluation of S-N fatigue resistance lines according to failure criteria *(a)* a) and *(b)* or *(c)* b).

Further developments

- For validating the Miner'rule in the field of low cycle fatigue design, tests based on amplitude variable loading history must be carried out.

References

- Calado, L., 2000. Fatigue Resistance Design. In F.M. Mazzolani e V. Gioncu (eds.) *Seismic Resistant Steel Structures,* Springer Wien New York, ch. 4, 381-399.

- Calderoni, B., De Martino, A., Ghersi, A., Landolfo, R., 1997. Influence of local buckling on the global seismic performance of light gauge portal frames. In F.M. Mazzolani & H. Akijama (eds), *Proceedings of STESSA '97*, 29-40, Edizioni 10/17, Salerno, Italy.

- Castiglioni, C.A., Bernuzzi, C. & Agatino, M.R., 1997. Progettare a fatica oligociclica (Low-cycle fatigue: a design approach). *Proceedings of the C.T.A. Congress, Ancona, 2-5 October 1997.*

- CEN (1995). ENV 1993-1-2. *Eurocode 3: Design of steel structures. – Part 1.1: General rules.*

- Formisano, A., 2003. Analisi teorico-sperimentale della fatica oligociclica di travi di acciaio in parete sottile. *Graduation Thesis*, University of Naples "Federico II", Faculty of Engineering.

Contacts *(include at least one COST C12 member)*

- Prof. Dr. Eng. Federico M. Mazzolani
  Department of Structural Analysis and Design
  University of Naples "Federico II" – P.le Tecchio, 80 – 80125 Naples (I)
  fmm@unina.it; Phone: (+39) 081 7682443 Fax: (+39) 081 5934792

INNOVATIVE BUILT-UP COLD-FORMED BEAMS

Description

- The use of cold-formed thin-walled structural systems is spreading more and more in the market of constructions.

- The improving technology of manufacture and corrosion protection together with the reduced costs of construction are the key aspects, which have led to the increased competitiveness of the resulting products with parallel developments in practical applications.

- The benefits of light gauge metallic structure include its potential for off-site and in-site prefabrication, a high quality product as well as the ability to achieve elevated structural and technological performances.

Field of application

- The cold-formed members, traditionally used as secondary beams in industrial buildings, started being widely employed in residential constructions as well. In this application field, the light gauge cold-formed steel structures have proved to be a suitable alternative to the traditional constructive technologies based on timber, reinforced concrete and masonry.

- On the other hand, several studies and a considerable amount of research have been carried out aiming at checking the reliability of the cold-formed steel systems also from a structural point of view: when adequately designed, they may perform remarkably well, even in seismic areas, both in terms of strength and ductility.

- Consequently, the potential application fields of innovative cold-formed beams appear to be manifold. Obviously, the key aspect is the competitiveness respect to equivalent hot-rolling members. From this point of view, the final cost is undoubtedly the crucial parameter and it must be considered as an additional target in the relevant design procedures.

Technical information

- The beams under investigation are obtained by connecting two cold-formed shapes, through mechanical fasteners, located in correspondence of the web and of the flanges, where reinforcing plates are located as well (see Fig. 1). These plates not only provide a connection between the cold-formed parts, but also improve the structural performance of the whole beam, so allowing the possibility of using different thickness and/or material.

- Shape and position of bends and folded grooves are designed to optimize the bearing capacity of the proposed built-up beams. The aim is to reduce the effects of local buckling phenomena in both the flange and web elements, leading to a very efficient structural member. The flanges are stiffened by edge and intermediate stiffeners (made of folds and folded grooves, respectively), while the web stiffeners are made of two semi-cylindrical curved grooves which, once the beam is assembled, form a sort of truss system, whose braces have a hollow cross-section.

- By combining successfully all the above mentioned typological aspects (reinforcing plates, stiffening devices and web holes), it is in fact possible to obtain a new generation of cold-formed members, which can be suitably used in several application fields. Of course, in order to be on this target, it is necessary to set-up an appropriate design procedure able to individuate, for each possible application field, the best combination of both structural and technological beam features.

- With regard to the first aspect, it is worth underlining that the structural performance of the proposed beams can be improved by modifying the properties of both the sheet material and the other beam components. In fact, the two halves, usually in mild steel, may be made, for special applications, of high-strength steels, in accordance with a general trend in increasing the use of such sort of materials in cold-formed structures. At the same time, the reinforcing plates – usually of steel – may be also made of different materials (aluminum, composite materials).

- Obviously, the outcoming structural performance such beams should be, in each case, guaranteed by effective and innovative connecting systems (bi-component and mono-component blind rivets, self-piercing rivets and press-joints), able to provide high structural responses together with a remarkable workability.

a)   b)

Figure 1: An example of innovative built-up cold-formed steel beams

Structural aspects

- Reference of relevant related codes
- CEN EN 10051 (2000). Continuously hot-rolled uncoated plate, sheet and strip of non-alloy and alloy steels-Tolerance on dimension and shape.
- CEN ENV 1993-1-3 (1996). Eurocode 3: Design of steel structures – Part 1-3:

General rules - Supplementary rules for cold formed thin gauge members and sheeting.
- ECCS n.49 (1987). European Recommendations for the design of light gauge steel members, (First edition).

Research activity

- The whole research activity has been subdivided in the following operative steps:

Step 1: Preliminary study (study of the relevant "state of the art" and critical examination of the specific international design codes);

Step 2: Characterization of the beam components (selection of the sheet material and of the connecting systems, optimization of both cross-section shape and stiffening devices);

Step 3: Experimental tests on single components (tensile tests on base material before and after cold-forming process, lap shear tests on selected connecting systems);

Step 4: Study of the assembly and competitiveness problems (design of the mechanical fasteners with regard to current connections and end joints);

Step 5: Experimental tests on beams (selection of different beam specimens, full-scale tests on the selected specimens under different load conditions, elaboration and critical interpretation of test results);

Step 6: Theoretical and numerical analyses by using computational models previously calibrate;

Step 7: Design of a specific beams in accordance with different application fields.

- The research is in progress and has reached step 4. The first two stages have allowed the preliminary definition of the beam shape, which shall be checked through the subsequent steps. Besides, a possible sheet material (steel S235JR) as well as some connecting systems (mono-components and bi-component bind rivets, self-piercing rivets and press-joints) have been selected. A briefly overview on the main obtained result is presented in the following. More detailed information about such aspects are reported in the relevant papers..

Method of analysis

- The complexity of the beams under examination has required the development of an appropriate research project aiming at the assessment of their structural behaviour. The whole study is on the following main directives:

1. Experimental investigations devoted to evaluate strength and ductility of both beam and its components under different loading conditions.

2. Theoretical and numerical analyses aiming at simulating the whole beam response starting the components modeling.

3. With regard to the first directive, the design assisted by test methodology will be adopted, in accordance with the provisions of the main European codes on cold-formed members (ENV 1993-1-3, 1996 and ECCS n.49, 1987). As far as the theoretical and numerical aspects are concerned, deterministic (finite strips and/or

nonlinear finite elements), not deterministic (soft computing) and mixed methodologies will be used.

- Thus, the results achieved through the experimental tests will be used both for

Results of analysis

- With reference to the beams under examination, the complete and correct identification of the "material mechanical model" has required the set-up of an appropriate methodology of analysis aiming at the assessment of all the imperfections arising from the whole manufacturing process. In particular, such a methodology is based on the separated evaluation of the effects on the material produced by rolling first (on the base sheet), then by cold forming (on the final shape).

- As far as the first aspect is concerned, a relevant experimental investigation has been preformed with reference to a S235JR steel sheet with nominal thickness equal to 2.00 mm. The study has focused on the evaluation of both mechanical and geometrical imperfections induced by the rolling process on the sheet material, owing to the interaction among rolls and flat sheets in the plastic and thermoplastic fields.

- With regard to mechanical imperfections, the tensile tests carried out on specimens taken out directly from the sheet under investigation (see Fig. 2a) have shown that they basically consist of the variation of the main material properties along the sheet. Such imperfections are more remarkable along the cross direction of the sheet, where a parabolic variation of the material strength parameters has been noticed.

- Figure 2b shows the tensile test results in terms of yield strength (fy). The data, for the sake of comparison, are normalized through the nominal strength values, equal to $f_{y,b}=235$ N/mm2. The obtained results well emphasize that such a variation is particularly considerable for fy, with a variation of about 12% between edge and middle values. On the contrary, in case of fu, the variation is equal to 4% only.

- As it is well known, the cold-working process of steel sheets, which is necessary for the manufacture of cold-formed members, is characterized by the plastic deformations of the base material performed through cold roll forming and/or die forming. Therefore, further mechanical and geometrical material imperfections arise, in addition to the ones induced by the previous rolling process, next to the folded zones.

- The mechanical imperfections basically consist of the variable distribution of the mechanical properties (mechanical heterogeneity) and by residual stress, both of them located next to the folded zones. The mechanical heterogeneity is due to both strain hardening and strain aging effects, generally determining both a strength increase and a ductility reduction.

- The study of the material alterations produced by cold working is particularly interesting for the beams under investigation. Such members, in fact, are characterized by a complex cross-section shape, having the folds with bend angle ranging from 0° to 180 °.

- In order to evaluate the influence of cold work, a specific experimental study has been developed. In particular, through tensile tests carried out on specimens taken out directly from the examined beam (see Fig. 3a), it has been evidenced that the yield strength is the material property mostly influenced by the cold working process. In fact, if compared with the relevant nominal value ($f_{y,b}$=235 N/mm2), it is possible to observe, near the folds, a rise in fy varying between 80 and 90%.

- This difference can be reduced to 60 ÷ 70% if the actual values of yield strength, by including also the rolling process effects, are considered (see Fig. 3b). On the contrary, the variation in terms of tensile strength has proved to be lower, with variations from 20 to 30% about the relevant nominal value.

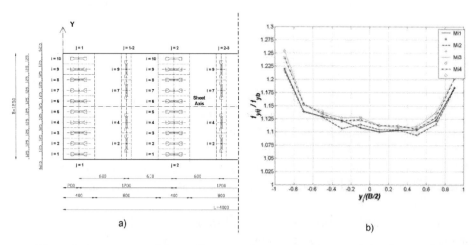

a)   b)

Figure 2: Arrangement of the tensile test specimens and experimental distribution of the yielding strength

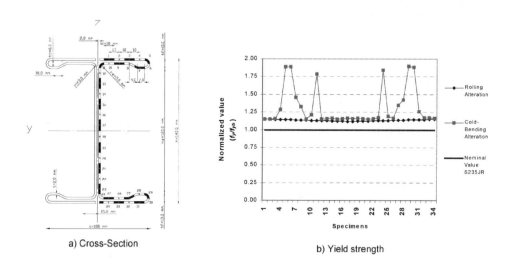

a) Cross-Section   b) Yield strength

Figure 3: Effects of cold work on material properties

Further developments

- All the aforementioned activities concerning the beams under investigation beam components have provided useful design information. In addition, the obtained results will be used to calibrate the theoretical and numerical models aiming at simulating the structural response of the whole beams.

References

- Davies J.M. (2000). Recent research advances in cold-formed steel structures, Journal of Constructional Steel Research, Vol. 55.
- Di Lorenzo G. and Landolfo R. (2002). Comparative study of new connecting systems for cold-formed structures. Proc. of the 3rd European Conference on Steel Structures, Coimbra.
- Di Lorenzo G. and Landolfo R. (2003). On the characterization of material properties for cold-formed members. Proc. of International Conference on Advances in structures (ASSCCA'03). Sydney (submitted paper).
- Landolfo R., Di Lorenzo G. (2001). Una nuova generazione di membrature leggere formate a freddo. XVIII Congresso CTA, Vol.2, Venice (in Italian).
- Landolfo R., Di Lorenzo G. and Fiorino R. (2002). Attualità e prospettive dei sistemi costruttivi cold-formed, Costruzioni Metalliche, n.1 (in Italian).
- Lawson R.M. (1996). Residential applications of light steel frames in the UK and Continental Europe. 5th International Colloquium Structural Stability Research Concilium SSRC, Chicago.
- Pedreschi R. F., (1999). Design and development of a cold-formed lightweight steel beam. 4th International Conference on Steel and Aluminum Structures - ICSAS'99, Helsinki.
- Schafer B. W., Pekoz T., (1999). Local and distortional buckling of cold-formed steel members with edge stiffened flanges, 4th International Conference on Steel and Aluminum Structures - ICSAS'99, Helsinki.

Contacts

- Raffaele Landolfo
  Address: Department of Constructions and Mathematical Methods in Architecture, University of Naples "Federico II", Via Monteoliveto, 80134 Napoli, Italia.
  e-mail: landolfo@unina.it
  Phone: ++39 081 7682447, Fax: ++39 081 5934792

COST C12 - WG2

Datasheet I.4.1.5
Prepared by: M. Manganiello[1], G. De Matteis[2]

[1] Dept. Constr. and Math. Methods in Arch., University of Naples
[2] Dept. Struct. Analysis and Design, University of Naples Federico II

INELASTIC BEHAVIOUR OF ALUMINIUM ALLOY STRUCTURES

Description

- The first difficulty that is usually encountered in theoretical analysis of plasticity and stability problems of aluminium alloy structures is connected with the idealisation of the tensile behaviour of the material.

- The stress-strain curve obtainable from a tension test cannot be simplified by means of an elastic/perfectly plastic schematisation usually adopted for mild steel, but it is necessary to consider more complex models. Among these, the Ramberg-Osgood continuous model and the exponential model [Manganiello, 2003] represent the most commonly adopted in the research field, since they give a very good agreement with experimental results.

- Likewise to the stainless steels, the aluminium alloys are characterised both by a remarkable strain hardening, commonly neglected in case of mild steels, and by a continuous constitutive law. Moreover, since they exhibit a limited ductility, it is necessary to verify, when inelastic analysis methodologies are implemented, if the ultimate deformation capacity of the material is reached.

- These two fundamental aspects pose an important question regarding the possibility to apply the conventional approach to collapse analysis, based on the well-known plastic hinge method (*PHM*). The kinematic theorem of the plastic collapse, in fact, is applicable when the stress-strain material behaviour shows a defined yield plateau with high deformation capacity.

Field of application

- Structural designers have been aware of aluminium's unique qualities for over one hundred years. As well as being one of the most abundant metals in the world, aluminium's formability, high strength-to-weight ratio, corrosion resistance, and ease of recycling make it the ideal material for a wide range of building applications.

- The aluminium and its alloys, from a structural engineering point of view, can be still considered as a 'young' material. Although first building structures made with aluminium alloys appeared in Europe in the early fifties of this century, their use remains still confined in a very small sphere [Mazzolani, 1995].

- Originally indispensable (for their lightweight) in the aeronautical industry, these materials are now widely used in cars, buses, coaches, trains, ferries, and bicycles. More than a quarter of all aluminium products are used in the transport sector.

- The offshore industry is surely the largest user of aluminium alloys for structural applications. Helicopter decks and living quarters of the offshore platforms are an example of structures where these materials have completely replaced the steel, which remains the most widely used material for structural civil applications.

- The main applications in field of civil engineering can be divided in the following groups: roofing structures (Figure 1), structures located in inaccessible places, moving parts, structures for special purposes, curtain-wall systems.

Figure 1. Desert Biome, Omaha, NE (USA)　　**Figure 2.** Helidecks support on offshore platform

Technical information

- The most important physical and mechanical properties can be as follows summarised: the mass density (ρ) and the Young's modulus (E) of aluminium and its alloys are approximately one third of the steel ones (for different alloys $\rho = 26 \div 28\,\mathrm{kNm^{-3}}$ and $E = 68500 \div 74500\,\mathrm{Nmm^{-2}}$); the thermal expansion coefficient (α) is twice that the one related to the steel ($\alpha = 19 \cdot 10^{-6} \div 25 \cdot 10^{-6}\,{}^{\circ}\mathrm{C^{-1}}$).

- To account for the reduced deformation capacity of the material some limit values for the cross-sectional curvature ($\chi_{\mathrm{ult.}}$) have been introduced in Eurocode 9. The codified values for $\chi_{\mathrm{ult.}}$ are $\chi_5 = 5 \cdot \chi_{0.2}$ for brittle alloys and $\chi_{10} = 10 \cdot \chi_{0.2}$ for ductile alloys.

- However, the effective limits assume extreme values very different from each other: they vary from a minimum less than $10\chi_{0.2}$ ($\chi_{\mathrm{elast.}} \equiv \chi_{0.2}$ since $f_{0.2}$ is the conventional elastic stress) to a maximum greater than 100 [Mazzolani, 1995].

Structural aspects

- The possibility to apply the plastic hinge method to continuous beam made with aluminium alloys was for the first time introduced in [Mandara & Mazzolani, 1995].

- A corrective factor η for the conventional yield stress $f_{0.2}$ ($f_y = \eta f_{0.2}$) was set up in order to take into account the limited deformation capacity of the different alloys that can be reached for values, more or less high, of load bearing capacity.

- Some simplified hypotheses were assumed: the first regarding the structural model and the second related to the load level on which to define the required ductility.

- It was utilised a model with concentrated plasticity, assuming a fixed value of the plastic hinge length, and the required ductility was evaluated for a load threshold that was not valuable *a priori* (but only comparing results of different methodologies of analysis).

- The rotational capacity necessary to attain certain ultimate load thresholds, which is an essential parameter to establish the cross-sectional dimensions for using simplified plastic methods of analysis (e.g. the plastic hinge method), was not investigated.

- This study, which has been carried out removing the basis hypotheses done in previous researches [Mandara & Mazzolani, 1995], has as primary aim the evaluation of the required ductility (measured using as parameter the rotational capacity) to implement the plastic hinge method.

- The post-elastic deformation demand will be conventionally evaluated as soon as the structural system attains the load thresholds equal to the one obtainable by using the plastic hinge method implemented with a plastic stress coincident with $f_{0.2}$.

- The post-elastic deformation available will be attained when fixed curvature limits will be reached (χ_i).

- Geometrical and mechanical parameters have been considered: the shape factor and the ultimate curvature limits - for the cross-section; the elastic bending moment distribution - for the structural configuration; the hardening degree - for the material.

- To generalise the study, the ultimate curvature has been varied between 5 and 30 times the elastic value. The shape factor and the hardening degree have been varied, respectively, between 1.10-1.30 and 5-30.

- The elastic stress distribution, which takes into account the structural configuration, has been evaluated by means of the ratio M_{max}/M_{min} related to the beam in which the plastic mechanism will form.

- The results obtained could be applied for different materials which constitutive law is roundhouse type (Ramberg-Osgood).

- A comparison with the EC9 [Eurocode 9, 1998] indications is proposed and commented.

Research activity

- This study is framed within the activities of CEN-TC250/SC9, the European committee in charge with the preparation of EC9 on aluminium alloy structures.

- The research is essentially devoted to provide the background data necessary to implement simplified inelastic design methodologies to structures made with strain hardening materials with reduced ductility.

- It will be defined the inelastic deformation required to reach certain load bearing capacity by means of to set up, for different hardening degrees, appropriate cross-sectional slenderness limits.

Method of analysis

- The proposed results are a part of a more general research [Manganiello, 2003] devoted to investigate the inelastic behaviour of aluminium alloy structures. It is organised in three main sections: *a*) the evaluation of the inelastic behaviour of structures made with strain hardening materials with reduced ductility; *b*) the investigation on the post-elastic behaviour of simple supported aluminium alloy beams under non uniform bending; *c*) the definition of a cross-sectional classification criterion.

- In this datasheet, part of the results related to the section *b*) will be briefly proposed.

- To analyse the inelastic behaviour of structural schemes the finite element codes ABAQUS/Standard_6.2 and SAP 2000_7.12 have been used.

- The rotational capacity (required and available) will be evaluated on a simple supported beam, which will be taken out from the considered structural configuration.

- The different numerical models implemented in non-linear FE codes have been calibrated on the basis of available experimental results.

Results of analysis

- The evaluation of the inelastic behaviour of aluminium alloy structures has been done on continuous beams. Successively, the validity of the obtained results has been verified on more complex systems (One story frames with one or more spans).
- The rotational capacity has been considered as the parameter representing the inelastic deformation capacity of the structure. The demand and the availability are defined according to the Figure 3.

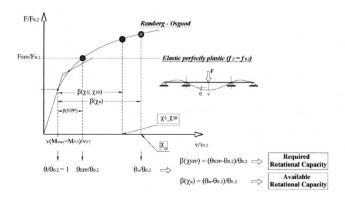

Figure 3. The (required and available) rotational capacity (β)

- The comparison between the demand (increasing curves) and the availability (decreasing curves) is proposed in Figure 4 for a symmetric (three span beam) and non-symmetric (two span beam) load condition. In both cases the load is applied at the midpoint of the loaded span.
- The influence of the other (secondary) parameters (distributed load, length of the spans and load position) has been also evaluated.

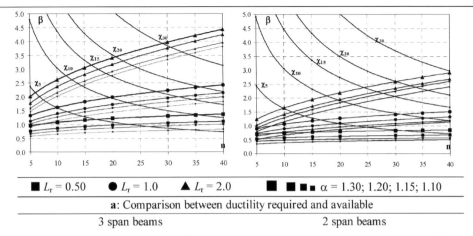

■ $L_r = 0.50$	● $L_r = 1.0$	▲ $L_r = 2.0$	■ ■■▪ $\alpha = 1.30; 1.20; 1.15; 1.10$

a: Comparison between ductility required and available

3 span beams	2 span beams

Figure 4. Required and available ductility for continuous (symmetrical and non-symmetrical) beams

- A value of 3 for the required rotational capacity (which has to be guaranteed for the class 1 cross-sections) seems to be the one more opportune.
- On the basis of the comparison between the rotational capacity (β) required and available, the coefficient η provided by the Eurocode 9 has been again calibrated.
- Combining the parameters α and L_r (α_{min} - L_{rmin}; α_{med} - L_{rmed}; α_{max} - L_{rmax}) it is possible to obtain the maximum, the minimum and the average values of the coefficient η.
- The formulations proposed (1) are enough different from the ones actually codified (2):

$$\eta = a \cdot EXP(bn) + c \cdot EXP(dn) \quad \text{with} \quad a, b, c, d = f(L_r, \alpha, \chi_{ult.}) \tag{1}$$

$$\eta = 1/(a + b \cdot n^c) \quad \text{with} \quad a, b, c = f(\alpha, \chi_{ult.}) \tag{2}$$

	a	b	c		a	b	c
	\multicolumn{3}{c}{$\alpha = 1.1 - 1.2$}		\multicolumn{3}{c}{$\alpha = 1.4 - 1.5$}
χ_5	1.15	-4 4	-0.66	χ_5	1.2	-5.0	-0.7
χ_{10}	1.13	-11.0	-0.81	χ_{10}	1.18	-8.4	-0.75

	a	b	c	d
	\multicolumn{4}{c}{(Lr, alpha)$_{min.}$}
χ_5	1.0260	-0.2201	1.0860	-0.0034
χ_{10}	1.7090	-0.2370	1.1730	-0.0030
	a	b	c	d
	\multicolumn{4}{c}{(Lr, alpha)$_{med.}$}
χ_5	0.9602	-0.2356	0.9859	-0.004564
χ_{10}	1.5720	-0.2615	1.1500	-0.006382
	a	b	c	d
	\multicolumn{4}{c}{(Lr, alpha)$_{max.}$}
χ_5	0.8830	-0.2369	0.8486	-0.004629
χ_{10}	1.2750	-0.2043	0.9317	-0.004741

Table 1: Coefficients given by Eurocode 9   **Table 2** Proposed coefficients

- In Figure 5, for different structural schemes (A…F), a comparison between the actual behaviour (points), the schematised one (line with points) and the one presently proposed by EC9 (line) is given.
- The proposed formulations obtained removing the plastic hinge concept and considering the values of rotational capacity necessary to attain a safe inelastic behaviour provide

analytical expression of η, which are different from the ones given by EC9 (Tables 1, 2).

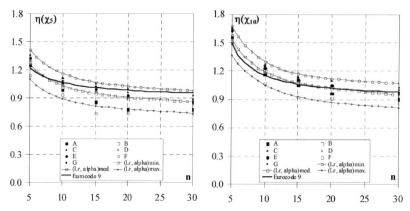

Figure 5. η coefficient for simple portal frames - Comparison between different proposed formulations

- They provide a range of variation for the numerical coefficient η, which contains the values given by actual code indications (Figure 6).

- On the basis of these considerations and using the results of a wide numerical investigation devoted to investigate the post-elastic behaviour of simple supported beam, the cross-sectional limit for the class 1 (the one utilisable when the inelastic resources of the structure is considered) has been established.

- In particular, the analytical formulations obtained interpolating the numerical results related to different cross-sectional shapes and hardening degrees have been solved [Manganiello, 2003] in order to obtain the values of local slenderness parameter b/t.

- The classification criterion provides class limits, which in case of strong hardening alloys are enough different from the ones previously provided from Eurocode 9.

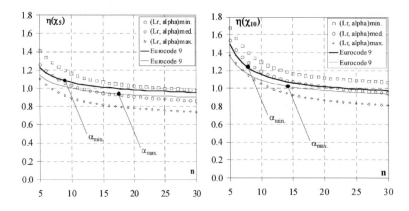

Figure 6. η coefficient - Comparison between the proposed formulations and the Eurocode 9 indications

Further developments

- Further studies should be devoted to analyse the influence of flexural-torsional buckling on plastic behaviour of aluminium alloys beams. They will provide appropriate formulae defining a reduction coefficient of the rotational capacity.
- Consequently, a reduction of the slenderness limits could be expressed as a function of global slenderness of the structural elements.

References

- European Committee for Standardisation ENV 1999. 1998. *"Eurocode 9 – Design of Aluminium Structures – part 1-1: General Rules and Rules for Buildings"*. CEN-TC250/SC9, Brussels, Belgium.
- Mandara, A., Mazzolani, F. M. 1995. *"Behavioural Aspects and Ductility demand of Aluminium Alloy Structures"*. Proc. of the Third International Conference on Steel and Aluminium Structures. Istanbul.
- Manganiello, M. 2003. *"On the Inelastic Behaviour of Aluminium Alloy Structures"*. PhD Dissertation. University of Chieti-Pescara "G. d'Annunzio", Pescara (ITALY).
- Mazzolani, F.M. 1995. *"Aluminium Alloy Structures"*, 2nd ed. Chapman & Hall. London, UK.

Contacts

- Federico M. Mazzolani
  Address: Department of Structural Analysis and Design, University of Naples "Federico II", P.le Tecchio, 80, 80125 Napoli, Italia.
  e-mail: fmm@unina.it; Phone: ++39 081 7682444, Fax: ++39 081 5934792
- Raffaele Landolfo
  Address: Department of Constructions and Mathematical Methods in Architecture, University of Naples "Federico II",
  Via Monteoliveto, 80134 Napoli, Italia.
  e-mail: landolfo@unina.it, Phone: ++39 081 7682447, Fax: ++39 081 5934792

COST C12 - WG2

Datasheet I.4.1.6.

Prepared by: G. De Matteis, B. Calderoni, C. Giubileo

University of Naples Federico II, Dept. Structural Analysis and Design

CAPACITY EVALUATION OF ANCIENT WOODEN BEAMS

Description

- Wood is a construction material which has been used for long time, and still plays a key role, for structural purposes, in many buildings.
- Ancient wooden beams are commonly used as structural elements in many historical buildings.
- Since wood is an organic material, it suffers deterioration due to biological causes and chemical, physical, mechanical attacks.
- Natural wood always presents a number of defects (nodes, shakes, shrinkage slits, shape imperfections, fibre deviations, etc.) and structural anomalies (cracking, deformations induced by long duration loads, connection degradation, etc.) which increase with time.
- According to the heterogeneity and behavioural anomalies of the basic material, any extrapolation of the local properties to the global response of the member and the whole structure present difficulties. Therefore, a comparison between full-scale loading tests and local investigations, by means of destructive (tests on small size specimens) and non-destructive analysis (resistographic analysis) are necessary, in order to evaluate the strength and ductility levels of such structures.

Field of application

- Ancient wooden slabs are an high vulnerability elements of buildings located in seismic areas. Therefore, it is necessary to know their structural reliability level.
- In the common historical buildings, floor slabs were constituted by principal roof beams, with a distance between centers of 70-80 cm. Over these, secondary wooden members carried mortar and pavement (figure 1).

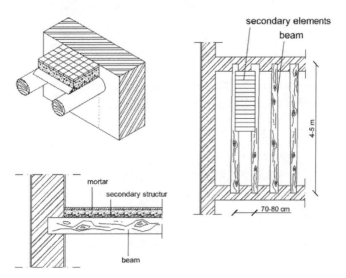

Figure 1. floor slabs in common historical buildings

- Wooden beams under consideration are employed as the supporting structure of floors and roofs in historical buildings.
- The beams were usually disposed without any previous manufacture. They present relevant shape defects, and a remarkable variation of cross sectional dimensions along the beams.

Technical information

- The ancient beams used in floor slabs of common historical building in Mediterranean area are usually made of chestnut.
- The beams are about 4.00-7.00 m long and usually they are only disbarked and present a roughly circular cross-sections. Diameters range is between 15 and 25 cm, varying along beams (figure 2).

Figure 2. Cross-sections at the two ends of an ancient wooden beam.

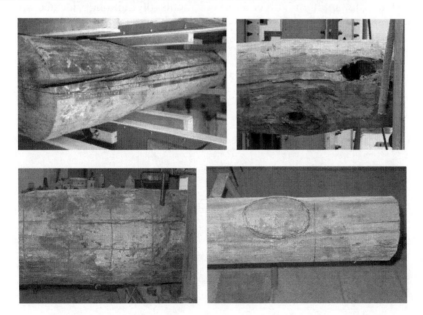

Figure 3. Typical imperfections in ancient wooden beams: longitudinal splitting, holes, lumps, degraded area of a beam with a shake (from top left, clockwise).

- Beams are generally characterised by different kinds of imperfections (longitudinal splitting, cracking, shakes, holes, lumps) and degradation phenomena, due to both age and the natural origin of the basic material (figure 3).

Structural aspects

- With reference to the elastic behaviour, acceptable estimation of the structural performance can be obtained on the basis of material data provided in the literature.
- On the other hand, with regard to the failure mechanisms, the high imperfection and degradation levels characterising the examined beams may favour the activation of slip surfaces induced by shear stresses.
- The collapse load corresponding to the activation of slipping between the upper and the lower part of the beam may significantly differ from the load corresponding to the bending-type collapse mechanism developing in the mid-span section.

Research activity

- A wide research activity is presently in progress at the Department of Structural Analysis and Design of the University of Naples Federico II (Italy).
- It aims at evaluating the elastic and post-elastic behaviour of simply supported ancient wooden beams.
- The attention is focused also on collapse mechanisms different from the mid-span bending-type failure.
- The experimental campaign is based on different levels:
  - tests on ancient wooden beams in actual dimension
  - tests on small size specimens extracted from ancient wooden beams
  - non-destructive tests by means of resistographic penetrations

Method of analysis

- Three-point and four-point bending tests have been carried out in the elastic and post-elastic regime on simply supported ancient wooden beams in actual dimensions, aiming at determining the structural response of the member to progressively increasing external loads (figure 4, 5).

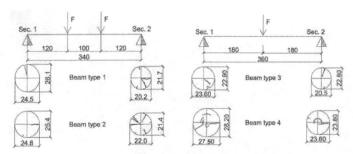

Figure 4. Test program on beams in actual dimensions: adopted static schemes and nominal dimensions of the end cross-sections for each of the analysed specimens.

Figure 5. The adopted experimental set-up for three-point (left) and four-point (right) bending loading.

- Compression (in longitudinal, radial and tangential direction), flexural and shear tests on small size specimens have been taken place out according to the ISO codes, in order to obtain basic mechanical properties and stree-strain behaviour.
- resistographic tests on ancient wooden elements carry out in longitudinal and transversal direction, in order to estimate a correlation between actual structural behaviour of ancient wooden beams and values obtained from non-destructive analysis (figure 6).

Figure 6. phases of the resistographic tests on wooden members

Results of analysis

- Despite the similarity of the beams in actual dimensions both from the geometrical and the mechanical point of view, collapse were not the same for all beams.

Figure 7. Shear-type failure at midspan of one of the tested beams (left); slipping between different parts of the beam at one of the end cross-section (right).

- In some cases, once attained the peak load, the beam experienced a sudden cracking with a significant drop of the applied load; a remarkable longitudinal slipping of the upper part of the beam with respect to the bottom one, due to shear stresses, then took place (figure 7). As displacement increased, the specimen exhibited further strength, even with strongly reduced flexural stiffness.
- In another case, the behaviour was linear-elastic up to the first breaking, but the post-peak behaviour was immediately characterised by total loss of bearing capacity. No appreciable slipping appeared throughout the beam, with an evident flexural failure taking place in the mid-span section, immediately developing a global kinematic mechanism (figure 8).

Figure 8. Bending-type failure at midspan of one of the tested beams.

- For the beams which experienced a shear failure mode with very significant slipping among the split parts of the members, the experimental values of ultimate loads were much lower than those corresponding to the nominal shear resistance, when this is evaluated using the mechanical properties of the basic material.

Further developments

- Simplified methods, normally adopted in order to evaluate the bearing capacity of ancient wooden beams, should be revised in order to take into account more accurately and more conservatively the behaviour of such beams under shear. In fact, tests have shown that this behaviour may be more influential than bending, in terms of collapse mode.
- For the considered wooden beams, defects and imperfections have a strong effect on shear resistance. Therefore, a noticeable reduction of the shear stress should be considered in the structural checking of the member. Besides, flexural resistance does not seem to be affected, even if flexural failure of the tested beams is always attained in the elastic field.
- Further investigations are necessary in order to set up reliable procedures for the adequate evaluation of the bearing capacity of ancient wooden structures. Such procedures should be based on non-destructive diagnostic analyses, capable of assessing the state of the beam.

References
- AITC, 1994. *Timber Construction Manual, American Institute of Timber*

Construction, Englewood, Colo, USA.

- Calderoni B., De Matteis G., Mazzolani F.M., 2002. Structural performance of ancient wooden beams: experimental analysis. In *Technological and technical cultures of European timber construction*, pp. 1-15. Editions scientifiques et medicales Elsevier SAS.
- Bodig J., Jayne B.A., 1982. *Mechanics of Wood and Wood Composites*, Van Nostrand Reinhold Co., New York, USA.

Contacts *(include at least one COST C12 member)*

Prof. Dr. Eng. Federico M. Mazzolani

Address: Department of Structural Analysis and Design – Faculty of Engineering, University Federico II of Naples, P.le Tecchio 80, 80125 Naples, Italy

E-mail: fmm@unina.it, Phone: 0039.081.7682444, Fax: 0039.081.5934792

COST C12 - WG2

Datasheet I.4.1.7

Prepared by: K. Gramatikov[1], T. Kocetov Misulic[2]

[1] St. Cyril & Methodius University, Skopje, Macedonia
[2] University of Novi Sad, Novi Sad, Serbia & Montenegro

TIMBER SHEAR WALLS
–SHEATING TO FRAMING JOINTS

Description

- Timber frame shear walls (TSW) comprise the vertical elements in lateral-force-resisting system. They support the horizontal diaphragm and transfer the lateral loads (wind or seismic forces) down into foundation, behaving like vertical bracing elements and are supposed to resist the vertical (dead and live) loads too.

- Typical timber-frame shear wall consists of timber frame with vertical – chords & studs, and horizontal members – struts (collectors), frame stiffening & blockings, and one or both sided sheathing materials fastened to frame generally with mechanical fasteners. Different tyings, anchors and holdowns are provided for interelements and to foundation connections.

- Under the assumption of uniform shear distribution, there is a number of items that should be considered in the design of timber shearwalls, such as sheathing thickness and shearwall nailing (depending of material and fasteners used in shear wall construction and unit shears), chord and strut design, as well as shear panel proportions and anchorage requirements (Faherty &Williamson, 1989).

- Modern design and behavioural prediction of TSW and timber frame buildings (TFB) is based on experimental investigation of component's materials and joints.

Field of application

- Sheathing to framing joints performances under different loading procedures are recognized as essential for TSW structural behaviour prediction and modelling.

- Results of static loading tests (monotonic and cyclic) conducted on nailed joints in TSW and full size TSW show good agreement (Wakashima & Sonoda, 2001), so it could be said that if hysteretic behaviour of sheathing to framing nailed joints is obtained, the earthquake response of TSW can be predicted without it's loading.

- As the sheet to frame joints refers as one of the most important items regarding the earthquake resistance of TSW, there is a constant need for data basis of their performance. The variety of joint design, mechanical properties of local timber and sheathing materials, as well as types of fasteners, demand an extensive worldwide testing campaign that enables comparison of behaviour of different joint configurations. The main task is possibility of potential generalisation of behavioural mechanisms and joints response, with the aim to improve the prediction models.

- Strength characteristics of applied timber fame / sheathing materials and mechanical fasteners represent input data for joint modelling.

Technical information

- Typical TSW in European market consists from solid timber frame and different sheathings (such as particleboard, oriented strand boards, cement-splinter boards or gipsum-fiber boards, etc.), mutually connected with nails or staples (Figure 1).

- Depending on structural system of timber frame buildings (TFB), skeleton or panel type, shear walls could be recognized as continuous or panel-type (small and large).

- Small panels width basic module is 1.2m, with height-to-width ratio of 2-2.5. Half-module panels are possible. For large panels width is in the range of 2.0-10.0m and height of 2.6-3.0m. In large panels and continuous TSW the studs distances are 0.4-0.7m, depending of chosen sheathing and nailing, as well as framing cross-section's dimensions depend on seismic, wind or climatic region (generally about 4/10 –4/15 cm). Main sheathing thickness is about 9-13mm.

- Nails (smooth, spiral, annular...), staples and screws are commonly used as fasteners. Nailed shear panel systems have very good ductile behaviour, as it stated in EC8, if they meet the proposed EC8 requirements about nail diameter d≤3.1mm, sheathing thickness t=min 4d, and nail length of l=4-6 t. Nail spacing is approx. c/c 10.0cm along the edges and approx. c/c 20.0cm along the middle of the panel.

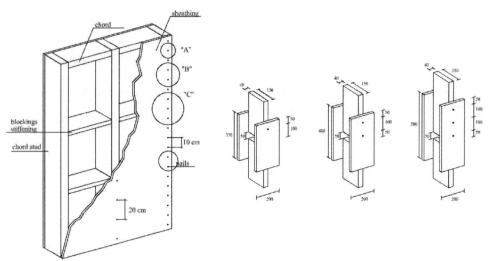

Figure 1. Typical TSW   **Figure 2.** TF / PB sheathing joints with nails
- test models

Structural aspects

- Worldwide structural codes give provisions and requirements for design and testing of TSW. For design purposes, the seismic behaviour of mechanical timber joints can be simplified as quasi-static behaviour under the cyclic loading, where, as a basic measures, are referred ductility and equivivalent viscous damping ratio. Structural detailing and conceptual design are very important issues in order to provide good seismic performance of TSW and TFB (Ceccotti, 2000).

- Testing models have to be designed according structural detailing requirements. It is necessary to provide uniform distribution of acting force and avoid eventual additional slip between specimen and testing device.

- Experimentally obtained data on monotone and cyclic loading response of joints could be referred as basic and valuable source of information needed for development, calibration and verification of inelastic computational models of TSW and consequently of TFB. Additional experimental data, i.e. under the dynamic, pseudo dynamic or "near fault" loading protocols, could give important information needed for correction and improvement of computational model. The choice of loading protocols is an issue what has to be carefully taken into account.

- Experimentally obtained load-slip loops and post failure visual inspection of tested joints are the main source of information regarding joint response under different loading protocols.

Research activity

- Research activities in wood based shear walls date back to the late 1920', but more recent are focused on experimental and modelling approaches in TSW research (EC5/Step 3 – Kallsner & Lam, 1995). Recent nailed joints and full size tests of TSW under dynamic loadings are based on majority of earlier research efforts focused on static case, regarding ultimate load carrying capacity of the system, structural behaviour of system components - connections and sheathing materials (Perkins, 1971; Polensek 1976; Hilton et al., 1976; Patton-Mallory et al., 1985; Polensek & Bastendorff, 1987; etc.). Hysteretic response of TSW gives important data about strength degradation and stiffness deterioration, dissipation of energy and ductility supply. Performed dynamic testing (Stewart, 1987; Dolan, 1989; Yasumura, 1999) and pseudo dynamic (Kamiya, 1988; Kaway, 1998; Wakashima, 2001) contributed to estimate the seismic performance of TSW. Many models, based on experimentally obtained data and with various approaches, have been proposed in order to analyse and predict the performance of TSW subjected to lateral loads (Foschi, 1977; Kamiya, 1981; Itani & Cheung, 1984; Stewart, 1987; Falk & Itani, 1989; Dolan & Fochi, 1991; etc). Finally, full scale testing of TFB was performed and a three-dimensional finite element model was developed to investigate the response of complete light-frame timber structures. 'Till today, the use of macroelements is more efficient than microelements for large structures.

- Current ongoing research activities are focused on gaining more information about ductility, stochastic nature of structural response and connections and sheathing materials in these systems as well as on use of innovative damping devices. Despite commonly used Code sets, limits & requirements (EC5, ASCE, ASTM, IBC, UBC, BCNZ,...), damages caused by recent severe earthquakes and high winds demand more detailed approach in analytical modeling and prediction of structural TSW behaviour, based on experimental investigations. One of the latest extensive research projects, fallowing the Northridge earthquake 1994, was CUREe Caltech Woodframe Project, 1998-2002. Researches were conducted in laboratory and in situ, considering developing of loading protocols, full-scale testing, data analysis and modelling. Results application had the aim to influence standards and codes, economical aspects and knowledge development.

- Constant need for data basis of sheathing / frame joint performance and potential possibility of response generalisation (Smith & Foliente, 2002) was initial demand for several European regional research projects on these topics (SLO-MAC, 2002; USA-MAC, 1990; YU-USA, 1985).

Method of analysis

- Experimentally obtained data are source of main connection properties: initial rigidity; load bearing capacity, ductility, strength degradation, stiffness deterioration and energy dissipation. Structural codes (EC5, EC8, EN26891, pr12512) recommend the testing procedure and methods for determining the relevant properties. As the result of data analysis, the sets of parameters, in the form of functional relationships (such as load degradation vs. joint slip trough cycles, "synthetic" load-slip curve, etc.) could be derived as input data for modelling of non-linear behaviour of TF-sheathing behaviour.

- Bilinear and three linear modelling of monotone and cyclic behaviour according different proposed theoretical models, are commonly used idealisation of experimentally obtained curves. The three linear envelope, fallowing the universal hysteresis rules, is usually defined as non-symmetric. In most cases, available and appropriate existing software influences the modelling tasks in a form of additional and independent parameters - such as strength deterioration after rupture point deformation, hardening, ductility, descending stiffness, stiffness degradation and slipping or pinching effect. Today, the mostly used actual software, suitable for different purposes, is DRAIN-2DX (Prakash, 1993). On the basis of hysteretic joint behaviour parallel and perpendicular to the grain, it is possible to develop spring element for simulation the inelastic nail connection behaviour (Dujic & Zarnic, 2001) and to implement it into DRAIN-2DX (Universal Longitudinal Spring element).

- The specific behaviour of TSW is guided by non-liner behaviour of mechanical fasteners i.e. deformability of TSW reflects in elastic deformation of sheathing material and framing members, and in inelastic deformation of fasteners from the very beginning. Therefore, the numerical analysis of cyclic connection behaviour as a result gives the hysteretic response of the wall. On the basis of this source, it is possible to derivate the mechanical characteristics of inelastic spring which simulate the TSW behaviour and replaces it in 3D modelling of timber building. Suitable software is DRAIN-3DX, DRAIN BUILDING and CANNY-E (Li, 1996).

- As the illustration, the experimental part of ongoing bilateral SLO-MAC project "Development of analytical methods for dynamic analysis of TFB seismic resistance" is presented. The experimental investigation of joints between the timber frame (TF) and particleboard sheathing (PB) under the different loading protocols (monotone, cyclic, dynamic and near fault - Krawinkler & alt., 2002) is done, with the aim to determine the meritory characteristics as the input data for prediction the mathematical model of joints and wall panel. The chosen models (Figure 2) were with 1, 2, 3 nails in a row (E25/50, single shear, parallel to the grain), in order of better simulation of continual sheathing – frame connection. The characteristic responses are presented in Figures 3-8.

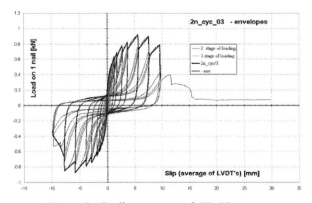 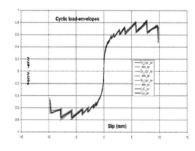

Figure 3. Cyclic response of TF_PB joint with 2 nails

Figure 4. Cyclic responses envelopes for all tested models

- Failure modes observed during the testing are plastic hinges deformation ending with pull-through of nail head (monotone, near fault and cyclic) and plastic hinge failure of nail due to fatigue in depth approximately equal to the PB thickness (dynamic and cyclic loading).

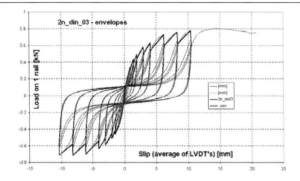

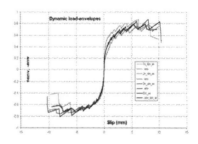

Figure 5. Dynamic response of TF_PB joint with 2 nails

Figure 6. Dynamic responses envelopes for all tested models

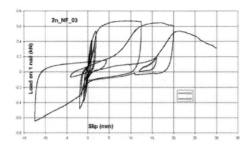

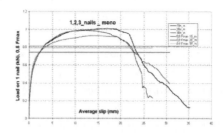

Figure 7. "Near fault" loading protocol response

Figure 8. Summarised mono diagrams

Results of analysis

- The ability to deform plastically is measured by ductility (D), defined as ratio of ultimate (v_u) and yield slip (v_y) in joint. Depending on chosen method for determine the yield point (EC5-8, USA, Japan) the ductility in the same joint could be referred in a wide range (for example from 5 to 15, EC value about 9), so it has to be carefully taken into account.

- The yield and ultimate strength in reversed cyclic loading could be from 5 to 35% smaller than those of the monotonic test. Ultimate displacements could be about 50% smaller. There is no significant difference in results using different loading protocols (CEN vs. ISO).

- Hysteresis equivalent viscous damping ratio (EC8) is referred as a ratio between the dissipated energy (E_d) and the available potential energy (E_p). Values about 8-10% have been evaluated for well-designed dowel type fasteners joints and even higher for nailed sheathing framed walls (15-20%). It could be almost constant regardless of the deformations (Yasumura, 1999).

- Reloading stiffness, stiffness deterioration, nail slip or pinching, ductility, descending parts of curves after rupture point deformation in compression and tension zone, etc., have to be determined numerically from the test diagrams in order to estimate the non-linear hysteresis behaviour and modelling the soft spring element for TSW.

- In modelling of TSW, each nail has to be represented by two springs – parallel and perpendicular to the grain. The framing members are considered as linear elastic beam elements with both edges plastic hinges.

Further developments

- As it was stated, more experimental data on different joint configuration with variety of local timber, sheathing materials and fasteners, obtained under the standardised loading protocols and analysed by similar numerical and analytical tools, could lead to potential generalisation.
- Influence the Structural codes by simplified methods for prediction of TSW behaviour.

References

- Ceccotti, A. 2000 *: Seismic behaviour of timber buildings – introduction;* Preliminary proceedings of the Workshop on seismic behaviour of timber buildings, COST E5 – Timber frame building systems, Venezia, Italy.
- Dujic, B., Zarnic, R. 2001: *Numerical model & Predictions – Team Slovenia;* CUREe Caltech Woodframe Project, Proceedings of International Benchmark Workshop, p.15, University of California, San Diego, USA.
- Faherty, K. & Williamson, T. 1989: *Wood engineering and construction handbook - Diaphragms & Shear Walls;* 68p., McGraw-Hill, Inc., USA.
- Gramatikov, K., Kocetov Misulic, T. 2002: *The experimental investigation of particleboard sheathing – timber stud connections in shear walls;* C12 COST meeting, p. 6, Timisoara, Romania.
- Kallsner, B. & Lam, F. 1995. Holzbauwerke nach Eurocode 5 – Step 3: *Diaphragms and shear walls;* 19p., Druckhaus Fromm, Osnabruck, Germany.
- Krawinkler, H., Parisi, F., Ibarra, L., Ayoub, A., Medina, R. 2001: *Development of a testing protocols for wood frame structures;* CURREe Caltech Woodframe Project - Final review draft, p.89, University of California, San Diego, USA.
- Prakash, V., Powell, G.H. 1993: *DRAIN-2DX, DRAIN-3DX and DRAIN BUILDING – Base program design documentation,* Rep.no UCB/SEMM-93/17, Berkeley, University of California, USA.
- Smith, I., Foliente, G. 2002: *Load and resistance factor design of timber joints: International practice and future direction;* Journal of Structural Engineering, Vol. 128, pp.1: 48-59,
- Wakashima, Y. & Sonoda, S. 2001: *Evaluation of seismic performance of nailed joints by pseudo-dynamic test;* Proceed. of Intern. RILEM Symp. – Joints in Timber Structures, pp.551-558, Stuttgart, Germany.
- Yasumura, M. 1999: *Static and Dynamic Performance of Wood-framed Shear Walls;* Proceed. of 1st Intern, RILEM Symp. on Timber Engineering, pp. 481-490., Stockholm, Sweden.

..

Contacts

- Kiril Gramatikov
    Address: Civil Engineering Faculty, University of St. Cyril & Methodius, Bul. Partizanski odredi 24, POB 560, Skopje, Rep. of Macedonia.
    e-mail: gramatikov@gf.ukim.edu.mk Phone/Fax: + 389 2 3117 367
- Tatjana Kocetov Misulic
    Address: Faculty of Technical Sciences, University of Novi Sad, Trg Dositeja Obradovica 6, 21000 Novi Sad, Serbia & Montenegro
    e-mail: tanya@uns.ns.ac.yu Phone: + 381 21 459 798 Fax: + 381 21 459 295

Prepared by: L. Krstevska[1], D. Ristic[2]

[1] Institute of Earthquake Engineering and Engineering Seismology (IZIIS),
University St. Cyril and Methodius, Skopje, FY Republic of Macedonia
[2] IZIIS, Skopje, FY Republic of Macedonia

NON-LINEAR STRESS-STRAIN RELATIONS OF PLAIN AND REINFORCED MASONRY INFILLS AS A PART OF RC FRAME STRUCTURES

Description

- Frame structures with masonry infill represent systems that are widely present worldwide, particularly in Europe. The nonlinear response of these systems to the effect of strong earthquakes represents one of the most complex physical phenomena, whereas the complexity of the characteristics of nonlinear behaviour is mainly due to the complex nonlinear behaviour and the complex failure mechanisms of the materials used for the structure.

- The complexity of modeling the nonlinear response of the frame is increased also due to the effect of the infill whose nonlinear response depends very much upon the specific characteristics of the applied materials. To provide relevant experimental data enabling realistic interpretation of the nonlinear behaviour of the brick masonry infill up to failure, a series of experimental tests were carried out at the Dynamic Testing Laboratory in IZIIS.

Field of application

- Within the frames of the international cooperative project INCO COPERNICUS PROJECT No. IC15-CT97-0203, "TOWARDS EUROPEAN INTEGRATION IN SEISMIC DESIGN AND UPGRADING OF BUILDING STRUCTURES", realized in the period October 1997 - January 2001, financed by the European Commission, complex investigation of reinforced-concrete frame structures with masonry infill has been performed from several aspects, by participation from seven renowned European institutions: University of Ljubljana, Slovenia, University in Bucharest, Romania, Institute for Structures and Architecture, Slovakia, ISMES, Bergamo, Italy, University in Bristol, Great Bretain, JRC, Ispra, Italy and Institute of Earthquake Engineering and Engineering Seismology (IZIIS), Skopje, Republic of Macedonia.

- The research activities within the project related to refined nonlinear numerical simulation of RC frames with infill have successfully been realized in IZIIS, Skopje, Republic of Macedonia. A new concept for nonlinear refined micro analysis of frame structures with infill of non-reinforced and reinforced masonry, with or without openings for doors or windows has been proposed.

- The verification of the applied analytical procedure through determination of the nonlinear response of these systems to earthquake effect showed that the nonlinear response of frame systems with masonry infill can successfully be simulated by use of the proposed concept for refined nonlinear micro-modeling and by application of the specially developed computer programme EURO NORA.

Technical information

- The masonry models on which experiments have been carried out were constructed in full scale by use of the same materials (hollow bricks, mortar, geogrids) that are used in practice. They were constructed of identical type of bricks with vertical holes (42% of the total cross-section area), with proportions 250x120x190 mm and compressive strength normal to the joints of σ_{bc} =13.3 MPa. The mortar was of category M3, with volume cement:mortar:sand ratio of 1 : 1 : 5 and average compressive strength of σ_{mc} = 5 MPa. The proportions of the models were b/h = 90/100 cm. Figure 1 shows the tested types of masonry models.

- Geogrids type TENSAR SS30 were used for strengthening, i.e., reinforcement of the models of reinforced masonry. The average yielding force under tension defined experimentally is F_{yt} = 29 kN/m (tested sample with a width of b = 100 cm), whereas the corresponding strain for that stage of yielding is ε_{yt} = 5% = 0.05.

Running bond (120mm) Flemish bond (250mm) Flemish bond (250mm) reinforced

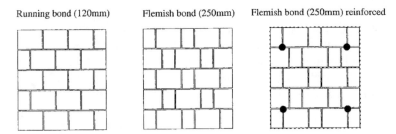

Figure 1. Types of masonry models tested experimentally

Structural aspects

- The experimental definition of stress-strain relationships of masonry is of an essential importance for the micro-modeling concept, enabling realistic simulation of the nonlinear behaviour of the infill within the frames of the RC frame structures under dynamic effect.

- Realization of basic experimental research activities was programmed within which frames the fundamental stress-strain relationships were experimentally defined for several different and representative types of infill, referring to both ordinary and reinforced infill. The type of failure and the collapse load were different for different types of tested masonry.

- For micro-modeling of an integral RC frame system with masonry infill, corresponding **models of nonlinear discrete components (NDC-models)** have been developed to represent the specific nonlinear characteristics of the different structural elements and materials of the system. Four different types of NDC models have been developed: 1) **NDC-FRAME model** - for nonlinear modeling of the behaviour of the structural elements of the frame - the columns and the beams; 2) **NDC-INFILL model** - for modeling of the infill elements; 3) **NDC-MORTAR model** - for modeling of behaviour of mortar in all the contact zones; 4) **NDC-REINFO model** - for modeling of the behaviour of reinforcement, in the case of reinforced masonry.

Research activity

- One of the main objectives of the complex INCO COPERNICUS European project was to investigate the possibility of increasing the ductility and capacity for energy dissipation of the masonry infill in RC frames. The proposed and adopted methodology (Sofronie, 1998) recommended to apply geogrid as confinement of the infill and also placing of geogrid in few horizontal mortar layers. A part of the complex experimental investigations for investigating the effectiveness of this methodology were the experimental research activities performed in IZIIS on masonry wall models.

- Based on the defined stress-strain relationships obtained by analysis of the measured values during the experiments, the quantities necessary for the comparative analysis of the advantages, i.e., differences among the tested types of masonry as well as the parameters necessary for further micro-analysis of the adopted analytical models were computed (Krstevska, Ristic, 2001). Based on the mean envelope curves (since two identical wall models were available for each test), idealization of the force-deformation $(P\text{-}\delta)$ i.e., $\sigma\text{-}\varepsilon$ and $\tau\text{-}\gamma$ relationships was further done for each tested type of infill whereby were obtained the parameters necessary for modeling of the nonlinear behaviour of the infill elements incorporated in the analysis.

- Taking into account the proposed analytical micro-model, the EURO-NORA computer programme has been developed and used in the analyses (Ristic, 2001). The programme is conceptualized such that linear and nonlinear static and dynamic analysis of two-dimensional structures can be performed.

- To verify the practical application of the formulated nonlinear micro-model for real numerical simulation of nonlinear behaviour of RC frame structures with masonry infill under earthquake excitation, applying the EURO-NORA programme, modeling and seismic analysis of two different plane frame models with masonry infill tested experimentally in laboratory conditions have been performed (Krstevska, Ristic, 2001).

Method of analysis

- Two characteristic types of experimental tests were performed in IZIIS: (1) tests with simulated diagonal pressure performed on 8 experimental full scale models, and (2) tests by simulated vertical pressure performed on 4 models, Figure 2. Tested were 4 different types of masonry and for each of these, 2 identical models were tested.

- In both types of tests, the compressive force was applied in semi-cycles by gradual application of force up to failure of the element via an actuator with a capacity of F_{max} = 1500 kN. To measure the physical quantities of interest, all the models were instrumented by corresponding gauges.

- Based on the experimentally defined stress-strain relationships, the quantities necessary for the comparative analysis of the advantages, i.e., differences among the tested types of masonry as well as the parameters necessary for further micro-analysis of the adopted analytical models were computed.

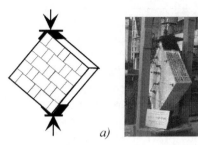 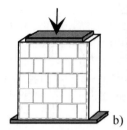 

Figure 2. Type of tests: (a) diagonal compression, (b) vertical compression

Results of analysis

- The type of failure experienced by the models was complex in all the tested elements whereat all the incorporated components (bricks, mortar and geogrids) affected the mode of failure (Figures 3, 4 and 5).
- Depending on the mode in which the models were tested (diagonal pressure or axial pressure), the strength characteristics of masonry are considerably different, which must be appropriately treated also in the formulation of the analytical models;
- The bearing capacity under diagonal pressure of elements constructed of ordinary masonry with a thickness of 25cm is about 55% greater than the bearing capacity of the elements with a thickness of 12.5cm, which means that there is no linear proportionality in the bearing capacity-wall thickness relationship. This is certainly due to the effect of the complex composition of the masonry. Referring to the capability for energy absorption, the models with thickness of 25 cm show considerably better capability. The greater ductility of this masonry is due to gradual loss of bearing capacity due to the complex participation of the different components in the failure process, while the loss of strength and the failure of the thinner element are very sudden.
- The plastering increased the bearing capacity significantly, but there was a very sharp drop in strength after the occurrence of cracks, which directly resulted in decrease of ductility.
- The reinforced masonry has increased bearing capacity, about 30% in diagonal compression and 15% under axial compression, significantly improved ability for energy absorption, greater ductility and protection of the element integrity, stable behaviour with controlled and uniform failure, properties preferred in earthquake conditions.
- The bearing capacity of masonry under axial compression is 4 to 6 times greater than the capacity under pure shear loading.
- Considering the complexity of masonry as a material and the possible variations in its basic strength parameters, it is important to perform a greater number of experimental tests for obtaining reliable data to characterize a certain masonry type. The analysis of the results should be carefully and critically carried out. In such a way, a qualitative analytical simulation of the nonlinear behaviour of the infill and its realistic interpretation as a part of the structure might be expected.

Figure 3. Failure mechanism of elements tested under diagonal compression

Figure 4. Failure mechanism, axial compression tests

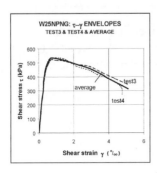
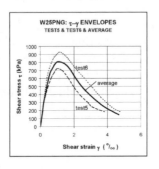
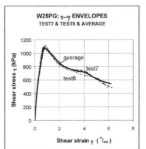

Figure 5. Stress-shear strain relationships, diagonal compression tests

Further developments

- There is a need for qualitative experimental base for definition of the basic mechanical characteristics of the materials, which means caring out a lot of elementary tests and taking a serious approach to analyzing the obtained results.

- There is a need for more intensive experimental and analytical investigations in order to enable development of more refined analytical models for simulation of the basic relationships of the built-in materials.

References

- Final Report (1.10.1997-31.01.2001) for the INCO-COPERNICUS Contract IC15-CT-97-0203 – *Towards European Integration in Seismic Design and Upgrading of Building Structures (EUROQUAKE)* 2001, University of Bristol, G. B.

- Krstevska, L., & Ristic, D., 2002. *Advanced micro-modeling of RC infilled frames-part 1: Basic stress-strain tests of masonry infills*, Conference on Structural Dynamics, EURODYN2002, Munich 2002, p.p. 949-954

- Krstevska, L., 2002. *Development and application of non-linear micro-models for evaluating of seismic behaviour of RC frames infilled with plain and reinforced masonry*. Doctoral dissertation, Skopje, Rep. of Macedonia

- Ristic, D., Yamada, Y., Iemura, H., Petrovski, J., 1988. *Nonlinear behaviour and stress-strain based modeling of reinforced concrete structures under earthquake induced bending and varying axial loads*, Kyoto University, School of Civil Engineering, Research report 88-ST-01

- Ristic, D., & Krstevska, L., 2002. *Advanced micro-modeling of RC infilled frames-part 2: Phenomenological non-linear discrete component models*, Conference on Structural Dynamics, EURODYN2002, Munich 2002, p.p. 1405-1410

- Severn, R., Conti, P., Fajfar, P., Franchioni, G., Juhasova, E., Pinto, A., Ristic, D., Sofronie, R., Taylor, C., Zarnic, R., 1998. *First Annual Report for the INCO-COPERNICUS EUROQUAKE IC15-CT-97-0203 Project*, University of Bristol, Bristol, G. B.

Contacts *(include at least one COST C12 member)*

- Lidija Krstevska
  Address: Institute of Earthquake Engineering and Engineering Seismology (IZIIS), Department Experimental Mechanics & Structural Control, University "St. Cyril and Methodius", Salvador Aljende 73, Skopje, Republic of Macedonia.
  e-mail: lidija@pluto.iziis.ukim.edu.mk, Phone: +389 2 3176 155, Fax: +398 2 3112 163

Datasheet I.4.2.1

Prepared by: G. Huber[1], U. Kuhlmann[2], M. Schäfer[2]

[1] Innsbruck/Austria, www.huber.net.tf
[2] Institute of Structural Design, Stuttgart/Germany, www. uni.stuttgart.de

ROTATION CAPACITY OF JOINTS

Description

- general background information on the derivation of available joint rotation (robustness) in comparison with required joint rotation
- objective general aspects and details for the case of composite joints
- field of application: joints under predominantly static loading

Definitions

- "*joint*" is the assembly of basic components; a beam-to-column joint consists of a web panel and either one or two "connections" (single or double sided)
- "*basic component*" is a specific part that makes an identified contribution to its structural properties
- main structural joint properties: resistance to internal forces and moments $M_{j,Rd}$, rotational stiffness $S_{j,ini}$, rotation capacity Φ_{pl} (Figure 2)
- *joint characterisation*: determination of the (available) joint's moment-rotation curves or key values (Figure 1a)
- *joint classification*: a possible but not obligatory tool for simplification of modelling (nominally pinned or nominally continuous)(Figure 1b)
- *joint idealisation*: conversion of more realistic moment-rotation curves into simplified (linearised) ones (Figure 1c)
- *joint modelling*: reproduction of joint behaviour within the structural analysis (Fig. 1d)
- *available rotation*: ϕ_{pl} is the overall rotation capacity; $(\phi_{pl} - \phi_{el})$ is the plastic rotation capacity (for global redistribution; corresp. to rotation capacity of members) (Figure 2)

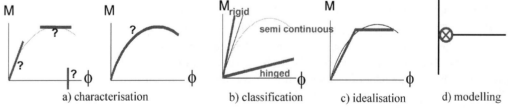

a) characterisation   b) classification   c) idealisation   d) modelling

Figure 1. Definitions

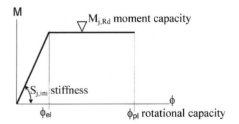

Figure 2. Characteristic values

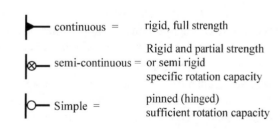

continuous =	rigid, full strength
semi-continuous =	Rigid and partial strength or semi rigid specific rotation capacity
Simple =	pinned (hinged) sufficient rotation capacity

Figure 3. Types of joint modelling

Method of global analysis

- three simplified joint models: simple, semi-continuous, continuous (Figure 3)
- applicability of a model as result of the joint classification
- beyond elastic global analysis local plastifications in the members and / or joints are progressing due to load increase leading to redistribution of internal forces
- in rigid-plastic or elastic-plastic global analysis based on plastic hinges it is assumed that the resistance keeps fully reliable despite of a continuous progress of deformations.
- deemed-to-satisfy criteria already exist for member classes; but such simple rules still have to be developed for joints; alternatively the designer has to compare required rotation values in the joints against the available rotation capacity.
- Attention! Component overstrength can lead to an unforeseen change of the decisive failure mode and therefore to a wrong calculation of the available joint's rotation capacity.

Available rotation capacity

- *JOINT CHARACTERISATION*: different possibilities to derive the available joint characteristics (here focus on rotation capacity):
    - i) experimental with full-scale joint tests
    - ii) numerical based on Finite Element calculations
    - iii) analytical based on component method as an every-day design tool

  Component method: macroscopic subdivision of the complex joint into basic components (modelled by translational springs); method is generally applicable for static and monotonic joint behaviour as well as for cold and fire state

  Steps of characterisation:
    - i) identification of contributing components
    - ii) component characterisation (force-deformation curves)
    - iii) assembly of the component properties or curves to those of the joint

- *COMPONENT IDENTIFICATION*: (Figure 4) distinguish between components loaded in tension (or bending), compression or shear; distinguish between components of the connecting elements, of the load-introduction into the column web panel and the component "column web panel in shear"; subdivision leads to the spring model (Figure 5); by the spring model (interplay of deformation influences) an economic joint construction can be consequently derived

- *COMPONENT CHARACTERISATION*: derivation of force-deformation curves of the individual contributing components (key values or full curves); different levels of accuracy (component tests - experimental approach; Finite Element simulations - numerical approach; analytical mechanical models)

  sophisticated mechanical models (developed at several research centres; validated against component tests and numerical simulations) -> reduced to easy-to-handle formats (by comprehensive parameter studies) which are integrated in the codes

  so far more focus to stiffness and strength values; little investigations and guidance available on the components' deformation capacity

- *COMPONENT ASSEMBLY*: transfer from all isolated components' force-deformation curves to the overall joint's moment-rotation curves fulfilling the requirements of equilibrium and compatibility; basing either on sophisticated or on simplified spring model (different modelling of the components' interplay) (Figure 5); assembly either only for the main key values (initial rotational stiffness, plastic moment resistance, failure mode) or for the full shape of the resulting M-ϕ curves; depending on these choices assembly either by iterative/computational procedure or by a simple hand calculation; especially in view of available rotation capacity a simplified analytical approach for a hand calculation could not yet be provided for all cases; only principle guidelines in the codes extrapolating from the failure mode, see for steel joints (Kühnemund, 2003). For more accuracy several software tools are available such as:

  i) (Silva/Coelho, 2001): formulae on the basis of an energy formulation

  ii) (Huber, 2000): computational tool (Figure 6) based on an iterative approach covering both the sophisticated "Innsbruck-model" (Figure 5a) as well as the simplified component model ENV 1993-1-8 (Figure 5b)

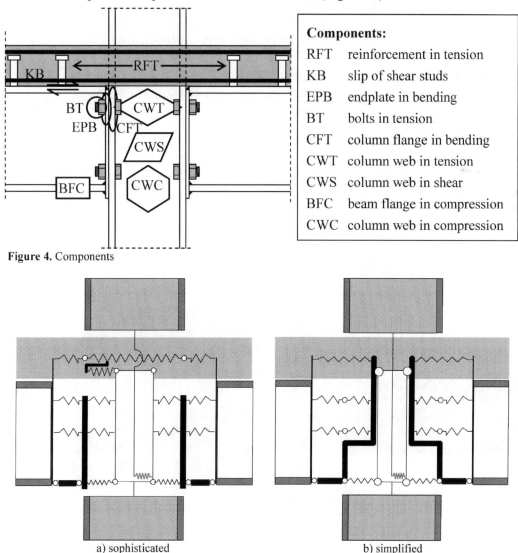

Components:

RFT	reinforcement in tension
KB	slip of shear studs
EPB	endplate in bending
BT	bolts in tension
CFT	column flange in bending
CWT	column web in tension
CWS	column web in shear
BFC	beam flange in compression
CWC	column web in compression

Figure 4. Components

a) sophisticated   b) simplified

Figure 5. Spring model: Accuracy levels of assembly

- *IMPROVEMENT OF DUCTILITY*
    - i) Failing components have to be ductile; ultimate strength and not yield strength is responsible for the failure mechanism; consider hereby additional unforeseen component strength due to strengthening and overstrength of material, membrane effects (EPB) and larger effective width (RFT) than calculated
    - ii) Only the plastifying components contribute remarkably to the rotational capacity
    - iii) BT: full threaded bolts, low material grade, avoid thread stripping (Steurer 1996+1999)
    - iv) EPB + BT: d_{bolt} /t_{plate} ductility criteria ENV 1993-1-1/A2 Annex J (J.49) may not avoid failure of the bolts, see Figure 9a, increase bolt diameter and arrange the bolt at a lower level (see Figure 9b)
    - v) CFT + BT: mode 1 failure (yielding of the plate), according to ENV 1993-1-1/A2 Annex J (J.12), see also Figure 9b
    - vi) CWC: reduce the calculative strength to consider the descending ductile branch of the component deflection curve, see Figure 7 (Kühnemund 2003) stiffener or filled in concrete to avoid failure
    - vii) RFT: Reinforcement properties $f_u/f_y > 1,13$, large spacing from first shear stud to the connection (Figure 8), full shear connection, (Kuhlmann/Schäfer 2003)

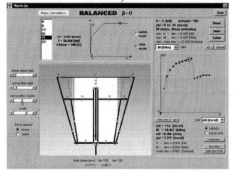

Figure 6. CoBeJo program for component assembly

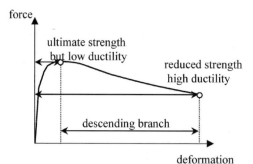

Figure 7. Typical force-deflection curve of CWC

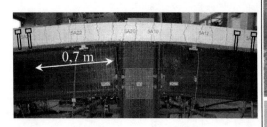

Figure 8. High-ductile composite joint

a) bolt: brittle failure   b) endplate: ductile failure
Figure 9. Failure mode

Required rotation capacity

see $\Phi_{j,req}$ in Figure 13

- Required rotation ("ductile" joint) has to be proven for plastic global analysis
- Required rotations may reach 35 to 70 mrad , see Figure 11
- In case of plastic hinge in the joint (due to partial strength $M_{j,Rd} < M_{Rd}$) $\Rightarrow$ joint has to provide sufficient rotation capacity ϕ_{pl} for further plastic redistribution within the beam (Figure 10)
- Plastifications of partial-strength joints occur under ultimate limit load
- For plastified joints, the required rotational capacity $\Phi_{j,req}$ is dependent on the plastic moment resistance of the joint $M_{j,pl}$ and not on the joints' stiffness $S_{j,ini}$, see Table 1
- For ultimate limit load (plastic hinge theory) the required joint rotation is linearly dependent on the beam's slenderness L/h, the steel grade (influence of yield strain ε_y), and the degree of restraint M_{hog}/M_{sag} (Kuhlmann/Schäfer 2003)
- Yield zones in the beam enlarge the required rotations, so plastic hinge theory underestimates the required rotations (Huber 2000, Kuhlmann/Schäfer 2003)
- Sway-frames require much higher rotations than non-sway ones (Kuhlmann/Schäfer 2003)
- Unpropped construction for composite structures increases the required rotations (Kuhlmann/Schäfer 2003, Kattner 1999

Errore. Il collegamento non è valido.Table 1. Required rotations (see Figure 10)

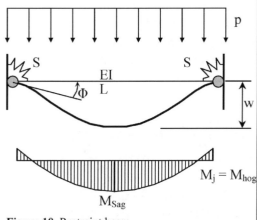

Figure 10. Restraint beam

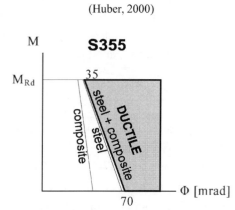

Figure 11. Proposal for required rotation boundaries

Verification procedure

- Separated verification of moment and rotation:

i) $M_{j,d} \leq M_{j,pl,Rd} / \gamma_{Mj}$

ii) $\Phi_{j,req} \leq \Phi_{j,avail} / \gamma_{M\Phi}$

Safety factor $\gamma_{M\Phi}$ has to be drawn from stochastic considerations (Steenhuis 2000)

- Beamline method, see Figure 12 (Kuhlmann/Schäfer 2003): The beamline illustrates the requirements; if the joint characteristic "meets" the beamline, no failure occurs

- Deemed-to-satisfy criteria: simple, suitable for practical application, are to be developed in the future

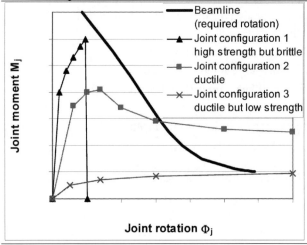

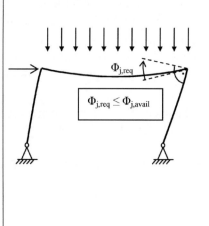

Figure 12. Beamline method   **Figure 13.** Verification of rotational cap.

Field of application

- fabrication of ductile partial-strength joints is economical

- ductile partial-strength joints allow for a moment redistribution with an optimised utilisation of the structural material especially for steel-concrete composite structures

- highly ductile joints can sustain large deformations that may occur under exceptional loading (e. g. earthquake, fire) and thus improve robustness and structural integrity

Research activity

- composite frames with partial strength joints (Kattner 1999, Kozłowski 1999, Kuhlmann/Schäfer 2003)

- semi-rigid joints (Weynand 1997, Jaspart 1991, ...)

- composite joints (Huber 2000, da Silva et al. 2000, Kuhlmann/Schäfer 2003)

- endplate in bending (Kuhlmann/Sedlacek/Stangenberg/Kühnemund 2002, Faella 2000, Steurer 1999)

- column web in compression (Kühnemund 2003)

- bolts in tension (Steurer 1996)

Method of analysis

Required rotations $\Phi_{j,req}$ of the structural system (Figure 13):

- plastic hinge theory underestimates the required rotations caused by plastifications
- yield zone theory calculates high required rotations when a plastification in the beam occurs

Available rotations $\Phi_{j,avail}$ of the joint:

- component model

Further developments

- Experimental studies on the deformation capacity of missing components
- Simple formulae for describing the deformation capacity of missing components
- Parameter studies on overstrength effects evaluating the joint ductility
- Stochastic simulations (Borges 2003) of required and available rotations, leading to deemed-to-satisfy criteria for joint classes

Technical information

- prEN-1993-1-8, 2003. *Design of steel structures, Design of joints.* November 2003.
- ENV-1993-1-1/pr A2, 1994. *Revised annex J (version Nov.94): Joints in building frames.*
- prEN-1994-1-1, 2003. *Design of composite steel and concrete structures, Section 8 Composite joints in frames for buildings.* 6 May 2003.
- ECCS, 1992. *Analysis and design of steel frames with semi-rigid joints.* ECCS TC8, report no.67.
- ECCS, 1999. *Design of composite joints for buildings (incl. Model code provisions for composite joints in building frames).* TC11, Brussels.
- Commission of the European Communities, 1997. *Frame design including joint behaviour* (Vol.1-3). Executive Committe Ref.no. 93-F6.05, ECSC no.7210-SA/212+320.
- COST C1, 1992. *Semi rigid behaviour of civil engineering structural connections.* Proceedings of the 1st state of the art workshop, Strassbourg.
- COST C1, 1994. *Semi rigid behaviour of civil engineering structural connections.* Proceedings of the 2nd state of the art workshop, Prague.
- COST C1, 1996. *Composite steel-concrete joints in braced frames for buildings.* Brussels/Luxembourg.
- COST C1, 1997. *Semi rigid behaviour of civil engineering structural connections.* Proceedings of the International Conference, Liège.
- COST C1, 1999. *Recent advances in the field of structural steel joints and their representation in the building frame analysis and design process.* Brussels/Luxembourg.

References

- Borges, L., 2003. *Probabilistic evaluation of the rotation capacity of steel joints.* Master thesis, Faculdade de Ciências e Tecnologia, especialidade de Esterturas. Coimbra.

- Faella, C., Piluso, V., Rizzano, G., 2000. *Structural steel semi-rigid connections. Theory, Design and Software.* CRC Press Boca Raton, London, New York.

- Huber, G., 2000. *Non-linear Calculations of Composite Sections and Semi-continuous Joints.* Ernst & Sohn, Berlin, ISBN 3-433-01250-4.

- IABSE, 1997. *International Conference "Composite Construction - Conventional and Innovative"* in Innsbruck. Zürich.

- Jaspart, 1991. *Etude de la semi-rigidité des noeuds poutre-colonne et son influence sur la résistance et la stabilité des ossatures en acier.* Doctoral thesis, Liège.

- Kattner, M., 1999. *Beitrag zum Entwurf von Rahmen mit Verbundknoten im Hochbau.* Thèse MK No. 2055, Ecole Polytechnique Fédérale de Lausanne.

- Kozłowski, A., 1999. *Kształtowanie szkieletów stalowych i zespolonych o węzłach półsztywnych.* Steel and composite frames with partial-strength joints. Oficyna Wydawnicza Politechniki Rzeszowskiej.

- Kuhlmann, U., Sedlacek, G., Kühnemund, F., Stangenberg, H., 2002. Research report: *Vorhandene Rotationskapazität wirtschaftlicher Anschlußkonstruktionen auf der Basis des Komponentenverhaltens für die Anwendung plastischer Bemessungskonzepte im Stahlbau.* Funded by Studiengesellschaft für Stahlanwendung e. V. Institute of Structural Design, University of Stuttgart. Proceedings No. 2002-3.

- Kuhlmann, U., Schäfer, M., 2003. Research report: *Innovative verschiebliche Verbundrahmen mit teiltragfähigen Verbundknoten.* Funded by Studiengesellschaft für Stahlanwendung e. V. Institute of Structural Design, University of Stuttgart. Proceedings No. 2003-46.

- Kühnemund, F., 2003. *Zum Rotationsnachweis nachgiebiger Knoten im Stahlbau.* PHD thesis, Institute of Structural Design, University of Stuttgart, Proceedings No. 2003-1.

- Schäfer, M., 2004: *Verification of the rotational capacity of composite sway frames.* Doctoral thesis, in preparation.

- Steenhuis, M., van Herwinen, F., Snijder, H., 2000. *Safety concepts for ductility of joints.* International AISC/ECCS workshop on connections in steel structures IV, Roanoke, Virginia, USA.

- da Silva, L., Coelho, A., 2001. *A ductility model for steel connections.* Journal of Constructional Steel Research, 57(1), 45-70.

- da Silva, L., Simoes, R., Cruz, P., 2001: *Experimental behaviour of end-plate beam-to-column joints under monotonical loading.* In Engineering Structures.

- Steurer, A., 1996, *Trag- und Verformungsverhalten von auf Zug beanspruchten Schrauben.* ETH Zürich, IBK-Bericht Nr. 217. Institut für Baustatik und Konstruktion, ETH Zürich. Birkenhäuser Verlag Basel, Mai 1996.

- Steurer, A., 1999. *Tragverhalten und Rotationsvermögen geschraubter Stirnplatten-verbindungen.* IBK Bericht Nr. 247, Institut für Baustatik und Konstruktion, ETH Zürich. Birkenhäuser Verlag, Basel, Dezember 1999.

- Weynand, K., 1997. *Sicherheits- und Wirtschaftlichkeitsuntersuchungen zur Anwendung nachgiebiger Anschlüsse im Stahlbau.* Doctoral thesis, Aachen.

Contacts

- Dr. Gerald Huber

  Address: Kohlsiedlung 9, A-6091 Götzens, Austria
  e-mail: geraldhuber@telering.at
  http://members.telering.at/geraldhuber/home.html

- Prof.-Dr.-Ing. U. Kuhlmann

  Address: Institute of Structural Design, University of Stuttgart, Pfaffenwaldring 7, D-70569 Stuttgart
  e-mail: u.kuhlmann@ke.uni-stuttgart.de
  http://www.uni-stuttgart.de/ke/f_kuhlmann.html

- Dipl.-Ing. M. Schäfer

  Address: Institute of Structural Design, University of Stuttgart, Pfaffenwaldring 7, D-70569 Stuttgart
  e-mail: martin.schaefer@ke.uni-stuttgart.de
  http://www.uni-stuttgart.de/ke/f_mitarbeiter.html

CYCLIC ROTATION CAPACITY OF BEAM-TO-COLUMN JOINTS

Description

- Actual design codes and recommendations for ductile Moment Resisting Steel Frames in seismic zones require to beam-to-column joints to sustain a total rotation of 0.035 rad (in Europe) to 0.040 rad (in United States). The problem is that in codes, there are no analytical methods to predict the rotation capacity of MR connections, not even for monotonic loading, which is simpler then compared with seismic action of cyclic type. Even if last years appeared some proposals for analytical or numerical based methods to evaluate rotation procedure for their qualification in terms of ductility, still remain the pre-qualification tests, both for static and seismic actions.

- In present datasheet, based on extended experimental studies, comparatively conducted under monotonic and cyclic loading on different types of connections, both for steel and composite MR frames, a reduction factor to be applied to monotonic rotation capacity is proposed in the purpose of evaluation of cyclic rotation capacity.

Field of application

- Four different experimental studies aiming to evaluate performances of both steel and composite beam-to-column joints under cyclic loading will be summarised.

Technical information

- **Rotation capacity for welded joints (INSA Rennes)**
  Aribert & Grecea (1998) have developed an experimental research program at INSA Rennes. This research program dealt with 8 beam-to-column welded joints under monotonic and repeated cyclic loading. The specimens were major axis joints with a symmetrical cruciform arrangement comprising an H or I column connected to two cantilever beams. No transverse stiffener was welded in the compression zone of the column web.

  The tests were performed according to the Recommended Testing Procedure of ECCS (1985). Each type of joint was subject first of all to a monotonic loading (CPP11, CPP13, CPP15 and CPP17, then to cyclic reversal loading (CPP12, CPP14, CPP16 and CPP18).

- **Rotation capacity for welded and bolted joints (UP Timisoara)**
  Investigations on beam-to-column joints, carried out at the laboratory of steel structures at the Civil Engineering Faculty of Timisoara are presented hereafter.
  Three typologies of beam-to-column joints have been tested from a total of 12 specimens. All the joints are double sided. For all the specimens the design steel grade was S235 (f_y=235 N/mm^2, f_u=360 N/mm^2), beams being IPE 360 and columns HEB 300.

  Two types of loading were applied: symmetrical and anti-symmetrical and three connection typologies were tested (Figure 1).

The joints (three different joint configurations and two load types) have been designed according to EN 1993-1-8. The loading history was made according to the ECCS Recommendations simplified procedure.

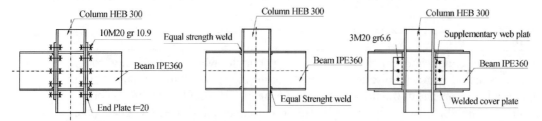

Figure 1. Connection configurations: (a) bolted, (b) welded and (c) with cover welded plate

- **Rotation capacity for bolted steel and composite joints (UP Timisoara)**

An alternative to the "standard" European column cross-section (hot-rolled I profiles) is the use of X-shaped cross-sections, built-up of two hot-rolled profiles welded along the median axis or built-up sections made out from welded plates, as shown in Figure 2.

The testing program comprised six specimens: three joints under symmetrical loading (BX-SS and BX-CS), and three joints under anti-symmetrical loading (BX-SU and BX-CU). Tests were performed in accordance with the ECCS Recommendations complete procedure (ECCS 1985). The first specimen from each series was tested monotonically, The load was applied quasi-statically, under displacement control.

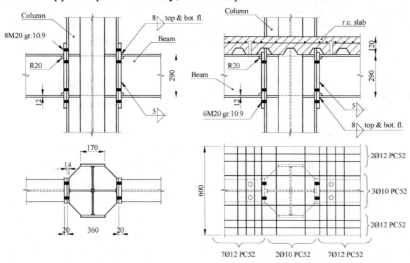

Figure 2. Connection configurations for BX-S series (a) and BX-C series (b) of joints.

- **Rotation capacity for bolted steel and composite joints (INSA Rennes)**

Ciutina (2003) has tested some composite connections at INSA Rennes (Figure 3) under monotonic and cyclic loads. G14 and G15 are composite joints under monotonic and cyclic loading, and G17 and G18 are composite joints with haunch under monotonic and cyclic loading.

Tests were performed in accordance with the ECCS Recommendations complete procedure (ECCS 1985).

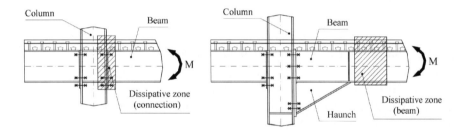

Figure 3. Tested joints

Structural aspects

•

Research activity

• Modern design codes characterise the behaviour of a connection by three basic parameters i.e. strength, stiffness and ductility. In MR connection the ductility is measured in terms of rotation capacity. From this point of view, connections may be classified, similar to the classification of sections (EN 1993-1-1) in terms of their ductility (EN 1993-1-8) as follows:

Class 1 joints:	*Ductile* joints. A ductile joint is able to develop its plastic moment resistance and to exhibit a sufficiently large rotation capacity.
Class 2 joints:	Joints of *intermediate ductility*. A joint of intermediate ductility is able to develop its plastic moment resistance but exhibits only a limited rotation capacity once this resistance is reached.
Class 3 joints:	*Non ductile* joints. Premature failure (due to instability or to brittle failure of one of the joint components) occurs within the joint before the moment resistance based on a full plastic redistribution of the internal forces is reached.

• The parameters of this classification are shown in Figure 4 (EN 1993-1-8).

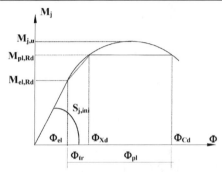

$M_{el,Rd}$ – elastic design moment resistance
$M_{pl,Rd}$ – plastic design moment resistance
$S_{j,ini}$ – initial rotational stiffness
Φ_{el} – elastic rotation
Φ_{tr} – another rotation necessary to get to the plastic design moment resistance $M_{pl,Rd}$
Φ_{pl} – plastic rotation ($\Phi_{pl} = \Phi_{Cd} - \Phi_{Xd}$)

Figure 4. Characteristic parameters for a beam-to-column connection

• For welded beam-to-column connection, in which the web is stiffened in compression, but unstiffened in tension, under monotonic load only, EN 1993-1-8 provide the following formula to estimate the design rotation capacity, Φ_{Cd}:

$$\Phi_{Cd} = 0.025\,h_c/h_b \qquad (1)$$

where h_b is the depth of the beam and h_c is the depth of the column.

- The deformation capacity of components has been studied by several researchers. Faella at al. (2000) carried out tests on T/stubs and derived analytical expression for the deformation capacity of this component. Kuhlmann and Kuhnemund (2000) performed tests on the column web subjected to transverse compression at different levels of compression axial force in the column. Some authors have tried to extract the information of the behaviour of single components from the tests on a whole joint. Bose et al. (1996) determined only the strength of the most important components, while da Silva et al. (2002) tried to determine all three important parameters, stiffness, strength and deformation capacity, at different levels of axial force in a beam.

- Recently, Beg et al. (2004) proposed a simple analytical method for calculation of rotation capacity of MR connection which is compatible with the "component method" used in EN 1993-1-8 to provide the stiffness and strength.

- Also Gioncu (2000) suggested to use the "local mechanism method" to evaluate the rotation capacity of top-and-seat cleat bolted MR connections. This method could be used for welded connections, but probably not for extended-end-plate bolted connections, or other types.

- However, even, as shown, there are tentatives to find analytical methods for evaluation of monotonic rotation capacity, some of these referred in this paper being promising, the basic procedure provided by design codes on this purpose still remain the prequalification tests.

- In case of seismic resistant connection the problem is much more complex, and if the relevant design codes have specific provisions to request precisely the lower bound values for the necessary rotation capacity, they do not provide a specific evaluation methodology, except, again testing.

Method of analysis

-

Results of analysis

- **Rotation capacity for welded joints (INSA Rennes)**
  Experimental values of ultimate resistance moment and rotation capacity obtained in both monotonic and cyclic reversal loading can be compared in Table 1.

  From the moment-rotation curves, it was observed that the ultimate moment and the initial stiffness of the joints are not strongly influenced by the repeated cyclic loading, so that in seismic design the corresponding formulae given in EN 1993-1-8 for the case of static loading can be used, as reasonable approximations. On the other hand it appears clearly that the rotation capacity of the joints is systematically reduced by a factor about 2.

Table 1. Comparison between joint characteristics under cyclic and monotonic loadings

	CPP11	CPP12	CPP13	CPP14	CPP15	CPP16	CPP17	CPP18
M_u [kNm]	230.0	253.0	166.9	180.3	349.2	368.5	467.5	486.2
Φ_u [rad]	0.064	0.031	0.045	0.023	0.045	0.020	0.052	0.030

- **Rotation capacity for welded and bolted joints (UP Timisoara)**

  Table 2 comprises the results of the experimental tests compared to that of EN 1993-1-8, in terms of joint bending moments, rotational stiffness and ultimate rotation attained.

Table 2. Comparison between computed and experimental joint characteristics

SPECIMEN	$M_{j,Rd}^{(exp)}$ [kNm]	$M_{j,Rd}^{(th)}$ [kNm]	$S_{j,ini}^{(exp)}$ [kNm/rad]	$S_{j,ini}^{(th)}$ [kNm/rad]	$\phi_y^{(exp)}$ [rad]	$\phi_u^{(exp)}$ [rad]
XS-EP 1	255.6	262.7	69539	142932.2	0.0038	0.033
XS-EP 2	288.9	261.3	44205	140886.8	0.0063	0.038
XS-W 1	305.6	309.3	333953	∞	0.0009	0.029
XS-W 2	277.8	317.6	321569	∞	0.0009	0.016
XS-CWP 1	316.7	452.1	366309	∞	0.0009	0.038
XS-CWP 2	**	449.0	**	∞	**	**
XU-EP 1	146.7	169.2	44081	43727.2	0.0033	0.060
XU-EP 2	157.8	169.1	49004	43718.2	0.0028	0.062
XU-W 1	113.3	163.6	63102	68792.1	0.0020	0.052
XU-W 2	131.1	164.1	49681	69062.1	0.0026	0.052
XU-CWP 1	131.1	178.6	60712	75597.2	0.0022	0.064
XU-CWP 2	164.4	177.4	58453	74963.1	0.0026	0.060

Comparing the experimental and computed values of joint moment capacity, it can be observed that generally, close values are obtained for the XS series. In the case of XU series, all experimental values are lower than the ones computed by EN 1993-1-8.

In what concerns the initial stiffness of the joints, numerical and experimental results agree fairly well for the XU series, while significant differences are noticed for XS series. Anyway, stiffness is much lower for the anti-symmetrical joints both from experimental and computed stiffness values. This fact is again given by the deformability of the panel zone.

Joint rotations are considerably higher for XU series, generally in the ratio of 1:2. Improved ductility in the case of XU joints is given by good rotation capacity and stable hysteresis loops of the web panel in shear. Anti-symmetrical joints have generally increased energy dissipation capacity with respect to the symmetrical ones.

- **Rotation capacity for bolted steel and composite joints (UP Timisoara)**

  Table 3 and Table 4 present the joint characteristics obtained from the tests and computed analytically in accordance to EN 1993-1-8 and EN1994-1-1 section 8, respectively.

Table 3. Comparison of test to the analytical results given by EN 1993-1-8 for steel joints

Specimen	Total Energy kNm rad	φ_{max}^+	φ_{max}^- mrad	M_{max}	M_{min} kNm	$S_{j,ini}^+$	$S_{j,ini}^-$ x10³kNm/rad	φ_y^+	φ_y^- mrad	M_y^+	M_y^- KNm
Symmetrically loaded joints											
EC3-full A_s	---	---		---		55.64		2.97		165.40	
EC3–red. A_s	---	---		---		55.64		2.97		165.40	
BX-SS-M	9.01	43.20		263.34		48.03		3.26		180.79	
BX-SS-C1	41.5	28.0	21.0	271.6	259.1	55.91	59.60	3.26	2.60	197.2	188.0
BX-SS-C2	23.4	17.4	18.1	261.8	259.8	71.24	63.5	2.66	2.39	194.8	206.8
Anti-symmetrically loaded joints											
EC3-full A_s	---	---		---		32.99		4.76		156.91	
EC3–red. A_s	---	---		---		25.19		4.24		106.77	
BX-SU-M	24.0	105.5		258.36		51.50		2.28		137.66	
BX-SU-C1	135.8	72.5	55.3	269.4	240.6	35.07	29.08	3.77	4.44	153.1	161.2
BX-SU-C2	88.37	39.2	46.8	240.1	236.6	27.82	40.53	5.54	3.37	179.8	161.2

full A_s – full shear area approach

red A_s – reduced shear area approach, according to EN 1993-1-8

307

Table 4. Comparison of test to the analytical results (EN 1994-1-1) for composite joints

Specimen	Total Energy kNm rad	φ_{max}^+	φ_{max}^- mrad	M_{max}	M_{min} kNm	$S_{j,ini}^+$	$S_{j,ini}^-$ x10³KNm/rad	φ_y^+	φ_y^- mrad	M_y^+	M_y^- KNm
Symmetrically loaded joints											
EC4 Mom -	---	---		---		57.11		2.34		133.80	
EC3* Mom +	---	---		---		99.68		2.30		230	
BX-CS-M1	16.60	86.7		244.25		98.47		1.27		155.51	
BX-CS-M2	8.4	32.7		316.09		105.27		2.52		231.32	
BX-CS-C1	32.0	44.8	14.1	305.5	196.9	102.5	75.05	1.60	1.73	195.2	149.9
Anti-symmetrically loaded joints											
EC4-comb.**	---	---		---		41.78		3.84		160.55	
BX-CU-M	15.2	90.1		198.24		47.19		2.34		126.23	
BX-CU-C1	104.5	60.5	53.8	187.1	193.3	36.87	37.92	3.49	3.38	142.67	137.3
BX-CU-C2	32.2	32.7	30.9	191.3	210.5	--	--	--	--	--	--

EC3 - computed according to EN 1993-1-8, by translation of the centre of compression*
*EC4-comb.** - mean value between the positive and negative values*

The experimental curves of the monotonic tests, as well as the envelopes of the cyclic tests, are compared to the analytical results given by the EN 1993-1-8. In the case of symmetrically loaded joints, it can be observed that the values of the experimental stiffness are close to those computed by EN 1993-1-8, but the values of the computed yielding bending moment is generally 10-20% smaller than the experimental ones. For the anti-symmetrically loaded joints, only the full-shear area approach gives close agreement to the analytical results given by EN 1993-1-8. The monotonic test is different from the cyclic tests in stiffness, maximum resistance and maximum rotation, as can be seen in Table 3. As usual, the maximum rotation in monotonic tests is 1.5-2 times greater than the maximum rotation attained under cyclic loading. The conventional values of yielding moment and of yielding rotation for the case of load reversal are closer to the computed values by EN 1993-1-8.

The analytical computations according to EN 1994-1-1 section 8 in the case of symmetrical positive bending leads to safer values in terms of resistance (15% approx.) and smaller stiffness. Due to slab degradation in cyclic tests, the resistance and stiffness values are smaller.

The EN 1994-1-1 section 8 does not give the possibility of computing composite joints subjected to positive moments. For this case, the values of stiffness and moment resistance presented in Table 4 are computed according to EN 1993-1-8 by a translation of the centre of compression from the upper beam flange to the middle of the concrete slab (considered without corrugated sheet). These assumptions lead to comparable values to the tests in terms of resistance and stiffness. For the case of cyclic loading, there can be observed a decrease in both resisting moment and stiffness due to rapid slab degradation.

Analytical values of moment resistance and stiffness for anti-symmetrical loading have been obtained by the mean values for the two connections subjected to anti-symmetrical loading, taking into account the full shear area approach. This prediction remains only an attempt of computing the composite joints under anti-symmetrical loading.

- **Rotation capacity for bolted steel and composite joints (INSA Rennes)**

Obtained results concerning the joint characteristics are presented in Table 5 for test specimens G14-G15 and Table 6 for G17-G18 respectively. In this case of tests, it is quite clear again that the rotation capacity of joints subjected to cyclic loading could be estimated as approximately 0.5 of the rotation capacity of joints subjected to monotonic loads.

Table 5. Experimental characteristics determined by tests

Test	$S_{j,ini}^+$ [kNm/rad]	$S_{j,ini}^-$ [kNm/rad]	M_{max}^+ [kNm]	M_{max}^- [kNm]	Φ_{max}^+ [mrad]	Φ_{max}^- [mrad]
Test G14	----	49900	----	342	----	91,8
Test G15	41098	53630	326	340	24	68

Table 6. Experimental characteristics determined by tests

Test	$S_{j,ini}^+$ [kNm/rad]	$S_{j,ini}^-$ [kNm/rad]	M_{max}^+ [kNm]	M_{max}^- [kNm]	Φ_{max}^+ [mrad]	Φ_{max}^- [mrad]
Test G17	----	54734	----	639	----	88
Test G18	101982	66587	740	665	40	32

A small difference should be remarked for G14 and G15, but this is due to the test conditions and the fact that the test could not be developed symmetrically. Anyhow, it can be seen that the sum of the negative and positive rotation is practically identical with the rotation capacity of the joint subjected to monotonic loads (24+68=92≈91.8 mrad).

In the second case, for G17 and G18, where negative and positive rotations are similar, it can be seen that 40+32=72≈88 mrad.

Further developments

- It is clear for all specialists of the field that the rotation capacity of beam-to column joints is one of the main characteristics which are influencing the seismic behaviour of the steel MR frames. That is the reason why in the last period, this characteristic has been introduced with some arbitrary values of reference for different types of frames. Unfortunately, formulae for evaluating this characteristic are very few. It is evident that the researches have to be continued in both directions experimental test research and numerical modelling, to establish new definitions for the evaluation of the rotation capacity of beam-to-column joints.

- Rotation capacity of joints is very difficult to establish except of test procedures. From analytical point of view, the evaluation in static range should be possible, using the method of components. Contrary, in cyclic range, Zandonini and Bursi (2002) have shown that the use of this method is impossible and have recommended the use of some macrocomponents, incorporating some of the interaction effects among elemental components. This method should be more adequate, because some macrocomponents are ductile ensuring the ductile behaviour and the rotation capacity, and some others are fragile, needing to ensure them a certain level of overstrength, in order to ensure the development of the rotation capacity. Anyhow the ductile components of the joint are the endplate and the column web panel. After Ciutina (2003), a sort of balance of equilibrium has to be done between the geometrical dimensions of the two ductile components, otherwise only one of them working for the joint ductility.

- An important observation in our opinion should be that taking into account the complex behaviour of the moment resisting joints, and the conclusions of Stojadinovici (2003) and Zandonini and Bursi (2002), concerning the pre-qualification and the introduction of macrocomponents, under seismic loads, it is more adequate to refer to *rotation capacity of the joint* using the concept of "macrocomponent method", instead of *rotation capacity of the connection* using the concept of "component method" utilised for monotonic loads.

- So, what could we do in order to establish the rotation capacity of the joints, without experimental cyclic tests? Analysing several tested joints under monotonic and cyclic loads, the authors of this datasheet have observed that the rotations under cyclic and monotonic loads are usually in the ratio of 1:2. Using this conclusion, the answer would be very simple.

- For the evaluation of the rotation capacity under cyclic loads, we could use an analytical method for evaluation under monotonic loads, or the well-known computer code DUCTROT of Gioncu and then affect the results with the correction coefficient for cyclic loads equal to 0.5.

- For the same evaluation, the correction coefficient of 0.5 could be used also to affect the values of rotation capacity of joints determined under monotonic loads. This proposal is made in order to use the large number of results, existing in literature, thinking also to some data bank like SERICON.

- From the same analytical point of view, we are reminding the importance of the macrocomponents method to be developed and used after some more numerical simulations.

- It is important also to remind the prequalification connections what FEMA 350 (2000) have made, in order to make a similar prequalification in Europe, with the usual connections utilised by designers in Europe.

References

- EN 1993-1-1. 2003. Design of steel structures. Part 1-1: General rules and rules for buildings. European standard.

- EN 1993-1-8. 2003. Design of steel structures. Part 1-8: Design of joints. European standard.

- EN 1998-1. 2003. Design of structures for earthquake resistance. Part 1: General rules, seismic actions and rules for buildings. European standard

- EN 1994-1-1. 2003. EUROCODE 4: Part 1.1. General rules and rules for buildings; Section 8: Composite joints in frames for buildings. European standard

- FEMA 350. 2000. Recommended Seismic Design Criteria for New Steel Moment-Frame Buildings, SAC Joint Venture.

- Aribert J.M., Grecea D. 1998. Experimental behaviour of partial-resistant beam-to-column joints and their influence on the q-factor of steel frames. The 11th European Conference of Earthquake Engineering, Paris, 6-11 September 1998.

- Beg D., Zupancic E., Vayas I. 2004. On the rotation capacity of moment connections. Journal of Constructional Steel Research 60 (2004) 601-620.

- Bose B., Youngson G.K., Wang Z.M. 1996. An appraisal of the design rules in Eurocode 3 for bolted endplate joints with experimental results. Proc Inst Civil Eng: Struct Build 1996; 116:221-34.

- Bose B., Sarkar S., Bahrami M. 1996. Extended endplate connections: comparison between three-dimensional nonlinear finite-element analysis and full-scale destructive tests. Struct Eng Rev 1996;8:315-28

- Ciutina A. 2003. Assemblages et comportement sismique de portiques en acier et mixtes acier-béton : expérimentation et simulation numérique. Thèse de Doctorat. INSA de Rennes, France.

- Dubina D., Ciutina A., Stratan A. 2001. "Cyclic Tests of Double-Sided Beam-to-Column Joints", Journal of Structural Engineering, Vol.127, No.2, Feb.2001, pp.129-136

- Dubina D., Ciutina A., Stratan A. 2002. "Cyclic tests on bolted steel and composite double-sided beam-to-column joints", Steel & Composite Structures, Vol.2, No.2, 2002, pp.147-160

- Faella C., Piluso V., Rizzano G. 2000. Plastic deformation capacity of bolted T-stubs: theoretical analysis and testing. Moment resistant connections of steel frames in seismic area, design and reliability, RECOS. London: E&FN Spoon, 2000.

- Gioncu V., Mateescu G., Petcu D., Anastasiadis A. 2000. Prediction of available ductility by means of local plastic mechanism method: DUCTROT computer program. Moment resistant connections of steel frames in seismic area, design and reliability, RECOS. London: E&FN Spoon, 2000.

- Grecea D. 1999. Caractérisation du comportement sismique des ossatures métalliques - Utilisation d'assemblages à résistance partielle. Thèse de Doctorat. INSA de Rennes, France.

- Kuhlmann U., Kuhnemund F. 2000. Rotation capacity of steel joints, NATO Advanced Research Workshop "The Paramount Role of Joints into the Reliable Response of Structures, From the Rigid and Pinned Joints to the Notion of Semi-rigidity", Ouranoupolis, Greece, 21-23 May 2000.

- da Silva L., Lima L., Vellasco P., Andrade S. 2002. Experimental behaviour of endplate beam-to-column joints under bending and axial force. Database reporting and discussion of results. Report on ECCS-TC10 Meeting in Ljubljana, April 2002.

- Stojadinovici B. 2003. Stability and low-cycle fatigue limits of moment connection rotation capacity. Engineering Structures 25 (2003) 691-700.

- Zandonini R., Bursi O. 2002. Monotonic and hysteretic behaviour of bolted endplate beam-to-column joints, Advances in Steel Structures, Vol.1, Edited by Chan, Teng and Chung, Elsevier Science Ltd., 81-94.

- ECCS. 1986. Recommended Testing Procedures for Assessing the Behaviour of Structural Elements under Cyclic Loads, European Convention for Constructional Steelworks, Technical Committee 1, TWG 1.3 – Seismic Design, No.45

Contacts

- Dubina Dan

  Address: Department of Steel Structures and Structural Mechanics, University Politehnica Timisoara, Ioan Curea 1, Timisoara, Romania.
  e-mail: dubina@constructii.west.ro, Phone: 00.40.256.40.39.32, Fax: 00.40.256.40.39.32

- Grecea Daniel

  Address: Department of Steel Structures and Structural Mechanics, University Politehnica Timisoara, Ioan Curea 1, Timisoara, Romania.
  e-mail: daniel@constructii.west.ro, Phone: 00.40.256.40.39.24, Fax: 00.40.256.40.39.32

[1] Politehnica" University of Timisoara, Romania, http://www.ceft.utt.ro
[2] Romanian Academy, Timisoara Branch/ Romania, http://acad-tim.utt.ro

WELDED JOINTS: EFFECT OF DETAILING AND STRAIN RATE

Description

- The poor behaviour of welding was the main cause of brittle fracture of MR steel frame connections during the important seismic events of Northridge 1994 and Kobe 1995

- Among the possible causes of brittle fractures in welded joints, have been identified:
  - workmanship (welding defects)
  - detailing (stress concentration at the root or the toe of welds)
  - design practices (larger beam and column sizes than those tested)
  - poor welding practice (low-toughness weld metal, poor quality control)
  - unusually high seismic input (high strain rates)

- The influence of strain rate on the performance of different types of welded joints was studied, both experimental and numerical

Structural aspects

- Mechanical properties of steel, such as yield strength and ultimate strength, and thus the ductility vary with strain rate (Figure 1). Modulus of elasticity is not influenced by the strain-rate variation and the upper yield stress is more strain-rate sensitive than the lower yield stress

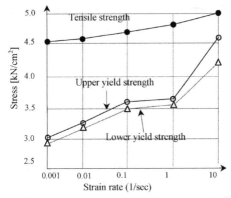

Figure 1. Yield stress and tensile strength vs. strain rate

- The behaviour of welded joint is more complex than that of steel material (three basic components: base material, deposited material and heat affected zone) ⇒ apart from strain rate, several other factors influence should be considered: cyclic loading, type of welding, welding defects

Research activity

- Strain rate effect during an earthquake was considered negligible (Mahin et al, 1982, Wallace and Krawinkler, 1989)

- After earthquakes of Northridge and Kobe, when recorded velocities have been very high, the strain rate was considered responsible for the unexpected poor behaviour of steel structures (SAC and COPERNICUS "RECOS" research projects, Kurobane et al, 1996, Kohsu and Suita, 1996, Dubina et al, 2001, Gioncu and Mazzolani, 2002).

Method of analysis

- Experimental program was mainly devoted to the behaviour of "T" assemblies, composed of an end plate and two flanges (Figure 2). A number of 54 specimens have been tested. Parameters:
  - three strain rates: $\dot{\varepsilon}_1 = 0.0001 s^{-1}$; $\dot{\varepsilon}_2 = 0.03 s^{-1}$; $\dot{\varepsilon}_3 = 0.06 s^{-1}$
  - two steel grades: S235 and S355
  - two types of loading: monotonic and cyclic pulsating
- Welding details are also considered in the program (Figure 2): fillet weld, double bevel butt weld – K type, single bevel butt weld – 1/2V type

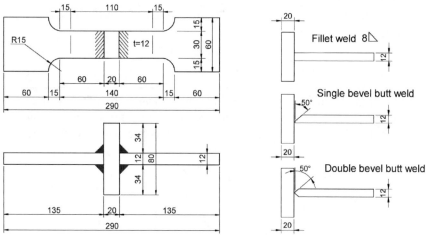

Figure 2. Welded specimens and edge preparation

- Experimental research is an important tool for the analysis of the behaviour of structural elements ⇒ disadvantage: substantial cost, difficult to simulate exactly on-site conditions, time consuming
- Numerical modelling, used together with experimental testing, could overpass these problems ⇒ advantage: greater flexibility in the analysis
- Finite element code NASTRAN 70.7 was used for the numerical modelling. Three numerical models for monotonically loading and three for cyclic loading, one for each weld type were calibrated, based on experimental results (see Figure 3)

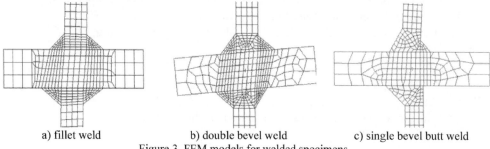

a) fillet weld b) double bevel weld c) single bevel butt weld
Figure 3. FEM models for welded specimens

Results of analysis - experimental

Tensile tests

- Tensile tests have been performed on base and deposited metals to determine the mechanical characteristics of the materials:
    - the lower yield strength (R_{el}) increases for higher strain rates
    - ultimate tensile strength (R_m) increases for higher strain rates but is less strain-rate sensitive than yield strength
    - maximum influence is observed for the mild steel (S235)
    - total elongation at fracture (A_t) is not influenced by strain rate, implying that strain rates of the magnitude of 0.03-0.06 s^{-1} do not reduce the ductility of the base and deposited metals

Tests on welded specimens

- Yield strength of welded specimens is less sensitive to strain rate than the yield strength of component materials

- Ultimate strength of the welded specimens (R_m) increases slightly with the increase of strain rate for the monotonically loaded specimens (maximum 10% for $\dot{\varepsilon}_3$) (Figure 3)

- Contrary to component materials, a higher strain rate does imply a reduction of the ductility for monotonically loaded welded specimens (Figure 4)

- In case of cyclic loading, the results are rather scattered. A possible explanation for the increase in ductility under high strain rate cyclic loading may be attributed to the specimen heating, as noted elsewhere

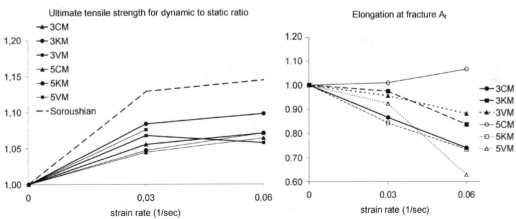

Figure 4. Influence of weld detail on the ultimate strength (R_m) of welded specimens

Figure 5. Ductility (elongation at fracture) vs. strain rate for monotonically loaded specimens

- Failure modes: two failure types for welded specimens: fracture in the base metal (BM) and in the weld (W):
    - for monotonically loaded specimens, the failure occurred in the base metal
    - for cyclically loaded specimens, the probability of failures in the weld increases for fillet and single bevel butt weld specimens. No weld failure occurred in double bevel butt weld specimens

$\Rightarrow$ *double bevel butt weld is less sensitive to imperfections and is recommended for shop welding*

$\Rightarrow$ *single bevel welds are adequate for site welding, but re-welding of the root is compulsory in order to eliminate defects at the root of the weld*

$\Rightarrow$ *fillet welds may be used for shop or site welding, with the condition of strict verification of weld size*

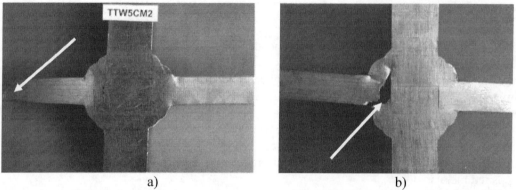

a)　　　　　　　　　　　b)

Fig.6 - Failure of fillet weld specimens: in the base metal (a), and in the weld (b)

Results of analysis - numerical

- Fillet weld: similar behaviour compared to the experimental one, both in terms of load-deformation curve and failure mode

- Double bevel butt weld: similar behaviour compared to the experimental one, both in terms of load-deformation curve and failure mode

- In case of single bevel butt weld, there are some differences, caused by the complex fracture mechanisms which develops at the root of the weld, where a lack of fusion was observed before the test

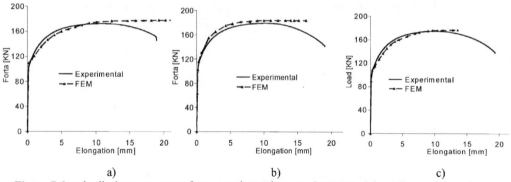

a)　　　　　　　　　　b)　　　　　　　　　　c)

Figure 7. Load - displacement curve from experimental tests and FEM models: a) fillet weld; b) double bevel butt weld; c) single bevel butt weld

- Numerical analysis has the capacity to simulate the effect of imperfections on the behaviour of welded joints (Figure 8) ⇒ further studies are needed

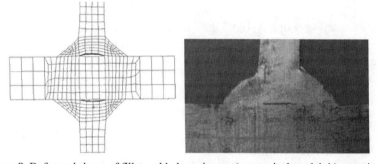

Figure 8. Deformed shape of fillet welded specimen: a) numerical model; b) experiment

Further developments

- parametric studies on the influence of imperfections $\Rightarrow$ to identify the critical imperfection for the three types of welding details

References

- Beg, D., Plumier, A., Remec, C., Sanchez, L., 2000. *Cyclic behavior of beam-to-column bare steel connections: Influence of strain rate*, Chapter 3.1 in: Moment Resistant Connections of Steel Building Frames in Seismic Areas, (Mazzolani F.M. ed.) E&FN SPON, London.

- Dubina, D., Stratan, A., 2002. *Behaviour of welded connections of moment resisting frames beam-to-column joints.* Engineering Structures, Vol. 24, No. 11, 1431-1440.

- Gioncu, V., Mazzolani, F.M., 2002. *Ductility of Seismic-Resistant Steel Structures. London*: SPON PRESS: 694 pp.

- Kaneko, H., 1997. *Influence of strain-rate on yield ratio. Kobe Earthquake Damage to Steel Moment Connections and Suggested Improvement.* JSSC Technical Report No.39.

- Kassar, M., Yu, W.W., 1992. Effect of strain-rate on material properties of sheet steel. Journal of Structural engineering, Vol.118, No.11, 3136-3150.

- Kohzu, I., Suita, K., 1996. *Single or few excursion failure of steel structural joints due to impulsive shocks in the 1995 Hyogoken Nanbu earthquake.* 11th World Conference on Earthquake Engineering, Acapulco, 23-28 June 1996, CD-ROM, Paper No.412.

- Leblois, C., 1972. *Influence de la limite d'elasticite superieure sur la comportament en flexion et tension de l'acier doux.* Ph.D. Thesis, Liege, Belgium.

- Manjoine, M.J., 1944. *Influence of rate of strain and temperature on yield stress of mild steel.* Journal of Applied Mechanics, No. 11, 211-218.

- Nagakomi, T., Tsuchihashi, H., 1988. *Fracture and deformation capacity of a welded T-shape joint under dynamic loading.* 9th World Conference on Earthquake Engineering, Tokyo-Kyoto, 2-9 August 1988, Vol.IV, 157-162.

- Nakashima, M., Tateyama, E., Morisako, K., Suita, K., 1998. *Full-Scale test of beam-column subassemblages having connection details of shop-welding type.* Structural Engineering Worldwide, Elsevier Science (CD-ROM), Paper Ref. T158-7.

- Soroushian, P., Choi, W.B., 1987. *Steel Mechanical properties at different strain rates,* Journal of Structural Engineering, Vol.113, No.4, pp.863-872.

- Wakabayashi, M., Nakamura, T., Iway, S. and Hayashi, Y., 1994. *Effects of strain rate on the behavior of structural members,* Proc. 8th World Conference on Earthquake Engineering, Vol.4: 491-498. San Francisco.

Contacts

- Prof. Dan DUBINA
  Address: "Politehnica" University of Timisoara
  Ioan Curea 1, Timisoara 1900 Romania
  Tel/Fax: +40 256 403 932;
  e-mail: dubina@constructii.west.ro http://www.ceft.utt.ro

FULL STRENGTH I BEAM SPLICE BOLTED STEEL CONNECTIONS

Description

- The present research work concerns the laboratory testing of full strength bolted splice connections for I profiles. Taking as reference results those obtained from a continuous intact beam subjected to a concentrated force at the middle of its span, s series of simply supported beams with full strength bolted splice connections have been tested. The obtained test results significantly diverge from the reference ones obtained from the intact beam.

Field of application

- The use of non-pretensioned bolts in full strength splice connections is a common practice in several countries. However, in such splices first due to the difference between the diameter of bolts and the respective holes and second due to a possible preliminary imperfection (rotation of the connected parts of the beam), an additional significant deflection is caused as soon as the loading is applied on the beam and before its value reaches the calculated value.

- Scope of the present research work was to experimentally define the relationship between the moment at the middle of the span and the deflection at the same point. As a matter of fact, the latter is tightly connected to the Serviceability Limit State of the beam, even before the total loading at the middle of the span has been applied (Kontoleon *et al*, 2003).

Technical information

- For the laboratory tests, I-beams have been selected because this profile is very often used as a structural member in a plethora of steel structures (cf. e.g. the purlins of steel roofs). Usually the material is Fe 360 and the non-pretensioned bolts M12-8.8 and M14-8.8. In the performed experiments, four groups of different connections were used, once with bolts M12 and once with bolts M14. Two steel plates having cross section 6x80 mm were used to connect the webs, whereas another two with cross section 12x100 mm to connect the flanges. The full strength bolted splices have been designed and constructed so that they exhibit at least the same load-bearing capacity as the intact beam (ENV 1993-1-1).

Structural aspects

- **TYPOLOGY-SPECIMENS B1 AND B2.** The suggested position of the bolts is shown in Figure 1 (diameters for the holes d_o =13 and d_o =15 mm) corresponding to specimens B1 and B2.

Figure 1. The position of the holes at the horizontal splice plates and flanges (specimens B1 and B2)

- **SPECIMENS B3 AND B4.** In an effort to add more stiffness to the area of the connection in order to obtain better results during deformation, it was decided that the distance between the holes to be greater in the specimens B3 and B4 than the distance in the specimens B1 and B2. The positions of the bolts in specimens B3 and B4 is shown in Figure 2.

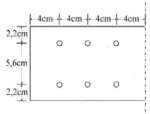

- **Figure 2.** The position of the holes at in the additional splice plates and the flanges (specimens B3, B4)

- **SPECIMENS B5 AND B6.**

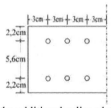

Figure 3. The position of the holes at the splice plates and the flanges (specimens B5 and B6)

- **SPECIMENS B7 AND B8.**

Figure 4. The position of the holes in the additional splice plates and the flanges for specimens B7, B8

Research activity

- All the experiments were performed at the Laboratory of the Institute of Steel Structures, Department of Civil Engineering, Aristotle University of Thessaloniki, Greece. All the specimens used were simply supported beams having a span of 1 m and a concentrated vertical force at the middle of their span.

- A composite hydraulic machine was used with an upper limit of force equal to 500 kN.

- A PC was used to record and store 1000 pairs of values corresponding to the force and the deflections at the middle of the span for each one of the specimens.

- In the sequel, the diagrams corresponding to each one of the specimens from B1 to B8 in parallel with the diagram of the reference beam B9 are depicted.

- Figure 7 shows the diagrams for all the 9 specimens including the reference one in order to give a general comparison of the splices test results with those of the intact beam.

Method of analysis

- The maximum moment and shear force that the selected reference profile HEB 100 can carry under the plastic limits of strength are equal to:

$M_{pl.y.Rd} = W_{pl.y} * f_y/\gamma_{M0} = 104,2 * 23,5/1,1 =$ **2221,818** kNcm and
$V_{pl.y.Rd} = 1,04 * h * t_w * f_y/(3^{1/2} * \gamma_{M0}) = 1,04*10*0,6*23,5 /(3^{1/2} * 1,1) =$ **76,966** kN

- For the connection of the two separated parts of the beam, two horizontal steel plates with cross section 10*1,2 cm were used for the flanges and two vertical ones with a cross section 0,6*5,5 cm for the web (Figure 5).

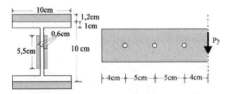

Figure 5. The splice connection (left) and the vertical additional plate (right)

The additional horizontal plates on the flanges were used to transfer the moment in the connection area, whereas the additional vertical ones were used to carry the shear forces. Because of the symmetry of both the beam and the loading, only half of a vertical plate was drawn and calculated. In Figure 1 one half of one of the two additional vertical plates is shown where the geometrical position of the center of the holes is the point of the load application. Note that in such splices the shear forces are not the principal problem.

- The same position and the same kind of bolts were used for all specimens. The experimental results showed that these vertical additional parts of the connection did not take or carry any force during the experiment.

- In the following equations, calculations and symbols from EC3 were used. A 6 bolt connection was used for M14-8.8 bolts for the half of the connection of a flange. The position of the bolts is shown for the first two specimens are shown in Figure 1. The horizontal distance between the two connected pieces of the beam was equal to 1,5 cm for all the specimens.
$P_y=79,966/2=38,48$ kN $M=9*P_y=348,347$ kNcm
P_y: $(V_{x,.p}=0$ $V_{y,.p}=P_y/3=12,828)$ $I_p=\Sigma(xi^2+yi^2)=2*5^2=50$
M $(V_{y,M}=M*x_i/I_p=346,347*5/50=34,635$ kN $V_{x,M}=My_i/I_p=0$ kN$)$
$V_x=0$ $V_y=12,828+34,635=47,463$ kN
$t=min(t_\wedge{}^\kappa=0,6$ $s/2=0,6/2=0,3)$ **t=0,3**
$\alpha=min(e_1/(3*1,5)=4/4,5=0,889$ $p_1/(3*1,5)-1/4=0,861$ $80/36=2,2$ 1) **α=0,861**
$F_{v.Rd}=0,6*f_{ub}*A/\gamma_{Mb}=0,6*80*1,54/1,25=$**59,136 KN** > 47,463 kN
$F_{b.Rd}=2,5*\alpha*f_u*d*t/\gamma_{Mb}=2,5*0,861*80*1,4*0,3/1,254=$**57,859** > 47,463
The cross section of 0,5*5,5 cm was not available in the market and for this reason, a cross-section 0,6*8 cm was used for the splice under investigation.

- Two groups of four different positioning of the bolts were used for the specimens. For the first group, bolts type M12-8.8 were used calculated for the connections on the flanges. For the second group, bolts type M14-8.8 were used in an effort to optimize the deformation of the connected areas. Calculations were not repeated because the change in the diameter of the bolts was in the safe side.

- **Calculation of the strength of the splice connection**

  $d_o = 13$ mm   A=11,3 mm f_{ub}= 80 kN/cm^2 f_{yb}=64 kN/cm^2

  $t = min$ (t=1 $t_A{}^\pi$=1,2) **t = 1 cm**

  $\alpha = min$ (4/(3*1,3)=1,026 8/(3*1,3)-1/4=1,8 80/36=2,22 1) **α=1**

  $F_{sd} = M_{sd}/z = 22,218/(10+0,6+0,6)*10^{-2}= 198,375$ kN

  $F_{v.Rd} = 0,6*f_{ub}*A/\gamma_{Mb}= 0,6*80*1,13/1,25=$**43,392** > 198,375/5=**39,675** kN

  $F_{b.Rd}=2,5*\alpha*f_u*d*t/\gamma_{Mb}=2,5*1*80*1,2*1/1,25=$**192** > **39,675** kN

- Splices B3-B8 are calculated in a similar way.

Results of analysis

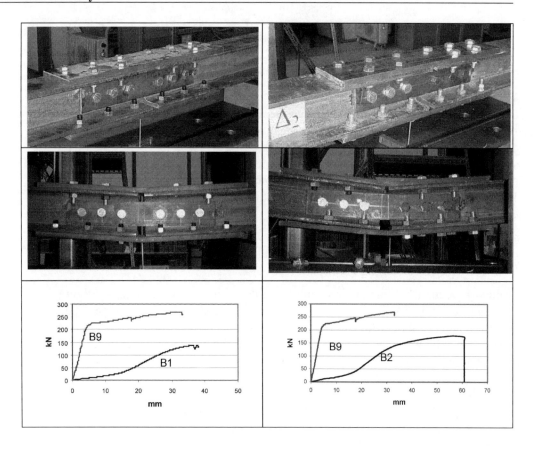

Figure 6. Figures of the specimens and the diagrams for B1, B2.

- The diagrams of the test results (Figures 6,7) for the connected beams gave a significant deflection that corresponds to a much lower force than this that corresponds to the reference curve of beam B9.
- Similar results have been obtained by the rest tests (cf. specimens 3-8) (Zygomalas and Baniotopoulos, 2004).
- In particular, only the 1/10 of the load used for reference beam B9 was used to give the same deflection to all the specimens 1-8.

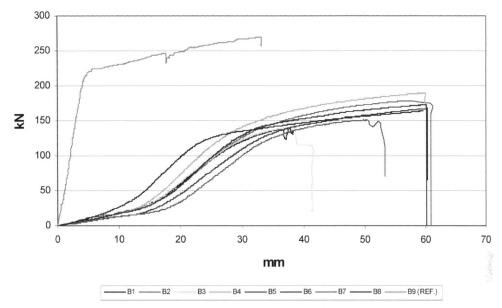

Figure 7. Force-deflection diagrams for the connected beams from B1 to B8 and the reference beam B9.

Figure 8. The holes (with the traces of the insertion of the thread into the steel mass) after the experiment at the upper horizontal splice plate and at the beam flange.

Further developments

- After the end of the testing program, it was observed that all the vertical additional pieces of the web remained undeformed (intact) due to the fact that they didn't carry any loads. The latter certifies the initial assumption taken into account in the calculation of the connection. Having in mind that it is always critical to check the serviceability limit state of such beams, the failure of the specimens due to excessive deflection around the area of the splice connection is obvious. The maximum acceptable percentage according to EC3 § 4.3.2 is equal to 1/200=0.05% for total loading (4). The experimental results gave a deflection of approximately 15/1000=1.5% below the 1/10 of the total load.

- Two were the reasons for the observed significant deflections around the area of the splice connections: The first is the difference of 1 mm between the diameters of the holes and those of the bolts. The existence of the clearance gives a kind of freedom of sliding as soon as a small (1/10) load is applied on the beam. At this first step, the force loaded only the two additional horizontal plates on the flanges which reacted as autonomous simply supported beams instead of the full cross-section of the beam. This is the reason for the existence of linear part at the beginning of the diagrams for the connected beams from 0 to near 15 mm.

- The second reason of the observed significant deflection is due to the insertion of the thread of the bolts into the mass of the steel in the vicinity of the contact area of the holes, as shown in Figure 9.
- As a conclusive remark it is noteworthy that the choice of the position of the splice connection must be very carefully chosen. It must be in a position along the length of the beam where the maximum moment is less than 1/10 of the maximum moment capacity of the profile HEB 100.

References

- Kontoleon, M. J., Kaziolas, D. N., Zygomalas M.D. & Baniotopoulos, C. C. (2003), *Analysis of Steel Bolted Connections by Means of a Nonsmooth Optimization Procedure*, COMPUTERS & STRUCTURES 81, 2455-2465.
- Zygomalas, M. J. and Baniotopoulos, C. C. (2004), *Four-Plate HEB-100 Beam Splice Bolted Connections: Tests and Comments*, Proceedings of the "Connections in Steel Structures V" Workshop, Amsterdam 3-4.6.04 (in print).
- Ivanyi, M. & Baniotopoulos C.C. (eds) (2000), Semi-rigid Joints in Structural Steelwork, Springer Wien, New York, pages 350.
- C.C.Baniotopoulos & F. Wald (2000) (eds), The Paramount Role of Joints into the Reliable Response of Structures. From the Classic Pinned and Rigid Joints to the Notion of Semi-rigidity, Kluwer, Dordrecht, pages 480.
- ENV1993-1-1 (1993). *Eurocode 3, Design of Steel Structures*. CEN, Brussels.

Contacts *(include at least one COST C12 member)*

- Michael Zygomalas, Ass. Prof. Dr.
    Address: Department of Civil Engineering, Aristotle University of Thessaloniki, University campus, Thessaloniki, 54124 Greece. e-mail: mzygo@civil.auth.gr, Phone: +30 2310 995816, 2310 995677 Fax: +30 2310 995642

- Charalambos Baniotopoulos, Prof. Dr.-Ing.
    Address: Department of Civil Engineering, Aristotle University of Thessaloniki, University campus, Thessaloniki, 54124 Greece. e-mail: ccb@civil.auth.gr, Phone: +30 2310 995753, 2310 995677 Fax: +30 2310 995642

STAINLESS STEEL BOLTED CONNECTIONS

Description

- In construction, stainless steel is commonly used for secondary elements. The stainless steel used for structural applications concern glass-steel facades as T-beams, tubes or ties. The structural design is covered by ENV 1993-1.4. The austenitic steel is the most widely used in construction.

- The stress-strain curve shows a non-linear and ductile behaviour with a high ratio of the ultimate to yield strengths.

- The plate-to-plate bolted connections are easy to use and the load transmission is made by plates in bearing and bolts in shear. The bolts pretension is not used because of the low friction coefficient between the plates and the difficulty to obtain a suitable pretension.

- Some experimental results are analysed to assess the resistance formulae for the cover-plate bolted connections and to obtain an estimation of the bearing deformation.

- The large ductility will give some advantages for the energy dissipation in case of cyclic loading and a large capacity of loads redistribution before failure.

- Attention could be given to the resistance capacity of the connection compared to that of the connected element.

- Stainless steel has a good fire resistance and large possibilities of strain hardening. However, for bolted connections, a design for the ULS loads may not be sufficient to avoid unacceptable plastic deformations in the SLS.

- The available design codes are focused on structural carbon steel. The formulae and design approaches would be modified to take into account the non-linear behaviour of stainless steel. This material behaviour is usually modelled by a one parameter Ramberg-Osgood law using a 0,2 % conventional elastic limit (Fig. 1).

Field of application

- The application concerns at first the connections submitted to direct tension load but it could be used to other kinds of cover plate connections. Beam-to-beam connection could be a realistic example.

- The analysis could take into account the connection deformation at different level of loading to allow the assessment of the global behaviour of the structure.

Technical information

- The connected elements are made of three types of stainless steel (austenitic, ferritic, and duplex).

- The thickness of connected plates are different for the three steel families.

- Three bolts diameters (M12, M16, M20) and four arrangements (2T, 2L, 3, 4) are used for each diameter.

- Bolts are made of austenitic stainless steel (A4L–80) with a shank thread on all their lengths. They were tested in tension and shear separately. For each diameter, four specimen were used for each testing type : direct tension, shear in tension and shear in compression.
- The stainless steel stress-strain curve is different from that of carbon steel. The curve is without a well marked elastic limit and it is nonlinear even for low levels of load. A conventional limit at 0,2 % of plastic strain is used.

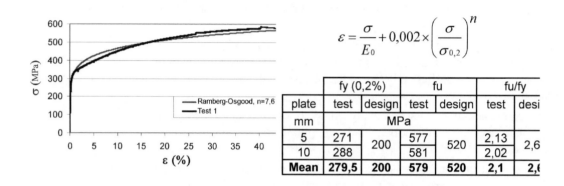

$$\varepsilon = \frac{\sigma}{E_0} + 0,002 \times \left(\frac{\sigma}{\sigma_{0,2}}\right)^n$$

plate	fy (0,2%)		fu		fu/fy	
	test	design	test	design	test	desi
mm	MPa					
5	271	200	577	520	2,13	2,6
10	288		581		2,02	
Mean	279,5	200	579	520	2,1	2,(

Figure 1. Non linear real stress strain curve for a 304 L austenitic stainless steel

Figure 2. Mechanical characteristics of the austenitic steel used in the tests

Structural and material aspects

- Under direct tension load in the connection, the bolts are subjected to shear load with a bending component at the ultimate state.
- The design formulae are in relation with the ENV – EC3 part 1.4 dedicated to the stainless steel. They are compared to the euroinox document (EuroInox, 1994). Theses formulae concern different failure modes : bolts in shear, plates in tension in net or gross section and bolts holes in bearing.
- Based on the experimental results, an evaluation of the hole elongation is made under SLS and other loads.
- The stainless steel has a non symmetrical behaviour in tension and compression and have a general tendency to show a non linear behaviour more marked in tension than in compression. So, an anisotropy exists and it is implicitly recognized in the codes which recommend to proceed to the characterization on test coupons taken in the transverse direction of rolling. However, the aspect which has potentially most influence for structural applications is the non linearity.
- As for certain metal alloys, in opposition to carbon steels, stainless steels are subject to significant creep at ambient temperatures. This incites to use stainless steels in a controlled way if a high level of stress must be maintained during a long period (limit the stress due to the long-term actions to 60 % of the conventional elastic limit for example).

- The austenitic tested steel (304L) of European designation (X2CrNi19-11 or 1.4306) is in the S220 resistance class, according to EC3-partie 1.4. Its measured elongation is close to 63% while the nominal one is equal to 45%.

Research activity and Method of analysis

- The design of carbon steel joints is usually made to the ULS, while the deformation criteria at the SLS is considered to be implicitly satisfied. This assumption is justified by the linear behaviour of carbon steel up to its elastic limit and by the low ratio between ultimate and yield strengths, which is generally comprised between 1,1 and 1,5. In the case of stainless steel, the stress-strain curve is non-linear and the ratio between the ultimate strength and the elastic limit may exceed 2,0. So, a design to the ULS may not guarantee that no excessive deformations will occur at the SLS, since the ratio between the ULS and SLS loads is usually comprised between 1,35 and 1,5.

- Another possible verification should concern the hole deformation in bearing, which is not easy to predict. However, the EC3 considers that the ULS verification is sufficient if a reduced bearing ultimate limit is used instead of the real ultimate limit.

- A ECSC project, covering all major aspects of the design of stainless steel elements in buildings, was initiated in 1998. An experimental study for the cover-plate joints, was conducted at Clermont Ferrand and the results were compared to the predictions of EC3-1.4 (Bouchair et al. 2001).

- The analysis is based on the simple failure mechanisms proposed by the EC3 taking account of the ductile deformation of stainless steel. A limitation is introduced for the net section resistance to limit the deformation at the SLS.

- The real stress-strain curve based on the Romberg-Osgood formula (Fig. 1) is used to calculate the deformation at the net section for a symmetrical connection.

- Experimental study with different cover plates configurations, with the estimation of the net section elongation, allow the bolt hole elongation due to bearing and shear bolt to be deducted.

- For bolted connections, a design for the ULS loads may not be sufficient to avoid unacceptable plastic deformations in the SLS. The formulae and design approaches for carbo steel must be modified to take into account the non-linear behaviour of stainless steel. The main idea is to define an analytical method to evaluate the real deformation around the bolt hole which will avoid the conservative limits.

- Test results show that the various ultimate state joint strengths are safely predicted with a very large ductility of all the connections under static loads. This will be a good guarantee of safety under cyclic or dynamic loads.

- A finite element approach is in progress to generalise the study to different configurations (beam-to-beam connection for example) and to evaluate the effect of the specific behaviour of stainless steel on the structural behaviour of a bending resistant connection.

- The FEM will also help to extend the study to other geometrical configurations and other types of connections (beam-to-column for example). Also, the evaluation of a beam deflections and a plastic loads redistribution need to be more analysed. For the members, the secant modulus could be used in substitution to the elastic modulus (Van Den Berg 2000). Also, attention has to be made to ensure that calculated plastic moment transmitted to the connections do not exceed their own strength.

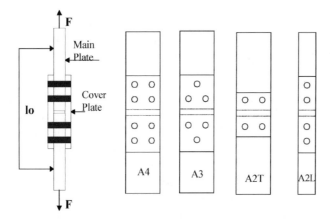

Figure 3. Schematic representation of the tested cover-plate connections

---

Results of analysis

- Two force-displacement curves, representing two distinct failure modes (Net-section and Bearing), are given for two different connections (Fig. 4-7). They show the high ductility of the connections, and the large difference between the test ultimate loads and the design values calculated according to the EC3 rules with a partial safety factor taken equal to 1,0.

- The major failure modes predicted by calculation according to the EC3 formulae are associated to the plastic limit resistance of the gross section. This limit is reached before the net section ultimate state because the ratio between the areas of the net and gross cross-sections is close to 0,7, while the ratio between the steel elastic limit and its ultimate limit strength is smaller than 0,5.

- The tests where failure occurred in the net cross-section show that, even when the 0,9 and k_r reducing factors are dropped, the real values of the connection resistance are always higher than those predicted by calculations. It was found that the ratio between the experimental and calculated resistances varied between 1,05 and 1,16.

- This fact is probably due to "notch" effects and it is confirmed by other experimental and numerical studies, using finite element models to analyse the elastic-plastic behaviour of a single plate with a hole, subjected to a tension load (bi-directional state of stress and using the Von Mises criterion).

- Test specimens were able to support high deformations, due to the high ductility of austenitic steel. At the ULS, the hole ovalisation seems to result more from the reduction of net cross-section than from bearing, except in the case of some particular connections (2T).

- An estimation of the elongation from each component of the joint is made. The global displacement measured on the length l_0 (Fig. 3) is divided into three components, corresponding to the elongation in the gross section components of the plates, in the net section components and in the holes bearing the loaded bolts. This method allows an analytical estimation of the contribution of each component, in order to isolate the one corresponding to the bearing load.

- During tests, the deformations corresponding to the maximum load for connections in bearing (A2T) reach relatively high values. However, at both the ULS and the SLS loading, the calculated deformations still relatively low, either in the case of bearing or net section failure modes.

- The calculated bearing deformations at the SLS remain relatively small (lower than 1,6 mm, and 1,1 mm, with a bearing resistance based on the ultimate limit (f_u), and its reduced value (f_{ur}), respectively). All the calculated elongations at the ULS are small (lower than 2,5 mm) except in the case of the A2T-20 connection.

- The net section elongation in the A2L and A2T connections are not very important, even when the less favourable hypothesis was considered. The maximum elongation at a hole of a net section is smaller than 0,8 mm (ULS) and 0,1 mm (SLS).

- It can be easily shown that, if the mean stress at the net section is about 1,5 times greater than the elastic limit, the net section elongation is still lower than 1 mm, for the three hole diameters used in the tested connections. The new condition introduced by the EC3-1.4, limiting the stress in the net section at the SLS to the yield strength does not seem to be necessary, because the length concerned by this condition is very limited (the hole diameter as a maximum).

- In the case of failures in bearing, which presented the most complex behaviour, the ultimate loads were 1,5 to 1,7 times greater than their calculated values, but failure occurred together with holes ovalisation. The main difficulty lies on making a realistic calculation of this ovalisation, which results not only from the bearing effects of the bolt but also from the elongation of the net section.

- The estimated bearing deformation at the hole is not very high and the net section elongation is even lower. Actually, the length concerned by the net section elongation is small and even if a plastic deformation occurs, its effect on the hole elongation is always small except at failure, when necking takes place at the net section.

- The ratio between the shear resistance in tension to that of direct tension gives mean values going from 0,69 to 0,62 for diameters from 12 mm to 20 mm. This ratio has a high value for the small diameter while direct tension gives comparable ultimate test resistances between diameters 12 and 20 mm.

- For shear in compression, the results dispersion is lower than that for shear in tension for the three diameters, while average resistance values are higher. In reality, the shear test in tension which gives lower mean values of resistance is closer to the connection loaded in tension.

- The test values of resistance in direct tension are higher than guaranteed values. For each diameter, the results of resistance show a low dispersion.

Figure 4. Net section failure mode

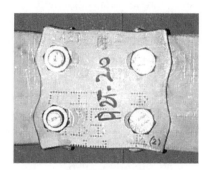

Figure 5. Bearing failure mode

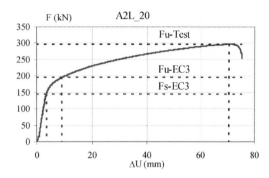

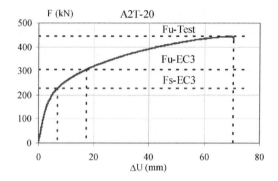

Figure 6. Global load-displacement curve for a connection with failure in net section

Figure 7. Global load-displacement curve for a connection with failure in bearing

Further developments

- Extension of the experimental results to analyse the behaviour of a beam-to-beam cover plate connection and a gusset connections.

- Some tests are in preparation for T-stubs with stiffeners. They will be compared to that of carbon steel and to the numerical models (FEM software CASTEM). This model is based on the 3D FEM developed for carbon steel T-stub reinforced by backing plates (Alkhatab, 2003).

- The evaluation of a beam deflections and a plastic loads redistribution needs to be more analysed taking account of the ductility of stainless steel. Also, attention has to be made to ensure that calculated plastic moment transmitted to the connections do not exceed their own strength.

- Despite interesting results for bolts in tension and shear, for actual connection, the bolts shear failure mode remains less interesting than the gross cross-section, the bearing or the net cross-section. For these failure modes it is possible to obtain relatively large deformation before failure.

- It is necessary to develop analytical approaches which allow to calculate in a precise way the elongation of the net cross-section and the deformations due to the bearing and to identify the cases for which the excessive deformations of holes have very important effects (inversion of load, etc.). So, it may be desirable to make a distinction between single and double shear connections.

- The analytical approach results could be compared to the corresponding FEM and experimental results to assess the level of accuracy of each of the proposed approaches.

Standard documents

- European Committee for Standardisation ENV 1993-1.4, *Eurocode 3 – Part 1.4 : General Rules – Supplementary Rules for Stainless Steels*, CEN/TC250/SC3, Brussels, Belgium, 1996.

- European Stainless Steel Development and Information Group, EuroInox, Design

Manual for Structural Stainless Steel, 1994 and new version in2003.

- American Society of Civil Engineers, ANSI/ASCE-8-90, *Specification for the Design of cold-formed stainless steel structural members*, 1991.

References

- Alkhatab, Z., 2003 : *Analyse de comportement d'assemblages métalliques renforcés par contre-plaques - Approche numérique et validation expérimentale.* PhD Thesis, Blaise Pascal University, Clermont-Ferrand, France, 282 pages.

- Arnedo, A., et al., *Deformations of flexural members of austenitic stainless steel*, 2nd World Conf. on Steel Constr., 1998, San Sebastian, Spain, paper 241, CD-Rom.

- ASCE, 1990. *Specification for the design of cold-formed stainless steel structural members,* ASCE Standards, 114 pages.

- Bouchair, A., Ryan, I., Kaitila, O. and Muzeau, J. P., 2001. *Some aspects of the analysis of bearing-type stainless steel bolted joints.* 9th Nordic Steel Const. Conf., Helsinki., 755-762.

- Burgan, .B. A, Baddoo, N. R. and Gilsenan, K. A., 2000, *Structural design of stainless steel members -comparison between EC3-1.4 and test results.* Journal of Constr. Steel Research, **54,** 51-73.

- CEN, Eurocode 3 - ENV 1993 – part 1.4, *Supplementary rules for stainless steels,* 1996.

- Kouhi, J., Talja, A., Salmi,P. and Ala-Outinen, T., 2000, *Current R&D work on the use of stainless steel in construction in Finland.* Journal of Constr. Steel Research, 54, pp. 31-50.

- NidL-EuroInox, 1994. *Design Manual for Structural Stainless Steel.* Nickel Develpt. Institute.

- Ryan, I., *Development of the use of stainless steel in construction (WP 4.2)- Bolted connections : Part 1. Cover plate connections,* 2000, Ugine - CTICM. 82 pages.

- SCI, 2000, *Development of the use of stainless steel in construction*, Contract 7210-SA/842, 903, 904, 327, 134, 425, Final Report to ECSC

- Van Den Berg, G. J., 2000. *The effect of the non-linear stress-strain behaviour of stainless steels on member capacity.* Journal of Constr. Steel Research, **54,** 135-160.

Contacts

- Dr. Abdelhamid BOUCHAIR

  Address: CUST- Civil Engineering Department, University Blaise Pascal, Rue des Meuniers, B.P. 206, 63174 Aubière-cedex, Clermont-Ferrand, France.
  e-mail: bouchair@cust.univ-bpclermont.fr
  Phone: (+33) 4 73 40 75 32
  Fax: (+33) 4 73 40 74 94

COST C12 - WG2

Datasheet I.4.2.6
Prepared by: G. De Matteis
       University of Naples Federico II, Dept. Structural Analysis and Design

T-STUB ALUMINIUM ALLOY JOINTS

Description

- T-stub aluminium joints consist of two plate elements assembled in a "T" configuration, usually by means of a weld connecting the web element with the flange element. An even number of holes, located symmetrically to the web, are punched or drilled on the flange, so that a bolted fastening system can be obtained (Figure 1).

- T-stubs may be subjected to axial load only, being the load applied in the direction of the web element.

- T-stubs may be either connected to each other by means of bolts joining the two opposite flanges, thus obtaining coupled aluminium T-stubs, or an isolated T-stub may be connected to a supporting structural element, usually having negligible deformability (Figure 2).

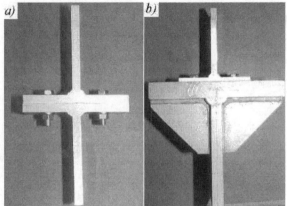

Figure 1. Isolated T-stub specimens   **Figure 2.** Coupled (a) and isolated (b) aluminium T-stubs

Field of application

- T-stub connection may be considered either as a stand alone connection (in case of joint of members under tensile forces) or as a part of more complex joints (in case of joint subjected to bending and/or normal forces), both under monotonic and cyclic loading (Figure 3).

- The single T-stub configuration is typical of end-plate beam splice joints (where the connected elements have the same flexural stiffness); conversely, the coupled configuration is representative of joints where the connected plate elements are different from each other, such as in the case of beam-to-column joints.

- The behaviour of T-stub is rather complex due to the fact that different collapse modes could be of concern, namely failure of bolts and failure connected flange elements.

- For aluminium joints, T-stub behaviour is more complicated with respect to steel due to the influence of material hardening and limited material ductility.

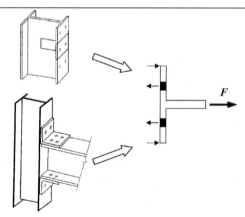

Figure 5. Modelling of more complex types of joints by means of T-stub models

---

Technical information

- Flanges and webs of T-stubs are made out of wrought aluminium alloy plates, joined by means of a welded connection, with thickness of few millimetres. The flange and the web elements may have the same thickness or not.
- In-plane flange dimensions, as well as web height, may vary.
- Bolts may be made either of steel or of aluminium alloy.
- Holes may be either punched or drilled.

Structural aspects

- The bolts are subjected predominantly to tension, as well as the web, while the flanges work in bending and shear. Therefore, the bearing capacity of this connecting system depends essentially on both the flexural resistance of the flange and the tensile strength of the bolts.
- Different collapse mechanisms, related to different ductility levels, may be achieved according to the geometrical configuration of the joint as well as the mechanical properties of the adopted materials.
- In particular, when calculating the bearing capacity and the ductility level of an aluminium alloy T-stub connection, attention must be paid to: (i) bolt strength and deformation capacity; (ii) influence of the Heat Affected Zone (HAZ) on the mechanical properties of the material in the vicinity of the weld; (iii) strain hardening and available ductility of the adopted aluminium alloy.
- Three basic failure mechanisms may be recognised for T-stubs (Figure 4): type-1 failure involves flange yielding, with the onset of four plastic hinges; type-2 failure occurs when bolt failure takes place together with yielding of the flange at the sections corresponding to the flange-to-web connection; type-3 failure consists in the sole bolt failure with the overall uplift of the flange.
- Since type-2 mechanism depends on both flange and bolt behaviour, three sub-mechanisms can be outlined in aluminium alloy T-stubs (Figure 5): type-2a is related to a plastic failure in the flange rather than in the bolt; type-2c reflects the bolt failure with limited plastic deformation in the flange; type-2b may be considered as an intermediate situation between the aforementioned ones.

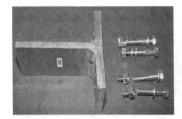

Figure 4. Basic failure mechanisms for aluminium alloy T-stubs (from left to rigth: type 1, type 2, type 3)

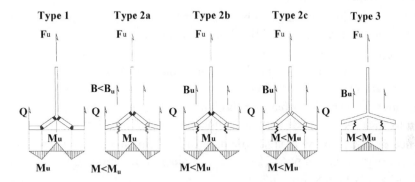

Figure 5. Failure mechanisms for aluminium alloy T-stubs

Research activity

- Research carried out within the Italian Research Project "Damage of Connecting Systems in Metallic and Composite Constructions" aims at verifying the codified approach for steel T-stubs when employing aluminium alloy plates and both steel and aluminium bolts.

- The research project also provides relevant corrections to the existing codified formulation regarding aluminium T-stubs.

- Research activities are also ongoing at the Department of Civil Engineering of the Aristotle University of Thessaloniki.

Method of analysis

- Both theoretical, numerical and experimental methods have been employed in the research project.

- In the experimental analysis, specimens have been drawn out of T-stub strips that have been pre-assembled by welding the corresponding flange and web plate elements. The basic plate elements are made of different wrought aluminium alloys and have different thickness (Figure 6). Fillet welds between the web and the flange have been made using the MIG welding process. Both aluminium alloy (type AA7075) and steel (grade 4.8 and grade 10.9) bolts have been adopted. In all cases the hole diameter is equal to 11 mm and the bolt diameter is equal to 10 mm.

- Monotonic and cyclic loading has been applied, up to failure, under displacement control, to the selected specimens. In particular, 26 welded T-stub specimens have been tested. In particular, while 20 specimens were isolated T-stubs, the remaining 6 were assembled to each other (coupled T-stubs).

- As analytical method, the bearing capacity of aluminium T-stubs has been evaluated by means of the component method.

- The strength of T-stub elements may be evaluated by following an approach similar to the one used for steel (EC3-Part 1.8), but accounting for the moment capacity of plastic hinges developing in the flanges (De Matteis et al., 2001).

- Finite element models have been calibrated on the experimental results, and parametric analyses have then been carried out in the "virtual laboratory" provided by finite element commercial codes.

- The monotonic behaviour of the T-stub aluminium joint, subjected to tensile force, has been simulated by means of the ABAQUS finite element code.

- Eight-node, octahedral, first-order, reduced-integration elements, with three active degrees of freedom per node (i.e., the translational ones), have been used for modelling both the T-stub flange and bolts.

- The aluminium alloy material behaviour (both in the plates and in the bolts) has been represented by means of the well-known exponential model (Hopperstad, 1993).

- The influence of HAZs in the area surrounding the welding depends on the welding process, the alloy and the temper. It has been considered in the numerical model by means of a resistance knock-down factor.

- In order to account for the unilateral interaction between different bodies, special contact elements with isotropic Coulomb model have been adopted in the analysis.

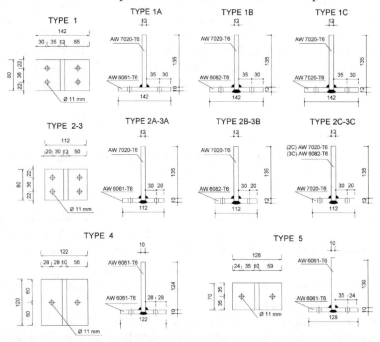

Figure 6. Nominal dimensions and adopted aluminium alloys for the tested T-stub specimens

Results of analysis

- From experimental results, the important role of bolt strength and deformation capacity on the response of T-stubs has been highlighted.

- Also, it has been verified that the effect of material strain-hardening has a paramount role in determining the ultimate bearing capacity of the joint.
- Numerical analyses are able to correctly predict any of the collapse mechanisms and sub-mechanisms, as illustrated above.
- By means of parametric analyses, the paramount influence of material strain-hardening, material ductility and the effect of HAZs (also in relation to the adopted weld detail) have been pointed out.
- The proposed analytical method for the evaluation of T-stub bearing capacity can be considered fairly reliable, conservative and, because of its inherent simplicity, suitable to design purposes, the interpretation of the actual joint failure mechanism being quite accurate.

Further developments

- In order to account for the effect of both strain-hardening and ductility on the ultimate resistance of aluminium T-stubs, a correction procedure to the usual formulation for steel joints, failing under type-1 collapse mechanism (see Eurocode 3), has been already proposed.
- Such a procedure has been profitably implemented in the Eurocode 9 for aluminium alloy T-stub connections.
- The extension of such an approach to different connection typology is strongly advisable.

References

- De Matteis G., Della Corte G., Mazzolani F.M., Experimental Analysis of Aluminium T-stubs: Tests under Monotonic Loading, *XVII Congresso del Collegio dei Tecnici dell'Acciaio (CTA)*, Venice, Italy, 2001.
- De Matteis G., Della Corte G., Mazzolani F.M., Experimental Analysis of Aluminium T-stubs: Tests under Cyclic Loading, *Advances in Structures: Steel, Concrete, Composite and Aluminium (ASSCCA 2003)*, Sydney, Australia, 2003.
- De Matteis G., Mandara A., Mazzolani F.M., Calculation Methods for Aluminium T-stubs: a Revision of the EC3 Annex J, *Proceedings of the 8th International Colloquium on Joints on Aluminium Structures (INALCO 2001)*, pp. 6.2.1-6.2.12, Munich, Germany, 2001.
- De Matteis G., Mandara A., Mazzolani F.M., Remarks on the Use of the EC3-Annex J for the Prediction of Aluminium Joint Behaviour, *Proceedings of the 6th International Colloquium on Stability and Ductility of Steel Structures (SDSS'99) –* D. Dubina and M. Ivanyi Editors, pp. 229-240, Timisoara, Romania, 1999.
- De Matteis G., Mandara A., Mazzolani F.M., T-stub Aluminium Joints: Influence of Behavioural Parameters, *Computers & Structures*, Vol. 78, pp. 311-327, 2000.
- De Matteis, G., Mandara, A., Mazzolani, F.M. 2002. Design of aluminium t-stub joints: calibration of analytical methods, *In Proc. of the Third European Conference on Steel Structures*, Coimbra (Portugal), 19-20 Sept., Vol. II 1017-1026 .
- Hopperstad O.S., *Modelling of Cyclic Plasticity with Application to Steel and Aluminium Structures*, Dr. Ing. Dissertation – Division of Structural Mechanics, The

Norwegian Institute of Technology, Trondheim, Norway, 1993.

- Mazzolani F.M., *Aluminium Alloy Structures*, E&FN SPON, London, United Kingdom, 1994.
- Mazzolani F.M., De Matteis G., Mandara A., Classification System for Aluminium Alloy Connections, *Proceedings of the International Colloquium on Semi-Rigid Structural Connections, IABSE* Vol. 75, pp. 83-94, Istanbul, Turkey, 1996.

Contacts

- Prof. F. M. Mazzolani
  Address: Department of Structural Analysis and Design, University of Naples 'Federico II', P.le Tecchio, 80, Naples, Italy.
  e-mail: fmm @unina.it
  Phone: (+39) 0817682443
  Fax: (+39) 0815934792

- Prof. Dr.-Ing. C. C. Baniotopoulos
  Address: Institute of Steel Structures- Department of Civil Engineering
  Aristotle University of Thessaloniki, 54124, Thessaloniki, Greece
  E-mail: ccb@civil.auth.gr
  Tel.: +302310-995753;
  Fax: +302310-995642

COST C12 - WG2

Datasheet I.4.2.7

Prepared by: G. De Matteis
> University of Naples Federico II, Dept. Structural Analysis and Design

ALUMINIUM ALLOY BOLTED CONNECTIONS IN SHEAR

Description

- Usually, connections between aluminium alloy structural systems are based on the use of bolts as a fastening system between plate elements. The connected plates may be subjected to either tension or bending/shear stress regime. Obviously, in both cases, shear stress regime does take place in the bolts as a consequence of the actions applied to the structural systems under consideration.

- In the simplest case, an isolated flat plate is connected to a supporting structure having negligible deformability; different boundary conditions may be applied on the four edges of the structural element. Otherwise, two aluminium alloy flat plates may still be joined by means of a bolted connection; in such a case, for the sake of the simplicity, only the case of symmetric arrangement of the plates with respect to the direction of the applied load will be analysed in the following.

- More usually, the two flat plates introduced above belong to more complex structural systems: for example, they can be either the internal or the outstanding elements of extruded aluminium alloy profiles.

Field of application

- Bolted connections joining aluminium alloy structural elements/structural systems in tension may be adopted to join tensioned members, both under monotonic and cyclic axial loads.

Technical information

- The connected structural elements/structural systems are made of wrought aluminium alloy material.
- The thickness of the connected plates may vary, as well as the geometry of the bolted connection.
- Bolts may be made either of steel or of aluminium alloy.
- Holes may be either punched or drilled.

Structural aspects

- As a consequence of the actions applied to the structural elements/structural systems under consideration, the bolts will be subjected to shear stresses.
- Different collapse mechanisms, related to different strength and ductility levels, may be achieved, depending on the geometrical configuration of the bolts, the mechanical properties of both the joined elements and the bolts, the thickness of the plates and, to a lesser extent, the remaining geometrical features of the connected plates. In particular, attention must be paid to: (i) the bolt strength and deformation capacity; (ii) the available ductility of the adopted aluminium alloy.

- Four basic failure mechanisms may be recognised: type-1 failure involves bearing ultimate resistance of the holes; type-2 mechanism occurs when shear ultimate resistance of the bolts is attained; type-3 failure takes place when the structural element/structural system net section ultimate resistance is reached; type-4 mechanism (the so-called "block-shear mechanism") is given by the extraction of the whole area containing the holes from the remaining part of the connected structural element/structural system.

- Block-shear failure may take place in structural elements/systems either under axial tensile loads or under bending/shear loading regime. In both cases, the extraction of the whole area containing the holes takes place under the combined effect of both normal tensile stress and shear stress. More precisely, the former are predominant along the yielding/failure lines perpendicular to the applied load direction (Figure 1a), while the latter are more important along the yielding/failure lines parallel to the applied load direction (Figure 1b).

- Prescriptions yielded by Structural Codes for aluminium alloy structures prove to be reliable and on the safe side with regard to failure mechanisms from 1 to 3; instead, as far as the block-shear failure mechanism is concerned, they either do not yield any indication (especially with regard to the axial tensile loading case) or they are non-conservative.

- Therefore, with reference to the block-shear failure type, either experimental or analytical methodologies should be adopted in order to evaluate correctly both the resistance and the ductility level of the connection. In particular, with regard to analytical approaches, reference should be made to the models provided by fracture mechanics, both in theoretical and in numerical studies

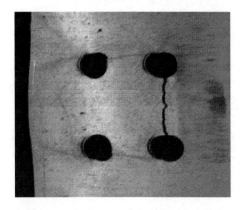

Figure 1. Block-shear failure mechanism in aluminium alloy extruded I-shaped beams: (a) under tensile axial load (Test A1); (b) under three-point loading (Test B3).

Research activity

- The research project illustrated in the following is currently in progress at the Department of Structural Analysis and Design of the University "Federico II" of Naples (Italy), in co-operation with the Department of Structural Engineering at NTNU of Trondheim (Norway).

- The research project aims at verifying the codified approach for block-shear mechanisms in aluminium alloy structural elements/structural systems.

- If necessary, the proposed research project will also suggest relevant corrections to the existing codified formulation regarding bolted aluminium alloy connections in shear.

Method of analysis

- Both theoretical, numerical and experimental methods have been employed in the research project.

- In particular, finite element models have been calibrated on the experimental tests highlighting block shear failure in aluminium alloy beams under tensile axial load.

- Parametric analyses will then be carried out in the "virtual laboratory" provided by finite element commercial codes.

- In order to verify the codified approach proposed for bolted connections between steel members in tension with regard to the block-shear failure mechanism, different aluminium alloy specimens have been designed in such a way to fail according to Type-4 mechanism, either under tensile axial loading (Figure 2) or under three-point loading on a simply-supported static scheme (Figure 3). Obviously, in both cases, shear stresses take place in the bolts.

- In particular, five different specimens have been tested in the former loading condition (Tests A1 to A5), three in the latter (Tests B1 to B3) (Figure 4).

- The experimental campaign has been carried out on HDR 0115 extruded beam, made of the AW-6082T6 aluminium alloy (0.2% offset proof stress and ultimate nominal tensile stress equal to 259 Nmm-2 and 276 Nmm-2, respectively).

- Connections with 4, 6 and 8 holes (Figure 4), located symmetrically on the connected plate elements, have been considered; hole spacing and hole distance from edges (both in the load direction and perpendicularly to the load direction) have been kept constant for all specimens.

- Steel bolts grade 8.8, having a nominal yielding stress of 640 Nmm-2 and a ultimate stress of 800 Nmm-2, have been adopted. In all cases (Figure 4) the hole diameter is equal to 15 mm, and the bolt diameter is equal to 14 mm.

- In order to apply the tensile axial load to the beam, a hydraulic actuator, able to apply a maximum axial load of 2000 kN and a maximum displacement of 2000 mm, has been employed. The displacement rate is equal to 2 mm/min, and it has been applied by means of two plates made of Weldox steel, with a thickness of 12 mm.

- With regard to the bending/shear regime applied to the beam, spanning 3200 mm, the load has been applied by means of a hydraulic actuator (maximum load 500 kN, maximum displacement 140 mm) placed at 335 mm from the connection to be tested.

Figure 2. *Bolted connection in shear in aluminium alloy extruded beam under axial tensile load (right-hand-side) and under three-point-loading (left-hand-side)*

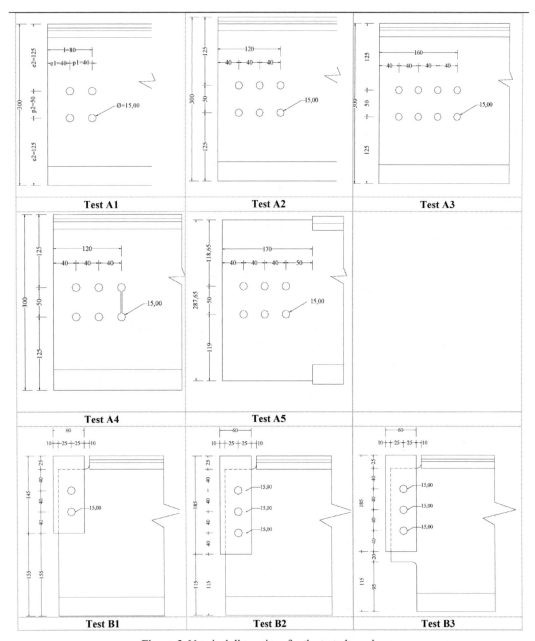

Figure 3. Nominal dimensions for the tested specimens

- The bearing capacity of bolted connections between aluminium alloy structural elements in tension has been evaluated by means of the limit-load approach, by adopting the rigid-perfectly plastic material model and the small displacement hypothesis.
- The monotonic behaviour of the T-stub aluminium joint, subjected to tensile force, is simulated by means of the ABAQUS finite element code.
- Due to symmetry, only half of the global model has been considered in the analysis.
- Four-node, quadrilateral, reduced-integration elements, with six active degrees of freedom per node, have been used for modelling the structural system subjected to

the block-shear failure mechanism.

- Bolts have been modelled as rigid surfaces, connected to a master joint and moving rigidly in the direction of the applied tensile load at the same rate as in the experimental tests.

- The edge of the structural system under consideration opposite the area where block-shear failure takes place has been fully clamped, for the sake of the simplicity.

- The joint response has been evaluated by means of a transient nonlinear dynamic analysis, based on the well-known Hilber-Hughes-Taylor implicit integration scheme. The load-displacement curve has been evaluated by calculating the reaction in the encastred end corresponding to each value of the applied master joint displacement. Steel bolts grade 8.8, having a nominal yielding stress of 640 Nmm-2 and a ultimate stress of 800 Nmm-2, have been adopted. In all cases (Figure 3) the hole diameter is equal to 15 mm, and the bolt diameter is equal to 14 mm.

- Tensile tests have been carried out on eight rectangular section specimens, cut out in different positions of the web of a HDR 0115 extruded beam, 3200 mm long.

- Due to the rolling process, residual stress and orthotropic material properties distribution take place in the specimens. Material tensile tests have been performed along different directions during the experimental campaign.

- The aluminium alloy material behaviour in the structural elements away from the area where crack onset and propagation takes place has been represented by means of the well-known exponential model [Hopperstad, 1993].

- In order to reproduce crack propagation effects on the strip connecting two adjacent holes, perpendicularly to the applied load direction, a modified material model has been implemented in the numerical model. In particular, recourse has been made to the necking stress-strain curve.

- The necking stress-strain curve is derived from the conventional engineering stress-strain curve by dividing, once attained the uniform strain, the displacement of the tensile specimen by a shorter gauge length rather than by the original one. In this way, necking effects are introduced, in the finite element model, at the material level.

- True stress-logarithmic strain coordinates have been adopted to describe the material behaviour on the whole structural element under consideration, thus allowing for a large-strain analysis. Instead, where the necking stress-strain curve was employed, engineering coordinates were adopted.

Results of analysis

- In all of the carried out experimental tests in the tensile axial load condition, a brittle collapse mechanism has taken place, failure has taken place corresponding to the net section failure, immediately followed by the plastic yielding lines connecting the edges of the holes, in direction parallel to the applied load.

- Block-shear failure of the tested specimens has been obtained in all cases, under strength and ductility levels depending on the tested specimen.

- Block-shear mechanism took place with rupture of the material on a thin strip connecting the centres of the innermost holes, in direction perpendicular to the applied load.

- Besides, in most cases yielding lines developed on the material of a thin strip connecting the centres of two adjacent holes, in direction of the applied load.

343

Yielding lines also developed between the outermost holes and the free (i.e., non-encastred) edge of the tested specimen.

- From the experimental campaign, it can be clearly seen that the design prescriptions proposed by some structural codes (EC9, CAN/CSA-157) are on the unsafe side, since they overestimate the block shear collapse strength. On the other hand, structural codes correctly model the collapse mechanism in the case of bending/shear stress regime in the beam.

- In case of a bolted connection between an extruded aluminium-alloy I-beam and a rigid support, it was proved that the contribution offered by the flanges of the beam is negligible.

- Numerical analyses are able to predict correctly both the strength and the ductility level related to the block shear failure mechanism, provided the criteria illustrated above are followed (Figure 4).

- Care are must be taken in adopting sufficiently refined meshes, in order to evaluate correctly the ductility level of the considered specimen.

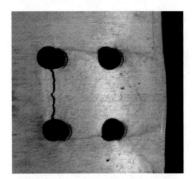 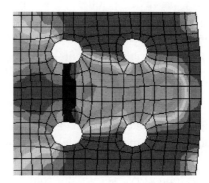

Figure 4. Comparison between experimental and FEM results in terms of stress contours

Further developments

- Parametric analyses should be carried out, accounting for different hole configuration, plate thickness and aluminium alloys.

- Once evaluated the strength and ductility level of each specimen in the virtual laboratory given by the numerical model, a correction procedure to the codified formulation for block-shear failure in steel structural elements could be proposed.

References

- European Committee for Standardisation ENV 1993-1.1, *Eurocode 3 – Part 1.1: General Rules and Rules for Buildings*, CEN/TC250/SC3, Brussels, Belgium, 1997.

- European Committee for Standardisation ENV 1999, *Eurocode 9 – Part 1.1: Design of Aluminium Structures*, CEN/TC250/SC9, Brussels, Belgium, 1998.

- Norsk Standard NS 3472, *Prosjectering av Stålkonstruksjoner - Beregnings -og Konstruksjonregler (Utgrave 3)*, Oslo, Norway, 2001.

- American Institute for Steel Construction, *AISC Manual of Steel Construction – Load and Resistance Factor Design (3rd Edition)*, Chicago, IL, USA, 1999.

- Canadian Standards Association CAN3-S157-FM83, *Calcul de la résistance mécanique des éléments en aluminium*, Mississauga, Ontario, Canada, 2001.

- Hopperstad O.S., *Modelling of Cyclic Plasticity with Application to Steel and Aluminium Structures*, Dr. Ing. Dissertation – Division of Structural Mechanics, The Norwegian Institute of Technology, Trondheim, Norway, 1993.

- Dexter R.J., Gentilcore M.L., 2002. Analysis of plane stress ductile failure propagation by simulating necking, *Journal of Structural Engineering*, Vol. 128, pp. 1003-1011.

- Kulak G.L., Grondin G.Y., Block Shear Failure in Steel Members – A Review of Design Practice, *Proceedings of the 4th International Workshop on Connections in Steel Structures*, Roanoke, VA, USA, 2000.

- Aalberg A., Larsen P.K., Strength and Ductility of Bolted Connections in Normal and High-Strength Steels, *Proceedings of the 7th International Symposium on Structural Impact and Plasticity (IMPLAST 2000)*, Melbourne, Australia, 2000.

Contacts

- Prof. F. M. Mazzolani
  Address: Department of Structural Analysis and Design, University of Naples 'Federico II', P.le Tecchio, 80, Naples, Italy.
  e-mail: fmm @unina.it
  Phone: (+39) 0817682443
  Fax: (+39) 0815934792

- Prof. Magnus Langseth
  Address: Department of Structural Analysis – Faculty of Engineering Science and Technology, Richard Birkelands vei 1A, 7491 Trondheim, Norway.
  Phone: 0047.73594782
  Fax: 0047.73594701
  E-mail: Magnus.Langseth@bygg.ntnu.no

COST C12 - WG2

Datasheet I.4.2.8

Prepared by: F.Wald[1], U. Kuhlmann[2], M. Rybinski[2], D.Gregor[1]

[1] Czech Technical University in Prague
[2] University of Stuttgart

STEEL TO CONCRETE JOINTS

Description

- Information concerning methods applicable to connection between steel and concrete members (e.g. steel beam to concrete column).

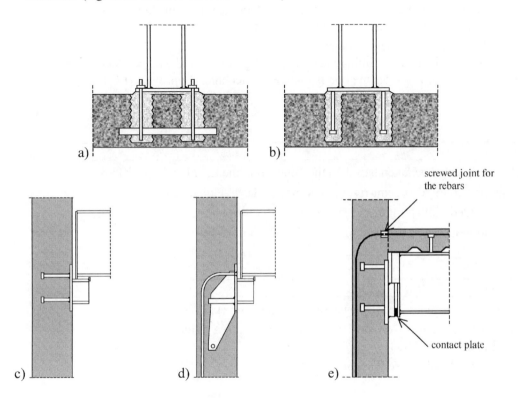

Figure 1 Exmaples of steel to concrete joints;
a) Column base with anchor bolts, b) column base with headed studs,
c) console with headed studs for steel beam, d) console with welded reinforcement for steel beam, e) console for composite beam

Field of application

Mixed building structures, steel structures with RC core. Simple and moment joints, see [Stark, Hordijk, 2001]

Prediction models

- Model by FEM
  Still difficult to define: contacts, concrete crushing and cracking, bond behaviour of anchors. Unknown imperfections. For scientific use only, calibrated on tests.

[Eligehausen, 2001]

- Semi analytical model
  Discretization of the joint into at least two parts (e.g. the tension and compression part for the joint loaded in flexion). Definition of force displacement (F-δ) relation for each part based on FEM analysis of these parts. Assembly procedure assuming the parts behave as springs (with their respective F-δ relationship) connected in joint by infinitely rigid.

- Analytical model
  Application of method of components. Description of the component F-δ relationship allows assembling complete M-φ curve of the joint. A bilinear relationship is available for prediction models. The prediction under cyclic loading is under development.

- Design model (Concrete capacity method)
  Developed for fastenings in unreinforced concrete [Eligehausen and Mallée, 2000]. For application of fixing devices this method is included in technical approvals [EOTA, 2001].

Global analysis

For the global design may be taken based on the connecting member bending stiffness as

- Nominally pinned, rigid or semi-rigid joints for the elastic global analysis.
- Nominally pinned, full-strength and partial strength for the plastic global analysis.
- Simple, continuous and semi-continuous model of joint.

Research activity

- New types of connections, see [Eligehausen, 2001], [Ando et al., 2001].
- Influence of reinforcement on steel to concrete joints, see [Kuhlman and Imminger, 2004], Kuhlmann et al., 2004].
- Models of cyclic loading, see [Gregor, 2004].

Component description

- Web panel in shear, see [prEN-1993-1-8: 2003].
- Anchor bolt in tension

Resistance is checked based on the basic failure modes: anchor steel failure, concrete cone failure, split of concrete, bursting and pull-out failure, see Figure 2. [ETAG 001 – Annex C] and/or [CEB, 1997].The ductile failure mode due to the steel part of the anchor bolt is preferred.

Stiffness may be predicted based on [prEN-1993-1-8: 2003].

Due to the required deformation capacity of joint, steel rupture should govern. Other modes of rupture should be avoided by appropriate embedment depth and edge distances and/or bar or stirrup reinforcement. In the case of low embedment depth of anchors (e.g. steel studs used as anchors) – stirrups reinforcement increase both joint ductility and strength [Kuhlmann, Imminger, 2004].

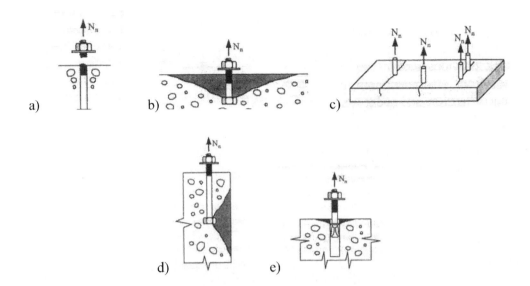

a)　　　b)　　　c)

d)　　　e)

Figure 2. a) Anchor steel failure, b) concrete cone failure, c) split of concrete,
d) bursting and e) pull-out failure

- <u>Anchor bolt in shear</u>

Two basic models are available.

Friction + tension of anchor [Bouwman et al., 1989] for long anchor bolts, see Figure 3.

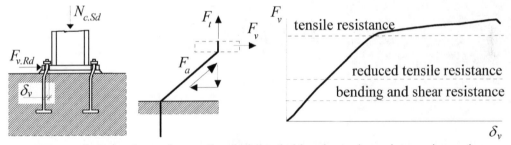

Figure 3. Behaviour of an anchor bolt loaded by shear, the resistance in tension,
[Bouwman et al., 1989].

Friction is calculated with coefficient 0.2 for standard grout (for 0.3 special grout). Anchor resists in shear even after formation of plastic hinge in bending and further behaves as a fibre in tension. Sufficient ductility of anchors is required. Applicability is limited to steel grade S235-S640.

Anchor in bending , see [CEB, 1997] Figure 4.

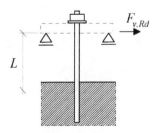

Figure 4. Model of anchor bolt in bending, [CEB, 1994]

Shear force is transmitted by anchor's bending only, friction is neglected, elastic force distribution between anchors is permitted.

- Concrete in compression and plate in bending, for resistance and stiffness prediction see [prEN-1993-1-8: 2003].

- Anchor bolt in tension and concrete in compression, for resistance and stiffness prediction see [prEN-1993-1-8: 2003].

- Reinforced concrete in tension

Tension force of anchor is transmitted to reinforcement by catch of the stirrup and length of the stirrup in the assumed failure cone [Eligehausen and Mallée, 2000]. Load capacity of concrete is not considered for tension.

Further developments

- Component tests
  - Tests of components in shear, see [Bouwman et al., 1989].
  - Tests of components concrete in compression and plate in bending, see [Gregor et al., 2002] Figure 5.
  - Tests of components anchor bolt in tension and plate in bending [Gregor et al., 2003].

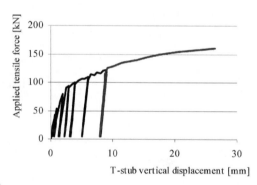

a)   b)

Figure 5. a) Test of component anchor bolt in tension and plate in bending [Gregor et al, 2002], b) force deformation curve

- Static behaviour
  - Stirrups increase load capacity and ductility of joints, see Figure 6.
  - Concrete capacity method highly underestimates load capacity as reinforcement is not taken into account.
  - Calculated load capacity by analytical model based on component method corresponds to test results for rigid anchor plates, failure of shear studs or concrete cone failure [Kuhlmann and Imminger, 2004].
  - Further studies are needed for failure mode yielding of the anchor plate to evaluate the effective inner lever arm and the deformability.

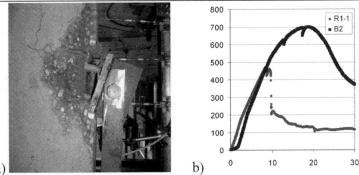

a)   b)

Figure 6. a) Anchor plate with welded studs [Kuhlmann and Imminger 2004], b) force deformation curve with and without reinforcement

- Cyclic behaviour
  - Models by FEM became more complex in the case of cyclic loads taking into account the deterioration of the material characteristic and the re-contact during cycling, see [Ádány, 2000].
  - For semi-analytical models, sophisticated material models for cyclic load are needed e.g. [Adány, 2000]. For assembly procedure of springs, commercial FEM software may be used. [Rassatti, 2000].
  - Analytical models define the stiffness of unloading path for all components and take plastic deformation of components into account. The initial stiffness is assumed as a unloading path stiffness for T-stub in tension and compression and for anchor bolts, see [Gregor, 2004] Figure 7.

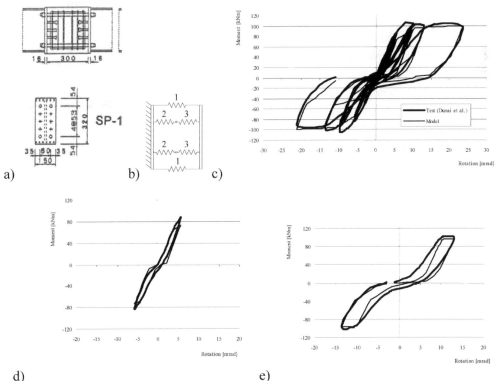

d)   e)

Figure 7. a) Test of whole joint [Dunai et al., 1994]; b) mechanical model [Gregor, 2004]; c) comparison of the model to test; d) comparison at second cycle, e) comparison at 6th cycle.

351

References

- Ádány, S., 2000: *Numerical and Experimental Analysis of Bolted End-Plate Joints under Monotonic and Cyclic Loading*, Ph.D. thesis, Budapest University of Technology and Economics, Budapest, Hungary.

- Ando N., Nishimura I., Kamo K. (2001) An experimental study on the connection joints between steel girder and reinforced concrete columns with various types of embedded load transferring plates, in Connections between steel and concrete, University of Stuttgart, ed. Eligehausen R., Stuttgart, pp. 1293-1302, ISBN 2-912143-25-X.

- Bouwman, L.P., Gresnigt A.M., Romeijn, A. Onderzoek naar de bevestiging van stalen voetplaten aan funderingen van beton. (Research into the connection of steel base plates to concrete foundations), in Dutch, Stevin Laboratory report 25.6.89.05/c6, Delft 1989.

- CEB, 1994: *Fastening to concrete and masonry structures*, State of the Art Report, p. 248, CEB, Thomas Telford Services Ltd., London 1994, ISBN 0 7277 1937 8.

- CEB, 1997: *Design of Fastenings in Concrete*, Design guide, p. 83, CEB, Thomas Telford Services Ltd, London 1997, ISBN 0 7277 2558 0.

- CEN, 2004: *Design of Fastenings for Use in Concrete*, Draft 10, Technical Specification, CEN / TC 250, March 2004.

- COST C1, 1999. Column Bases in Steel Building Frames, Brussels/Luxembourg.

- Dunai L., Ohtani Y., Fukumoto Y., 1994: *Experimental Study of Steel-to-Concrete End-Plate Connections under Combined Thrust and Bending*, Technology Reports of Osaka University, Vol. 44, No. 2197, Osaka.

- Eligehausen, R. ed., 2001: *Proceeding to International Symposium on Connections between Steel and Concrete*.

- Eligehausen R., Mallée R., 2000: *Befestigungstechnik im Beton- und Mauerwerksbau (Fastenings for use in concrete and masonry)*, in German, Ernst & Sohn Verlag, Berlin, ISBN 3-433-01134-6.

- EOTA, 2001: *ETAG 001 –Annex C*.

- Fuchs W., Eligehausen R., Breen J.E.,1995: *Concrete Capacity Design: Approach for Fastenings to Concrete*, ACI Structural Journal, Vol. 92, No. 1.

- Gregor D., 2004: *Analysis of Cyclically Loaded Mixed Building Joints by Method of Components*, Ph.D. thesis, CTU Prague, Prague, in preparation (in Czech).

- Gregor D., Wald F., Sokol Z. 2003: *Experiments with End Plate Joints for Mixed Building Technology*, v Experimental Investigation of Building Materials and Technologies, ed. Konvalinka P., Luxemburg F., ČVUT in Prague, p. 65-82, ISBN 80-01-02835-6.

- Gregor D., Wald F., Eliášová M., Jírovský I., 2002: *Joints for mixed building technology with view to experiments of component steel plate in bending and concrete in compression*, in Eurosteel, Coimbra, p. 977-986, ISBN 972-98376-3-5.

- prEN 1993-1-8, 2003: 2003, Eurocode 3, *Design of Steel Structures*, Part 1.8: *Design of Joints*, European Norm, CEN, Brussels.

- prEN-1994-1-1: 2003, Eurocode 4, *Design of composite steel and concrete structures*, *Section 8 Composite joints in frames for buildings*, European Norm, CEN, Brussels.

- Kuhlmann U., Imminger T.; 2004: *Ankerplatten und Einbaudetails zur Kraftübertragung im Stahlbau*, in German, Forschungsbericht Nr. 1/2004, Deutscher Ausschuß für Stahlbau DASt, Düsseldorf.

- Kuhlmann U., Imminger T., Rybinski M.; 2004: *Zur Tragfähigkeit von Ankerplatten in bewehrtem Beton,* in German, Stahlbau 72, Ernst & Sohn Verlag, Berlin, p. 270-275.
- Rassati G.A., Noè S., Leon, R.T., 2000: *PR Composite joints under cyclic and dynamic loading conditions: The component model approach.* Proc. 4th AISC International Workshop on Connections in Steel Structures, Roanoke.
- Stark J., Hordijk D.A., 2001: Where structural steel and concrete meet, in Connections between steel and concrete, University of Stuttgart, ed. Eligehausen R., Stuttgart, pp. 1-11, ISBN 2-912143-25-X.
- Penserini P., 1991: *Caracterisation et modelisation du comportement des liaisons structure métallique-fondation, Ph.D. theses,* l'Université Pierre et Marie Curie, Paris 6.

Contacts

- František Wald
  Address: Department of Steel structures, Czech Technical University in Prague, Thakurova street 7, Prague, Czech Republic.
  e-mail: wald@fsv.cvut.cz,
  Phone: +42024354757, Fax: +420233334766
- Ulrike Kuhlmann
  Address: Institute of Structural Design, University of Stuttgart, Pfaffenwaldring 7, 70569 Stuttgart, Germany.
  e-mail: u.kuhlmann@ke.uni-stuttgart.de,
  Phone: +497116856245, Fax: +497116856236
- Markus Rybinski
  Address: Institute of Structural Design, University of Stuttgart, Pfaffenwaldring 7, 70569 Stuttgart, Germany.
  e-mail: markus.rybinski@ke.uni-stuttgart.de,
  Phone: +497116856245, Fax: +497116856236
- Dalibor Gregor
  Address: Department of Steel structures, Czech Technical University in Prague, Thakurova street 7, Prague, Czech Republic.
  e-mail: dalibor.gregor@fsv.cvut.cz,
  Phone: +42024354763, Fax: +420233334766

Chapter II

New Building Technologies

COST ACTION C12

Chapter II - New Building Technologies

Introduction
II.1 Datasheets on New Buildings
    II.1.1 General definition
    II.1.2 New design concepts
        II.1.2.1 Integrated life cycle design of structures
        II.1.2.2 A method for analyzing the IFD performance of floor systems
        II.1.2.3 LC(I) A comparison of two alternative building structures

II.2 Datasheets on Existing Buildings
    II.2.1 General definition
    II.2.2 Consolidation levels
        II.2.2.1 Safeguard
        II.2.2.2 Repairing
        II.2.2.3 Reinforcing
        II.2.2.4 Restructuring

II.3 Datasheets on Suitable Combinations
    II.3.1 New buildings
        II.3.1.1 Steel and concrete
        II.3.1.2 Timber and glass
        II.3.1.3 Steel and FRP
        II.3.1.4 Concrete and FRP
        II.3.1.5 Concrete and timber
        II.3.1.6 Steel profiles and panels
        II.3.1.7 Technical textiles/FRP and timber
        II.3.1.8 Glass and steel
    II.3.2 Existing buildings
        II.3.2.1 Concrete and steel
        II.3.2.2 Masonry and steel
        II.3.2.3 Timber and concrete
        II.3.2.4 Masonry and concrete
        II.3.2.5 Concrete and FRP
        II.3.2.6 Masonry and FRP

II.1 NEW BUILDINGS

II.3.2.5 Concrete and FRP

II.3.2.6 Masonry and FRP

COST ACTION C12

Introduction to Chapter II

The following survey, in the format of compiled datasheets, gives an overview of the work of the WG 1–Mixed building technologies. The aim of this datasheet-project was to reveal the knowledge as well as the specific working areas of each working group's members. The research and development of new types of construction as well as the methods in order to improve existing buildings formed the center of their occupation.

For this reason the datasheets have been divided up into parts. While the first one deals with new buildings, the second one focuses on rebuilding and upkeep.

Appropriate combinations, which were selected according to the interaction of the individual materials used, had to be taken into consideration.

However, the term Mixed-Building Technology has to be defined in a much wider sense, as the mixing does not only refer to basic materials.

The definition exceeds the understanding of hybrid constructions and relates the combinations of materials to the material-specific types of construction as well as to the design-methods. The possibility to combine different materials along with their specific types of construction, developed over history, facilitates that new developments can be applied practically and more quickly.

The technological competition, which is intensifying more and more, wants also the Constructive Civil Engineering to think about the goals and chances concerning the research and development of improved and new qualities of building-materials and prefabricated parts.

There is such a complex spectrum of demands, concerning the buildings to be constructed, that, in the sense of an integrated solution, a monolithic type of construction could never reach the goal. Only by using modern mixed - building technologies, that is to say by means of a convenient combination of materials, could the ideal compressing of several building-technical qualities be achieved.

The different systems of the material parameters have to be selected and used constructively in such a way that their weaknesses are reciprocally balanced and that they could benefit from each other - a behaviour which is called symbiosis in biology.

Beside the traditional types of construction, mixed-building technologies offer a wide range of new possibilities concerning the design and the construction of buildings. The resulting wing assemblies are characterised by a high load capacity with small measurements of the prefabricated parts and a great width of the beams. The low cross-section measurements will lead to a higher flexibility in use and moreover, they are advantageous, as far as the later running costs of the building are concerned.

In view of the selection of the material combinations, the compiled datasheets cannot be claimed to be complete. They were selected according to the available knowledge of the individual participants of the working group. However, they demonstrate perfectly how research co-operation within Europe could work in future.

Christian Schaur
WG1 Chairman

NEW BUILDINGS

General definitions

- This part of the documentation in form of datasheets is dealing with

  Residential Apartments

  Office buildings

  Industrial buildings

- For these kinds of structures currently new structural members as well as structural systems are developed. From the numerous possibilities of new mixed building technologies suitable combinations of material and different construction methods are introduced within this publication.

- In mixed building technologies structural elements can be constructed of different materials including steel, concrete, timber, glass or any kind composites.

- Starting with so called classical mixed building technologies using steel and concrete new developments like timber and technical textiles or masonry and FRP are also shown

- The new possibilities for structural systems were analysed with respect to their economical, social and structural requirements. Special attention has been paid to the improvement of serviceability and load bearing capacity.

COST C12 – WG1

Datasheet II.1.2.1
Prepared by: ir. Rijk Blok[1], prof. ir. Frans van Herwijnen[1]

[1] TU/e Technical University Eindhoven Netherlands

INTEGRATED LIFE CYCLE DESIGN OF STRUCTURES

Introduction

- INTEGRATED LIFE CYCLE DESIGN is one of four important fields of attention in Life time engineering

- LIFE TIME ENGINEERING:
    - Life time investment planning and decision making
    - **Integrated life cycle design**
    - Integrated life time management and maintenance planning
    - Recovery, reuse, recycling and disposal
- Design considerations taken into account when applying Integrated life cycle design for structures are summarised below

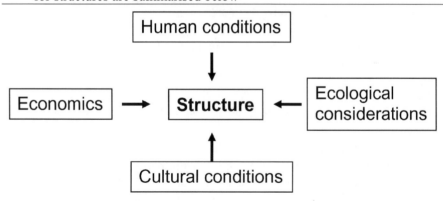

Figure 1. Design considerations for structures

Technical Information

- Examples of important design considerations in Integrated Life Cycle Design

Human conditions: functionality / usability, safety, health and comfort
Economics: construction cost, (projected) cost for operation, maintenance and repair, financial returns, costs for disposal.
Cultural conditions: common materials and construction methods used in different cultures, historical value.
Ecological considerations: environmental aspects, sustainability (consumption of raw materials, fossil fuels, production of pollutants, land use etc)

With the notion that our natural resources become limited and may become exhausted ecological considerations become more and more important. To asses and compare environmental impacts of building structures, **Life cycle impact assessment (LC(I)A) methods** are developed.

See Data sheet *"Towards sustainability, Life cycle assessment of building products"*.

Terms and definitions

Regarding the Life of structures the following terms are defined:

- **Design life**
  The Design life is the assumed period (target life) for which as structure is to be used for its intended purpose (with anticipated maintenance but without major repair).
- **Technical service life**
  The technical service life is the period for which a structure can actually be used for its intended purpose (possibly with necessary maintenance but without major repair)
- **Functional working life**
  The functional working life is the period for which a structure can still meet the demands of its (possibly) changing) users (may be with repairs and/or adaptations)
- **Economic service life**
  The economic service life is the period for which the structure can still provide a sufficient level of financial returns as required by its oners.
- **Technical durability**
  The ability / capacity of a structure in its environment to remain fit for its intended use during its technical working life (with anticipated maintenance but without major repair)
- **Functional durability**:
  The ability / capacity of a structure to fit the requirements, despite changes in use.

Performance based structural design.
The technical, functional and economical aspects mentioned above (with their mutual interactions) play an important role in the decision process regarding the ending of the service life of a building structure. These factors should already be taken into account at the design stage. (Social and cultural factors for example historic value can sometimes also play a significant role.)

With regard to the **Technical durability** the Ultimate limit states and Serviceability limit states define the minimum requirements:
- Ultimate limit state
  Minimum level of safety: Failure probability: $Pf(t) = P\{ R(t) < S(t)$ for all $t <= td$
- Serviceability limit state
  Maximum values for deflections and cracking

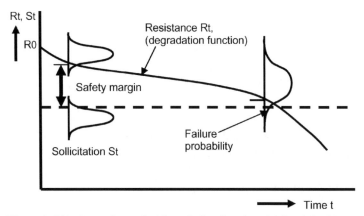

Figure 2. Life time safety using degradation functions (After A.Sarja)

The prediction of technical service life of building materials and components is subject of numerous studies. A guide and bibliography is given by "Service Life and Durability Research for Building Materials and Components" by the joint CIB W80 / RILEM TC 140-TSL report publication 295 march 2004.

With regard to the **Functional durability** the minimum requirements and demands of the users during the Structures Functional Working Life have to be met with regard to:

- spatial needs (column free floor area, not obstructed by bearing walls and stability elements, structural free storey-height)
- Accessibility for maintenance, repair and replacement and adaptation of the structure as well as of other building components such as building facade, building services and installations

Integrated Life Cycle Design of Structures

From the viewpoint of economics and sustainability, optimisation of Technical service Life with Functional working life is needed.

Despite the fact that their Technical Service Life has not ended, there is a large amount of buildings and building structures that no longer fullfill the functional needs of their users (today or in the near future). (For example a large part of the high rise housing in our new E.C. member country's). On the other hand, demands of building occupants become more and more dynamic. Changing demands increase the probability that the functional requirements of the building and its structure are no longer met. If so, the structure either needs to be adapted, dismantled or even demolished.

Integrated Life cycle design is way of improving the quality of future buildings and building structures. An example of optimisation of the Functional working life with the Technical service life at the design stage is the XX-office building Delft. (See also the Data sheet *"LC(I)A comparison of two alternative buildingstructures for an office building (XX-office in Delft, Netherlands), each as a result of a different Life-cycle design approach".*)

Integrated Life cycle design of structures aims to design for possible changes.

In practise, when changes in functional requirements are anticipated, two options can be distinguished:

1. Long life structures with flexibility and/or adaptability
2. Short time structures with possibilities for re-use of components, elements or materials (recycling)

IFD building

IFD is an example of a building concept which combines sustainability with functionality and at the same time aims at a higher quality level of the building.

IFD: Industrial Flexible and Demountable Building

Industrial: Increase in quality of the components, reduction in the amount of energy for production and construction and reduction in the amount of waste on the building site: Less waste, less energy

Flexible: possibilities for adaptation of the building and its structure increases the functional working life.

Demountable: re-use of elements and components as well as restructuring becomes possible: Loose fit , less waste.

Relations between IFD criteria and design considerations:

Industrial	Economy	Long life
Flexible	Ecological consideration	Less energy, less waste
Demountable	Human condition	Loose fit

Integrated design

The IFD-philosophy leads among others to integration and independence of disciplines. The accommodation of technical services and installations within the structure leads to new structural designs.

Examples:
- **IFD structure Delft Tech Office**   See Data sheet IFD structure Delft Tech Office
- **Integrated Floor design**   See Data sheet IFD floor design
- **Other projects: see SEV (below)**

Field of Application and End Users
- The **Integrated life cycle design of buildings and structures** affects vast parts of our society. Building users, building developers as well as owners, architects, structural engineers and other parts of the construction industry all need to be involved to improve on (Mixed) Building Techniques.
- Many demonstration projects in the Netherlands have been realized, many are still under way. The projects range from Feasibility studies to a very wide range of existing as well as new Building projects.
- The SEV organisation, Rotterdam, Netherlands can provide information on these

projects.

www.sev.nl

www.sev.nl/ifd

Structural Assessment

- not applicable

References

- A. Sarja; *Integrated Life cycle design of structures; Spon press London/New York 2002*
- R. Blok, F. Van Herwijnen e.o.; Service life and life cycle of building structures; *Proceedingsof the internationmal seminar Lisbon 2002; Inprovement of buildings' struct.. etc.Cost C12; European Comm. Eur 20728*
- *Service Life and Durability Research for Building Materials and Components* by the joint CIB W80 / RILEM TC 140-TSL report publication 295 march 2004.

Contacts

- Rijk Blok; Frans van Herwijnen
   Address: TU/e, University of Technology Eindhoven
       PO. Box 513
       5600MB Eindhoven, Netherlands
       email R.Blok@bwk.tue.nl ; F.v.Herwijnen@bwk.tue.nl
       Tel. + 31 2472619
       Fax. + 31

COST C12 – WG1

DatasheetII.1.2.2

Prepared by: . S.F.A.J.G. Zegers[1]

F. van Herwijnen[1]

R.Blok[1]

[1] TU/e University of Technology Eindhoven, Netherland

A METHOD FOR ANALYZING AND DETERMINING THE IFD PERFORMANCE OF FLOOR SYSTEMS

Description

- Analyzing and determining IFD performance of (integrated) floor systems is an important step in achieving better performance of new (integrated) floor designs thus achieving better performance of building structures as a whole.

- IFD means Industrial, Flexible and Demountable building.

  For explanation of the IFD concept is referred to the datasheet "Integrated Life Cycle Design of Structures".

  Performance based engineering and -design consider also the functional user requirements. These functional requirements of buildings and their structures are certain to change during the life time of a building. To prevent early dismantling and or demolishing of building structures these possible changes in use have to be taken into consideration.

  Integrated Life cycle design of structures needs to design for these possible changes.

- IFD Floor systems aim to provide the building with the necessary flexibility in order to increase the Functional Working Life of a building. Integration of the floor system and at the same time providing independency of the other building disciplines are important factors. By reducing floor thickness and at the same time creating accessible space, for example in the neutral zone of the floor system, integration of building services such as heating and cooling systems, data systems etc. can be achieved. The accessibility ensures independence of the structural qualities, a large degree of independence of the installation and construction process, and ensures easy future maintenance, repair or renewal. This is important because the Design life and Technical Service Life of building structures, including the floor system, is usually much longer (for example 50-100 years) than the service life of building services and installations (for example 20-30 years).

- Lately a number of different integrated floor designs have become available to the Building industry and have already been used in different projects.

  Examples:

  - Infra + floor
  - Wing floor
  - IDES floor (Integrated Deck Extra Space)

  More information: see contacts

- Development of an IFD floor system is currently under research at the University of Technology Eindhoven, Netherlands.

- A method for analyzing and evaluating these new floor designs, determining the strengths and weaknesses of these designs regarding IFD characteristics has been developed by S.F.A.J.G. Zegers, University of Technology, Eindhoven, Netherlands.

Technical Information

- Most important features of the Evaluation method are:

  o Subdivision of the floor system into parts and provisions.

  A floor system has to perform various functions, which are performed by the parts that make up the floor system. These parts, e.g. beams, plating etc, are the physical parts of the floor system. Besides these physical parts, a floor system can contain provisions. These provide the floor system with added functionality without being a physical part. Examples of provisions found in a floor system are: free space in a floor for installations and services, hollow tubes, etc.

  o Assigning contribution level to functional aspects

  The parts and provisions, which were defined in the step described above, are assigned to functional aspects. The functional aspects are categorized into three groups namely, Structural- (S), Building physics- (B), Infra-structural (I) aspects. Each functional aspects group is subdivided into one or more functional aspects that characterize floor systems, as shown in Table 1.

Table 1, List of functional aspects of a floor system

Code	Name	Description
S1:	Load-bearing function	carrying off vertical loads acting on the floor system to the main load-bearing structure.
S2:	Stabilizing function	carrying off horizontal loads due to wind forces, second order and tilting.
S3:	Supporting function	transferring loads from the floor system to the load-bearing structure.
B1:	Sound insulation	sound insulation between adjacent rooms.
B2:	Vibration	effect of dynamic loads.
I1:	Gas, water, electricity	holding ducts that have a small diameter (< 30 mm) and a long time interval for change, replacement or maintenance
I2:	Computer network, cable network	holding low voltage installation that has a short time interval for change, replacement or maintenance
I3:	Air-conditioning	holding ducts that have a big diameter (> 100 mm) and a long time interval for change, replacement or maintenance
I4:	Sewerage	holding ducts that have medium diameter (> 30 mm and < 100 mm), are sloping and have a long time interval for change, replacement or maintenance.

  o Assigning contribution levels

  The second step of the evaluation requires that the contribution level of the previously de-fined parts and provisions have to be assessed for each functional aspect. Each part or provision is assessed and rewarded a relative contribution level from 0 to 10 in which a contribution level of 10 means that the part or provision contributes most to the regarded functional aspect, and 0 means no contribution at all. The contribution levels, c, with the score base of 10, are normalized by Eq. (1), so that the sum of all contribution levels with regard to one functional aspect equals one:

$$c_{f,PnP}^{1} = \frac{c_{f,PnP}^{10}}{\sum_{PnP=1}^{n} c_{f,PnP}^{10}}$$

where: f is the regarded functional aspect; PnP is the regarded part or provision; n is the total number of parts and provisions,; the superscript 1 means normalized; superscript 10 means on a 1-10 scale o   Assigning scores on each part or provision on every characteristic mentioned

The third step of the evaluation requires that each part or provision be given a score with regard to the assessed characteristics. These characteristics reflect the main areas of interest concerning the Industrial, Flexible and Demountable way of building. The characteristics are grouped into Industrial, Flexible and Demountable characteristics.

Table 2. Definition of IFD characteristics

Phase Characteristic	Description of positive characteristic of part or provision
Industrial	
[D] [Standardized parts]	consists of subparts that are manufactured in series
[Modular measurement system]	has a measurement system that is modular
[D] [Number of parts]	consists of a small number of parts
[A] [Controlled manufacturing environment]	has been manufactured under controlled manufacturing conditions
[A] [Reproducible production process]	is manufactured or assembled by means of a reproducible production process
[A]	[Simple assembly protocol] can be assembled on site by means of simple actions and lightweight equipment
[A] assembly on site	[Little waste] produces little waste during manufacturing and
[S] [Changing of parts]	can be changed during service life by standardized parts
Flexible	
[D] [Freedom of design]	can be changed according to new requirements during the entire design process
[A] [Adaptable during assembly]	can be changed according to new requirements during the assembly process
[A] [Freedom of assembly]	is not depending on a strict assembly planning
[S] [Changing of function]	supports the possibility of changing the functionality of the room with little disturbance to other parts of the building.
[S] [Changing of layout]	supports the possibility of changing the layout of the floor field with little disturbance to other parts of the building.
[S] [Mentally]	has been designed in such a way and is provided with such good documentation that the user can be aware of the possibilities of flexibility.
Demountable	
[D] [Reuse from other buildings]	can be used in another building without alterations after demounting the original building
[A] [Dry connections]	are connected to the rest of the floor system and construction through dry connections
[S] [Demounting of parts]	can be demounted without disturbing this or other parts so it can be used in other buildings
[M] [Demounting without waste]	can be demounted without creating waste materials.
[M] [Reuse of materials]	can be treated in such a way that the materials of this part can be used as new raw materials
[M] [Reuse of building parts]	supports the reuse of the entire floor system in other buildings.

o   Rating system

375

A matrix, [Rc,PnP], is made in which horizontally the set of characteristics is listed and verti-cally all the parts and provisions. Subsequently every part or provision is given a rating on each of those characteristics.

o   Processing of scores

Three lists have been defined:

- A list of functional aspects (Table 1)
- A list of parts and provisions (Unique for every floor system)
- A list of Industrial, Flexible and demountable characteristics (Table 2)

The first two lists are linked through the set of contribution levels, $[c^1_{f,PnP}]$, and this matrix is linked to the last list by the ratings matrix, $[R_{c,PnP}]$. A table that lists ratings, ranging from $+2$ to -2, for the functional aspects on each IFD characteristic is constructed by calculating the rating of a functional aspect on the regarded characteristic by summing up the contributions of every part and provision considering their contribution levels. Because the contribution levels have been normalized the contribution level can be multiplied with the rating and are added for all the parts and provisions. In matrix formulation:

$$\left[R_{f,c} \right] = \left[c^1_{f,PnP} \right]^T \cdot \left[R_{PnP,c} \right]$$

o   Sorting of results

The results can be grouped in a number of different tables in order to provide a better insight in functional aspects and IFD characteristics. Graphical representation is used to enhance characteristics.

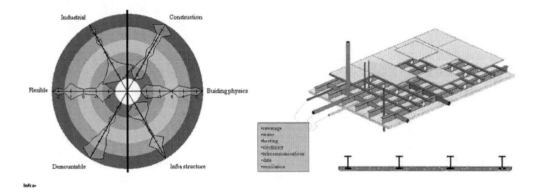

Figure 1. Example of Infra + floor system, (left rating characteristics)

Field of Application and End Users

- The evaluation tool will provide building designers insight in the different floor system characteristics. Architects, Structural Engineers and Designers of building services can create new integrated (IFD) solutions. The performance requirements can be matched with performance characteristic of the floor system. Different options

and solutions can easily be compared.

Structural Assessment

Not applicable

Economical Aspects

- By a better match between functional working life and technical service life through the use of integrated floor systems the service life of buildings will increase. This will result in a better use of materials and resources, thus achieving a better economic performance. The evaluation method of floor systems described above is an important step in this direction.

Contacts

- Ir. S.F.A.J.G. Zegers
  Address: TU/e, University of Technology Eindhoven, Netherlands
  Faculty of Architecture, Building and Planning
  Department Structural Design
  De Wielen Vertigo
  P.O. Box 513
  5600 MB Eindhoven Netherlands
  s.f.a.j.g.zegers@bwk.tue.nl
  Tel. + 31 40 2473963
  Fax. + 31 40 2450328

[1] TU/e University of Technology Eindhoven, Netherlands

LC(I)A COMPARISON OF TWO ALTERNATIVE BUILDING STRUCTURES FOR AN OFFICE BUILDING (XX-OFFICE IN DELFT, NETHERLANDS), EACH AS A RESULT OF A DIFFERENT LIFE-CYCLE DESIGN APPROACH

Description

- The actual build Structure of the XX-office building in Delft Netherlands has been the result of a new design approach of balancing the Technical Service Life (TSL) of the building and its structure with the estimated and foreseen (shorter) Functional Working Life (FWL). TSF = FWL = XX =20(-25) years, while at the same time minimizing the environmental impact. To achieve this the structure of the XX-office has been build in reusable and recyclable, untreated, **mainly wooden (Swedlam) materials and components.**

- The environmental impact of this XX-office Structure has been compared with a more traditionally designed Reference Structure with a Design Life of 50 years.

- It shows that from an environmental point of view, the design approach of the XX-office Structure proofs to be successful.

- To evaluate the design approach an alternative **Reference Structure** for the building has been designed. For this Reference Structure the minimum Design Life of 50 years was used. The Reference Structure uses the same building dimensions. The structure has been designed as a **braced steel construction with prefabricated pre-stressed concrete hollow-core floor slabs**

- **Life Cycle (Impact) Analyses** (LC(I)A), were carried out for the two different building structures, taking into account different possible life-time scenario's of 25 and 50 years for both structures. To evaluate the different approaches in the design of the structure the results for an assumed fifty year period of use have been compared using **different Life scenario's** and **different disposal scenario's**. (fig. 1.) The environmental consequences of this new design approach could thus be evaluated.

- The used LC(I)A is based on the method developed by the Centre of Environmental Science Leiden (ref. [2]) It clearly identifies a number of different steps. To analyse the materials and the processes involved for both structures the computer program Simapro (see [3]) has been used. The negative impact or damage assessment of the used materials and processes has been quantified in different categories using the Eco-indicator. The Eco-indicator is a damage oriented method for LC(I)A. It uses the following main classification of categories: Damage to Human Health, Damage to Ecosystem Quality and Resources (each in turn divided in sub categories).

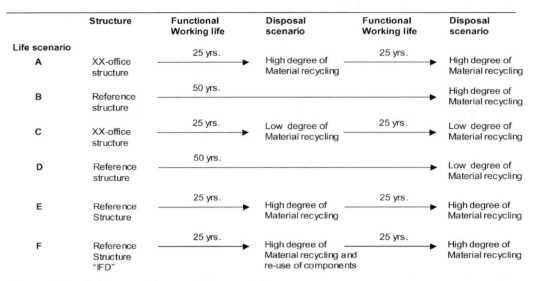

Life scenario	Structure	Functional Working life	Disposal scenario	Functional Working life	Disposal scenario
A	XX-office structure	25 yrs. →	High degree of Material recycling	25 yrs. →	High degree of Material recycling
B	Reference structure	50 yrs. →			High degree of Material recycling
C	XX-office structure	25 yrs. →	Low degree of Material recycling	25 yrs. →	Low degree of Material recycling
D	Reference structure	50 yrs. →			Low degree of Material recycling
E	Reference Structure	25 yrs. →	High degree of Material recycling	25 yrs. →	High degree of Material recycling
F	Reference Structure "IFD"	25 yrs. →	High degree of Material recycling and re-use of components	25 yrs. →	High degree of Material recycling

Figure 1. The six life scenario's over a 50 year period

Results of the Scenario's A to F on Human Health, Ecosystem Quality and Resources

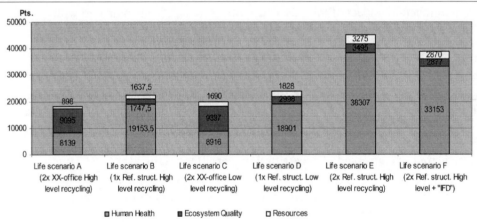

Figure 2. Single score results scenario's A to F

Because of the high amount of resulting output only the most striking and summarized results will be shown here. Although the shorter assumed FWL and TSL of the XX-office structure requires the use of two structures over a 50 year period, this scenario (A) still shows lower impact figures (a total of 18.132 Points in single score).

Some results

- On **Human Health** the XX-office Structure shows the best performance with the lowest impact figures (Life scenario A and C)
- The effects of **respiratory inorganics** make up a large contribution to the relatively high figures on Human Health for the Reference Structure. This is mainly caused by the extraction, production etc. of cement-containing materials and elements, such as concrete floors.

- With regard to **Ecosystem Quality** the XX-office Structure shows higher impact figures than the Reference Structure. Reason for this is the large negative impact contribution on Land-use. Reduction in Ecosystem Quality is measured in the loss of biodiversity in a certain area per time-unit. By this definition of Land use, the commonly used methods of **mono-culture forestry** causes these negative contributions in Ecosystem Quality. The correctness of this outcome requires further investigation.

- Contrary to this effect is the relatively low impact of the XX-office Structure on **Resources**, mainly due to the large use of (renewable!) timber.

- The Reference Structure shows larger impact figures on resources mainly because of the extraction of iron-oar, needed for the steel production. Also the used amount of fossil fuels needed for production and transportation contribute to the higher impact figures on resources.

- The results show fairly minor differences between the applied **disposal scenario's** as described by "High Degree of Material Recycling" and "Low Degree of Material Recycling" for both the XX-office Structure and the Reference Structure. (High degree of material recycling resulted in about 5% lower environmental impact for the Reference structure and 10% lower for the XX-office structure) The reason for this is, that according to current practise in the Netherlands, most steel components are recycled at material level anyway. The timber elements can be recycled at the level of wood pulp and paper production, or can be decomposed.

- The design approach used in the XX-office design shows to be successful in minimizing environmental impact effects if the life scenario of a Functional Working Life of about twenty five years is assumed. Using these scenario's, the used LCIA assessment (Eco-indicator 99) result in considerably lower environmental impact figures compared with the Reference Structure.

- It can be seen that decisions taken at the design stage inparticular the choice of building materials show a large influence on the environmental impact .

Conclusions

- This example shows that LC(I)A calculations can increase the awareness of engineers and designers in their responsibility to incorporate sustainability in building design and construction.

- Integrated Life Cycle Engineering and Design is needed to incorporate and further optimise the many different aspects involved, ranging from Human (technical, safety) , to Economical and Ecological considerations.

- LC(I)A calculations are elaborate and time-consuming. Further development of simplified but reliable methods is needed to make easy comparison at the design stage possible and allow them to become common practise.

- The impact results depend directly on the length of the assumed Functional Working Life. Improvement of techniques to take into account possible differences in the Life span of a structure is needed. For example a method to evaluate differences in functional quality of a structure, and therefore a possible longer expected Functional Working Life, is needed.

References

- [1] Herwijnen, F. van: (In Dutch) Twintig duurt het langst, Houtblad, vol. 11, No. 2 1999, pp 32 - 37
- [2] Centre of Environmental Science Leiden (CML), Einsteinweg 2, 2333 CC Leiden.
- Goedkoop, M. [et al]: The ECO-indicator 99,Methodology report, Amersfoort, 2001
- [3] R. Blok, F. Van Herwijnen e.o.; Service life and life cycle of building structures; *Proceedingsof the internationmal seminar Lisbon 2002; Inprovement of buildings' struct.. etc.Cost C12; European Comm. Eur 20728*

Contacts

- Rijk Blok; Frans van Herwijnen
  Address: TU/e, University of Technology Eindhoven
  PO. Box 513
  5600MB Eindhoven, Netherlands
  email R.Blok@bwk.tue.nl ; F.v.Herwijnen@bwk.tue.nl
  Tel. + 31 2472619
  Fax. + 31 249999999999

EXISTING BUILDINGS

General definitions

- This part of the documentation in form of datasheets is dealing with

   Common buildings

   Historical buildings

   Monuments

- The use of new techniques and new combinations, so called mixed building technologies, have become more and more important also in the field of upgrading. Therefore also this part is subdivided according to suitable combinations of material and construction methods.

- Strengthening of existing buildings can be necessary for the following reasons :

   damage or degradation of material and structural elements

   new requirements for the use of the building

   change of the structural system due to subsequent placing of openings

   lack of structural properties adequate to comply with current structural codification

   no adequate structural properties to offer resistance to those actions not considered at the design stage

   the structural scheme and the used materials do not allow to resist significant exceptional actions

CONSOLIDATION LEVELS: SAFEGUARD

Description

- Safeguard is, in cronological order, the first level of consolidation of existing buildings
- It consists of a set of provisional interventions able to assure an adequate safety during the transitory phase for both the public and the site, which precedes any definitive consolidation operations (Mazzolani, 1985; Mazzolani & Mandara, 2001)
- This provision is used both for protecting the site and for avoiding partial or total collapse of the building when damage conditions require urgent recovery actions
- Peculiar characteristics of the safeguard systems are:

    a. speed of execution

    b. flexibility of the constructive systems

    c. adaptability to narrow and hardly accessible work areas

    d. reversibility of interventions

Field of application

- Temporary support of facades during the reconstruction of a new building between two existing ones, by means of reticular space structures (Figure 1)
- Steelwork structure which supports the façade during the demolition of the internal part of the building (Figure 2)
- Temporary support of building facades during the post-earthquake emergency, by

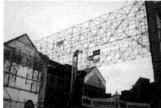

Figure 1.

Figure 4 a.

Figure 2.   **Figure 3.**   **Figure 4 b.**

using tubular steel scaffoldings (Figure 3) which realise contrast reticular structures, allowing the transit at the street level (Mazzolani, 1985)

- Provisional roofing (Figures 4a and 4b) for providing adequate protection of the site area from the atmospheric agents during the restoration operations

Technical information

- Structural steel elements offer, under form of "scaffolding systems", "ad hoc" solutions designed to achieve optimum results and tailored to specific requirements (Mazzolani, 1990)
- Heavy steel structures (welded or bolted profiles) and light steel ones (pipes with bolted joints) are effectively adopted in safeguard active and passive operations
- Prerequisites of the above structural steel systems are (Mazzolani, 1994):
  a. lightness
  b. high prefabrication
  c. facility of transport and assemblage
  d. economic convenience, due to their re-use
- Traditional alternative materials are masonry (made of tuff or local stones) and timber elements

Example of application

- In the field of safeguard passive interventions, structural masonry elements, called *"barbacani"* (Figures 5a and 5b), are sometime used for supporting the facades of buildings in Mediterranean Countries

Figure 5a.   **Figure 5b.**

- Timber struts, of several types and shapes, are inserted in masonry for creating a more continuous distribution of vertical loads before consolidation (Figure 6a) or for providing also lateral support and access facilities (Figure 6b).
- Bamboo scaffoldings are used in Asia as service systems during refurbishment operations (Figure 6c)

Figure 6a.   Figure 6b.   Figure 6c.

- A funny solution, among hundred, to provisionally sustain the Pisa's Tower for avoiding an immediate danger of collapse in view of its definitive consolidation has been proposed under form of "girello" (Figure 7)
- Steel elements both for active and passive actions have been effectively used in many cases.

1. *Active support during the restoration of the York Cathedral* (Figure 8)

The building lateral facade, which presented an out-of-plumb of 635 mm, was provisionally substained to allow under-foundation operations. Safeguard intervention consisted on the use of steel reticular trusses, erected on adequate r.c. foundations, which contrasted a load control device, made of a couple of hydraulic jacks with constant pressure, which compensate the effect of termal variations. The jack system guaranteed a given constant load on the masonry structure during the under-foundation realization (Defez, 1997).

Figure 7.   Figure 8.

2. *Consolidation of Palazzo Carigliano in Turin* (Figure 9)

Hot-rolled steel sections were designed for the temporary shoring up of the stone columns at the entrance hall of the building during restoration works, consisting on the substitution of old floors with new ones. Steel sections with corrugated steel sheets acting as formwork for a concrete cast slab were effectively employed (Mazzolani, 1990).

3. *Rebuilding of the Waring and Gillow's Store in London* (Figure 10)

Steel columns and girders acting as a bracing system are used for preserving the old facade. These elements, once assembled together, create an "ad hoc" structure which supports the facade and a 5 m zone behind it during demolition of the remaining part of the building. After the internal rebuilding, according to a new functional lay-out, steel elements were gradually removed.

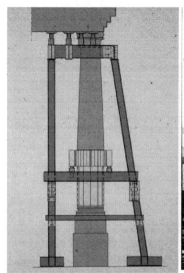

Figure 10 b.

Figure 9.   **Figure 10 a.**   **Figure 10 c.**

4. *Re-use of the old Moller theatre in Darmstadt* (Figure 11)

Horizontal rings of steel sections fixed to the walls are used as a temporary bracing of perimetrical wall in this building, transformed in a State-record office.

Figure 11.   **Figure 12.**

5. *Restructuring of a building in the centre of Lisbon* (Figure 12)

A big fire produced the total empting of a large masonry building. The remaining external facades and internal core were provisionally supported by means of steelwork and perimetral ties, before the starting of the restoration and consolation operations.

6. *Max Plank Institute in Rome* (Figure 13)

A complex refurbishment operation of an old building in Rome required the provisional support of the façade by means of appropriately designed steelworks, before a degutting operation (Marimpietri et al., 2004).

7. *Temporary support of walls in the site area of a new underground station in Naples* (Figure 14)

Tubular steel struts were erected for retaining the perimetral walls and for allowing the archeological discovery of the ruins of the old roman harbour during the execution of the new underground line.

Figure 13a.   **Figure 13b.**   **Figure 14.**

References

- Defez, A., 1997. *Il consolidamento degli edifici*, Liguori Editore, Napoli, Italy.

- Mazzolani, F. M. 1985. L'acciaio ed consolidamento degli edifici. *Acciaio, n.12-1985*

- Mazzolani, F. M. 1990. *Refurbishment*, Arbed, Luxemburg.

- Mazzolani, F. M. 1994. Il consolidamento strutturale. In *La Progettazione in acciaio,* Consorzio CREA Editore, Massa, Italy.

- Mazzolani, F.M. & Mandara, A. 1991. *L'acciaio nel consolidamento*, Assa Editore, Milano, Italy.

- Marimpietri, F.M., Mezzi, M. & Parducci, A. 2004. *Constraints and construction procedures to insert a new building in a historical centre,* Proc. of the 8th Int. Conf. on Modern Building Materials, Structures and Techniques, Vilnius, Lithuania, 19-21 May 2004.

Contacts

- Prof. Dr. Eng. Federico M. Mazzolani
  Address: Department of Structural Analysis and Design,
  University of Naples "Federico II",
  P.le Tecchio, 80, 80125 Naples, Italy.
  e-mail: fmm@unina.it,
  Phone: (+39) 081 7682443, Fax: (+39) 081 5934792

CONSOLIDATION LEVELS: REPAIRING

Description

- Repairing is, in cronological order, the second level of consolidation of existing buildings

- It involves a series of operations carried out on the building to restore its former structural efficiency before the damage occurred. (Mazzolani, 1990, 1994a)

- Repairing, differently from safeguard, represents a definitive operation used in case of damage caused by simply individuable factors which normally produce their effects during a long period of time and, therefore, do not require urgent interventions

- It provides a simple restoration of the structural performance, under safety point of view, without introducing any strengthening of the construction

Field of application

- Building structures damaged by atmospheric agents, age and revage of time (Figures 1, 2, 3 and 4)

Figure 1. Masonry building damaged by age

Figure 2. Deteriorated old wooden floor

Figure 3. Damaged wooden structure

Figure 4. Corrosion phenomena on a steel structure

Technical information

- Within the repairing stage, there are numeorus technological consolidation systems based on the use of steelwork, improving the structural behaviour of both masonry, reinforced concrete and timber building structures (Mazzolani, 1990)

- Steel structural elements offers, by means of "prefabricated" types of technology, "ad hoc" solutions designed to achieve optimum results and tailored to specific requirements (Mazzolani, 1994b)

- Prerequisites of the above described components are:
  a. lightness, which allows ease of transport and erection, which are important conditions when it is necessary to work in narrow operational spaces of the hystorical centers of the old towns.
  b. reversibility, thanks to the use of bolted joints, which allows the re-use of the structure after dismanteling
  c. speedy erection, useful during the repairing emergency, when the damage quickly progresses
  d. economic convenience, thanks to the potential re-use
  e. modern feature, clearly identified with the additional advantage of the use of tecnologies and materials which can be removed any time, without producing damage to the building

- Recently, fibre reinforced polymers (FRP) are ever-increasing used in the reparation of existing buildings

- FRP are used as reinforcing bars for concrete and masonry structures or like strips and sheets for strengthening existing elements

- The useful properties of FRP materials are:
  - excellent strength/self-weight ratio (approximatively 40-50 times better than structural steel);
  - easily formed into any shape;
  - largely corrosion-free;
  - highly resistant to fatigue.

- The FRP development is mainly motivated by the problem of corrosion which affect both the re-bars in r.c. structures and the profiles in steelworks

Example of application

- ***Masonry structures***
  Vertical load-bearing capacity of this kind of constructions can be improved by the following methods (Mazzolani & Ivanyi, 2002):
  - encircling damaged masonry columns with vertical steel L profiles and cross batten plates (Figure 5);
  - insertion of new steel columns along or inside the wall to be consolidated (Figure 6);
  - insertion of steel frames around the wall opening for restoring the original strength of the masonry structure (Figure 7);
  - reparation by means of FRP elements under form of strip and/or issues.

  Horizontal load-bearing capacity is restored as follows:
  - Securing façade walls with horizontal steel beams connected to the transversal walls by means of tie;
  - Securing building corners through vertical steel elements connected together by means of girders or tie-beams.

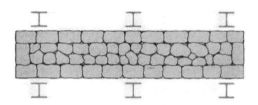

Figure 5. Consolidation of masonry column

Figure 6. Consolidation of masonry walls

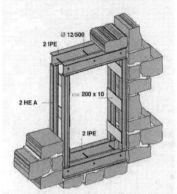

Figure 7. Steel frame used for a wall opening

Figure 8. Repairing of existing r.c. column by means of steel angles and tie bars

- ### *R.C. structures*

    Columns can be repaired in several way for increasing their resistance both to horizontal and vertical actions (Mazzolani, 1992):

    - r.c. columns can be consolidated by means of steel angles which are connected among them throught steel profiles and ties (Figure 8) or welded batten plates, giving rise to an encircling effect (Figure 9);
    - when the intervention must be visible, steel sections can be bolted to the existing column (Figure 10a);
    - cold-formed steel profiles can be assembled together around the r.c. column by means of bolted connection (Figure 10b);

    Beams can be consolidated in the following way:

    - Steel angles and batten plates are used for improving the flexural and shear resistance of r.c. beam-to-column nodes. Welding and epoxidic resin can be used as connection system for joining the new structural components;
    - Coupled vertical ties, connected to an upper concrete slab and bolted to a bottom steel plate, act like additional stirrups for the existing r.c. beam;
    - Cold formed steel profiles and hot rolled plates can be fixed to the inferior part of r.c. members by means of rivets or bolts and epoxidic resin (Figure 11).

    Within the modern repairing systems, the interventions based on FRP elements are effectively used to provide an adequate over-strength to existing damaged r.c. beams (Figure 12) and columns (Figure 13).

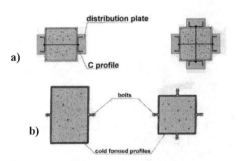

Figure 9. Steel angles used as repairing system for r.c. columns

Figure 10. Increasing the cross-section of r.c. columns by means of hot rolled (a) and cold-formed (b) steel profiles

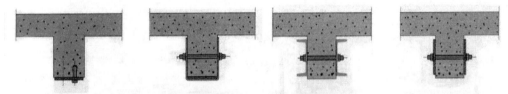

Figure 11. Steel plates and profiles fixed at the bottom of the beam by means of bolts and epoxidic resins

Figure 12. Repairing intervention of a beam by means of CFRP

Figure 13. Repairing intervention of a column using CFRP

- ### *Steel structures*

    For repairing steel structures the following systems can be used:

    - increasing of the member cross-section by welding or bolting new structural steel elements (Figure 14);
    - stiffening of structural nodes by introducing steel profiles, plates and stiffeners.

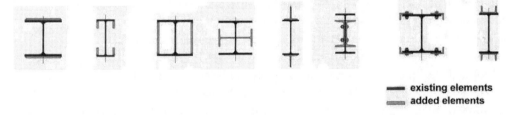

Figure 14. Repairing of steel sections

- **_Floor structures_**

The floor types susceptible to be consolidated are the following (Mazzolani & Mandara, 1991):
1) Floors with wooden beams;
2) Floors with steel profiles:
3) Floors with reinforced concrete beams and tiles.
As an alternative to the complete substitution, the first type of floor structure may be reinforced by two constructional methods:
- working from the bottom upwards, a pair of steel sections (double T, channel, cold-formed) are located side by side to each wooden beam (Figure 15), in some case they must be integrated by metal sheets or ferro-cement coating for supporting the secondary elements (Figure 16);
- working from the top downwards, when the wooden beams are in good conditions, they are joined to I-beams located on their top and linked by suitable connection system (Figure 17). More complex alternative solutions can be achieved by creating a multiple composite action: the cast reinforced concrete slab on corrugated steel sheets is connected to the bottom part by means of stud shear connectors welded on top of the steel profile which cooperates with the old wooden beam (Figure 18).

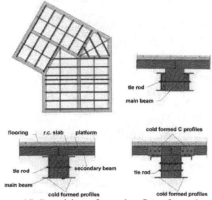

Figure 15. Repairing of wooden floor from the bottom side

Figure 16. Upwards view of the intervention of reparation of wooden beams

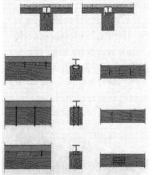

Figure 17. Repairing of a wooden floor by connectors and steel beams for creating a composite steel-wood system

Figure 18. Reparation of a wooden floor by means of a multi-composite system

The floors of type 2), very frequently used at the beginning of the 20th century, are made of I-beams in combination with brick or tile vaults plus concrete. Due to the degradation of the restraint conditions, an increased rigidity is usually required; the following three procedures are used:

- working from the bottom upwards, the section modulus of the beams can be increased by welding to their bottom flange steel sections, under form of plates, upside-down T, double T, box sections, etc. (Figures 19 a and b);
- working from the top downwards, a r.c. slab, linked to the beams by suitable connectors welded to their top flanges, can be realised;
- increasing the section modulus by gluing C-FRP strips.

Mixed concrete and tile floors of type 3) can be repaired by the following methods:
- repairing the r.c. beams by new re-bars and new special concrete after eliminating the degraded parts;
- adding glued steel plates or FRP strips for plating the bottom parts of the concrete beams (Figure 20);
- inserting steel I-beams in suitable cavities between the concrete beams.

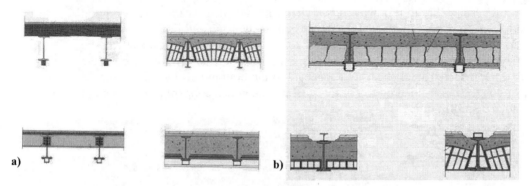

Figure 19. Increasing of the steel beam resistance by means of welded steel sections

- ### *Roofing structures*

Roofing of masonry buildings are generally made of timber trusses, often deteriored, being at direct contact with atmospheric agents.

When wood is not in bad conditions, a possible solution for repairing can be done by dressing the wooden elements with steel plates, connected to the structure through bolts or rivets, integrated by resin (Figure 21).

The use of C-FRP for repairing wooden elements is becoming more and more popular under form of strip, sheet and poltruded bars.

Contrary, when the roofing structures appear not recuperable, the optimal solution is obtained by substituting the old wooden truss with a new steel one, integrated by trapezoidal sheeting and concrete casting. In many cases this system has been adopted for roofing church buildings damaged by earthquake (Figures 22 a and b).

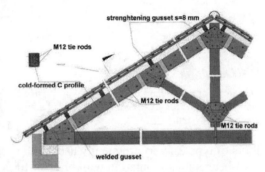

Figure 20. Repairing intervention by means of steel plates

Figure 21. Repairing of wooden roof structure by means of steel plates

a)  b)

Figure 22. New steel roof of a restored church

References

- Mazzolani, F. M. 1990. *Refurbishment*, Arbed, Luxemburg.
- Mazzolani, F. M. 1992. *The use of steel in refurbishment,* Proceedings of the 1st World Conference on Constructional Steel Design, Acapulco.
- Mazzolani, F. M. 1994a. *Il consolidamento strutturale,* La Progettazione in acciaio, Consorzio CREA Editore, Massa, Italy.
- Mazzolani, F. M. 1994b. *Transformation and Repair*, Lecture 16.2, ESDEP.
- Mazzolani, F.M. & Ivanyi, M. (editors) 2002. *Refurbishment of buildings and bridges (CISM Course)*, Springer - Verlag Wien - New York.
- Mazzolani, F.M. & Mandara, A. 1991. *L'acciaio nel consolidamento*, Assa Editore, Milano, Italy.

Contacts

- Prof. Dr. Eng. Federico M. Mazzolani
  Address: Department of Structural Analysis and Design,
  University of Naples "Federico II",
  P.le Tecchio, 80, 80125 Naples, Italy.
  e-mail: fmm@unina.it,
  Phone: (+39) 081 7682443, Fax: (+39) 081 5934792

CONSOLIDATION LEVELS: REINFORCING

Description

- Reinforcing is, in cronological order, the third consolidation level of existing buildings

- It involves the improving of the structural performance in order to enable the building to fulfil new functional requirements or environmental conditions. (Mazzolani, 1990, 1994a)

- This consolidation level does not produce a significant change of the structural scheme, but introduce new elements with the function to statically integrate the existing ones without substantially altering both the mass and the stiffness distribution of the building

- Contrary to the simple reparation, the reinforcing can be carried out with various degree of intensity according to the different increase of strength required for the building by the new conditions, also considering the previous level of damage when it exists.

- Under the seismic point of view, the strengthening operation can be distinguished in two levels: simple improvement and upgrading interventions

- The improvement intervention is faced to obtain a higher safety degree. In this case the reinforcing intervention acts on a single part or on the whole, but without excessively modifying its static scheme and, therefore, the global behaviour. (Figures 1 and 2)

- Improvement interventions can be also done on single structural elements when they are affected by design mistakes or bad execution

- The seismic upgrading intervention is characterised by a set of operations necessary to make the structure able to withstand the new earthquake design actions. It can also requires a large revision of the structural scheme, with a complete modification of the global seismic behaviour. In such a case, this intervention has to be classified from the structural point of view within the restructuring operations (Figures 3 and 4).

Field of application

- Buildings subjected to heavier loading conditions, because of a change of use requiring an increase of live loads

- Existing constructions which are located in an area recently included in a new seismic zone, therefore subjected to more severe loading conditions

- In some codes there is a rigorous distinction between the simple improvement and the seismic up-grading interventions (Mazzolani, 1994b)

- The improvement operations can be adopted in the following cases:
  - when any variation of the use destination occurs
  - when design and/or execution defects must be eliminated
  - when the consolidation operations are applied on monumental buildings

- The seismic upgrading is compulsory in the following conditions:
  - superelevation or amplification of the construction, with increase of volume and areas
  - load increasing due to the destination change
  - when transformations substantially modify the structural resistant scheme in comparison with the original one or, in general, when they affect the global behaviour, as in restructuring cases

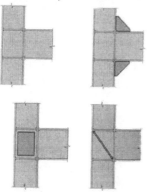

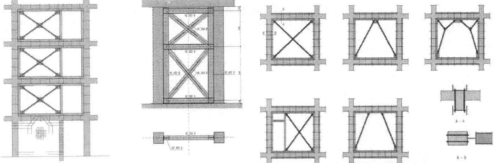

Figure 1. The improvement intervention carried out by adding welded parts on beam-to-column steel connections

Figure 2. The improvement intervention carried out on r.c. nodes by means of steel angles, plates and battens

Figure 3. Shear walls made by steel bracing systems for seismic strengthening r.c. and masonry structures

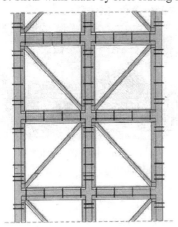

Figure 4. The intervention of seismic upgrading of a r.c. structure based on the use of steel St. Andrew's cross bracing with a two storey mesh (Mexico City)

- The different levels of reinforcing, from simple improvement to upgrading, can be done by using the same technological consolidation systems as for repairing, but with a more consistent intensity
- Steelworks are commonly used for improving the static behaviour of both masonry and reinforced concrete buildings (Mazzolani, 1992)
- More recently, reinforcing systems based on FRP have been successfully applied on masonry and reinforced concrete structures
- Bracing systems are often used for seismic upgrading of masonry (Figures 5 and 6) and reinforced concrete structures (Figures 7 and 8)
- Innovative bracing systems are based on the use of steel eccentric bracing (EB) (Figure 9), steel buckling restrained bracing (BRB) (Figure 10), shape memory alloy bracing (SMA-B) (Figure 11), low yield steel or pure aluminium stiffened panels (Figure 12) (Mazzolani et al., 2003)

Figure 5. Electrical Power Station in Hungry   **Figure 6.** A church near Avellino (Italy)

Figure 7. An apartment building in Santa Monica (California, USA)   **Figure 8.** A building in Tessaloniki (Greece)

Figure 9. R. c. structure reinforced by eccentric braces   **Figure 10.** R. c. structure reinforced by buckling restrained braces

Figure 11. R.c. structure reinforced by SMA braces   **Figure 12.** Stiffened panel for reinforcing r.c. structures

- The technologically acceptable combinations among structural types to be consolidated and materials suitable for consolidation are given in the following matrix, giving rise to new composite materials:

NEW COMPOSITE MATERIALS					
		MATERIALS FOR CONSOLIDATION			
		STEEL	CONCRETE	MASONRY	WOOD	FRP
DAMAGED STRUCTURES	STEEL	★				★
	CONCRETE	★	★			★
	MASONRY	★	★	★	★	★
	WOOD	★			★	★

- It can be observed that not all combinations are acceptable. Contrary, some of them give rise to synergetic composite systems
- Among the different consolidation techniques, not all possess the feature of reversibility, which is an important prerequisite in structural restoration of monumental buildings

Example of application

1. Sigma Coating Building in Agnano, Naples

The re-use of an industrial building required to modify the load bearing capacity of the structure, because the live load increased from 2 KNm^{-2} to 20 KNm^{-2} for a change in use. The original structure was made of two storeys r.c. frames with direct isolated foundations. The reinforcing operations consisted on (Mazzolani & Mandara, 1991):
- transformation of the isolated foundations into a r.c. beam grid;
- increasing of the axial load bearing capacity of r.c. columns by means of cold-formed steel sections (Figure 13);
- increasing the flexural load bearing capacity of r.c. beams by adding 20 cm of collaborating r.c. slab on the top of the existing one and by gluying a steel plate at the bottom of the existing beam, connected to the top part by means of vertical ties (Figure 14)

2. R.C. buildings of the University of California, Berkeley (USA)

After the Loma Prieta earthquake in 1989 in California, many buildings in Berkeley were upgraded by means of steel bracings (Mazzolani & Ivanyi, 2002).

The University Hall is a significant example where the use of steel bracings with a mash covering two floors contributes both to increase the seismic strength of the existing r.c. structure and to give rise to an improvement of the aesthetic of the original façade (Figure 15). Similar systems were used for upgrading the seismic performance of the r.c. structures of the apartment buildings for students (Figure 16) and also in the case of a car parking building (Figure 17)

Figure 13. Consolidation of r.c. foundations (Sigma Coating,Naples)

Figure 14. Consolidation of r.c. beams(Sigma Coating,Naples)

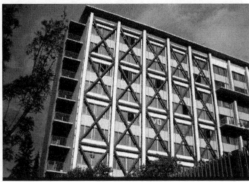

Figure 15. University Hall in Berkeley (California), seismically up-graded by steel bracings

Figure 16. Apartment building in Berkeley (California)

Figure 17. Car parking building in Berkeley (California) and detail of the base hinge

Figure 18. A steel mill in Bagnoli (Naples)

3. ILVA continuous casting mill in Bagnoli, Naples

The earthquake of Novembre 1980 happened when the building was under construction (Figure 18). After this event, the area was included into a new seismic zone and, therefore, it was compulsory to seismically up-grade the structure according to the italian regulations.
The only structural intervention consisted on connecting the column bases of the bracing structures by means of steel links (Figure 19).

Figure 19. The connecting structures at foundation level

References

- Mazzolani, F. M. 1990. *Refurbishment*, Arbed, Luxemburg.
- Mazzolani, F. M. 1992. *The use of steel in refurbishment,* Proc. of the 1st World Conference on Constructional Steel Design, Acapulco.
- Mazzolani, F. M. 1994a. *Il consolidamento strutturale.* In La Progettazione in acciaio, Consorzio CREA Editore, Massa, Italy.
- Mazzolani, F. M. 1994b. Strengthening of Structures, Lecture 16.1, ESDEP.
- Mazzolani, F.M., Calderoni, B., Della Corte G., De Matteis, G., Faggiano, B., Panico, S., Landolfo, R. 2003. *Full-scale testing of different seismic upgrading metal techniques on an existing RC building,* Proceedings of the Fourth International Conference STESSA '03, Naples, Italy.
- Mazzolani, F.M. & Ivanyi, M. (editors) 2002. *Refurbishment of buildings and bridges (CISM Course)*, Springer - Verlag Wien - New York.
- Mazzolani, F.M. & Mandara, A. 1991. *L'acciaio nel consolidamento*, Assa Editore, Milano, Italy.

Contacts

- Prof. Dr. Eng. Federico M. Mazzolani
  Address: Department of Structural Analysis and Design,
  University of Naples "Federico II",
  P.le Tecchio, 80, 80125 Naples, Italy.
  e-mail: fmm@unina.it,
  Phone: (+39) 081 7682443, Fax: (+39) 081 5934792

Department of Structural Analysis and Design
University of Naples "Federico II"

CONSOLIDATION LEVELS: RESTRUCTURING

Description

- Restructuring represents, in hyerarchical order, the more general consolidation level of existing buildings

- It consists of the partial or total modification of the functional distribution, lay-out and volumetric dimensions, together with the change of other original features of the building, including a drastic change of the structural system (Figures 1 and 2)

- There are four different kinds of restructuring interventions: degutting, insertion, addition and lightening (Mazzolani, 1990, 1994)

- *Degutting* is the total or partial substitution of the internal part of a building with new structures of different type. It is carried out when architectural and/or town-planning reasons require the complete conservation of the building facades, whilst the interior layout is changed for functional reasons (Figure 3).

- *Insertion* represents the introduction of new structures or structural elements into the existing volumetric dimension. Additional intermediate floors or mezanines are created in order to increase the usable area within the limits of a given volume (Figure 4). The self-supporting frames inserted in museums for containing special exhibition show-cases visible on several levels as well as the staircase and lift cages can represent some more examples of the insertion operations.

- *Addition* is carried out for new functional requirements, involving the increasing of the original volume of the building, which can be extended in horizontal or vertical direction (Mazzolani, 2001).

  - *Horizontal addition* consists on the introduction of new lateral volumes to add to the original structure. In these cases, aestetich aspects, more than the structural ones, play a great role due to the necessity to correlate different architectural languages. (Figure 5a)

  - *Vertical addition* requires to increase the building height of one or more stories above the existing structure (Figure 5b). Depending on the amount of the new additional masses, it is necessary to re-check the load bearing capacity of the original structure in order to take into account possible consolidation interventions preliminary to the superelevation. This problem is particularly important in seismic zone, where the global behaviour of the building is strongly influenced by the addition of new masses particularly in the upper part. The necessity to minimize the added structural weight makes steel the most suitable material, thanks to its high strength/specific weight ratio.

- *Lightening*, opposite to vertical addition, can include the demolition of one or more storeys at the top due to the need to reduce the stress state in the structure. This scope can be achieved by means of operations based on the substitutions of the original floors, roofs or other structural elements with new lighter materials. The replacement of heavy wooden floors with light steel I-section and corrugated steel sheets as well as the old roofs with steel trusses is very common (Figures 6 a and b).

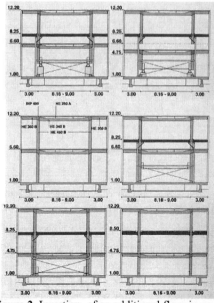

Figure 1. A new steel structure inserted in the original lay-out of a r.c. building (Cantù, Como, Italy)

Figure 2. Insertion of an additional floor in an existing steel building (Amstelveen, The Netherland)

Figure 3. Example of degutting intervention (Zurich, Switzerland)

Figure 4. Example of insertion intervention (Genoa, Italy)

a)

a)

b)

b)

Figure 5. Examples of horizontal (a. La Villette, Paris) and vertical (b. Toronto, Canada) additions

Figure 6. Example of lightnening intervention by new steel roofing (a. Museum of Rivoli, Turin, Italy; b. Borbons' Iron Mill of Mongiana, Italy)

Field of application

- When the modification of the functional lay-out of a building requires the introducion of new volumes and new areas
- When the respect of the code provisions requires to modify the resistant structural scheme
- For highly damaged buildings which require the complete revision of the structural scheme and its upgrading

Technical information

- When the building to be consolidated is of historical interest, the use of reversible mixed technologies must be chosen in its restoration (Mazzolani, 2001)
- Conservation of the existing building and its integration with new clearly distinguishable and reversible works represents the criteria upon which restructuring operations must be based, according to the modern theory of restoration
- A logical application of this principle undoubtedly shows that steel and its technology have the necessary prerequisites of being a modern material with "reversible" characteristics, particularly suited to harmonize with the ancient materials, so to form integrated structural system

Example of application

- ***Degutting (The Law Court Palace in Ancona, Italy)***

The restoration of the Law Court in Ancona is an emblematic example of degutting. The building was degutted and restructured, while the masonry facades were preserved, showing their neo-renaissance style. Four r.c. towers 9 x 9 m, containing stairs, lifts and floor services, were located in the corners of the covered courtyard in order to withstand the vertical loads (roofing system and suspended floors) as well as the seismic actions (Figure 7). The suspension system consists of 4 pairs of steel truss girders located in the inside edge of the 4 r.c. towers, along the courtyard perimeter (Figure 8). The five suspended floors, composed by steel beams and corrugated sheets with a cast concrete slab, are located in the four zones of approximately 9 x 20 m between the four towers. (Mazzolani & Mandara, 1992).

Figure 7. The tower and the suspended floors of the Law Court of Ancona, Italy

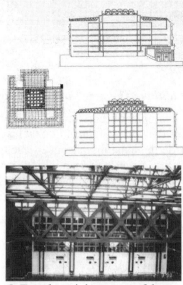

Figure 8. Truss box girders on top of the towers in the Law Court of Ancona, Italy

- *Insertion (The Ducal Palace in Genoa, Italy)*

 In the past centuries this building suffered drastic variations of destination of its use and heavy degradation of its structure. In fact, the fires of 1591 and 1977 and the damage due to the bombing in 1942 have made necessary the static restauration of the building. The large use of steel has been required by the necessity to differentiate the new parts from the old ones by means of a series of interventions able to characterise the building with a modern aspect. In particular, the insertion of new steel structural elements have been done. Among them, the ramp connecting the "Loggia degli Abati" with the "Torre di Palazzo" is suspended to the roof (Figure 9). An amphitheater entirely made of curved steel beams, as well as the articulated system of inside service staircases and intermediary mezanines represent further insertion interventions (Figures 10 a and b) (Mazzolani & Mandara, 1992).

Figure 9. The suspended ramp in the Ducal Palace in Genoa, Italy

Figure 10. The amphitheater (a) and the staircase-mezanine system (b) in the Ducal Palace of Genoa, Italy

- *Horizontal addition (The new Faculty of Economy and Commerce in Turin, Italy)*

It represents an example of horizontal extension of an existing building, the former hospice "House for the Old" (Figure 11). Steel work has been used for creating mezanines and staircases inside the existing masonry structure, in the full respect of the reversibility characteristics of the intervention. In addition, it was necessary to create new volumes for locating the new auditoriums. The new buildings are characterised by modular façade panels keeping the same configuration of the windows in the original building (Figure 12) (Mazzolani & Ivanyi, 2002).

Figure 11. View of the "House for the Old" before the intervention

Figure 12. The new steel structure erected in adherence to the existing one

- *Vertical addition (Country Club in Briatico, Catanzaro, Italy)*

The building was an ancient sugar factory, a relevant example of industrial archaeology. The restoration operation was finalized to transform it for social activities. The building suffered large damage from the previous earthquakes and from the sea effects which determined erosion of its inferior level (Figure 13). A new level has been created above the existing masonry structure which transfer vertical loads only to the masonry below, thanks to special support devices. (Figure 14)

Figure 13. The old building in Briatico before the restoration

Figure 14. The Country Club of Briatico after the restoration

- *Vertical addition (Building in Toronto, Canada)*

The original building was a 6 storeys building made of reinforced concrete frames (Figure 15). It was designed to be superelevated with more 4 storeys, always in r.c. After some years the superelevation started, but changing the material from r.c. to steel. Due to the advantages of the steel technology, instead of 4 storeys, they were able to add 8 storeys more, giving rise to a superelevated building of 14 storeys (Figure 16).

Figure 15. The original building

Figure 16. The building after superelevation

- ***Lightening (Cultural centre in Succivo, Caserta, Italy)***

The ex Carabinieri Barrack of Succivo was transformed into an antiquarium and Cultural Centre. The change of use of the building, together with static improvement demands, required the creation of additional new spaces with a contemporary lightening of the structural conditions. In order to fulfil both statical and functional requirements, the original roof structure has been replaced by a new metallic mansard (Figure 17), composed by a series of Vierendeel trusses, with four vertical elements, sufficiently tall to entertain a new suitable space (Figure 18). Despite a little increase of the overall building volume, due to a slight increase of the high of the roofing level, a sensible weight reduction has been obtained thanks to the use of light gauge steelworks.

Figure 17. The new steel roof of the ex Cultural Centre of Succivo, Italy

Figure 18. The new internal space

References

- Mazzolani, F. M. 1990. *Refurbishment*, Arbed, Luxemburg.
- Mazzolani, F. M. 1994. Il consolidamento strutturale. In *La Progettazione in acciaio,* Consorzio CREA Editore, Massa, Italy.
- Mazzolani, F.M. 2001. *Die Anwendung von Stahl bei der Restaurierung von Gebäuden in Italien*, Bauingenieur, Band 76, May 2001.
- Mazzolani, F. M. & Mandara, A. 1991. *L'acciaio nel consolidamento*, Assa Editore, Milano, Italy.
- Mazzolani, F. M. & Mandara, A. 1992. *L'acciaio nel restauro*, Assa Editore, Milano, Italy.
- Mazzolani, F.M. & Ivanyi, M. (eds.). 2002. *Refurbishment of buildings and bridges,* CISM Course N.435, Springer Wien New York.

Contacts

- Prof. Dr. Eng. Federico M. Mazzolani
    Address: Department of Structural Analysis and Design,
        University of Naples "Federico II",
        P.le Tecchio, 80, 80125 Naples, Italy.
        e-mail: fmm@unina.it,
        Phone: (+39) 081 7682443, Fax: (+39) 081 5934792

STEEL-CONCRETE MOMENT CONNECTION

Description

- beam-to-column joints; semi-continuous (semi-rigid; partial strength)
- moment transfer between composite columns and composite slim floors
- composite column: filled hollow section or encased I-section
- principles also applicable to steel columns (hollow and I sections)
- composite slim floor: steel beam integrated into the slim concrete slab

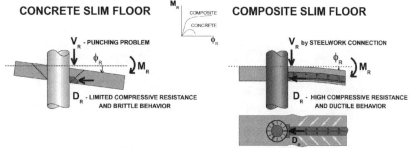

Figure 1. Conventional concrete against modern steel-concrete slim floor connection

Figure 2. Joint layout before concreting

Figure 3. Advantage in view of beam sag

Technical Information

- especially for high-rise buildings: column design is mainly determined by normal forces; additionally applied restraint moments from the floors do not considerably influence the column design but significantly help in view of ULS and SLS design of the slabs/beams

- two problems solved by one solution

    i) punching of concrete slab solved by T-shaped steel beam as continuous support of the slab; the collected vertical forces then are transferred to the column by a combined fin-plate / bracket support (cold and hot stage)

    ii) moment connection resistance of slim concrete slab would be limited by brittle concrete failure due to concentrated compression contact; here high local compression resistance due to contact between lower beam flange and load introduction bracket

- joint configuration: hogging connection moment results from lower compression force (transferred from lower beam flange via clearance filling shims into the column bracket) and upper tension force (within the reinforcement loops) in a vertical distance; a welded positioning saddle for the reinforcement ensures this very crucial lever arm

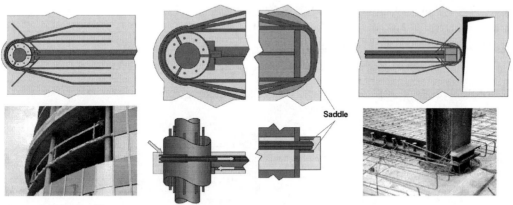

Figure 4. Joint detailing

Fields of Application and End Users

- generally applicable solution to create restraint beam end moments to improve the ULS and SLS behaviour of the beams/slabs
- Attention: where these additional moments in comparison with the normal forces in the columns are not negligible this effect has to be considered for the column design!

Structural Assessment

- Advantages:

  reduced sagging span leads to improved ULS and SLS; resulting in very slim floor depths; therefore reduced dead load; this effects the column dimensions and the foundation design; the structural efforts for these benefits are very low (easy erection, no considerable increase of erection time and costs); due to the vertical continuity of the columns a very clear load transfer of the normal forces is guaranteed; erection speed

- Disadvantages:

  column design for normal force and moment interaction (only decisive for small normal forces; in case of low rise buildings)

- Potentials:

  architectural benefit: slender vertically continuous columns in combination with very slim floors

  structural benefit: reduced construction height ⇒ gain of commercial height and reduced dead load (foundation design!)

- Risks:

  taking into account such beam restraints requires the guaranteed lever arm between the lower beam flange and the hogging reinforcement loops (positioning saddles, control on site)

Economical Aspects

- material savings
- fast erection

Robustness and Testing

- analytical joint characterisation by means of the component method (assembly of individual component springs according to the spring model)
- the analytical characterisation has then been supplemented by a full-scale joint test which impressively proved the calculated results

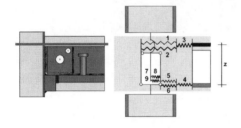

Figure 5. Component method, spring model

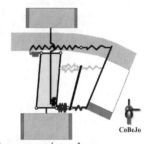

Figure 6. Component interplay

Figure 7. Advantage: very slim floors

Examples

- **MILLENNIUM TOWER** (Vienna, Austria) – 1999 (IStHM, Univ. Innsbruck)

  50 upper floors, 202m, 8 months of construction (2½ - 3 floors of 1080m² per week)

  concrete core (tower back) + composite tower around

  moment connections between inner columns + slim floor beams + external columns lead to radial composite frames ⇒ minor influence on columns but considerable reduction of beam sagging (moment, deflection, vibration); frames not used to transfer horizontal wind/earthquake forces (passed further to the concrete core)

Figure 8. Millennium Tower 1998/1999

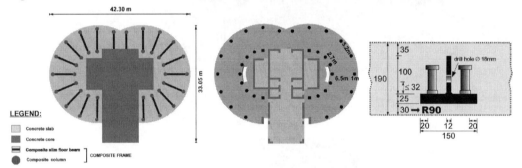

Figure 9. Plan of concrete core and composite frames; slim floor cross section

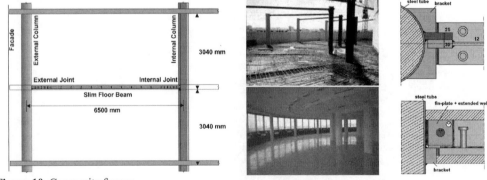

Figure 10. Composite frames

414

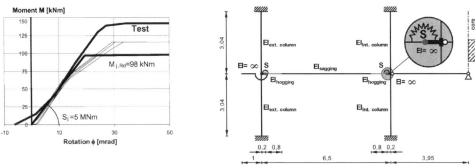

Figure 11. Joint characteristics (calculation / test)   **Figure 12.** Global frame analysis

- **DEZ CAR PARK** (Innsbruck, Austria) – 1999 (A-MBT, Univ. of Innsbruck)

  composite columns with foldable beam stubs in four directions for moment connection and against punching shear (no continuous steel beams spanning from column to column but only beam stubs in the hogging area); floors as conventional concrete slabs

  speciality: 5m slab cantilever

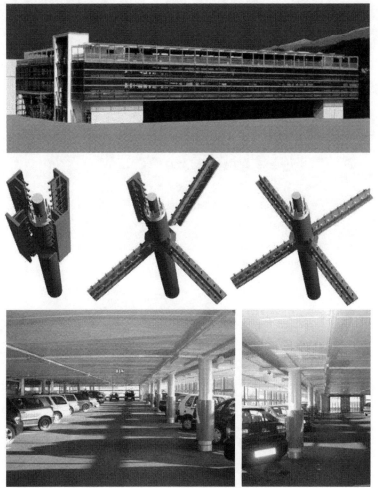

Figure 13. Foldable beam stubs at the Park Deck DEZ in Innsbruck

- **AK OFFICE** (Innsbruck, Austria) – 2002/2003 (aste konstruktion)

  full refurbishment of an old brick-concrete four storey office building; removal of main walls and replacement by supporting beams on slender columns (gain of space; more light); implementation of two new slabs in slim floor technology with moment connection to the tube columns; composite slabs with re-entrant steel sheeting

  further advantages: reduction of noise disturbance; reduced floor depth; reduced load on existing foundation; frame action for horizontal forces

Figure 14. Refurbishment of the AK office building in Innsbruck

Codification

- EN-1993-1-8. *Design of steel structures, Design of joints.*
- EN-1994-1-1. *Design of composite steel and concrete structures, Section 8 Composite joints in frames for build-ings.*

References

- Anderson, D., 1999, *Design of Composite Joints for Buildings*, ECCS Publication No. 109, ISBN 92-9147-000-52, Brussels.

- Huber, G., 1999, *Non-linear calculations of composite sections and semi-continuous joints*, Doctoral thesis, Ernst & Sohn, Berlin, ISBN 3-433-01250-4.

- Huber, G., Michl, T., 1999, *Beispiele zur Bemessung von Riegel-Stützen-Verbindungen (Example calculation for a beam-to-column joint)*, Fachseminar und Workshop Verbundbau 3, Fachhochschule München/Munich, Germany.

- Huber, G., Rubin, D., 1999, *Verbundrahmen mit momententragfähigen Knoten beim Millennium Tower (Composite frames with semi-continuous joints at the Millennium Tower)* Stahlbau 68, Ernst & Sohn, Berlin, p. 612-622.

- Müller, G., 1998, *Das Momentenrotationsverhalten von Verbundknoten mit Verbundstützen aus Rohrprofilen (The moment-rotation response of composite joints with composite tubular hollow columns)*, Doctoral thesis, IStHM, University of Innsbruck, Austria.

- Taus, M., 1999, *Neue Entwicklungen im Stahlverbundbau am Beispiel Millennium Tower Wien und Citibank Duisburg (New developments in composite construction at the Millennium Tower in Vienna and the Citibank in Duisburg)*, Commemorative publication Prof. Dr. Ferdinand Tschemmernegg, IStHM, University of Innsbruck, Austria, ISBN 3-9501069-0-1.

- Tschemmernegg, F., 1999, *Innsbrucker Mischbautechnologie im Wiener Millennium Tower (Mixed building technology of Innsbruck at the Millennium Tower in Vienna)*, Stahlbau 68, Ernst & Sohn, Berlin, p. 606-611.

Contacts

- Dr. Gerald Huber
  Address: Kohlsiedlung 9, A-6091 Goetzens, Austria
  e-mail: geraldhuber@telering.at
  http://members.telering.at/geraldhuber/home.html

COST C12 – WG1

Datasheet II.3.1.1.2

Prepared by: G. Huber[1]

[1] Institution: Huber, Innsbruck/Austria, www.huber.net.tf

H. Beck[2]

[2] Institution: Hilti AG, Schaan/Liechtenstein, www.hilti.com

SHEAR TRANSFER BETWEEN STEEL AND CONCRETE WITHIN COMPOSITE TUBES

Description

- new type of shear connector
- powder-actuated fastener: Nail X-HVN32P10 (as conventionally used for fastening purposes)
- mainly for steel hollow sections filled with concrete

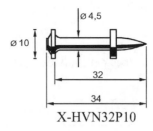

Figure 1. Powder-actuated fasteners (nails) for shear transfer

Technical Information

- in frame structures the vertical support forces of the beams are handed over to the column steel tubes; parts of these concentrated forces then have to be passed to the chamber concrete and eventually further to a steel core profile; instead of conventional welded studs shot-fired nails and bolts can been applied.

- according to Eurocode 4 shear connectors have to provide sufficient resistance against uplift; in the actual application case of a chambered concrete such uplift is automatically prevented and the nails/bolts can be applied without further measures.

- despite of high strength steel material of the nails they proved to behave very ductile due to the chamber effect within the hollow steel section.

- Nail type X-HVN32P10 (Hilti)   (former X-DSH32P10)

- design resistance values according to the submission for a "Allgemeine bauaufsichtliche Zulassung" at the DIBt (Deutsches Institut für Bautechnik, Berlin): design resistance dependent on concrete grade: for example $R_{d,ULS} = 12,0$ kN for C35/45. Application boundaries:

  i)   tubes (pipes) without dimension restriction
  ii) wall thickness 5,6 to 12,5 mm
  iii) d/t-ratio acc. to E-DIN 18800-5
  iv) maximum aggregate size 16 mm
  v)   minimum distance to concrete construction joint: 20 cm
  vi) minimum transverse nail distance: 50 mm
  vii) minimum vertical/longitudinal nail distance: 100 mm

Fields of Application and End Users

- general shear transfer within hollow steel sections filled with concrete
- in general nails are applied before concreting
- but even strengthening of an existing composite tube is possible, providing enough energy to drive the fastener

Structural Assessment

- Advantages:

  fast, cheap, easily applicable from the outside of the hollow section, possibility for later strengthening, no welding, use of conventional powder-actuated tools to drive the nails
- Disadvantages:

  relatively low resistance per individual nail $\Rightarrow$ several nails per load introduction point

Economical Aspects

- fast, cheap

Robustness and Testing

- load-slip push-out tests at the University of Innsbruck, at Hilti Technical Center/Schaan and at the University of Wuppertal

Examples

- **MILLENNIUM TOWER** (Vienna, Austria) – 1999 (IStHM, Univ. Innsbruck)

  first application

  for all external columns (about 1000 beam supports with about 16 nails and 16 bolts per connection - in this case also the shear connection of the inner solid steel cores was made with powder-actuated bolt fasteners)

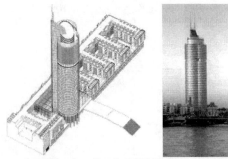 

Figure 2. Millennium Tower 1998/1999

Figure 3. Driving the nails in the shop

- **<u>DEZ CAR PARK</u>** (Innsbruck, Austria) – 1999 (A-MBT)

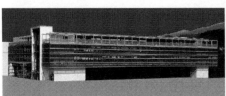

Figure 4. Car Park Deck DEZ in Innsbruck

- **<u>AK OFFICE</u>** (Innsbruck, Austria) – 1999 (aste konstruktion)

Figure 6. Refurbishment of the AK office building in Innsbruck

Codification

- EN-1994-1-1. *Design of composite steel and concrete structures.*

References

- Angerer, T., Rubin, D., Taus, M., 1999, *Verbundstützen und Querkraftanschlüsse der Verbundflachdecken beim Millennium Tower (Composite columns and vertical support connection of the slim floor beams at the Millennium Tower)* Stahlbau 68, Ernst & Sohn, Berlin, p. 641-646.

- Beck, H., 1999, *Nailed shear connection in composite tube columns*, Conference Report Eurosteel '99 in Praha, ISBN 80-01-01963-2.

- Taus, M., 1999, *Neue Entwicklungen im Stahlverbundbau am Beispiel Millennium Tower Wien und Citibank Duisburg (New developments in composite construction at the Millennium Tower in Vienna and the Citibank in Duisburg)*, Commemorative publication Prof. Dr. Ferdinand Tschemmernegg, IStHM, University of Innsbruck, Austria, ISBN 3-9501069-0-1.

- Taus, M., 1999, *Verbundkonstruktion beim Millennium Tower – Fertigung, Montage, neue Verbundmittel (Composite construction at the Millennium Tower – production, erection, new shear elements)* Stahlbau 68, Ernst & Sohn, Berlin, p. 647-651.

- Tschemmernegg, F., Beck, H., 1998, *Nailed shear connection in composite tube columns*, ACI-Paper, Houston Convention.

- Tschemmernegg, F., 1999, *Innsbrucker Mischbautechnologie im Wiener Millennium Tower (Mixed building technology of Innsbruck at the Millennium Tower in Vienna)*, Stahlbau 68, Ernst & Sohn, Berlin, p. 606-611.

- Hanswille, G., Beck, H., Neubauer, T., 2001, *Design Concept of nailed shear connections in composite tube columns*, RILEM Conference Proceedings "Connections between steel and concrete" Stuttgart 2001, ISBN 2-912143-27-6

Contacts

- DI. Hermann Beck
  Address: Hilti AG, Feldkircherstr. 100, FL-9494 Schaan, Liechtenstein
  phone +423 234 2290
  e-mail: beckher@hilti.com
  http://www.hilti.com
- Dr. Gerald Huber
  Address: Kohlsiedlung 9, A-6091 Goetzens, Austria
  e-mail: geraldhuber@telering.at
  http://members.telering.at/geraldhuber/home.html

STEEL TRUSS TO CONCRETE CONNECTION

Description

- steel truss-to-concrete connection
- transfer from localised connection forces into a concrete section
- pre-stressed docking devices
- for dominating monotonic tension forces

Figure 1. Pre-stressed docking devices placed into the formwork

Technical Information

- load transfer into concrete walls with mutual pre-stressing cables (interior tendons); from one docking point to the opposite one or with internal cable anchorage
- prestressing after hardening of the concrete and before connecting the steel sections; dimensioning/number of strands depending on the tension forces; conduits then filled with injection grout against corrosion
- additional concentrated rebars in the local load introduction zones to cover the bursting forces and for crack distribution
- slotted puch-over hollow steel section welded on site to the cantilevering steel docking elements (four longitudinal ribs); necessary welding length decisive for bracket size

Figure 2. Slotted hollow sections welded to the docking brackets

Fields of Application and End Users

- rising structures, industrial buildings, bridges (quasi-monotonic)

Structural Assessment

- Advantages:

  clear separation between concrete structure and steel cage during erection

  minimum disturbance of the climbing formwork

  load transfer of high tension forces
- Disadvantages:

  no later accessibility of the tendon heads

  only for static loading
- Potentials:

  supports the combination of a light-weight pre-fabricated steel cage in combination with a conventional concrete building
- Risks:

  pre-stressing after sufficient hardening period of the concrete; locally concentrated load introduction forces have to be covered by adequate rebar reinforcement; corrosion protection of the tendon heads has to be assured; welding has to be applied in several layers not to heat up the tendon heads beyond tolerable limits

Economical Aspects

- fast erection of pre-fabricated steel trusses
- clear separation of steel and concrete parts in case of dismantling / recycling

Robustness and Testing

- eventual negative influence of the high welding temperatures on the end anchorage of the pre-stressing strands could be dispelled by a test specimen; the maximum heat increase was measured to be only 50 degrees

Example

- **<u>SKI JUMP BERGISEL</u>** (Innsbruck, Austria) – 2001/2002 (aste konstruktion)

  ski jump facilities fully renewed; architectural landmark by Zaha Hadid/London; concrete tower + docking steel cage at the top; steel cage of rectangular hollow section trusses is cantilevering up to 12 m in a height of 35 m above ground; transfer from the localised truss tension forces into the concrete box section by pre-stressed docking devices within the climbing formwork

  tower: height 60 m above the foundation; 49 m above ground; concrete box type section 7 x 7 m at the bottom and 7 x 3,7 m at the top levels; wall thickness 40 cm; fair-faced concrete with climbing formwork

  tower top: three-level steel cap; café-restaurant and observation platform; transparency and elegance of the facade is supported by the fact that diagonal bars within the front could be avoided and huge glass elements were placed into the facade

Figure 3. Panoramic view of Bergisel facing north

Figure 4. Sketch of steelwork construction   **Figure 5.** Concrete tower with docking devices

Figure 6. Cantilevering steel cage during construction

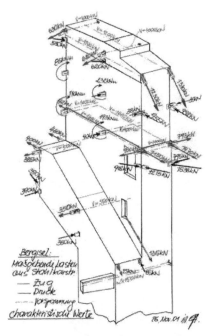

Figure 7. Characteristic docking forces

Figure 8. Finalised steel top

References

- Aste, Ch., Glatzl, A., Huber, G., 2002, *Ski jump „Bergisel" – A new landmark of Innsbruck*, 3rd Eurosteel Conference, Lisbon, ISBN 972-98376-3-5.
- Aste, Ch., Glatzl, A., Huber, G., 2002, *Schisprungschanze „Bergisel" – Ein neues Wahrzeichen von Innsbruck*, Stahlbau Heft 3/02, S.171-177, ISSN 0038-9145, Ernst&Sohn, Berlin.
- Aste, Ch., Glatzl, A., Huber, G., 2003, *Steel-concrete mixed building technology at the ski jump tower of Innsbruck, Austria*, International Journal on Steel and Composite Structures, Technopress/Korea, ISSN 1229-9367.
- Aste, Ch., Glatzl, A., Huber, G., 2003, *Innsbruck ski jump: a triumph of mixed building technology*, Concrete Journal (The Concrete Society, UK), Berkshire.
- Aste, Ch., Glatzl, A., Huber, G., 2003, *Schisprungschanze „Bergisel" – Ein neues Wahrzeichen von Innsbruck*, Zement + Beton Heft 3/03, vöz, www.zement.at, Wien.

Contacts

- Dipl.-Ing. Christian Aste, Dipl.-Ing. Andreas Glatzl
  Address: aste konstruktion, A-6020 Innsbruck, Austria
  tel. +43 512 580330
  e-mail: aste@utanet.at
  http://www.aste.at
- Dr. Gerald Huber
  Address: Kohlsiedlung 9, A-6091 Goetzens, Austria
  e-mail: geraldhuber@telering.at
  http://members.telering.at/geraldhuber/home.html

COST C12 - WG1

Datasheet II.3.1.1.4

Prepared by: M. Ferrer, F. Marimon, F. Roure
Barcelona School of Engineering. Technical University of Catalonia (ES)

FINITE ELEMENT SIMULATION OF SHEAR TRANSFER ON COMPOSITE SLABS

Description

- Resistant failure of composite slabs is generally due to **longitudinal slip** between steel sheet and concrete. This slip is associated to shear forces of simple bending.
- In order to resist shear forces, current designs of cold-formed steel sheet present a repeated **embossment pattern**, of many shapes and sizes, all along the span of the composite slab. The wedge effect of these embossments on steel, transforms the longitudinal slip to different actions on cold-formed steel sheet, such as **transversal bending** –especially important because of its very low stiffness– and posterior vertical release of both elements.

Figure 1 Eurocode-4 standard bending test of a composite slab

- The predominant physical phenomenon is sliding contact with friction between steel and concrete surfaces, with pressures induced by wedge effect of embossments.
- A validated procedure for building **finite element models**; in particular, the **interaction** between steel sheet and concrete, has been established to analyse and predict shear resistance.
- A new design has been developed and optimised.

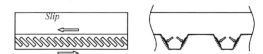

Figure 2 Longitudinal slip causes transversal bending of sheet

Field of application

- This research work is devoted to **improve and optimise steel sheet designs** by first analysing and understanding the mechanical behaviour of interaction between steel and concrete. The established procedure is a very efficient method for steel sheets designers, to innovate, develop and optimise current sheets, and, so, making lighters and more effective composite slabs.
- The long-term aim is to propose new ideas in order to achieve full connection.

- Other derived conclusions are obtained, for reviewing EC-4 standard test conditions and specifications.

Parametrical analysis

- Numerical models have been parametrically built in order to evaluate the dependency of shear resistant mechanics on geometrical design parameters.
- Sensibility of shear resistance is analysed, in relation to: friction coefficient, embossing depth and slope, sheet thickness, inclination angle and length of embossment, profiling angle of rib shape (from trapezoidal to dovetailed) for different embossment patterns.

The physical mechanism being modelled: the pull-out test

- The test specimen consists of two confronted forms of composite slab. A thick steel sheet joints and stiffs both forms. Upper traction load is applied to the cold-formed steel sheet prolongation. Steel screwed bars are inserted in concrete and used as bottom traction elements. The Pull-out test (Daniels 1988, Crisinel 2004) was designed to reproduce the longitudinal slip behaviour exclusively.

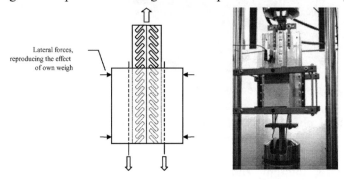

Figure 3 The pull-out test

Description of the finite elements model

- Preliminary finite element models treated concrete (20 node solids) and steel (8 node shell) as linear elastic materials (Ferrer 2002). Obtained results showed that steel must be implemented using a multilinear elastoplastic model. On the other hand, concrete block stiffness was proved to be much higher than steel sheet (Ferrer 2003).
- Therefore, the concrete is treated as a perfectly rigid contacting surface with friction. The elastic solid elements mesh has been removed resulting a much simpler model. Two consecutive embossment patterns are needed for concrete to let steel sliding over it. A multilinear stress-strain curve, obtained from cold-formed steel sheet traction tests, has been introduced for steel. Residual stresses of steel sheet are not taken into account.
- Contact elements with friction (Baltay 1990) have been implemented between both materials. Several values have been considered, maximum coefficient $\mu = 0,6$

(Veljkovic 1996), minimum coefficient $\mu = 0,2$ (Shuurman 2001) and null coefficient has been calculated as well, to completely evaluate its influence.

- The boundary conditions are those for double longitudinal symmetry of the pull-out test specimen, and the cyclic symmetry due to the embossment patterns.

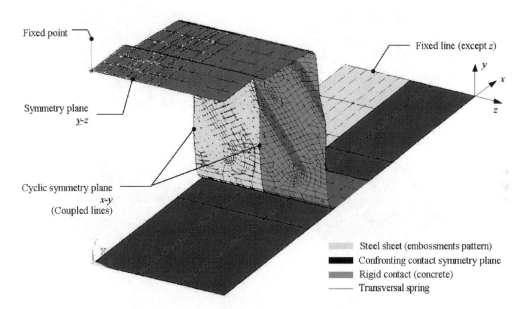

Figure 4 Finite element model for HB profile

- The external loads consist of prescribed displacement in slip direction (Z), on both two cyclic symmetry lines. Therefore, these two lines remain flat and equidistant. In other words, longitudinal (z-direction) length change of the embossment pattern is neglected and assumed to be independent of transversal bending.

Examples of existing analysed designs

- The following figures show a concrete side perspective view of some commercial designs to be analysed and their identification names (company names omitted):

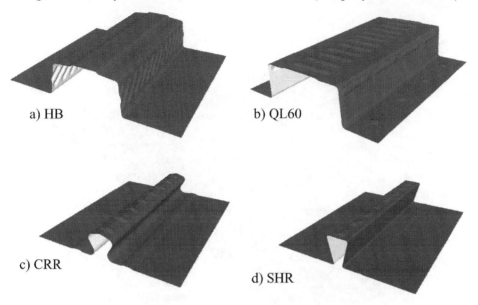

a) HB

b) QL60

c) CRR

d) SHR

Figures 4a-4d Images of analysed profiles

FEM results images

- The following figures show a concrete side perspective view of the deformed Von Mises stress maps for all above designs, at the maximum shear force point:

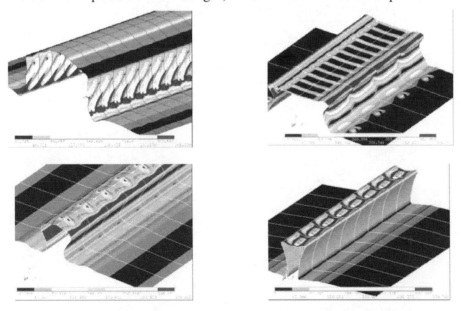

Figures 5a-5d Deformed equivalent stress maps

Example of a new developed design

- The authors of this paper have designed T80 profile. A new optimised design based on T80 is being developed according to conclusions of this research work and will be presented shortly.

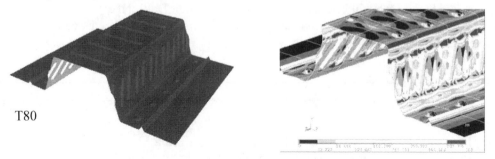

T80

Figure 6a-6b New "T80" profile and results

Results of analyses

- Complete and easy comprehension of any sliding resistant mechanism.
- Easy evaluation of any changes on geometrical or physical parameters, so complete and easy comprehension of their influence.
- Optimisation of T80 new profile, doubling its initial shear strength by very simple changes.
- Derived conclusions are being obtained (final paper is near to be finished) about the necessity of reviewing EC-4 standard test conditions and specifications in order to reduce scattering on tests results.
- Summary of recommendations for designing steel sheets.

Further developments

- Inclusion of crackable concrete mesh to reproduce concrete peel-off failure mode.
- Inclusion of profiling and embossing residual stresses.
- Extension of the model to the whole span (one rib), therefore including flexion and also avoiding the hypothesis inherent to boundary conditions, such as equal behaviour of all embossment patterns along the shear span.

References

- Baltay, P & Gjelsvik, A. 1990. *Coefficient of Friction for Steel on Concrete at High Normal Stress*, J. of Materials in Civil Engineering, Vol. 2, No. 1, 46-49.
- Crisinel, M & Marimon F. 2004. *A New Simplified Method for the Design of Composite Slabs*, J. of Constructional Steel Research, Vol. 60, 481-491.
- Daniels, B.J. 1988. *Shear Bond Pull-Out Tests for Cold-Formed-Steel Composite Slabs*, Rapport d'essais, ICOM 194. École Polytechnique Fédérale de Lausanne, CH.
- Ferrer, M; Marimon, F. & Roure, F. 2002. *Losas Mixtas (I): Modelado Mediante Elementos Finitos del Fallo por Deslizamiento Longitudinal*, XV National Congr. on Mech. Eng. Cádiz, ES.

- Ferrer, M; Marimon, F. & Roure, F. 2003. *Composite Slabs (II): Finite Element Modelling of Longitudinal Shear Failure*. Proc. of 7th Int Research/Expert Conf. Trends in the Development of Machinery and Associated Technology. Lloret de Mar, Barcelona, ES.
- Schuurman, R.G. 2001. *The Physical Behaviour of Shear Connections in Composite Slabs*, Doctoral Thesis. Technische Universiteit Delft. NL.
- Veljkovic, M. 1996. *Behaviour and Resistance of Composite Slabs*, Doctoral Thesis. Luleå University of Technology, SW.

Contacts

- Miquel Ferrer
  Dep. of Strength of Materials and Structural Engineering
  Barcelona School of Engineering, Technical University of Catalonia
  Avda. Diagonal, 647. 08028-Barcelona (ES)
  e-mail: miquel.ferrer@upc.es, Phone: +34 934010861, Fax: +34 934011034
- Frederic Marimon
  idem
  e-mail: frederic.marimon@upc.es, Phone: +34 934016537
- Francesc Roure
  idem
  e-mail: francesc.roure@upc.es, Phone: +34 934016533

COMPOSITE ELEMENTS IN TIMBER AND GLASS

Description

- Timber and glass in greenhouses, facades and other structures
- Timber-glass composite element has been developed where the wooden frame is directly glued to the glass-plate

Figure 1. composite elements

Technical Information

- composite timber-glass elements
- float-glass, fully tempered glass or laminated glass
- solid wood or wood based materials
- polyurethan adhesive

Field of Application and End Users

- Residential housing and industry buildings
    - i) windows
    - ii) facades
    - iii) structures

Structural Assessment

- light structure:
  A comparison between the traditional and the composite structures shows that the wooden sections are much smaller with timber-glass composite structures.
- prefabricated elements:
  Timber-glass composite elements can be mounted quickly on the building site.
- an optimum use of material:
  timber under tension, glass under compression
- constructive wood preservation
  Timber is protected by the glass-plate on the outside.
- The timber-sections prevent the brittle-failure of the composite section

Economical Aspects

- prefabricated elements
- optimum use of matirals
- light structure

Robustness and Testing

The glue

- To find the right glue for the timber-glass composite structure, several small specimens (fig.1) were tested under shear compressive loading.

- the test specimens were submitted to several climatic cycles. One cycle consisted of 4 hours at -30°C, 4 hours at +70°C, 16 hours at +30°C and 80% humidity

- cycle was repeated up to seven times and specimens were tested after 0-cycle, 2-cycles, 5-cycles and 7-cycles to evaluate the possible degradation. Of the four glues which were tested, only one exhibited adequate resistance.

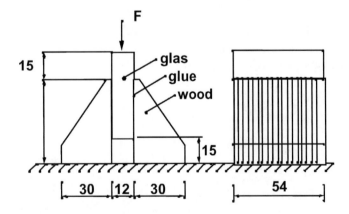

Figure 1: specimen for shear compression test [mm]

Timber-glass composite plates

After the right glue was found, 4-point loading tests on timber-glass composite plates were carried out.

- Load was introduced on the glass over the wooden sections, so that the composite behavior was tested.
- results of these tests showed that the efficiency of the glue was nearly 100%
- Failure was caused by breakage in the wood under tensile stress (on average: 66,24 N/mm^2)
- After the formation of cracks in the timber section, the level of the neutral axis was shifted higher, so that the glass plate broke under tensile strength. The pattern of the broken glass of two different glues showed the influence of the rigidity of the glue on the formation of cracks in the glass plate (fig.3 and 4).

Figure 2: formation of cracks with a rigid glue   **Figure 3:** formation of cracks with a soft glue

- The load-deformation curve of a timber-glass composite plate shows the different stages of the load capacity of these elements (fig.5).

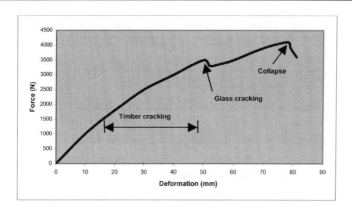

Figure 4: load-deformation curve of a timber-glass composite plate

I-beam

- The I-beams were also tested under 4-point loading.

- All test specimens showed a similar behavior.

- Long before the failure of the test beam, the glass began to crack in several places.

- The cracking occurred in regular time intervals as the load increased. The first cracks were registered at a load of approx. 4 kN. An average of nine successive cracking steps developed before failure occurred (fig. 6).

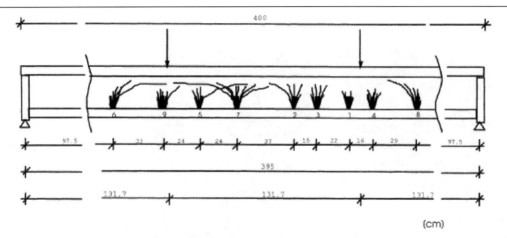

Figure 5: Groups of cracks in a tested I-beam

- The first cracks developed between the places where the forces were applied. Only later did cracks develop outside of this area.
- Towards the end of the loading tests the groups of cracks had an average distance of 25 cm from each other.
- The force-deformation curve shows a kink when the cracks occurred. This phenomenon can be attributed to the fact that with the development of cracks the total flexural rigidity of the laminate beam decreases (fig. 7).

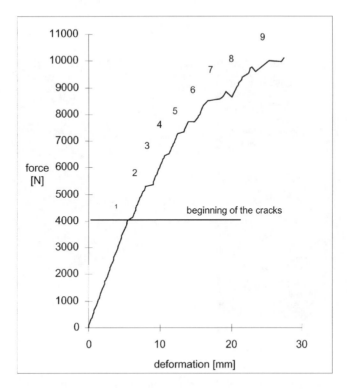

Figure 6: load-deformation curve of a timber-glass composite I-beam

Examples

Figure 7: *Facade and sliding door in timber, glass composite section Etoy, CH, 1997*

Figure 8: Handrail with load-bearing function: Ecole Polytechnique Fédéralede Lausanne, CH, 1999

Figure 9: *House, Shizuoka (Japon), 2000 Architect: Yoshiaki AMINO*

Figure 10: composite window (Switzerland)

Codification

- EOTA (eta)-Richtlinie: Guideline for Technical Approval for Structural Sealant Glazing Systems (SSGS), *1997*

References

- Natterer,J. ;Herzog, T.; Volz, M.; Construire en bois 2; Presses polytechniques et universitaires romandes, Lausanne (1991)

- Hamm, J.; Tragverhalten von Holz und Holzwerkstoffen im statischen Verbund mit Glas; Thèse No 2065 (1999); Ecole Polytechnique Fédérale de Lausanne (Switzerland)

- Kollmann, F.F.P; Principles of Wood Science and Technology; Volume I, Springer-Verlag, Berlin (1984)

- Hoeft, M.; Zur Berechnung von Verbundträgern mit beliebig gefügtem Querschnitt; Thèse No 1213 (1994); Ecole Polytechnique Fédérale de Lausanne (Switzerland)

- Kerkhof, F.; Grundlagen der Festigkeit und des Bruchverhaltens von keramischen Werkstoffen; Handbuch der Keramik; Verlag Schmid GmbH Freiburg im Breisgau (1982)

Contacts

- Dr. Jan Hamm

  Address: School of Architecture, Civil and Wood Engineering HSB, Solothurnstrasse 102, 2504 Biel, Switzerland
  jan.hamm@hsb.bfh.ch
  Tel. +41-(0)32/3440319
  Fax. +41-(0)32/3440391

COST C12 – WG1

Datasheet II.3.1.3.1
Prepared by: H. Trumpf[1] *and H. Gürtler*[2]

[1] Institute of Steel Construction, RWTH Aachen, Germany
[2] Cmposite Construction Laboratory, EPF Lausanne, Switzerland

FRP-STEEL COMPOSITE GIRDERS

Description

- Composite structures made of FRP and steel, Figure 1
- Lightweight FRP-deck elements in composite action with the steel girder
- Shear connection realised with headed shear studs, bonding and combined stud-bonding action
- Solutions for temporary bridges are feasible, Figure 2
- FRP-deck is lightweight, maintenance free and fatigue resistance
- Pre-fabrication of the deck elements in the plant

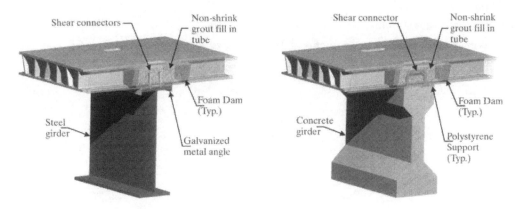

Figure 1. FRP-deck in composite action with the steel girder [1]

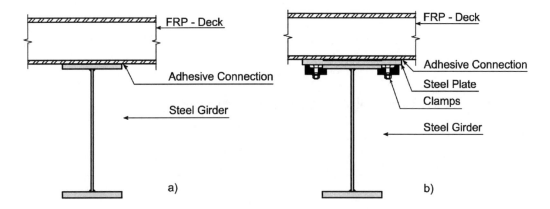

Figure 2. a) Adhesively bonded connection, b) solution for temporary connection [3]

Technical Information

- FRP-deck very light-weighted (γ = 17,95 kN/m³), orthotropic stiffness E_x and E_y can be determined in advance before pultruding the deck elements, resistance in pultrusion direction up to $f_{t,II}$ = 300 N/mm² but low shear resistance f_τ = 30 N/mm², corrosion-/decay- and ultraviolet resistance, inhibited fire properties, very low maintenance costs, very easy handling, economic life-cycle costs
- Materials:
  - i) Conventional steel girder S235 or S460
  - ii) FRP e.g. E23 acc. EN 13706
  - iii) Conventional shear studs St37-3 or alternative shear connectors
  - iv) Adhesive: thixotropic two-component impregnation resin on epoxy resin basis
- Determination of the stiffness and strength of fibre-reinforced laminates by the classical laminate theory (micro-mechanics and macro-mechanics) and advanced strength hypotheses.
- Design of the composite structure according to the Eurocodes
- Different geometries for the deck elements – Warren truss, trapezoidal profile and rectangular hollow section

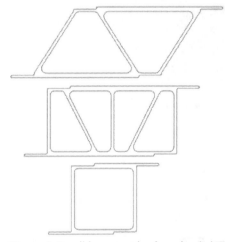

Figure 3. Possible geometries for pultruded FRP-decks

- Shear connection realised by shear studs, bonding and combinations

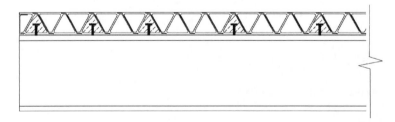

Figure 4. Headed shear studs as shear connectors in the openings filled with mortar

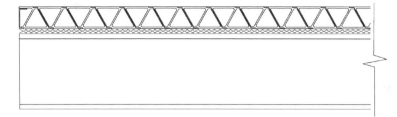

Figure 5. Thixotropic two-component impregnation resin on epoxy basis for the shear connection joint

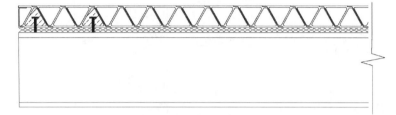

Figure 6. Combination with headed shear studs and bonded shear connection

- High grade of pre-fabrication by assembling the deck plate in the plant

Figure 7. Assembling of the DuraSpan FRP-deck element in the plant [2]

- Push-Out test to determine the static and dynamic load deformation behaviour, stiffness and strength according to Eurocode 4

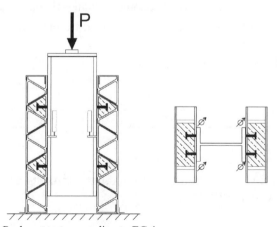

Figure 6. Test set-up for Push-out tests according to EC 4

- System tests to evaluate the in-plane compression and shear stiffness of FRP bridge decks as well as the maximum transmittable forces

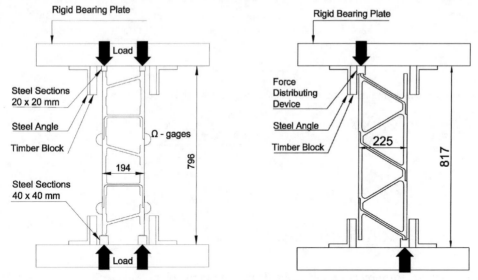

Figure 7. Set-up for the evaluation of the system properties of GFRP bridge decks [3]

Field of Application and End Users

- Industry and office buildings
    - i)   lightweight and flexible structures
    - ii)   chemical industry, purification plants
- Infrastructure
    - i)   bridges (cable-stayed, truss, arch, girder)

Structural Assessment

- Advantages: light-weight, high strength, corrosion resistant, easy handling, economic
- Disadvantages: design of the structure has to be adapted to material characteristics
- Potentials: optimizing structures by using the preferences of different materials to a hybrid design
- Risks: possible creeping effects have to be considered

Economical Aspects

- easy and fast erection
- less maintenance costs, no corrosion coating
- widening of bridges one or two lanes without re-building abutments and piers by replacing the concrete deck with a FRP deck

- System tests as shown in Figure 7 and full scale-tests on FRP-Steel-Composite Structures have already been performed at the École Polytechnique Féderal de Lausanne and will be published soon [3]. The setup and some results are shown in Figures 8 and 9. The tests were carried out for the DuraSpan (trapezoidal section, Fix 1, Fix 2) and the ASSET (triangular section, Fix 3, Fix 4) bridge deck

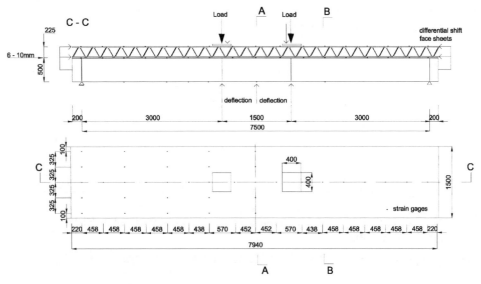

Figure 8. Set-up of four point bending tests [3]

- Main results tests with composite girers (see also Figure 9)
  - o Bonding a GFRP bridge deck on a steel girder enhances the deflection and failure behavior considerably
  - o The decrease of deflection of the girder depends strongly on the cross-section of the bridge deck
  - o Ten million fatigue cycles did not influence the girder behavior

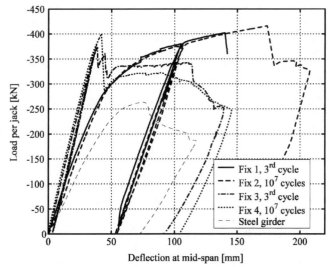

Figure 9. Load deflection behavior of two different composite girders and a reference steel girder for comparing reasons [3]

- The ASSET Project [4] performed by eight European partners have considered static and dynamic tests on FRP-decks supported by numerical investigations.

Examples

- West-Mill Bridge in Oxfordshire, [4]
- Bridge over Woodington Run, Ohio, USA [5]

Codification

- EN 13706; *Reinforced plastics composites - Specifications for pultruded profiles, Part 1,2 and 3; CEN /TC 249*

References

[1] Solomon, G. et al: „Expanded Use of Composite Deck Projects in USA", SEI, Volume 12, Number 2, May 2002

[2] Martin Marietta Composites, DuraSpan© Fiber-Reinforced Polymer Bridge Deck Systems

[3] H.W.Gürtler, PhD Candidate: „Composite Action of Adhesively Bonded Connections Between FRP Bridge Decks and Steel Main Girders", Swiss Federal Institute of Technology Lausanne, in preparation

[4] Canning, L. et al: „Advanced Composite Bridge Decking System – Project ASSET", paper, Structural Engineering International, February 2002

[5] Patrick Cassity et al: Final Technical Report; Advanced Composites for Bridge Infrastructure Renewal – Phase II DARPA Agreement No. MDA972-94-3-0030; Volume IV Task 16 – Modular Composite Bridge

Contacts

- Johannes Naumes and Heiko Trumpf
  Address: Institute of Steel Construction, RWTH Aachen, Mies-van-der Rohe Str. 1, D-52074 Aachen, Germany
  www.stb.rwth-aachen.de
  Tel. +49/ (0) 241 – 80-25177
  Fax. +49/ (0) 241 – 80-22140
  e-mail: stb@stb.rwth-aachen.de
- Herbert Gürtler
  Address: CCLAB, Swiss Federal Institute of Technology Lausanne, BP Ecublens, 1015 Lausanne, Switzerland
  www.cclab.ch
  Tel. +41/ (0) 21 – 693 6263
  Fax. +41/ (0) 21 – 693 6240
  e-mail: herbert.guertler@epfl.ch

STEEL AND FIBRE REINFORCED POLYMER

Description

- dismountable Fibre-Reinforced Polymer (FRP) and Steel Connections, Figure 1
- strengthening of the FRP for high internal forces by bonded (stainless) steel plates
- bonding by a solvent free, thixotropic two-component impregnation resin in epoxy resin basis
- load transfer by shear in the overlapping joint
- corrosion resistant and durable connection

Figure 1. Dismountable FRP-Steel Connection

Technical Information

- FRP-structures very light-weighted (γ = 17,95 kN/m³), resistance in pultrusion direction $f_{t,II}$ = 300 N/mm² but low shear resistance f_τ = 30 N/mm², corrosion-/decay- and ultraviolet resistance, inhibited fire properties, very low maintenance costs, very easy handling, economic life-cycle costs
- strengthening of the FRP is necessary, because traditional joining techniques like bolts don't meet special requirements of fibre-reinforced materials
- more sophisticated connections are in the development but need special fasteners/ connection elements and tools to meet the material characteristics
- Materials:
    - i) Steel plates: stainless steel 1.4571
    - ii) Bolts: Stainless steel, 8.8
    - iii) FRP: Quality E23 acc. EN 13706
    - iv) Adhesive: thixotropic two-component impregnation resin on epoxy resin basis

- FRP acc. E23 should have a ratio of longitudinal- to transverse fibres of appr. 4:1, glass-fibre reinforcement with a weight proportion of 50-60%, perfectly executed ratio of procedural- and functional additives to resin
- approach for the determination of the maximum shear transfer in an overlapping joints with hybrid materials by VOLKERSEN, which can be also used for FRP/FRP-, FRP/steel- and FRP/stainless steel-connections:

$$\frac{\tau(x)}{\tau_m} = \frac{\rho}{2}\left[\frac{\cosh\left(\frac{\rho}{l}\cdot x\right)}{\sinh\left(\frac{\rho}{2}\right)} - \frac{(1-\psi)\cdot \sinh\left(\frac{\rho}{l}\cdot x\right)}{(1+\psi)\cdot \cosh\left(\frac{\rho}{2}\right)}\right]$$

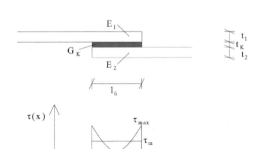

$$with \quad \rho = \sqrt{\frac{(1+\psi)\cdot G_k \cdot l_{\ddot{u}}^2}{E_1 \cdot t_1 \cdot t_k}} \quad and \quad \psi = \frac{E_1 \cdot t_1}{E_2 \cdot t_2}$$

$$peak \quad factor \quad at \quad x = -l/2:$$

$$\frac{\tau_{K\,max}}{\tau_{Km}} = \frac{\rho}{2}\left[\coth\frac{\rho}{2} + \frac{(1-\psi)}{(1+\psi)}\tanh\frac{\rho}{2}\right]$$

- highest resistance is reached if the stiffness ratio is
  $\psi = (E_1 * t_1)/(E_2 * t_2) = 1,0$
- further information for adhesive joints regarding modes of action, linkage forces, calculation models, test specimens are given in HABENICHT
- selection of the thixotropic two-component impregnation resin on epoxy resin basis is based on special requirements:
    - i) high thixotropic behaviour on vertical and overhead surfaces
    - ii) easy thin application (pasty, not flowing)
    - iii) good handling in process
    - iv) pot life of about 60-90 minutes for 15°C and 15-30 minutes for 40°C
- substrate is prepared by grinding followed by removing dust or loose particles by means of an industrial cleaner; substrate must be clean, free from grease and oil and dry; mixed adhesive is applied by a towel or brush; required thickness for a maximum ultimate resistance is about 0,2mm, applied resins must be protect from rain and sun for at least 12 hours
- tension strength of the adhesive joint has to be determined in tests according to DIN 53455/ EN ISO 527-1
- complex connections with combined bolt-adhesive action cannot be calculated by simple equations but have to be tested, Figure 2
- internal forces in the bolts are transferred by shear and bearing in the stainless steel plate and are transmit by means of shear in the adhesive joint into the FRP-profile, Figure 2 – maximum peak of the strain difference between both materials in the joint is the maximum shear strength of the adhesive
- FEM-methods show a good accordance to the load-deformation-behaviour if elements are well chosen, three dimensional effects are considered and plastic resources of the adhesive are given in the material law

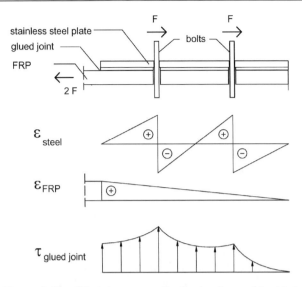

Figure 2. Simplified shear stress distribution for combined bolt-adhesive connections

Field of Application and End Users

- Residential apartments
    - i) façade and housing
    - ii) structural thermal insulation elements
- Industry buildings
    - i) cell cooling towers
    - ii) chemical industry
    - iii) purification plants
- Infrastructure
    - i) bridges (cable-stayed, truss, arch, girder)
    - ii) service platforms and roofs for railway
    - iii) metal-free structures

Structural Assessment

- Advantages: light-weight, high strength, corrosion resistant, easy handling, economic
- Disadvantages: design of the structure has to be adapted to material characteristics
- Potentials: optimizing structures by using the preferences of different materials to a hybrid design
- Risks: possible creeping effects have to be considered

Economical Aspects

- easy and fast erection
- less maintenance costs, no corrosion coating

- small- and full-scale tests have been performed with different stainless steel plate thicknesses of t = 2, 3 and 4 mm, Figure 3

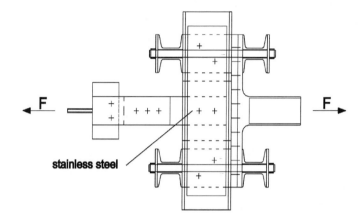

Figure 3. Full-scale test specimen truss joint

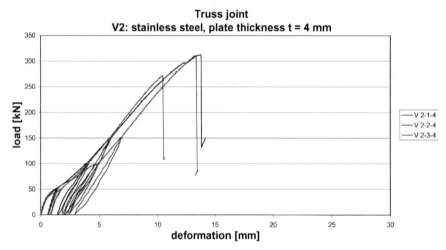

Figure 4. Load-deflection-curve truss joints

- several pre-load steps up to 50% of the calculated ultimate load, Figure 4
- connections show an elastic-plastic load-deformation curve, Figure 4
- first displacements always from the slip in the clamping devices
- stiffness and ultimate load is depending on the stainless steel plate thickness
- always failure by delamination of the FRP-profile but not in the adhesive joint
- very ductile load-deflection behaviour
- stiffness ratio for stainless steel plate thickness t = 2 mm:

$$\psi = (E_1 * t_1)/ (E_2 * t_2) = 2,28$$

Examples

- Mobile light-weight FRP-bridge for humanitarian out of area cases
    - i) Truss bridge with a dead weight of the structure 5,4 t
    - ii) load class MLC 12, exceptional load MLC 14 (test load MLC 40!)
    - iii) span l = 30000 mm and width b = 3436 mm
    - iv) dismountable connections, module length of 4000 mm
    - v) launching of the bridge by helicopter, crane or by a launching nose

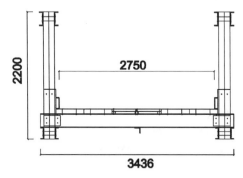

Figure 5. U-frame cross-section

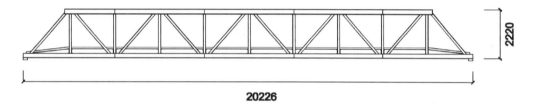

Figure 6. Elevation of the bridge

Figure 7 and Figure 8. Proving tests of the mobile light-weight bridge

Codification

- EN 13706; *Reinforced plastics composites - Specifications for pultruded profiles, Part 1,2 and 3; CEN /TC 249*

- DIN 53452, *EN ISO 527-1; Plastics - Determination of tensile properties -, Part 1: General principles; German version EN ISO 527-1: 1996*

References

- Habenicht G.; *Kleben - Grundlagen, Technologie, Anwendung; 3rd Auflage,* Springer Berlin Heidelberg New York 1997

- Volkersen; *Die Schubkraftverteilung in Leim-, Niet- und Bolzenverbindungen, Energie und Technik;* 1958

- Sedlacek, G., Trumpf, H., Geßler, A., Castrischer U.; *Abschlussbericht und Statische Berechnung: Pionier-Leichtbau E/K 41C/V 0085/V 5115, Konstruktive Entwicklung und statische Bemessung einer mobilen Leichtbau-Festbrücke aus pultrudierten glasfaserverstärkten Polymerprofilen mit Hybridverbindungen Lehrstuhl für Stahlbau;* RWTH Aachen 2002 – unpublished –

Contacts

- Heiko Trumpf

  Address: Institute of Steel Construction, RWTH Aachen, Mies-van-der Rohe Str. 1, D-52074 Aachen, Germany
  www.stb.rwth-aachen.de
  Tel. +49/ (0) 241 – 80-23595
  Fax. +49/ (0) 241 – 80-22140
  e-mail: tru@stb.rwth-aachen.de

Prepared by: D. Serdjuks, K. Rocens

Institution: Institute of Structural Engineering and Reconstruction
of Riga Technical University, Riga/Latvia, www.bf.rtu.lv

EVALUATION OF BEHAVIOUR OF HYBRID COMPOSITE CABLE IN SADDLE-SHAPED ROOF

Description

- New type of hybrid composite cable on the base of steel, glass fiber reinforced plastic (GFRP), and carbon fiber reinforced plastic (CFRP).

- Hybrid composite cable with increased, in comparison with carbon fiber composite cable (CFCC), breaking elongation and decreased, in comparison with the steel cables, dead weight.

- Mainly for prestressed saddle-shaped cable roofs.

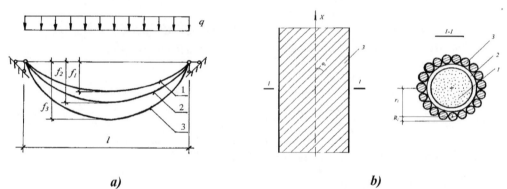

a)   *b)*

Figure 1. *a)* Scheme of hybrid composite cable work:1 - steel wire, GFRP and CFRP work commonly; 2 - GFRP is excluded from the work; 3 - GFRP with CFRP are excluded from the work and steel wire works alone; q - design vertical load, acting at the cable; f_1 - deflection of the cable, which corresponds to the stage, when steel wire, GFRP and CFRP work commonly; f_2 - deflection, which corresponds to the stage, when GFRP is excluded from the work and steel wire works commonly with CFRP; f_3 -deflection, which is corresponds to the stage, when GFRP and CFRP are excluded from the work and steel wire works alone; l - span of the cable.
b) Scheme of hybrid composite cable made from the CFRP, GFRP and steel: 1 – CFRP core, 2 – GFRP distributional layer, 3 – steel wire strands, α_i – angle of steel wire strands twisting, X – longitudinal axis of the cable, r_i – distance between the center of separate steel wire strand and whole cable, R_i – radius of separate steel wire strand.

Technical Information

- Hybrid composite cable contains three layers: carbon fiber reinforced plastic (CFRP) core, glass fiber reinforced plastic (GFRP) distributional layer and steel wire strands.

- All layers of hybrid composite cable take up tension stresses, acting in the cable during exploitation. But GFRP has second function – distribution of transversal pressure of steel wire strands at CFRP core.

- Volume fractions of steel and carbon were evaluated basing on the assumption, that in an emergency, when the strain of carbon fiber exceeds the ultimate value and this fibers are disrupted, strands of steel wire must be able to take up significantly decreased tension stresses. the decrease of the tension forces, acting in the cable, is joined with the growing of defletion after excluding from the work of GFRP and CFRP components.
- Design resistance of steel component depends on the steel grade and should be found as ultimate resistance, divided by 1,6 (in accordance with SNIP II-23-81*).
- Angle of steel wire strands twisting should not exceeds 20 degrees.
- GFRP (E-glass and epoxy matrix at 60% fiber content) and CFRP (AS4/3501-6 graphite fibers and epoxy matrix at a 60% fiber content) can be considered as materials of a hybrid composite cable.

Fields of Application and End Users

- Hybrid composite cable on the base of steel, GFRP and CFRP can be used for tension and suspension cables, which are the heaviest loaded units of saddle-shaped cable roofs.
- Hybrid composite cable on the base of steel, GFRP and CFRP can be used for hierarchic cable structures, where saddle-shaped cable roof is typical element.

Structural Assessment

- Advantages:
  increased safety, decreased (in comparison with the steel cables) dead weight, decreased maximum vertical displacements of structures.

  Disadvantages:
  small breaking elongation of CFRP core.

Economical Aspects

- Increased (in comparison with the CFCC cables) cost.

Robustness and Testing

- Tension tests of steel and CFCC cables at the Chalmers University, Göteborg.

Examples

- **Tension cable of saddle-shaped cable roof** with dimension in plan 30x30 m (Conceptual design).
  The cable has rational from the point of view of materials consumption initial
  deflection $f_1 = 5,7$ m. The uniformly distributed load with intensity q = 21 kN/m loads

the cable. Volume fractions of steel wire, GFRP and CFCC are 0,4 ; 0,2 and 0,4 respectively. Total area of the cable cross-section is 0,00097m².

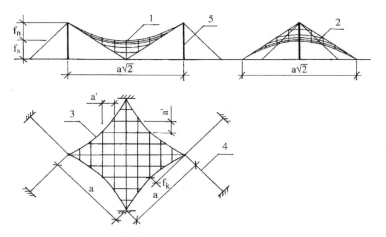

Figure 2. Scheme of saddle-shaped cable roof:
1-suspension cables; 2-stressing cables; 3-tension cables; 4-anchors; 5-columns; a-dimension of structure in plan; a'-step in plan of suspension and stressing cables; f_k-initial deflection of tension cables; f_n, f_s–initial deflections of suspension and stressing cables respectively.

- **<u>Supporting contour of saddle-shaped cable roof</u>** with dimensions in plan 50x50 m. Supporting contour consists on the two hybrid composite cables, which are joined together by the tie-bars. Initial deflection of top chord of compliant supporting contour was equal to 3,13 m. Distances between the support points of tie-bars were equal to 7 m. (Conceptual design).

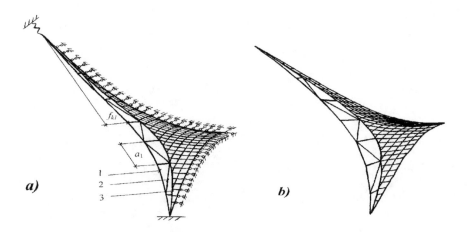

Figure 3. *a)* Quarter of cable roof, supported by compliant supporting contour:
1 – top cable of supporting contour, 2 – tie-bar, 3 – bottom cable of supporting contour,
f_{kl} – initial deflection of top chord of compliant supporting contour, a_l – distance between the support points of tie-bars. *b)* Shape of prestressed cable net after vertical design load application.

Codification

- SNIP II-23-81*. *Steel structures*
- SNIP 2.01.07-85 *Loads and actions*

References

- Bengtson, A., 1994, *Fatique Tests with Carbon-Fiber-Reinforced Composite Cable as Nonmetallic Reinforcement in Concrete*, Göteborg, p. 1-14.
- Peters, S. T., 1998, *Handbook of composites*, London .
- Otto, F., Shleifer, K., 1970, *Tent and Cable Building Structures*, Stroyizdat, Moscow.
- Kumar, K., CochranIr.I.E., 1997, *Closed Form Analysis for Elastic Deformations of Multylayred Strands,* Journal of Applied Mechanics, ASME Vol.54.

Contacts

- Dr. Dmitrijs Serdjuks

  Address: Institute of Structural Engineering and Reconstruction, Riga Technical University, Kalku street-1, Riga, Latvia, LV-1048.

  e-mail: dmitrijs@bf.rtu.lv

- Prof. Karlis Rocens

  Address: Institute of Structural Engineering and Reconstruction, Riga Technical University, Kalku street-1, Riga, Latvia, LV-1048.

  e-mail: rocensk@latnet.lv

COST C12 – WG1

Datasheet II.3.1.4.1

Prepared by: Zbigniew PLEWAKO[1]

[1] Rzeszów University of Technology, Poland

STIFFNESS OF BEAMS PRESTRESSED WITH FRP TENDONS

Description

- Key words: Reinforced and prestressed concrete structures, non-ferrous reinforcement, structural behaviour

- Materials used for prestressing must have very high tensile strength, which allows producing appropriate prestressing force, and ultimate strains able to resist elongation resulting from bending caused by additional loads. High durability (corrosion resistance) is also required. Most commonly used is high strength steel, which strength reach up to 1900 MPa, and ultimate (plastic) strain $\geq 3.5\%$. Interesting alternate materials are fiberbased composites made of aramid, carbon and glass fibers impregnated in a synthetic resin.

Technical Information

- Keywords: Fibre Reinforced Polymers, strength, elongation, modulus of elasticity,

- Fibre Reinforced Polymers materials (FRP) consist of uniaxially oriented parallel thin and strong fibres of different chemical origin embedded in a matrix. Respectively, there are:
    - Aramide FRP (AFRP)
    - Carbon FRP (CFRP)
    - Glass FRP (GFRP)

- FRP reinforcing bars, tendons and strips, which recently have been used in particular numbers of concrete structures, show significant differences in mechanical properties comparing to steel elements (**Table 1**). Opposite to steel, FRP products have anisotropic mechanical properties, and different resistance to environmental media. Basic mechanical differences are modulus of elasticy constant in full range of strains and various values of ultimate strength (**Figure 1**).

Table 1. Basic properties of reinforcement products

Property	Fibre			Prestressing Steel
	Glass	Aramide	Carbon	
Tensile strength	+	+	++	+
Long-term strength	O	O	+	+
Fatigue strength	-	++	+	+
Multiaxial strength	-	-	O	+
Durability in alkaline environment *(like concrete)*	-	O	+	++
Durability in acidic environment	+	+	++	-
Durability in carbonated concrete	++	++	++	-
Weight	+	++	++	-

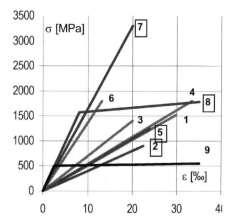

Legend

	Producer Name/Country		Type	E [GPa]	f [MPa]
1	POLYSTAL	GER	GFRP	51	1520
2	KHS KROSNO	POL	GFRP	40	900
3	FIBRA	JAP	AFRP	70	1400
4	TECHNORA	JAP	AFRP	54	1800
5	ARAPREE	NED	AFRP	52	1250
6	CFCC	USA	CFRP	137	1800
7	BBR CFP	SUI	CFRP	165	3300
8	Prestr. Steel			190	1900
9	Reinf. Steel			200	500

Figure 1.. Range of steel and FRP Reinforcement elasic behaviour

Field of Application and End Users

- residential housing

  Structures and elements reinforced of prestressed with FRP tendons can be applied especially in objects subjected to electromagnetic fields coming from
  - i) transformer or switching station rooms
  - ii) aerial masts and towers located on roofs
  - iii) radio/TV transmitter rooms

- infrastructure

  These elements can be applied especially in industrial objects subjected to particular environmental factors
  - i) Chemical media attack
  - ii) Electromagnetic fields (as described above)
  - iii) Off-shore structures subjected to sea water and tides

- bridges
  - i) Structures subjected to de-icing salts (chloride attack)
  - ii) Sea-shore routes

Structural Assessment

- Stiffness of beams is correlated with its bearing capacity, and implicates deflections under service loads. Deflection increase as visible signal of reaching the bearing capacity is a very important safety alert. The problem is, how different of steel mechanical properties of FRP tendons, especially various modulus of elasticity, constant if full range of stresses, influence on stiffness of prestressing steel. To solve this problem, some theoretical considerations and experimental tests were carried on.

- The theoretical analysis of concrete cross-section material models (**Figure 2**) and calculating procedures according to EUROCODE 2 was carried out. This analysis was performed for unit geometrical dimensions and acting forces. Its results allowed determining universal relationships between strain and stress distribution in analysed cross-sections. Using received results, numerical method of dimensionless strain and displacement analysis of prestressed cross-sections and beams was worked out. Engaging this method, influence of tendon types, their bearing capacity and pre-tensioning on properties of cross-section and beams prestressed with and without bond, was obtained.

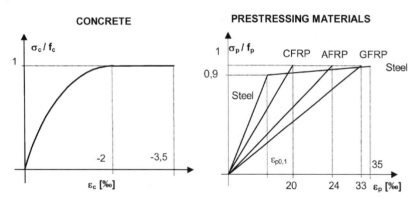

Figure 2. Stress-strain curves of materials applied for analysis

- **Figure 3** below presents deflection vs load behaviour of theoretical models of four beams. All assumed models have equal geometrical dimensions and concrete strength, are prestressed with tendons of different origin and cross-section area, but developed the same prestressing force (pre-tensioned to equal force) and the same bearing capacity (breaking load). The only actual difference in beam models lies in various modulus of elasticy of tendons as an effect of different ultimate strains. Analysis of plots leads to following general conclusions:
  - beams subjected to loads developing bending moment M which value is lower than cracking moment M_{cr} of concrete cross-section, have the same stiffness independently of applied prestressing tendon type.
  - for cracked beams, $M > M_{cr}$, stiffness of beams strongly depends in proportion on tendon stiffness. For resulting strain increase of tendons lower than yielding strain of steel, stiffness of beams prestressed with FRP is smaller than prestressed with steel tendons. Oppositely, when strain increase involve yield strains in steel, stiffness of beams prestressed with FRP is smaller than prestressed with steel tendons ones. Consequently, bearing capacity of beam prestressed with FRP tendons could be equal as for beam prestress with steel tendons.

Figure 3. Deflection vs bending moment M curves for theoretical beam models

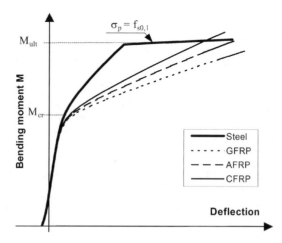

- Lower and constants in full range of stress modulus of elasticy of FRP tendons and described above course of beam bending under load caused large deformations

accompanying to failure. In beam with relatively smaller FRP prestressing reinforcement, failure could be caused by sudden tendon break, differently from beam prestressed with equivalent steel. This situation is determined by ratio of prestressing ω_{p0} = (prestressing force P_0)/(breaking load F_p), which is approximately irrelevant to bearing capacity of beams with steel tendons (**Figure 4**).

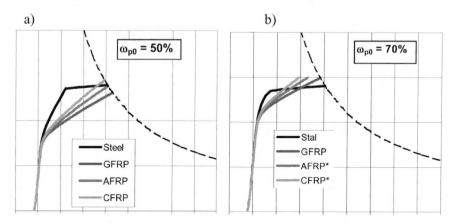

Figure 4. Course of beam bending example for theoretical models with various ratio of prestressing ω_{p0} = 50% and ω_{p0} = 70%
(symbols * following tendon type denotes tendon break failure mode)

Economical Aspects

- Up till now there no sufficient data to obtain real economical profits due to applying FRP reinforced/prestressed structures. It is due to low scale of current applications which have rather experimental meaning.

- Production costs of FRP products designated for reinforcing of engineering structures are high relating to steel.It is determined by low scale of commercial production

- Commercial wide range of application which enforce decrease of costs require more experimental tests. Special problem for prestressed structures is developing reliable anchoring devices.

Robustness and Testing

- To verify results obtained in theoretical analysis, experimental tests were provided.

Table 2. Basic data of prestressing tendons in tested beams

Beam symbol	Tendon type	Area A_p [mm^2]	Breaking Force $F_{p,min}$ [kN]	Modulus of Elasticy E_p [GPa]	Prestr. Force P_0 [kN]
S1	Steel strand 1∅12.5 mm (∅4.5+6∅4.0)	91.3	162.7	195.0	32.8
GFRP	GFRP 2∅8 mm	2x50.0	48.0	39.0	28.4
AFRP	AFRP 2∅8.5 mm	2x56.7	81.6	47.5	22.6

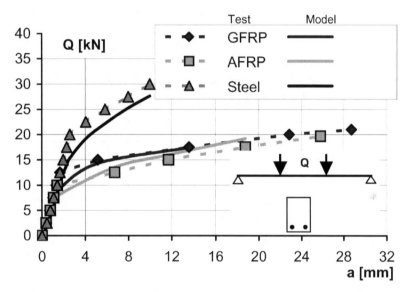

Figure 5. Tested beams apperance

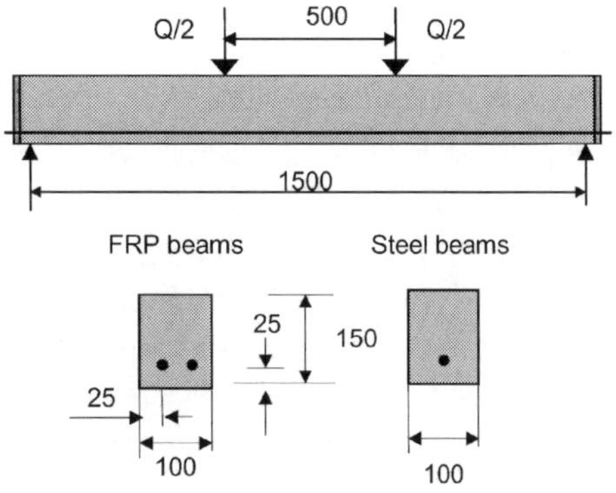

- Stress-strain curves show good conformity between theoretical and experimental results.

Figure 6. Deflection vs load curves for beams prestressed with different type of tendons. Test and model results

- Verification of theoretical analysis by test results proved that assumption of material models and calculation procedures according to EUROCODE 2 ensure good conformity with experiment in full range of loading. This points, that EUROCODE 2 could be applied for analysis and designing of beams prestressed with FRP tendons.

- Assuming safety of structure as most important requirement, bearing capacity and pre-tensioning of applied tendons should assure failure mode by concrete compression (over-reinforced beams). In this mode large deflection, as a signal of beam overloading occurs. Fulfilling this postulate require application of FRP tendons with bearing capacity F_p higher than of steel once, and with significantly lower pre-tensioning P_0 (for FRP $P_0 = 40\% \div 50\% F_p$, for steel $P_0 = 70\% F_p$).

- Direct relation between tensile strength and ultimate strains for FRP tendons results from constant value of modulus of elasticity. This relation requires exact calculation of structure deformation and corresponding tendon elongation in ultimate limit state of beam subjected to bending.

References

- *FIP Recommendation: Practical design of structural concrete*, Fédération Internationale de la Précontrainte SETO, London 1999

- *prEN 1992-1-1 EUROCODE 2: Design of Concrete Structures – part 1: General Rules and Rules for Building,* CEN Brussels, March 2003.

- *Proceedings of "Non-Metallic (FRP) Reinforcement for Concrete Structures":*
  - 1st Int. Symposium (FRPRCS-1), Vancouver, Mar. 1993, editor Antonio Nanni and Charles W. Dolan; American Concrete Institute SP-, Detroit, MI, 1993.
  - 2nd Int. RILEM Symposium (FRPRCS-2), Ghent, Aug. 1995, editor L. Taerwe, E&FN SPON, London, Glasgow, Weinheim, New York, Tokyo, Melbourne, Madras 1995.
  - 3th Int. Symposium (FRPRCS-3), Sapporo Japan, Oct. 1997, editor Japan Concrete Institute, Tokyo, 1997.
  - 4th Int. Symposium (FRPRCS-4),), Baltimore, MD, Oct. 31 - Nov.5 1999, American Concrete Institute Special Publication (ACI SP-188) 1999
  - 5th Int. Symposium (FRPRCS-5),), Cambridge, UK, Jul. 2001, Edited by Chris Burgoyne, Cambridge Univ. Press 2001
  - 6th Int. Symposium (FRPRCS-6),), Singapore Jul. 2003, Edited by K.H. Tan, National University of Singapore, 2003

Contacts

- Zbigniew PLEWAKO

  Rzeszów University of Technology,
  W. Pola 2
  Rzeszów,Poland
  plewakoz@prz.edu.pl
  Tel. +48178651626
  Fax. +48178542974

CONCRETE-TIMBER CONSTRUCTION

Description

- Wooden tubes reinforced by textile reinforced concrete

The reinforcement of wooden structures by technical textiles provides a universal technology for the solution of problems related to anisotropy and preservation of wood. A combination of load- and shape-adapted reinforcement of joints is achieved using stitch bonded structures.

fibre orientation	material	density	textile weight per area
[°]		[tex]	[kg/m^2]
0	VETROTEX-AR-Glass	640	180,89
90	VETROTEX-AR-Glass	640	

Table 1. Used non-woven textile structure

Figure 1. Wooden tube reinforced by textile reinforced concrete

Technical Information

- wooden tube
    - i) cement based matrix
    - ii) textile reinforced by non-woven fabric (4 layers)

Figure 2. Wooden tube reinforced by textile reinforced concrete

Field of Application and End Users

- Potential field of application are multi storey buildings with enhanced fire behaviour

Structural Assessment

Economical Aspects

- use of cheap cement instead of synthetical resins

Robustness and Testing

Codification

References

- Haller, P.: *Formholzprofile und textilbewehrter Beton – Ein neuer Verbundquerschnitt.* IN: BETON- UND STAHLBETONBAU. VERLAG ERNST & SOHN, BERLIN, 2004 (IN PRESS)

Contacts

- Name: Peer Haller
  Address: TU Dresden
  Institute of Steel- and Timber Structures,
  Eisenstuckstraße 33
  01069 Dresden
  Germany
  Peer.Haller@mailbox.tu-dresden.de
  Tel. + 49 – 351 / 46 33 55 75
  Fax. + 49 – 351 / 46 33 63 08

COST C12 – WG1

Datasheet II.3.1.5.2
Prepared by: Szczepan WOLINSKI
Rzeszow University of Technology, Poland

.

PROBABILISTIC DESIGN AND EVALUATION OF COMPOSITE – TIMBER - CONCRETE FLOOR

Description

- Keywords: Mixed building structures, reliability of elements and systems, probabilistic design and evaluation
- All stages of the structural design and evaluation process of mixed building elements and structures involve various uncertainties and because of these uncertainties loads and resistance (load-carrying capacities) are random variables.
- Structural systems of buildings consist usually of many interconnected elements and components often made of different materials using different connections and technologies.
- Present day design practice is focused on dimensioning and/or evaluation of individual sections and sometimes simple structural elements, however the reliability of an individual element may or may not be representative for the entire structural system (prEN 1990, 2002, ISO 2394, 1998).
- Mixed building or composite elements and structures are usually complex hybrid systems (combined systems) with brittle and ductile, usually partially intercorrelated members.
- A structure is a system of elements described by finite number of failure functions; each function can described the state of one structural element or connection, and several failure functions can describe the state of an entire structure.
- The probabilistic design of the structures and structural elements consists in such a choice of the structural system, dimensioning procedure and detailing, that during the required design working life the probability of failure will not exceed an assumed target value P_{fd}.

Technical Information

- Keywords: Load carying capacity, statistical parameters, Monte Carlo method, mechanisms of structural failure
- Because a present generation of design codes is based on probabilistic models of loads and load-carrying capacities, concepts investigated in this paper can be applied to the design of new mixed building elements and structures as well as to the evaluation of existing ones.
- The probabilistic measures of structural reliability and safety can be considered as a rational basis for design of a more reliable structures for a given costs, for design of a more economical structures for a given reliability and for decisions about repair, rehabilitation and replacement of existing structures.
- Mechanisms of structural failure or inadequate serviceability of mixed building elements and structures are usually correlated, their failure functions are highly nonlinear or/and not given in explicit form, so in most practical cases only Monte Carlo method makes possible to obtain approximate solutions of the probabilistic design and evaluation.

Field of Application and End Users

- e.g. residental housing
  Composite (mixed building) wood – concrete floors can be applied in reconstructed historical and residental housing as well as new auxiliry service and agricultural buildings.
- e.g. infrastructure
  These structures can be also applied in small industrial objects.
- e.g. bridges
  Structures of small (short span) composite wood and reinforced concrete bridges.

Structural Assessment

- The load-carrying capacity of a mixed building element or structure depends on its components, connections and interface between materials, and is typically a function of material properties, section, element and structure geometry and dimensions.
- The levels of reliability releted to structural resistance and serviceability can be achived by combinations of measures relating to: design calculations, quality management, accuracy of the mechanical models used, detailing, adequate inspection, etc. (prEN 1990, 2002).
- The probabilistic methods of design and evaluation calculations for mixed building elements and structures are used in this work to achive required levels of their reliability.
- Three levels of probabilistic methods can be distinguished:
  - semi-probabilistic (Level I; Partial Factors Method),
  - simplified probabilistic (Level II; First Order Reliability Method FORM and Second Order Reliability Method SORM),
  - full probabilistic (Level III).
- In both the simplified probabilistic and full probabilistic methods the measure of reliability is identified with the survival probability $P_s = 1 - P_{fd}$, where P_{fd} is the target failure probability which depends on various design situations, failure modes and reference periods.
- Mixed building structures can be considered as a combination of series and parallel systems (combined systems) with uncorrelated, perfectly correlated or partly correlated, brittle and/or ductile elements.
- The failure or decommissioning probability of combined systems with intercorrelated elements cannot be calculated analytically and usually is assesse using the lower and upper bounds on the probability of failure (Thoft-Christensen and Baker, 1982, Ditlevsen and Bjerager, 1986, Nowak and Collins, 2000).

Economical Aspects

- There are no sfficient data to obtain the real economical profits due to applying composite wood – concrete floors.
- Application of wood – concrete floors in reconstructed historical buildings where the original wood beams can be reused is economically profitable.

- The statistical parameters for material properties and dimensions are quantified by tests and observations and are available for the basic structural materials: concrete, steel, timber, aluminum and different types of reinforcements, but the variability of the resistance components, connections and interfaces are available only to some extent.
- The probability distribution types of random material properties are usually assumed as: normal, lognormal and Weibull, probability distributions of loads: normal for dead loads, gamma for long term live loads (sustained), exponential for short time live loads, Weibull or Gumbel for wind speed, etc., and probability distributions of model factors are usually assumed as normal (e.g.: JCSS Model Code, 2001, Nowak and Collins, 2000).
- Due to the limited experimental results identification of the probability distribution and random variation of mixed (composite) elements and structures resistance is difficult and only few test results are available (Ellingwood et.al. 1980, Nowak and Collins, 2000, Murzewski, 1989).

Example

- The joist floor structure consists of eight simply supported solid wood beams ($n = 8$) joined with the continuous one way, nine-span reinforced concrete slab ($m = 9$) is exposed to dead, long lasting and short lasting live loads: $q = g + p_l + p_s$ (Figure 1).
- The problem consists in the determination of the required dimensions of wood beams and an area of tension reinforcing steel at the critical cross-sections of the RC slab in order to obtain the assumed reliability $P_s = 0.999$, or the probability of failure of the entire floor structure $P_{fd} = 0.001$.

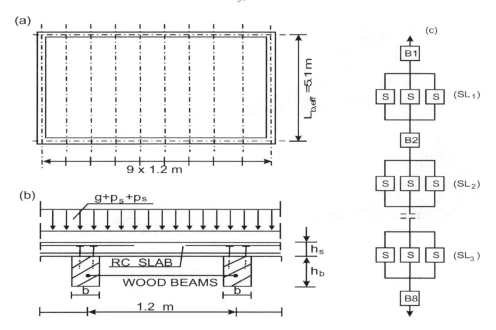

Figure 1. Composite wood-concrete floor: (a) layout, (b) cross-section, (c) system reliability analysis model

- The probability distributions and statistical parameters of random variables considered in analysis: dead load $g \Rightarrow N$ (2.7 kN/m^2, 0.063), long term live load $p_l \Rightarrow \Gamma$ (0.6 kN/ m^2, 0.25), short term live load $p_s \Rightarrow E$ (0.4 kN/m^2, 0.50), compressive strength of concrete (element 1; RC slab) $f_c \Rightarrow LN$ (28 N / mm^2, 0.15), yielding stress of reinforcing steel $f_y \Rightarrow LN$ (280 N /mm^2, 0.057), tensile strength of wood (element 2; timber beam) $f_w \Rightarrow LN$ (20 N / mm^2, 0.25), where: N – normal, Γ - gamma, E – exponential, LN – lognormal distributions.
- Assumed distributions of random variables correspond to the following design values of these variables, considered in the semi-probabilistic Partial Factors Method: g_d=3.3 kN/ m^2, p_{ld}=1.2 kN /m^2, p_{sd}=1.7 kN /m^2, f_{cd}=13.3 N /mm^2, f_{yd}=210 N /mm^2, f_{wd}=10.9 N /mm^2. The geometrical dimensions that were taken into account in the analysis are considered to be deterministic: effective spans of the beam and slab $L_{b,eff}$=5.10 m, $L_{s,eff}$=1.20 m, the width of the beam b =0.15 m, and the depth of the slab h_s =0.06 m.
- The depth h_b of solid wood beam and the amount of tension reinforcing steel A_s in concrete slab were calculated according to principals given in Eurocodes 2 , 5 (prEN 1992-1, 2001, prEN 1995-1-1, 1993) were calculated using formulas:

$$h_b = \sqrt{\frac{6(g + p_l + p_s)L_{b,eff}^2}{8bf_w}} \quad , \quad A_s = \frac{17bd\xi(0.85f_c)}{21f_y}$$

$$\xi = \frac{98}{66d}\left(\frac{17}{21} - \sqrt{\left(-\frac{17}{21}\right)^2 - 4\left(\frac{33}{98}\right)\frac{(g + p_l + p_s)L_{s,eff}^2}{11bd^2(0.85f_c)}}\right)$$

- Using the Monte Carlo simulation method and the M-Star computer program (Marek and Gustar, 1999) histograms of random variables h_b and A_s were calculated (Figure 2).

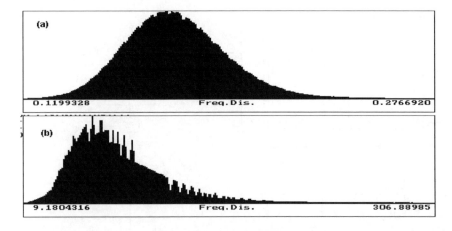

Figure 2. Histograms of random variables: (a) depth of wood beam,(b) amount of tension reinforcing steel.

- The entire floor structure can be considered as a combination of series system consists of eight identical elements B_i (wood beams) and of eight identical elements SL_i which next consist of three identical elements S (cross-sections of RC slab) connected in parallel (Figure 1c).
- To obtain the assumed reliability of the entire system $P_s = 1 - P_{fd} = 0.999$, the depth of each wood beam has been calculated for the reliability $P_s(B) = 0.9999$ and $P_{fd}(B) = 10^{-4}$ is $h_b = 260\ mm$, and the amount of tension reinforcing steel has been calculated for the reliability $P(S) = 0.97$ and $P_f d = 0.03$ is $A_s = 155\ mm^2$. Reliability of the system is:

$$P_s = 1 - P_{fd} = 1 - \{1 - (0.9999)^8 \times [1 - (0.03)^3]^8\} \approx 0.999 .$$

- Assuming that the load-bearing capacity of cross-sections of RC slab are equally correlated and $\rho_S = 0.3$ (correlation coefficient), and that the load-bearing capacity of each wood beam and the three neighboring cross-sections of slabs are also equally correlated and $\rho_{B,SL} = 0.8$, the probability of failure of the entire floor structure can be calculated using the Stuart's formula. The failure probability of the entire floor structural system with equally correlated load-bearing capacity of all wood beams and RC slab is $P_f = P_{fd} = 0.001$. This result was achieved by a process of trial and error for equal values of $P_f(SL) = P_f(B) = 5.8 \times 10^{-5}$, and the corresponding depth of wood beams $h_b = 260\ mm$, and the amount of tension reinforcing steel in slab $A_s = 301\ mm^2/m$.
- The reason why more conservative results were obtained for the system with correlated elements is that the reliability of parallel subsystems SL_i, composed of three correlated elements, is considerably smaller than the reliability of the same subsystems with uncorrelated elements

Codification

- prEN 1990 : 2002 (E). *Eurocode – Basis of structural design*, CEN, Brussels, Belgium, 2003.
- ISO 2394 : 1998 (E), International Standard. *General principles on reliability for structures*, ISO, Geneve, Switzerland, 1998.
- Joint Committee on Structural Safety, 2001. *Probabilistic Model Code*. JCSS Internet Site: http://www.iabse.ethz.ch/lc/jcss.html, 2002.
- prEN 1993-1-1, 2003. *Design of steel structures. General rules and rules for buildings*. CEN, Brussels, Belgium, 2003.
- prEN 1995-1-1, 1996. *Design of timber structures. General rules and rules for buildings*. CEN, Brussels, Belgium, 1996.

References

- Nowak A.S, Collins K.R., 2000. *Reliability of Structures, McGraw – Hill, Int. Edition, 2000.*
- Ellingwood B., Galambos T., MacGregor J.G, Cornell C.A., 1980. *Development of a Probability Based Load Criterion for American National Standard A 58. NBS Special Publication 577, Washington DC, USA, 1980.*
- Murzewski J., 1989. *Reliability of Engineering Structures. (in Polish), Arkady, Warszawa, 1989.*
- Thoft – Christensen P., Baker M.J., 1982. *Structural Reliability Theory and Its Applications. Berlin, Springer-Verlag, Germany, 1982.*
- Ditlevsen O., Bjerager P., 1986. *Methods of structural system reliability. Structural Safety, No. 3, pp. 195 – 229, 1986.*
- Marek P., Gustar M., 1999, *Monte Carlo simulation programs for PC. AntHill, M-Star, Res-Com, DamAc, LoadCom, ARTech Prague, Czech Republic, 1999.*

Contacts

- Assoc. Prof. Szczepan Wolinski

  Address: Faculty of Civil and Environmental Engineering,
  Rzeszow University of Technology,
  Ul. Poznanska 2, 35-040 Rzeszow, Poland
  e-mail: szwolkkb@prz.rzeszow.pl

TIMBER CONCRETE COMPOSITE STRUCTURES

Description

- Composite of timber and concrete
- Timber in the tension zone; concrete in the pressure zone
- Cracked concrete zones are replaced by the timber
- Connection between timber and concrete done by variations of typical connection devices, e.g.

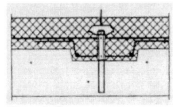

Figure 1: SFS-Screws

Figure 2: Flat steel lock

→ Screws (comp, Figure 1)

→ nails

→ nail-plates

→ groove (comp. Figure 3)

→ flat steel locks (comp. Figure 2)

→ headed studs

Figure 3: Groove

Figure 4: Tecnaria fastener

Figure 5: glued-in flat steel

→ glued-in steel connection devices such as in Figure 5

→ connection of timber and concrete by glueing them

- Except of glueing timber and concrete, all connections are flexible.

- → The internal forces with respect to the deformation of the connection and the different materials can be determined by the design method of Eurocode 5 Annex B.
- Creep and shrinkage can not be neglected
    - → Increase of the deflection due to shrinkage: ~70% of the elastic deformation
    - → Increase of the deformation due to creep: ~150 % – 450% of the elastic deflection
- Different rheological behaviour
    - → Within the first 3 years, concrete creeping is stronger than timber creeping
        - ▶ The internal forces of the concrete are reduced. Therefore stresses in the timber cross-section are increased.
    - → Within the year 3 to 7 after erection, both materials are creeping in comparable ways.
        - ▶ No stress redistribution within the cross-section
    - → After 7 years concrete has finished most of its creep deformation.
        - ▶ Timber stresses are reduced due to creeping
    - → As a consequence at the point in time of 3 to 7 years after erection the stresses of the timber can reach a maximum value. Therefore this point in time has to be considered in the structural design.
    - → Effective creep coefficients have to be determined due to the different temporal development of creep and shrinkage and due to the composite interaction.
        - ▶ according [Schaenzlin 2003] or
        - ▶ according

$$\phi_{Composite} = \psi \cdot \phi_{Material} \tag{1}$$

where $\phi_{Composite}$ composite creep coefficient

       ψ factor describing the influence of the composite interaction and the influence of the different temporal developement

       $\phi_{Material}$ material creep coefficient according to the standards

Table 1: Factor desribing the influence of composite interaction and the influence of the different temporal development

Time	ψ_{Timber}	$\psi_{Concrete}$
0	0	0
3-7 years	0,5	1,9
∞	1	2

- ▶ Shrinkage can be taken into account by fictive, external loading; see following equations

$$p_{sID,d} = C_{p,sID} \cdot \Delta\varepsilon_{sID,d}$$

and

$$C_{\mathsf{p,sID}} = k_{s,res} \cdot \frac{\pi^2 \cdot E_2 \cdot A_2 \cdot E_1 \cdot A_1 \cdot (h_1 + h_2) \cdot \gamma_1}{2 \cdot l^2 \cdot (E_1 \cdot A_1 + E_2 \cdot A_2)}$$

$$\Delta\varepsilon_{\mathsf{sID},d} = \varepsilon_{H,d,\infty} - \varepsilon_{B,d,\infty}$$

Definitions according Eurocode 5 Annex B

▸ The influence of shrinkage on the effective bending stiffness has to be taken into account by

$$(E \cdot J)_{\mathsf{eff,sID}} = C_{\mathsf{J,sID}} \cdot (E \cdot J)_{\mathsf{eff}}$$

and

$$C_{\mathsf{J,sID}} = \frac{C_{\mathsf{p,sID}} \cdot \Delta\varepsilon_{\mathsf{sID},d} + q_d}{\frac{E_1 \cdot A_1 + E_2 \cdot A_2}{E_1 \cdot \gamma_1 \cdot A_1 + E_2 \cdot A_2} \cdot C_{\mathsf{p,sID}} \cdot \Delta\varepsilon_{\mathsf{sID},d} + q_d}$$

Definitions according Eurocode 5 Annex B

Technical Information

- General technical information
    - → NATTERER, J.: Concepts and details of mixed timber-concrete structures. In: *Composite construction - conventional and innovative*, Conference Report, Innsbruck, 1997, p. 175 – 180
    - → KREUZINGER, H.: Die Holz-Beton-Verbundbauweise. In: Fachtagungen Holzbau 1999_2000. Informationsdienst Holz. p.70-83
    - → BRAUN, H.-J. ; SCHAAL, W. ; SCHNECK, F.: Der Brettstapel als Verbundelement – Vorschlag für einen Berechnungsansatz. In: *Bautechnik* 75 (1998), No. 8, p. 539 – 547
- German approvals for fasteners for timber-concrete-composite structures:
    - → DEUTSCHES INSTITUT FÜR BAUTECHNIK: Z-9.1-557: Holz-Beton-Verbundsystem mit eingeklebten TC-Schubverbindern, Bathon & Bahmer GbR (2003) (comp. Figure 5)
    - → DEUTSCHES INSTITUT FÜR BAUTECHNIK: Z-9.1-473: Brettstapel-Beton-Verbunddecken mit Flachstahlschlössern, Dipl.- Ing. W. Bauer (2002) (comp. Figure 2)
    - → DEUTSCHES INSTITUT FÜR BAUTECHNIK: Z-9.1-445: Timco II Schrauben als Verbindungsmittel für das Timco Holz- Beton-Verbundsystem, Wieland Engineering AG (2000) (comp. Figure 6)

Figure 6: Timco screws

→ Deutsches Institut für Bautechnik: Z-9.1-342: SFS-Verbundschrauben VB- 48-7,5 x 100 als Verbindungsmittel für das SFS-Holz-Beton-Verbundsystem, SFS Unimarket AG (1998) (comp. Figure 1)

→ Deutsches Institut für Bautechnik: Z-9.1-331: EW-Holz-Beton-Verbundelement, Hedareds Sand & Betong (1998)

→ Deutsches Institut für Bautechnik: Z-9.1-474: Dennert Holz- Beton-Verbundelemente, Veit Dennert KG (2003)

- Connection systems

  → Blass, H.-J. ; Ehlbeck, J. ; Linden, M. L. R. ; Schlager, M.: Trag- und Verformungsverhalten von Holz-Beton-Verbundkonstruktionen. In: *bauen mit holz* (1996), 5, p. 392 – 399

  → Bathon, L. ; Bahmer, R.; Fritzen, K.: Mut zu Neuem. In: bauen mit holz 2003, No. 1 , p. 21-26

  → Kuhlmann, U. ; Michelfelder, B.: *Kerven mit Schlüsselschrauben als Verbindung bei Brettstapel-Beton-Verbunddecken,* Institut für Konstruktion und Entwurf, Universität Stuttgart. 2001. – Zwischenbericht zum Forschungsvorhaben AiF 13 204

- Long and short term behaviour of timber concrete composite structures

  → Kenel, A. ; Meierhofer, U.: *Holz / Beton-Verbund unter langfristiger Beanspruchung.* EMPA Dübendorf (CH), Abteilung Holz, 1998 (report No. 115/39)

  → Schänzlin, J.: *Zum Langzeitverhalten von Brettstapel-Beton-Verbunddecken,* Institut für Konstruktion und Entwurf, Universität Stuttgart, phd-thesis, 2003

  → Höhmann, R. ; Siemers, M.: Untersuchungen zum Trag- und Verformungsverhalten von Holz-Beton-Verbundträgern. In: *Bautechnik 75* (1998), No. 11, p. 922–929

  → Grosse, M.; Hartnack, R.; Lehmann, S.; Rautenstrauch, K.: Modellierung von diskontinuierlich verbundenen Holz-Beton-Verbundkonstruktionen. In: *Bautechnik 80* (2003) p. 534-541 and p. 693-701

  → Schmidt, J.; Schneider, W.; Thiele, R.: Zum Kriechen von Holz-Beton-Verbundkonstruktionen. In *Beton- und Stahlbetonbau 98* (2003), p.399-407

  → Fragiacomo, M.; Schänzlin , J.: Modelling of timber-concrete floor structures. In: Cecotti , A. (Ed.) ; Thelandersson , S. (Ed.): *Timber constructions in the new millenium,* Cost E5, September 2000

  → Fragiacomo, M.: *Comportamento a lungo termine di travi composte legno-calcestruzzo,* Universität Trieste, phd-thesis, 2000

→ KUHLMANN, U. ; GEROLD, M. ; SCHÄNZLIN, J.: Trag- und Verformungsverhalten von Brettstapel-Beton-Verbunddecken. In: *Bauingenieur* 77 (2002), 1, No. 1, p. 22–34

→ GEROLD, M. ; KUHLMANN, U. ; RISIO, T. D. ; SULZBERGER, L. ; SCHÄNZLIN, J.: *Verformungs- und Dehnungsmessungen von Brettstapel-Beton-Verbunddecken.* Institut für Konstruktion und Entwurf, Universität Stuttgart, 2001 (report 2001-5X)

Field of Application and End Users

- e.g. residential housing
    - i) Tübingen: 6 storey multiple family dwelling

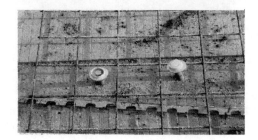

Contacts

- Joerg Schaenzlin

    Institute of Structural Design, Pfaffenwaldring 7, Stuttgart, Germany
    email: joerg.schaenzlin@ke.uni-stuttgart.de
    Tel. +49 711 685 6240
    Fax. +49 711 685 6236

- Ulrike Kuhlmann

    Institute of Structural Design, Pfaffenwaldring 7, Stuttgart, Germany
    email: Sekretariat@ke.uni-stuttgart.de
    Tel. +49 711 685 6245
    Fax. +49 711 685 6236

DEVELOPMENT OF GLUED TIMBER -CONCRETE-COMPOSITE STRUCTURAL ELEMENTS

Description

- Timber-concrete-composite structural elements with an adhesive connection
- Stiff connection with no slippage between timber and concrete
- Stiffening of the structure
- Suitable for use as a bending element, floor deck

adhesive application over full contact area

Figure 1. Production of timber-concrete-beams for bending tests

Technical Information

- The connection between the timber and the concrete is achieved with an adhesive based on epoxide.
- Stiff adhesive connection without slippage between the timber and the concrete: less deflections than occurs when mechanical connectors are used.
- Material properties:
    - i) Concrete: C20/25 (SIA 262 resp. EN 206-1) or better quality, maximum size of aggregate 16mm
    - ii) Timber: C24 (SIA 265 resp. EN 338) or better quality
    - iii) Adhesive: solvent-free, thixotropic 2-component adhesive on epoxide basis, produced by SIKA AG, Zürich
- Two production processes are possible:
    - i) Wet process: after the adhesive has been applied to the timber surface, the concrete is poured onto the still wet adhesive.
    - ii) Dry process: a prefabricated concrete element is attached to the timber with the adhesive.
- Specifications for the concrete process:

i)   Pouring of concrete 90 minutes after the adhesive has been applied to the timber surface.

ii)   Dropping height when pouring the concrete: maximum 50cm

iii)   Careful pouring of concrete at many rather than just one point

Field of Application and End Users

- New structures and also in the rehabilitation of old buildings
- Deck systems in housing projects, industrial buildings, wall elements, bridge slabs

Structural Assessment

- Advantages:
  i)   Structural efficiency: less deflections and greater stability. Thanks to the adhesive connection with no slippage, the structural elements are stiffer than those which use mechanical connectors.

  ii)   Better vibration behaviour because of the greater mass and stiffness - in particular the distribution of the stimulation due to the greater lateral stiffness

  iii)   Building physics: improved fire resistance and noise protection

  iv)   Better stress distribution in the adhesive line. There are no stress concentrations as is the case when mechanical connectors are used.

  v)   Optimized use of the different materials: timber under tension, concrete under compression

- Disadvantages:
  i)   Reliability of the adhesive bond: the pouring of the concrete has to be done very carefully in order to prevent a dislocation of the adhesive.

  ii)   The time interval of about 90 minutes between the application of the adhesive and the pouring of the concrete should be adhered to strictly because otherwise the adhesive may be still too wet or it may have already started to set.

  iii)   Assembly of the prefabricated elements involves moving large weights

Economical Aspects

- Since many work processes such as drilling screw holes or cutting indentations in the timber can be omitted, the costs are reduced in comparison to systems using mechanical connectors.
- Thanks to the greater protection against fire hazards, noise and vibration problems, no further measures are necessary.

Robustness and Testing

- The dislocation of the adhesive during the pouring of the concrete was studied with a visual analysis as well as the testing of small specimens in shear and in bending. The results helped to determine the necessary parameters for the materials and the process of pouring the concrete.

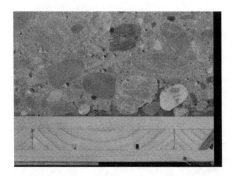

Figure 2. Visual analysis of the adhesive line

Figure 3. Bending Test

- Large-scale test specimens were prepared for bending tests at the HSB in Biel. The test results agree quite well with the calculated values. The bending tests have met all expectations, thus demonstrating the usefulness of the parameters determined in the test series concerning the dislocation of the adhesive.

Figure 4. Bending Test with large-scale specimens

Pilot structure

- To the best of our knowledge, so far no major structures have yet been constructed involving timber-concrete-decks with an adhesive connection. A pilot project is envisaged.

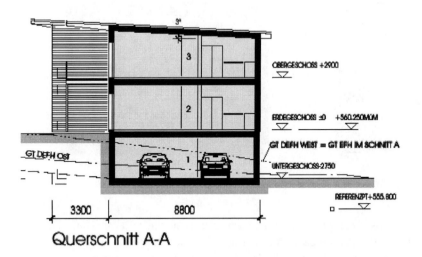

Querschnitt A-A

Figure 5. Pilot project in Kanton Obwalden, Switzerland: Cross section through a block of flats

Standards

- Standard SIA 262; *Betonbau (Concrete Structures); Zurich, 2003*
- Standard SIA 265; *Holzbau (Timber Structures); Zurich, 2003*

References

- Blass H.J., Schlager M.: "Trag- und Verformungsverhalten von Holz-Beton-Verbundkonstruktionen", Bauen mit Holz, Hefte 5 & 6, 1996.
- Brunner M., Gerber C.: "Holz-Beton-Verbunddecken mit Klebeverbindung", 31. SAH-Fortbildungskurs Weinfelden, 1999.
- Brunner M., Gerber C.: "Holz-Beton-Verbundelemente: Entwicklung von Holz-Beton-Verbundelementen durch die Anwendung von Klebesystemen", Interner Bericht der SH-Holz, Biel, März 2000.
- Gisel Peter: "Holz-Beton-Verbund-Elemente: alternative Anwendungen mit Klebesystemen", Diplomarbeit der SH-Holz, 1997.
- Kenel A.: "Zur Berechnung von Holz-Beton-Verbundkonstruktionen", Forschungs- und Arbeitsbericht, EMPA-Debendorf, 1999.
- Kuhlmann U., Gerold M.,. Schänzlin J: "Brettstapel-Beton-Verbund - Berücksichtigung von Kriechen und Schwinden", Bauingenieur 77, 6/2000, S. 281-288
- Kuhlmann, U.; Gerold, M. und Schänzlin, J.: "Trag- und Verformungsverhalten von Brettstapel-Beton-Verbunddecken", Bauingenieur, Band 77, Januar 2002, S. 22-34
- Natterer J., Hoeft M: "Zum Tragverhalten von Holz-Beton-Verbundkonstruktionen", Forschungsbericht CERS Nr. 1345, IBOIS, EPFL, 1987

- Versch.: "Holz-Beton-Verbundbau", Fachtagung Universität Stuttgart, 13.10.2003

Contacts

- Dr. Maurice Brunner, Marco Schnüriger
  - Address: School of Architecture, Civil and Wood Engineering HSB, Solothurnstrasse 102, Biel, Switzerland
  - email: maurice.brunner@hsb.bfh.ch
    - marco.schnueriger@hsb.bfh.ch
  - Tel. +41 32 344 03 41
  - Fax. +41 32 344 03 91

COST C12 – WG1

Datasheet II.3.1.5.5....
Prepared by: Dr. Jan Hamm[1]

> [1] *School of Architecture, Civil and Wood Engineering HSB*

COMPOSITE TIMBER-CONCRETE ELEMENTS

Description

- *Timber*

The timber for the composite system can be choosen as follows:
- o planks
- o square-sawn timber
- o round wood

These timber sections can be glued, nailed or connected with dowels.

- *Concrete*

The concrete used normally for the wood-concrete composite system is a standard C20/25 concrete. At the institute of Timber Construction at the Swiss Federal Institute of Technology of Lausanne (Switzerland) tests with wood-lightweight concrete and fiber reinforced wood-lightweght concrete showed encouraging results.

- *Composite connectors*

A detail of a post-tensioned HILTI-dowel is shown in fig.1. The dowel is glued in the timber section from above and protected by a plastic head and tube, to avoid bond so that the dowel can be post-tensioned after the concrete has set.

Figure 1. Detail of a post-tensioned HILTI – dowel glued in a groove

Technical Information

For residential buildings with live loads less than 3.5 kN/m^2 the following design rules can be applied:

- $h_C + h_W \cong \dfrac{l}{24}$ (depending on the material quality, loads and span)
  - l: span of the composite deck
  - $h_C \approx 0.6 \cdot h_{tot}$
  - $h_W \approx 0.4 \cdot h_{tot}$
- mesh reinforcement for shrinkage: $\varnothing$ 5mm, spacing 15cm
- strutting lines for decks:
  - 1 strutting line in the middle   (for l ≤ 7m)
  - 3 strutting lines   (for l > 7m)
- grooves for post-tensioned dowels:
  - 2 grooves for   l ≤ 7m
  - 3 grooves for   7m < l ≤ 10m
  - 4 grooves for   l > 10m

Field of Application and End Users

This kind of deck is normaly used in residential buildings but it's nowadays also often used in walls, bridges, commercial-building, schools or other public buildings.

- *Structural analysis*

The degree of composite action that can be achieved depends on the effectiveness of the shear transfer between the timber section and the concrete deck.
The stiffness of the timber-concrete decks using this kind of connectors depends essentialy on the span and how often the planks of laminated timber sections are interupted.

The rigity of the composite section can be calculated as follows:

$$B = \Sigma\ E_i \cdot I_i + \eta \cdot S$$

$\eta = 0.80$: length of the planks equal span
$\eta = 0.85$: length of the planks less than the span, planks fingerjointed and glued
$\eta = 0.90$: length of the timber sections equal the span, round wood used

- *Building physics (fire/sound)*

Fire resistance of F30, F60 or F90 AB, as well as noise insulation criteria up to 60 db for walls and decks can be reached.
In conclusion the present experience suggests that this kind of system is still in a developing stage, but it is advanced enough to be applied in different ways. Timber-concrete composite systems show great flexibility in their application. This technique has both the advantage of timber structures and the advantage of concrete structures (fig.2.). All the advantages of massive timber construction as fire resistance are conserved. Moreover concrete with its self-weight improves thermal inertia, a better noise insulation and gives a greater stiffness to the system.
Finally it is very important to note that this technique can be transfered to the small and medium construction company without too much financial investment. This system can be used in a broad variety of construction elements.

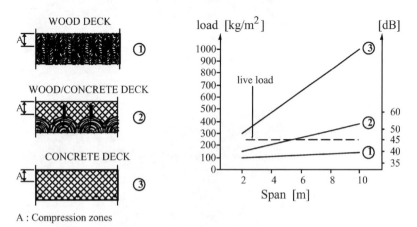

A : Compression zones

fig.2: Comparison of different decks

Economical Aspects

Wood-concrete decks including a load-bearing concrete slab present new advantages, especially for houses, schools and public buildings. With these techniques the long term behaviour and the bending properties of structures with small dead loads can be fulfilled economically.

- *Costs*

Approximatly 100€/m2 in visible quality of the vertical laminated timber deck and without a floating floor. The costs depend of the span and the timber quality of the timber-concrete deck.

Examples

fig.8. A ceiling as composite structure made with vertical, nailed planks

fig.10. Woodelements and dowels for a composite timber-concrete slab: Schule Triesenberg (FL).

fig.11. Composite structures for a restoration: Kloster Wurmsbach, Bollingen (CH).

References

1. Hoeft, M.: Zur Berechnung von Verbundträgern mit beliebig gefügtem Querschnitt, Thèse N° 1213, Ecole Polytechnique Fédéral de Lausanne, 1994

2. Götz, K.-H.; Hoor, D.; Möhler, K.; Natterer, J. Holzbau Atlas, Institut für internationale Architektur-Dokumentation GmbH, München, 1978.

3. Götz, K.-H.; Hoor, D.; Möhler, K.; Natterer, J. Timber Design and Construction Sourcebook, McGraw-Hill, Inc., 1989.

4. Natterer, J.; Herzog, T.; Volz, M. Holzbau Atlas Zwei, Institut für internattionale Architektur-Dokumentation GmbH, München, 1991.

5. Natterer, J.; Hoeft, M.: Zum Tragverhalten von Holz-Beton Verbundkonstruktionen Forschungsbericht CERS Nr. 1345, Ecole Polytechnique Fédéral de Lausanne, 1994

Contacts

- Dr. Jan Hamm
  Address: School of Architecture, Civil and Wood Engineering HSB, Solothurnstrasse 102, 2504 Biel, Switzerland
  jan.hamm@hsb.bfh.ch
  Tel. +41-(0)32/3440319
  Fax. +41-(0)32/3440391

COST C12 - WG2
Datasheet II.3.1.6.1
Prepared by: L. Fulop[1], D. Dubina[2]

[1] Politehnica University of Timisoara
[2] Politehnica University of Timisoara

LOAD BEARING CAPACITY OF LIGHT GAUGE STEEL SHEAR WALLS

Description

- In Light Gauge Steel (LGS) house structures the wall panels are the stuctural subasablies that are supposed to resist lateral (i.e. from wind and eartquake) loads. Under lateral forces Light Gauge Steel Wall Panels (LGSWP) are loaded mainly in shear. In low rise buildings the walls are supposed to resist vertical as well as horizontal actions.

- A LGSWP is defined as the assembly of a LGS skeleton and the corresponding sheeting panels, which are usually OSB, corrugated sheeting or gypsum wall board (GWB). The sheeting panels are connected to one, or both, sides of the skeleton with self drilling screws following a connection pattern.

Field of application

- LGS house structures are relatively new on the market in Europe. For earthquake (or wind) design there is no available design specification. Furthermore, there is are yet not sufficient research results in order to draft a code for design.

- The application of LGSWP as load bearing elements for lateral (ie. horizontal) loads finds its application mainly in house structures, but also in case of light steel structures for offices, schools, social buildings.

- The other possible application of LGSWP is in combination with classical steel frames structures, where they, as infill elements, contribute to the load bearing of the frame under moderate earthquake (under strong earthquakes they are destroyed) helping to achieve the serviceability criteria imposed on frames.

Technical information

- The skeleton of LGSWP is made of cold formed profiles (usually C, Σ and U) connected using self drilling screws. The sheeting panels are connected to one or both sides of the skeleton with self drilling screws following a connection pattern.

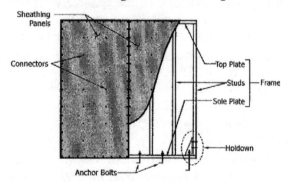

Figure 1. Main components of a wall panel

- Basic components of a typical wall panel are the: (1) skeleton, (2) sheeting, (3) sheeting to skeleton connections, (4) connections between sheeting panels (seam type connections if there is any) and (5) anchors (Fig. 1).

Structural aspects

- There is the need for a set of simple rules and formulas to determine the rigidity and the load bearing capacity of different wall panel configurations with diverse connection schedules.

- The simplest approach is the evaluation of the load bearing capacity based on the observation that the capacity per unit length of sheeted wall is constant for the same configuration. Therefore, if the capacity per unit length is known it only has to be multiplied by the sheeted length of the wall panel to determine its total capacity. This approach has been developed in the US and it is oriented towards practical design (AISI, 1988). The capacity per unit length values are determined from experiments on wall panels and are included in tables.

- To take into account the effect of openings empirical formulas have been developed by Bradford and Sugiyama (Bradford, 2001), which relates the load bearing capacity of the wall panel with opening to the load bearing capacity of the fully sheeted wall.

- For the case of walls equipped with corrugated sheeting the calculation procedure recommended by the ECCS (ECCS-P88, 1995) can be used with minor adaptations. Comparative calculations have shown very good correlation between these calculation method and experiments in terms of initial rigidity. In terms of load bearing capacity the method proved to yield to a very conservative evaluation.

Research activity

- Research in the USA (AISI, 1988) has been focused mainly on experimental testing of shear walls, typical to their home practice, in order to produce practical racking load values. Load bearing capacities were derived both from monotonic push-over curves, envelope and stabilised envelope curves from cyclic tests. Findings of these studies suggest a conventional elastic stiffness for a wall panel at 0.4 of the ultimate load (Serrette, 1998). Different frame typologies with various cladding materials were tested, studies being conducted to determine the influence of length/height ratios as well as the effect of openings.

- Testing and numerical simulation was combined in order to account for hysteretic characteristics in an attempt to provide evidence on the possible values of response modification factors (Kawai, 1999). Vibration tests of steel-framed houses were conducted and relatively large damping ratios were found due to finishes.

- The same issue was analysed by Gad et al. (Gad, 2000) who proposes a new analytical approach to evaluate the ductility parameter (R_μ), and found values between 1.5 and 3.0 to be suitable. The same research briefly assesses inherent structural overstrength and finds it to be very important factor as far as earthquake resistance is concerned. The quantitative evaluation of overstrength is more difficult, but an empirical evaluation attempt is performed.

- Experimental tests and FE modelling was employed by De Matteis (De Matteis, 1998) to asses shear behaviour of sandwich panels both in single story and multi-story buildings. According to the conclusions diaphragm action can replace classical bracing solutions only in low-rise buildings, and in areas of low seismicity.

- In the light of the above described research activities and results, the problem of earthquake performance of LGS houses has been studied, starting from the wall panels, trough experimental and numerical studies. The evaluation of the hysteretic shear force vs. displacement curves for wall panels has been attempted only by *experimental testing*.

- The experimental program was based on six series of full-scale wall tests with different cladding arrangements based on common solutions used in practice. Each series consisted of identical wall panels tested statically, both monotonic and cyclic.

- The main frame of the wall panels (3600x2440mm) was made of cold-formed steel elements, top and bottom tracks were U while studs were C profiles, fixed at each end to tracks with self-drilling self-taping screws. In specimens using corrugated sheet as cladding the sheets were placed in a horizontal position (Fig. 2.a.b.d.) tightened with seam fasteners. Corrugated sheet was fixed to the wall frame using self-tapping screws. Additionally on the 'interior' side of specimens Series II, gypsum panels were placed vertically and fixed to the studs.

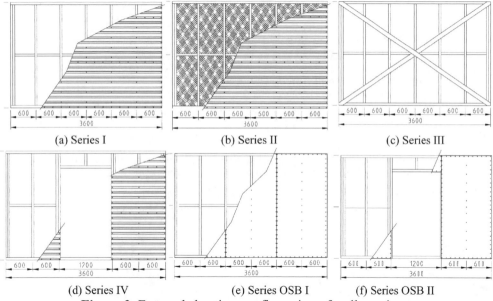

Figure 2. External sheeting configuration of wall specimens

- Bracing was used in three specimens (Fig 2.c), by means of straps on both sides of the frame. Steel straps were fixed to the wall structure using self-drilling screws, the number of screws being determined to avoid failure at strap end fixings.

- OSB panels were placed vertically (Fig. 2.e.f) only on the 'external' side of the panel and fixed to the frame using bugle head self-drilling screws.

Method of analysis

- The earthquake performance of the panels has been evaluated, based on the experimental results, using *Finite Element Modelling* (time history analysis).

- In order to model the complex hysteretic behaviour of the panel different techniques can be employed, depending on the desired accuracy, possibilities ranging from simple bilinear to highly complex nonlinear models.

- A nonlinear model can be based on the proposal of Della Corte *et. al.* (Della Corte, 2000), which is based on a Richard-Abbott type curve, and has very good capability to predict all aspects of the panel behaviour.

- During modelling it is important to take into account the main characteristics of the panel behaviour; (1) pinching, (2) overstrength (as difference between allowable "elastic" design limit and actual capacity) and (3) plastic deformation capacity.

- A model based on DRAIN-3DX computer code, and used in the studies, is easy to be included in a larger structural model. A hinged frame with a "fiber hinge" accommodating the desired hysteretic behaviour has been constructed and calibrated using the experimental results. According to this approach the wall panel can be replaced by an equivalent pinned frame with dissipative diagonals (Fig 3.).

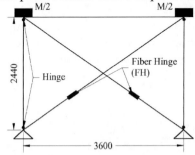

Figure 3. Equivalent hinged frame model

Results of analysis

- Top horizontal load vs. displacement curves are nonlinear from the beginning of the loading. Wall panels posses medium to high ductility and the hysteretic behaviour is characterised by low energy dissipation due to pinching (Fig. 4).

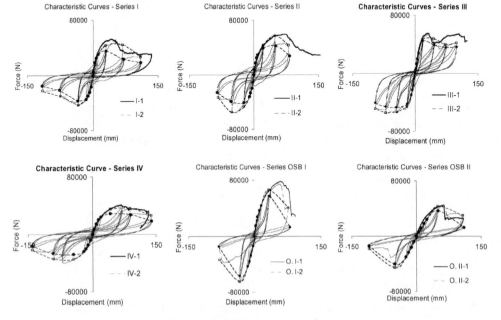

Figure 4. Characteristic force vs. displacement curves

494

- Shear-resistance of wall panels is significant both in terms of rigidity and load bearing capacity. The hysteretic behaviour is characterised by very significant pinching and reduced energy dissipation.

- Failure is starting at the bottom track in the anchor bolt region. The ideal shape of corner detail is so that uplift force is directly transmitted from the brace (or corner stud) to the anchoring bolt, without inducing bending in the bottom track. Failing to strengthen wall panel corners has important effects on the initial rigidity of the system and can be the cause of large sway and premature failure for the panel.

- The seam fastener represented the most sensitive part of the corrugated sheet specimens; damage is gradually increased in seam fasteners, until their failure causes the overall failure of the panel. Much of the post elastic deformation of the wall panel is in the region of seam fasteners.

- Starting from these conclusions, and using the DRAIN-3DX model described earlier, a number of incremental dynamic analysis runs have been performed using historical earthquake records.

- Three performance levels were associated with corresponding lateral displacement of the panels (Fig. 5) and 'partial behaviour' factors have been identified for the panels (as ratio of the earthquake levels corresponding to the reaching of each displacement limits) based on time-history analysis results (Fig. 6). The way of choosing D_{el}, D_{yield} and D_{ult} can be done using different methods and interpretations (Kawai, 1997; ECCS, 1985), and is crucial from the point of view of the results.

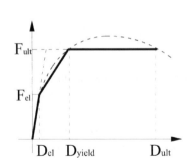

Figure 5. Performance criteria based on lateral displacement

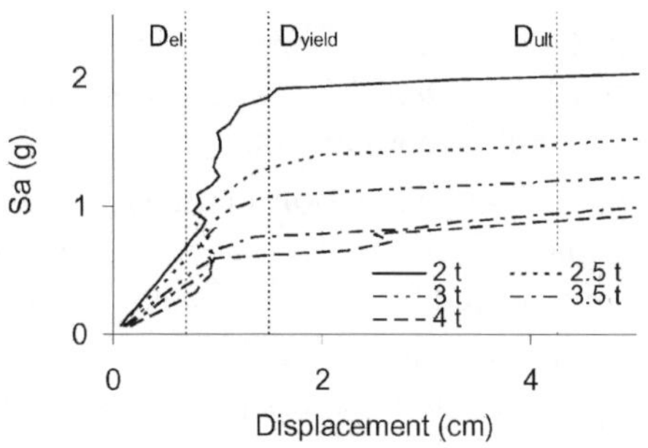

Figure 6. The displacement criteria applied for IDA curves

- The effect of over-strength is identified to be an important in the post elastic behaviour of panels and source of a possible design earthquake-force reduction. The resulting factor (2.2-2.6). The possibility of design force reduction due to ductility and energy dissipation seems to be more limited (1.4-1.6) probably due to low energy dissipation capacity of the hysteretic loops.

Further developments

- The following research directions can be anticipated to be of interest for the future:

- (1) The development of more general methods for the evaluation if the rigidity and load bearing capacity of wall panels in a general case. The method may be based on the experimental determination of the behaviour of components but should not use full scale experimenting on walls. A simple to use, jet powerful design method is required if these structural solutions are ever to be used on a large scale.

- (2) The assessment of the effect of the hysteretic behaviour on the earthquake performance of LGS wall panel structures has to further be investigated.

- (3)The effect of damping and the presence of secondary (non structural) elements may have important impact on the seismic performance of a structure.

References

- AISI, *Shear Wall Design Guide*, Publication RG-9804, American Iron and Steel Institute, February 1998

- Bradford K.D., Sugiyama H., *Perforated shear wall design approach*, American Forrest and Paper Association, 2001

- Della Corte G., De Matteis G., Landolfo R.. *Influence of Connection Modelling on the Seismic Response of Moment Resisting Steel Frames*, Moment Resistant Connections of Steel Frames in Seismic Areas, E & FN Spoon, 2000

- De Matteis G, *The Effect of Cladding in Steel Buildings under Seismic Actions*, PhD Thesis, Universita degli Studi di Napoli Federico II, 1998

- ECCS, *Recommended Testing Procedure for Assessing the Behaviour of Structural Steel Elements under Cyclic Loads*, 1985

- ECCS P88, *European Recommendations for the Application of Metal Sheeting Acting as a Diaphragm*. Publication No. 88, European Convention for Constructional Steelwork, Brussels, 1995.

- L.A. Fülöp, D. Dubina, *Performance of wall-stud cold-formed shear panels under monotonic and cyclic loading Part I: Experimental research*, Thin-Walled Structures, Volume 42, Issue 2, February 2004

- L.A. Fülöp, D. Dubina, *Performance of wall-stud cold-formed shear panels under monotonic and cyclic loading Part II: Numerical modelling and performance analysis*, Thin-Walled Structures, Volume 42, Issue 2, February 2004

- Gad E.F., Duffield C.F., *Lateral Behaviour of Light Framed Walls in Residential structures*, 12th World Conference on Earthquake Engineering, Auckland, New-Zeeland, February 2000

- Kawai Y, Kanno R, Hanya K, *Cyclic Shear Resistance of Light-Gauge Steel Framed Walls*, ASCE Structures Congress, Portland, 1997

- Kawai Y, Kanno R, Uno N, Sakumoto Y, *Seismic resistance and design of steel framed houses*, Nippon Steel Technical report, No. 79, 1999

- Serrette R.L., *Seismic Design of Light Gauge Steel Structures: A discussion*, Fourteenth Int. Speciality Conference on Cold-Formed Steel Structures, St. Louis, Missouri, Oct.15-16, 1998

Contacts *(include at least one COST C12 member)*
- Ludovic Alexandru FÜLÖP
- Address: str. Ioan Curea nr. 1, 300224 Timisoara, ROMANIA
- E-mail: lala@constructii.west .ro
- Phone/Fax: ++40-256-403-932

COST C12 – WG1

Datasheet II.3.1.6.2
Prepared by: L. Fulop[1], D. Dubina[2]

[1] Politehnica University of Timisoara
[2] Politehnica University of Timisoara

PERFORMANCE OF SCREWED CONNECTIONS IN LIGHT GAUGE STEEL SHEAR WALLS

Description

- In case of Light Gauge Steel Wall Panels (LGSWP) loaded in shear (i.e. under horizontal actions from wind and earthquake) the critical components of the wall panel are the sheeting-to-skeleton connections (and seam connections if there are any), with the condition that uplift of the corners is adequately restrained.

- The design of the wall panels for earthquake loads has to take into account the behaviour characteristics of the sheeting-to-skeleton connection both in terms of rigidity and load bearing capacity. Furthermore, the main source of the ductility of wall panels is exactly these connections. Connectors are usually self-drilling screws, which connect either the corrugated sheet or the OSB panel to the steel skeleton.

Field of application

- LGSWP are the usual structural elements that are supposed to resist lateral loads in small Light Gauge Steel (LGS) structures (ex. houses, small offices, schools).

- There is need for analytical design methods to determine the load bearing capacity and rigidity of wall panels with different sheeting and connector schedules. Even if the way of experimentation is always open, it is both time consuming and costly to make full scale tests on wall panels of different typologies.

- As the failure of the LGSWP is governed by the failure of the connections between sheeting and skeleton, it is evident that relationship exists between the performance of connections and that of the entire wall panel.

- If reliable transformation relationships (or suitable methods) between the connection and wall panel behaviour can be established, instead of wall panel experiments tests on connections (and a few other components) would suffice for the determination of the performance of LGSWP.

Technical information

- In LGSWP sheeting is connected to the skeleton with self drilling screws. The connection is loaded in shear due to the different deformation pattern of the sheeting and skeleton. The differential displacements of the sheeting and skeleton have, in fact, to be accommodated by the connecting screws.

- The behaviour of the screwed connections depends on the type of the connector and the type, and properties, of the connected materials.

Structural aspects

- There are two different failure mechanisms observed in LGSWP (Fulop, 2004), depending on the sheeting typology (Fig. 1). In both case there are certain connection groups that are loaded, and their failure governs the failure of the entire panel.

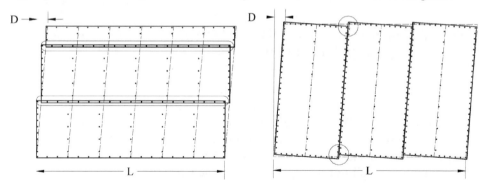

Figure 1. Typical deformation pattern of corrugated sheet and OSB sheeted LGSWP

- The behaviour of steel-to-steel connections using self drilling screws is a well studied. There are code provisions (EC3, 2001) for calculation, even if not all thickness combinations are covered by the standard (ex. no thick-to-thin sheet connections). Unfortunately EC3 does not provide the means to calculate the rigidity of steel-to-steel connections.

- Other possibility for the evaluation of the load bearing capacity is offered by an ECCS document (ECCS, 1985). In this document the rigidity of some connection typologies is also given (without a method to calculate them).

- Using recent research results, (Fan, 1996, 1997) there is a proposal to improve the Eurocode 3 calculation method and to extend its validity to more connection typologies. With this methodology the capacity of thick-to-thin connections can also be calculated, and there are formulas for the calculation of connection rigidity.

- If codes do not cover the desired connection typologies, the possibility of testing exist, testing methodologies being codified (ECCS, 1983).

- The calculation of OSB-to-steel connections is not codified and no standard testing methodologies exist (according to the author's knowledge), therefore the approach for this typology of connections is more difficult.

Research activity

- Experiments were carried out to determine the mechanical properties of the connections that were identified to have a crucial effect on the behavior of LGSWP.

- Self-drilling screws connecting the sheeting material to the steel skeleton and seam connections were of the following typologies: (1) connection between the corrugated steel sheeting and steel profiles; (2) sheeting-to-sheeting seam connections (Fig. 2).

- Connections between the OSB panels and steel profiles (skeleton) were also tested (Fig. 3). Specimens were chosen in a way that they represent connection typologies used in tested wall panels.

- Two loading velocities were applied, $V_1=1mm/min$ for quasi static loading conditions and $V_2=420mm/min$ for high velocity tests (in the case of OSB specimens only V_1 was used). In case of thin-to-thick (sheeting-to-skelton) connections the failure mode observed during the wall panel tests was due to tearing of thinner sheet. In order to force the connection specimens to fail in the desired mode of failure additional clips were provided to prevent the distortion of the thinner sheets as this is not possible in the wall specimen due to the presence of the corrugations.

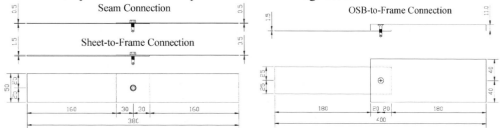

Figure 2. Dimensions of the steel-to-steel connections

Figure 3. OSB to steel skeleton connection specimens

- In case of thin-to-thin sheet connections, which were modeling seam fasteners, the failure mode was due to tilting and pull-out of the screw. The same failure mode was observed for seam connections during the wall panel experiments.

- Load versus slip curves are presented for all specimens at both loading velocities (Fig. 4, 5, 6, 7).

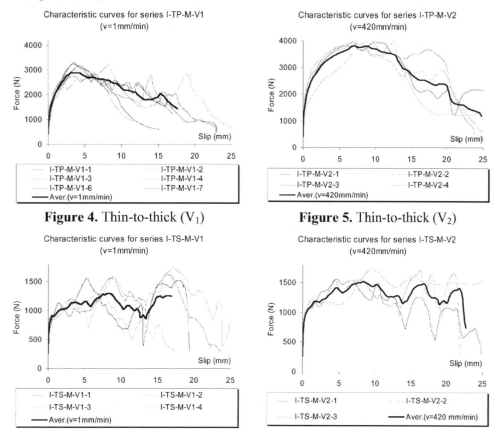

Figure 4. Thin-to-thick (V₁)

Figure 5. Thin-to-thick (V₂)

Figure 6. Thin-to-thin (V₁)

Figure 7. Thin-to-thin (V₂)

- Load versus slip curves are presented for all specimens at both loading velocities (Fig. 4, 5, 6, 7).

- The testing of OSB specimens yielded very inhomogeneous results depending on the direction and density of fibers in the vicinity of the screw and between the screw and

499

the margin of the OSB panel. No generalizing conclusion can be drawn from these experiments besides that OSB connections possess less ductility.

- After establishing an acceptable level of deformation at connection level this can be related to the overall deformation of the wall panel. The following acceptable deformations in the seam fasteners are suggested: (1) Slip does not exceed the elastic limit (D_e, Fig. 8), corresponding to $0.6F_{max}$ of the connection. The integrity of the cladding is fully preserved (serviceability conditions); (2) Slip is limited to the diameter of the screw (D_r=4.8mm, Fig. 8) the cladding requires some repair. There is damage, but not excessive and can be repaired (immediate occupancy); (3) In case of life safety criteria any kind of damage is acceptable, without endangering the safety of occupants. This criterion corresponds to the attainment of the ultimate force (F_{ult}) and the starting of the downwards slope.

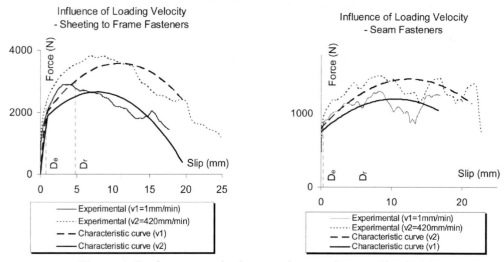

Figure 8. Performance criteria at steel-to-steel connection level

- In case of OSB-to-steel connections, which are characterised by a fragile behaviour, the design has to be controlled by the *elastic* limit only (D_e). In such a case multiple performance levels can not be applied.

Method of analysis

- Based on the component characteristics determined experimentally, a Finite Element (FE) modelling was made in ANSYS to reproduce the behaviour of the wall panels with corrugated sheet.

- The bars of the skeleton were modeled as elastic beam elements. The corrugated sheet was modeled as an equivalent orthotropic plate in order to take into account the different mechanical properties in the two principal directions and the distortion of the corrugated sheet when loaded in shear. Connections were modeled using link elements taking deformation properties of the tested connection loaded at V_1. The uplift deformation characteristics of the wall corners has also een taken into account.

- The FE model was subjected to increasing horizontal loading at the upper part of the panel similarly to wall-panels during the full scale test. The deformation pattern and the non-linear behavior curve obtained with the FEM model was compared with the monotonic curves obtained from wall panel experiments (Fig. 9, 10).

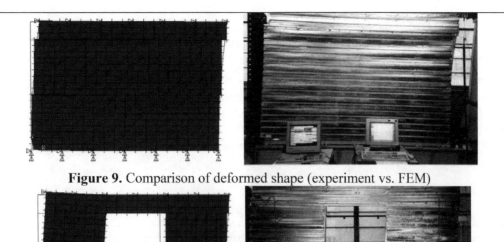

Figure 9. Comparison of deformed shape (experiment vs. FEM)

Figure 10. Comparison of deformed shape (experiment vs. FEM)

Results of analysis

- Based on the acceptability assumptions at connection level the following performance criteria are suggested for wall panels clad with corrugated sheet: (1) fully operational ($\delta<0.003$); (2) partially operational ($\delta<0.015$); (3) safe but extensive repairs required ($\delta<0.025$). (Fig. 11)

- The first performance level does not provide ductility, because shear panel work is elastic. This could be the design criteria for frequent, but low intensity earthquakes. In case of rare but severe earthquakes, the last two design criteria can be used and some ductility will be available.

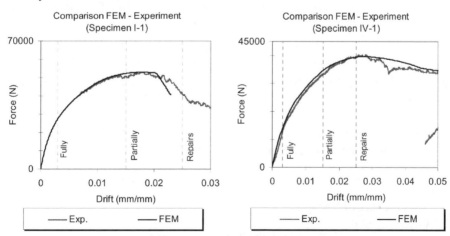

Figure 11. Comparison of deformed shape (experiment vs. FEM)

- The FEM proved to be able to replicate the characteristic curves and the deformation pattern of the panels, and can be used to partially replace experiments in evaluating the performance of the panels.

- An attempt was made to establish performance criteria on the level of the wall panel based on the behavior characteristics of the seam connection.

- An important conclusion of the tests on steel-to-steel connection is that at higher velocity loading connections gain in terms of load bearing capacity without loosing ductility. This observation justifies the use of quasi static loading for wall panel experiments as at higher velocity load transmission rates (ex. near field pulse type earthquakes) the panels can be expected to have better performance than in case of low rate static loading.

Further developments

- Even if the FEM presented here gave satisfactory results compared to the experiments, the effort to set up such a model is beyond the possibility of any designer, so it is still not very practical. Further simplification in the methodology of calculating rigidity and load bearing capacity has to be searched.

- The behaviour of the OSB-to-steel connection has to be further studied. It seams that these types of connections are very inhomogeneous in behaviour and there are no reliable calculation procedures yet.

- The FEM modelling can be extended to other configurations of LGSWP (ex. with different openings, width to height rations, connection schedules) but using the same base materials and connection typologies.

References

- L.A. Fülöp, D. Dubina, *Performance of wall-stud cold-formed shear panels under monotonic and cyclic loading Part I: Experimental research*, Thin-Walled Structures, Volume 42, Issue 2, February 2004

- Eurocode 3 Part 1.3., *Supplementary rules for cold formed thin gauge members and sheeting*, 20 July 2001

- ECCS, *P21 - European Recommendation for Steel Construction: The Design and Testing of Connections in Steel Sheeting and Sections*, ECCS 1983

- ECCS, *P88 - European Recommendations for the Application of Metal Sheeting acting as a Diaphragm, European Convention for Constructional Steelwork*, 1995

- L. Fan, J. Ronda, S. Cescotto, *Finite element modeling of single lap screw connections in steel sheeting under static shear*, Thin Walled Structures 1997; 27(2):165–85.

- L. Fan, *Contributions to Steel Sheet Connections of Screws, Blind Rivets and Cartridge Fired Pins*, PhD. Thesis, University of Liege, 1996

Contacts *(include at least one COST C12 member)*

- Ludovic Alexandru FÜLÖP
- Address: str. Ioan Curea nr. 1, 300224 Timisoara, ROMANIA
- E-mail: lala@constructii.west .ro
- Phone/Fax: ++40-256-403-932

Datasheet II.3.1.7.1.
Prepared by: P.Haller[1] T.Birk[1]

[1] Institute of Steel- and Timber Structure, TU Dresden, Germany

TEXTILE REINFORCED TIMBER STRUCTURES

Description

- Textile reinforced dowel type fasteners
- Textile reinforced timber connection
- Load-adapted textiles (multiaxial stitch bonding technique, biaxial weft knitting and flat knitting technology)
- Fully fashioned reinforced wooden structures

Timber joints are prone to early shear-, split- and compression failure due to excess of the allowed tension perpendicular to the grain or the embedding stress (Fig. 1). With the help of biaxially reinforced knitted or multiaxial stitch bonded structures made by glass this early failure can be reduced.

Technical Information

- the use of knitted or stitch bonded technical textiles provides substantial improvement of the load bearing behaviour of dowel type fasteners, the reduction of weight and an increase of strength, stiffness and ductility of wood constructions
- the used textile structures varied with respect to type of fibre (reinforcement fibre: glass; mesh fibres: glass, aramid), yarn count, ratio of the textile reinforcement and manufacturing process

  i)   biaxially reinforced flat knitted structure (Fig. 1b and Fig. 3a)

  ii) multiaxial stitch bonded structure (Fig. 1a and Fig. 2)

  iii) loops, made of glass, aramid or carbon (Fig. 1c and Fig. 3b)

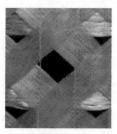

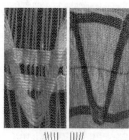

| Multiaxial stitch bonded machine | Flat knitting machine | Stitch bonded machine, variable oriented fiber placement |

Figure 1. Reinforcement structures

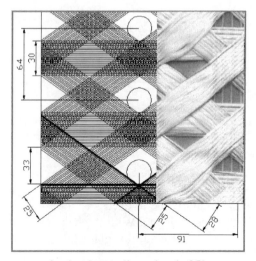 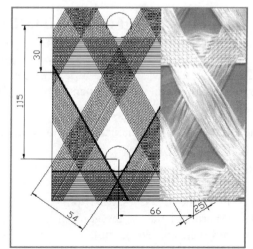

Angle of warp thread: + / - 35°
Textile surface density: 1302 g/m²
Yarn count: 1200 tex

Angle of warp thread : + / - 60°
Textile surface density: 1063 g/m²
Yarn count: 1200 tex

Figure 2. Geometric parameters of multiaxial stitch bonded structure

(a)   knitted fabric made of glass and aramid (spiral)   *(b)*   *Transversally reinforced loop made of carbon*

Figure 3. Tested specimen

Field of Application and End Users

- enginered wood construction
- wood buildings with better seismic behavior
- lightweight structures

Structural Assessment

- Advantages

  easily applicable, significant increase of load bearing capacity

Economical Aspects

- Cheap, efficient use of high strength fibers

Robustness and Testing

- investigations of specimens and test evaluation are carried out according to DIN EN 383

Figure 4. Test setup and specimen

- The experiments were performed using a servo mechanical testing machine with the full-scale load cell capacity of 1000 kN
- The determination of the embedding strength parallel to the grain was executed with a test rate of 200 N/s and 600 N/s
- The sampling rate of measurement was chosen as 2 Hz
- The deformations were measured using 2 or 4 independent displacement transducers

Examples

Codification

- DIN EN 383. *Determination of embedding strength and foundation values for dowel type fasteners, 1993*

References

- Birk, T.; Haller, P.: *Textile Verstärkung von Holzbauteilen.* IN: 1. FACHKOLLOQUIUM TEXTILBETON. RWTH AACHEN, 2001, S. 293–304
- Offermann, P.; Cebulla, H.; Diestel, O.: *Modelling and Production of Fully Fashioned Biaxial Weft Knitted Fabrics.* 1ST AUTEX-CONFERENCE, POVOA DE VARZIM, PORTUGAL 2001. - S. 263-269

- Haller, P; Birk, T.; Putzger, R.: *Physikalische und mechanische Untersuchungen an textilbewehrtem Holz und Holzbauteilen.* TECHNISCHE UNIVERSITÄT DRESDEN: EIGENVERLAG, 2002 – ARBEITSBERICHT DES SFB 528, S. 283-322

Contacts

-     Name: Peer Haller
      Address: TU Dresden
                  Institute of Steel- and Timber Structures,
                  Eisenstuckstraße 33
                  01069 Dresden
                  Germany
                  Peer.Haller@mailbox.tu-dresden.de
                  Tel. + 49 – 351 / 46 33 55 75
                  Fax. + 49 – 351 / 46 33 63 08

TEXTILE REINFORCED TIMBER STRUCTURES

Description

- Textile reinforced wooden tubes connection (bifurcation joint)

A combination of load- and shape-adapted reinforcement of joints is achieved using reinforced knitted and stitch bonded structures. Load-adapted joints consist of a solid joint and wooden tubes reinforced by knitted textile structures.

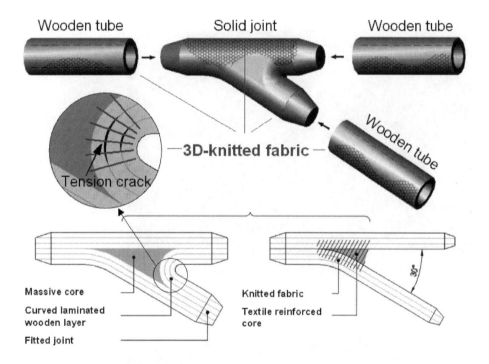

Figure 1. Bifurcation joint (schematically)

Technical Information

- wooden tube and bifurcation joint

        i)   epoxy matrix
        ii)   textile reinforced by knitted fabric (2-4 layers)

Material	Strength [N/mm²]	Failure strain [%]	Youngs-modulus [N/mm²]	Yarn count [tex]
Roving material				
warp thread (E-glass)	1 335	4,7	75 000	1 200
mesh thread (E-glass)	1 335	4,7	75 000	68

	Strength [N/mm²]	Failure strain [%]	Youngs-modulus [N/mm²]	Density [g/cm³]
Epoxy resin				
LN-1 VOSSCHEMIE	19	4	3 500	1,2

Table 1. Material characteristics

Figure 1. Wooden tube, diameter 150 mm thickness 16 mm

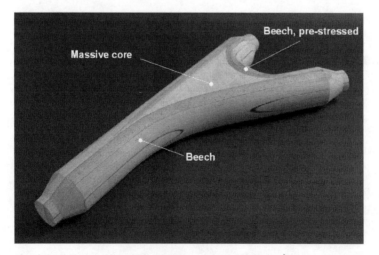

Figure 2. Bifurcation joint (cross section: 200 x 200 mm; length: 1300 mm)

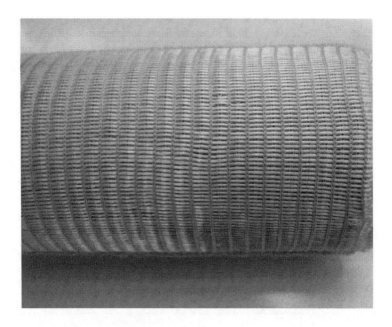

Figure 3. Example of monoaxialy reinforced woven fabric

Field of Application and End Users

- Load-adapted joints consist of a bifurcation made of beach wood and wooden tubes reinforced by knitted textile structures. This design concept may lead to new architectural design.

Structural Assessment

- The shear- and tension strength as well the durability of wood is improved. The load adapted orientation of the wood and textile fibers leads to a reduction and dissipation of stress concentration.

Economical Aspects

- Cheap, efficient use of high strength fibers

Robustness and Testing

Codification

References

- Haller, P; Birk, T.; Curbach, M. (Editor): *Der Einsatz von multiaxialen Nähgewirken und Biaxial-Gestricken zur Verstärkung von Holzkonstruktionen*. IN: PROCEEDINGS OF THE 2ND COLLOQUIUM ON TEXTILE REINFORCED STRUCTURES (CTRS2), Technische Universität Dresden: Eigenverlag, 2003 S. 235-246

Contacts

- Name: Peer Haller
  Address: TU Dresden
  Institute of Steel- and Timber Structures,
  Eisenstuckstraße 33
  01069 Dresden
  Germany
  Peer.Haller@mailbox.tu-dresden.de
  Tel. + 49 – 351 / 46 33 55 75
  Fax. + 49 – 351 / 46 33 63 08

COST C12 – WG1

Datasheet II.3.1.7.3
Prepared by: P. Haller[1]; J. Wehsener[1]
    [1] Institut of Steel- and Timber Structure, TU Dresden, Germany

Load Bearing Behaviour of Textile Reinforced and Densified Timber Joints

Description

- glass-fibre-reinforced timber joints
- densified wood with high density
- higher embedding strength for dowel type fasteners
- improvement of stiffness, strength and ductility

Figure 1. wood before and after the densification   timber beam with orthogonal glass fibre structure

Technical Information

Densified Wood

Densified wood was made of kiln dried spruce of good quality. A grading regarding knots, grain deviation and other characteristics was not done. The geometry of the sawn cross section was 2500 x 140 x 100 mm. Moisture content before and after densification was 12 to 15% and 7 to 9% respectively. The initial density ranged from 380 to 530 kg/m^3. The densification procedure took place in a conventional hot press.

The manufacturing of compressed wood was realized in three steps: heating, densification and cooling.

As wood remains still plastically deformable at 100 °C, it has to be cooled down under decreasing pressure to at least 80 °C. The densified laminations had no visible checks or deformations with a final thickness of about 50%. Figure 2 shows the densification process in the multiple daylight press.

    i)   multiple daylight (figure 2)

textile structure

Orthogonal glass fibre fabric with a weight of 300 g/m² were use as reinforcement.

Figure 2 Compression of the wood in a multiple daylight press Becker & Van Hüllen

Table 1 Glass fibre properties

Material	yarn count [tex]	Young´s modulus [N/mm^2]	modulus of rupture [N/mm^2]
E-glass-roving	2400	75000	3100-7000
woven fabric (E-glass)	138 (300g/m²)	12000	2400

unreinforced
0% densified

glass fibre reinforced
0% densified

glass fibre reinforced
50% densified

Figure 3 timber joints with 6 mm steel plates and 8 mm dowel typ fasteners

The load-carrying behaviour of the joints was tested for three different kind of loadings – tensile, shear and bending - according to DIN EN 26981 and DIN EN 408. The geometry of samples and test set-up for bending joints are shown in figure 4.

The bending specimen measured 120 x 200 x 3000 mm and consisted of gluelam made from ordinary and densified boards with an optional glass fibre reinforcement. The joint itself presented two steel plates of 6 mm thickness fixed with dowels of Ø 10 mm arranged in a rectangular shape.

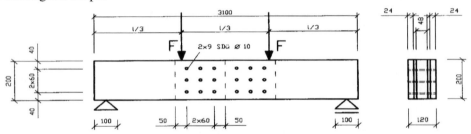

Figure 4 bending test set up and specimen

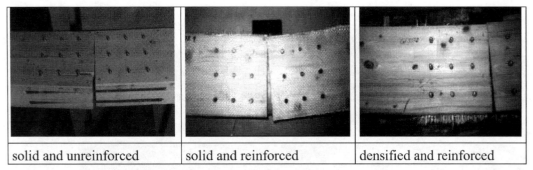

| solid and unreinforced | solid and reinforced | densified and reinforced |

Figure 5 specimen of bending test

The test on glass fibre reinforced and densified timber joints has shown that the load carrying capacity could be increased up to two times. Apart from the ultimate load, stiffness and ductility are improved considerably so that construction elements and parts could be prevented from sudden failure.

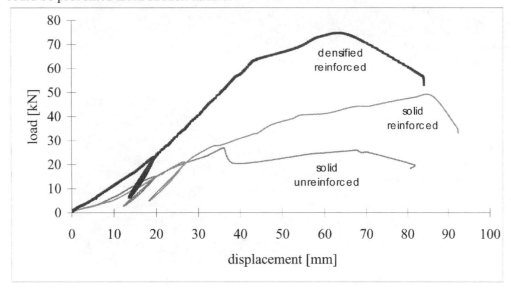

Figure 6 Load - displacement - relationship of bending stress

Field of Application and End Users
▪ Engineered timber structures
▪ Strengthening and repair of existing structures

Structural Assessment

▪ Advantages

higher strength and rigidity of timber joints

▪ Disadvantages

Avoid direct contact with water or high humidity especially in outdoor applications or control swelling effects

Economical Aspects

- reduction of cross section and weight in design
- reinforcement easy to handle in SME's

Robustness and Testing

- tensile, bending, compression and embedding tests on textile reinforced wood

Examples

Figure 7 bridge at the river "Simme" (CH) undensified with glass fibre structure

Figure 8. bridge "Baselerbieterstab" near Bubendorf (Siibedupf, CH) with glass fibre structure [Bauen mit Holz, 12. 2003, s.p. 10-13]

Codification

- DIN EN 383 embedding strength
- DIN EN 26891 timber structures

References

- Haller, P. 1997: Technische Textilien im Holzbau und ihre Möglichkeiten in der Verbindungstechnik. Bauen mit Textilien, Berlin: Ernst & Sohn
- Haller, P., Wehsener, J. 1999 *Use of technical and densified wood for Timber joints* Proceedings, 1[St] International RILEM Symposium on Timber Engineering, Stockholm
- Birk, T., Haller, P. 2001: *Textile Verstärkung von Holzbauteilen.* Textilbeton – 1. Fachkolloquium der Sonderforschungsbereiche 528 und 532, Aachen
- Haller, P.; Wehsener, J., 2003: Entwicklung innovativer Verbindungen aus Pressholz und Glasfaserarmierung für den Ingenieurholzbau, Fraunhofer IRB Verlag Stuttgart

- Zeuggin, N., 2003: Temporäre Rautenfachwerkbrücke mit glasfaserarmierten Verbindungen, Bauen mit Holz 12 / 2003; Karlsruhe, Bruderverlag
- Haller, P.; Wehsener, J., Glass fibre Reinforcement on Timber Joints, in proceeding: Joints in Timber Structures, [ea. Aicher, S. ;Reinhardt H.W.), Publisher RILEM 2001, p 243-252

Contacts

Prof. Dr.-Ing. Peer Haller
Dresden University of Technology
Faculty of Civil Engineering
Mommsenstraße 13   Tel.:   ++49 351 463 36305
D - 01062 Dresden   Fax.:   ++49 351 463 36306
Germany   e-mail: peer.haller @mailbox.tu-dresden.de

COST C12 – WG1

Datasheet II.3.1.7.4

Prepared by: M. Brunner, M. Schnüriger[1]

[1] School of Architecture, Civil and Wood Engineering HSB,
Biel / Switzerland, www.hsb.bfh.ch

STRENGTHENING OF TIMBER BEAMS WITH PRESTRESSED FIBRES: DEVELOPMENT OF A SUITABLE ADHESIVE AND TECHNOLOGY

Description

- Timber beams with prestressed fibres glued on the tensile side

- Suitable for long span structures. Reduction of the displacements due partly to the higher stiffness of the timber construction and mainly to the prestressing of the fibres

- The delamination problem has prevented to date the application of high prestressing forces. The authors have envisaged two possible solutions:

    i. Prestressing the fibres with the gradiented anchoring system developed by the EMPA (Swiss Federal Materials Testing Institute) to strengthen existing concrete structures. This project marks the first time that this system has ever been used to strengthen timber elements.

    ii. Developing a special, "ductile" adhesive

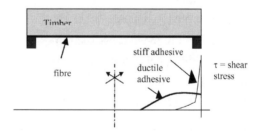

Figure 1. Detail of the prestressing device during the prestressing process of a timber beam

Figure 2. Spreading of the shear stresses over a larger area by using a ductile adhesive

Technical Information

- Timber beam: solid wood or glulam

- Carbon fibres

- Prestressing of carbon fibres with the gradiented anchoring system developed by EMPA, Switzerland. After the fibre has been prestressed with the full force, it is attached in stages starting from the centre of the beam by activating a specially designed epoxide through the local application of heat. At each step the prestressing force is slightly reduced, so that the prestressing force is in effect anchored over the entire beam length.

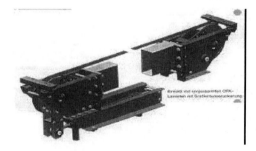

Figure 3. EMPA Prestressing device with a gradiented anchoring system

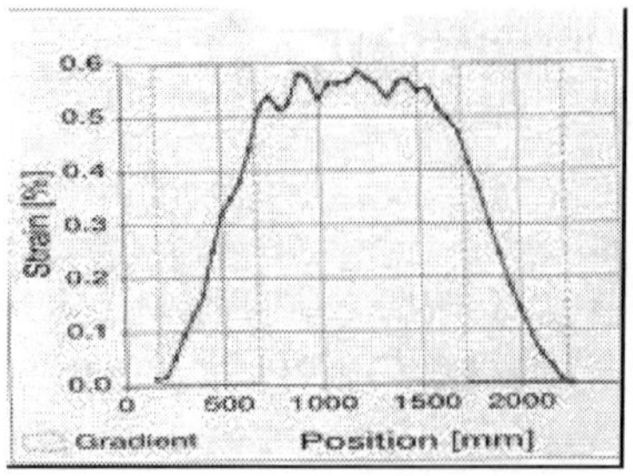

Figure 4. Strain distribution of a prestressed CFRP strip showing the gradiented prestressing force along the beam length

- Prefabrication of timber elements or prestressing in situ
- Development of a "ductile" adhesive. After some positive preliminary research work, the newly developed adhesive was used to attach a prestressed CFRP strip to a timber beam. Unfortunately, it failed to maintain the prestressing force, which was reduced to zero within a few days. More research works is necessary before this second system can be used.

Field of Application and End Users

- Residential and industrial constructions:
    i.   Beams or floors
    ii.   Elements with long spans
    iii. Elements with reduced cross section
    iv.   Elements with special displacement requirements
- Bridge structural elements, eg. cross beams
- Strengthening of structural elements

Structural Assessment

- Advantages:
    i)   Light structure: the wooden section of a prestressed element is smaller in comparison with a normal timber beam with the same load and span
    ii)   Longer spans can be attained than with a normal timber beam with the same cross section and the displacements are smaller because of the higher stiffness of the elements and the negative displacement induced by the prestressing force
    iii) Higher loads are possible
    iv)   The prefabrication of elements under factory conditions or the prestressing on site are possible with the gradiented anchoring system
- Disadvantages:
    i)   Ductile adhesive: at the moment not applicable. More research is necessary.
    ii)   Measures against fire and corrosion substances are necessary
    iii) The prestressing force may decrease with time. More research is necessary to find the relevant parameters and to develop appropriate calculation models.

Economical Aspects

- Savings in material
- Prefabrication, fast erection

Robustness and Testing

Developement of a ductile adhesive

- Analysis of adhesives mixes and tensile/shear tests:
    i.   Two main components on epoxide and polyurethane basis were mixed in different proportions and their characteristics, in particular the hardening times, were determined.
    ii. Three different glue mixes were chosen and applied, first in thin layers (less than 0.1mm) as used in industry, and then in thicker (0.6mm) layers. The specimens were used for shear tests.
- Pull-off tests:
    i.   The glue mix which exhibited the best results for the shear tests was selected and used to attach a slack laminate of carbon fibre to a timber board.
    ii. Due to the favorable results of the tensile shear tests, there were high expectations of the chosen mix. Unfortunately, the initial promise could not be confirmed in the pull-off tests

Figure 5. Set-up for shear tests according to German standard DIN prEN 205

Figure 6. Set-up of tensile-shear tests

- Bending tests with prestressed glulam beams
    i.   Due to the positive results of the preliminary tests, the best glue mix was chosen to attach a prestressed laminate to a glulam beam. Unfortunately, the attempt failed: the prestressed force of 30kN could not be maintained and was reduced to zero within a few days.
- It was not possible to find the reason for the premature failure within the framework of the project. More research work is necessary before the system can be safely applied.

Gradiented anchoring technique

- Bending tests with prestressed glulam beams
  - i. In a first series of tests, glulam beams were strengthened with a carbon laminate prestressed to 60kN.
  - ii. In a second series, glulam beams were strengthened with an unstressed carbon laminate of the same size. For control purposes, some normal glulam beams of the same size were also prepared.
  - iii. The results demonstrate how a glulam beam could be strengthened with a carbon fibre laminate both unstressed and prestressed. The tests also demonstrate that the strengthening with a prestressed laminate is more effective than the use of a slack laminate: the bending resistance of the naked glulam beam was increased by 32% when strengthened with a prestressed laminate as against 21% when the laminate is not prestressed.

 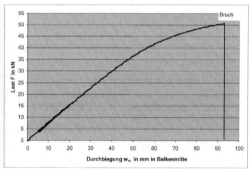

Figure 7. Four-point bending test according to German standard DIN EN 408

Examples

- Prestressed timber beams have not yet been used in practical structures

Figure 8. Prestressing process of a concrete beam

Figure 9. Bending test of a prestressed timber beam

Codification

- EN 205; *Adhesives - Wood adhesives for non strucural applications - Determination of tensile shear strenght of la joints; German Version EN 205/2003*
- prEN 408; *Timber structures - Structural Timber and glued laminated timber - Determination of some Physical and mechanical properties, German Version pr EN 408/2000*

References

- Tingley D.,*FIRP Reinforcement Technology Information Packet*, Science and Technology Institute, Corvallis OR, USA 1995

- Romani M, Blass HJ, *Design model for FRP reinforced glulam beams*, International Council for Research on Innovation in Building and Construction, Working Commission W18 - Timber Structures, Meeting 34, Venice, Italy, August 2001

- Stöcklin & Meier U., *Strengthening of concrete structures with prestressed and gradually anchored CFRP strips*, IABSE International Conference, Malta 2001

- Luggin W. F., *Die Applikation vorgespannter CFK-Lamellen auf Brettschichtholzträger*, Dissertation, Universität für Bodenkultur, Vienna, Austria, 2000.

- Holzenkämpfer P., *Ingenieurmodelle des Verbunds geklebter Bewehrung für Betonbauteile*, Deutscher Ausschuss für Stahlbeton, Heft 473, Beuth Verlag 1997.

- Triantafillou T.C., Deskovic N., "Innovative prestressing with FRP-sheets: Mechanics of short-term behaviour", *Journal of Engineering Mechnanics*, Vol. 117, Nr. 7, July 1991, pp1652-1672

- Triantafillou T.C., Deskovic N., "Prestressed FRP-Sheets as external reinforcement of wood members", *Journal of Structural Engineering*, Vol 118, Nr. 5, May 1992, pp1270-1284

- Gustafsson J., *Tests and Test Results on Mechanical Properties of Adhesive Bond Lines*, Chapter 2 of Final Report, COST E13, Version 4, Jan. 2000

Contacts

- Dr. Maurice Brunner, Marco Schnüriger

    Address: School of Architecture, Civil and Wood Engineering HSB, Solothurnstrasse 102, Biel, Switzerland

    email: maurice.brunner@hsb.bfh.ch
    marco.schnueriger@hsb.bfh.ch

    Tel. +41 32 344 03 41
    Fax. +41 32 344 03 91

MATERIAL BEHAVIOUR OF PVB-INTERLAYER IN LAMINATED SAFETY GLASS

Description

- polymer/glass laminates are used for load bearing elements in buildings (transparent and filigree constructions)
- load bearing capacity of the PVB-interlayer (polyvinyl butyral) should be considered in static calculation by short time loading
- laminated safety glass are comprised of not less than two glass (float glass, toughened glass or heat strengthened glass) planes bonded with an interlayer of plasticized polyvinyl butyral, Figure 1
- load transmission of laminated safety glass results from shear force transfer by PVB-interlayer → shear modulus as a mechanical property of shear behaviour
- the stiffness of PVB-interlayer is dependent on parameters as temperature, loading time, rate of loading and amplitude of loading

PVB-interlayer

glass pane

Figure 1. Structure of laminated safety glass

Technical Information

- Material
    - i) glass pane
    - ii) PVB-interlayer

- glass shows a linear-elastic material behaviour, Young's modulus = 70000 N/mm², Poisson's ratio = 0.23, destiny = 25 kN/m³, compressive strength > 800 N/mm², flexural tensile strength up to 120 N/mm² (toughened glass)
- PVB-interlayer is a amorphous thermoplastic resin (polymer basic material)
- amorphous thermoplastics show a visco-elastic material behaviour
- PVB-interlayer can behave either linear or non-linear visco-elastic depending on the amplitude of loading
- glass transition temperature ranges between 12°C and 16°C

- material behaviour is dependent on parameters as temperature, loading time, rate of loading and amplitude of loading
- the thermal and time-dependent material behaviour of amorphous thermoplastic resin can be describe with the mathematical model of relaxation (extended Maxwell-model), creep (extended Kelvin-Voigt-model) and time-temperature superposition (Williams-Landell-Ferry equation)
- relaxation tests by torsional loading by different temperatures to assess the shear modulus of PVB-interlayer
- shear modulus of the PVB-interlayer can be calculated from torsional rigidity of the composite section
- more over the rotation angle of the beam as well as the torsional stress can be calculated as follows

rotation angle of the beam:

$$\vartheta_T(x) = \frac{M_t}{G \cdot S_{11}}\left(1 + \frac{S_{12}^{\,2}}{S_{11} \cdot \tilde{S}_{22}}\right) \cdot x$$

torsional stresses:

$$\tau_T = \frac{M_t}{S_{11}}\left(2y + 2y \frac{S_{12}^{\,2}}{S_{11}} \frac{1}{\tilde{S}_{22}} + \frac{1}{2}\frac{S_{12}}{\tilde{S}_{22}}\right) \qquad \text{für } y > 0$$

$$\tau_T = \frac{M_t}{S_{11}}\left(2y + 2y \frac{S_{12}^{\,2}}{S_{11}} \frac{1}{\tilde{S}_{22}} - \frac{1}{2}\frac{S_{12}}{\tilde{S}_{22}}\right) \qquad \text{für } y < 0$$

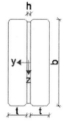

$$S_{11} = \frac{2}{3}b \cdot t\left(4t^2 + 6th + 3h^2\right)$$

$$S_{12} = -b \cdot t(t + h)$$

$$S_{22} = \frac{1}{2}b \cdot t$$

$$S_{22,F} = \frac{1}{12}\frac{b^3}{h}$$

$$\tilde{S}_{22} = -\frac{S_{12}^{\,2}}{S_{11}} + S_{22} + r_G \cdot S_{22,F}$$

- the relaxation tests can be simulated with finite element method by modelling the PVB-interlayer as a linear visco-elastic material
- the stiffness of PVB-interlayer can be considered in static calculation by short time loading

Field of Application and End Users

- Residential housing
    - i) load bearing elements (glass beam, glass panel, glass roof etc.)
    - ii) facade
- Industry buildings
    - i) load bearing elements (see above)
    - ii) facade

Structural Assessment

- Advantages: transparent and filigree construction, aesthetic aspects
- Disadvantages: glass is a brittle construction material, risk of fracture, stiffness of PVB-interlayer depends on temperature and loading time
- Potentials: usage alternative intermediate layer by laminated safety glass with higher stiffness as PVB-interlayer by elevated temperature

Economical Aspects

- economic static calculation with involvement of stiffness of the PVB-interlayer by short time loading

Robustness and Testing

- relaxation tests by torsional loading to assess the shear modulus of PVB-interlayer, Figure 2 und 3
- dimensions of the laminated safety glass: 1100 mm x 360 mm x thickness (6, 8 or 10 mm)

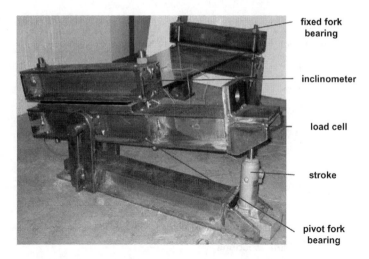

Figure 2. Experimental equipment under torsional loading and the assembly of measuring instruments

Figure 3. Static system of torsion tests

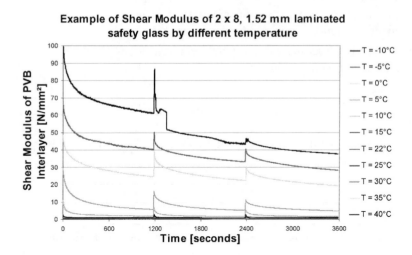

Figure 4. Shear modulus by temperatures between -10°C and 40°C

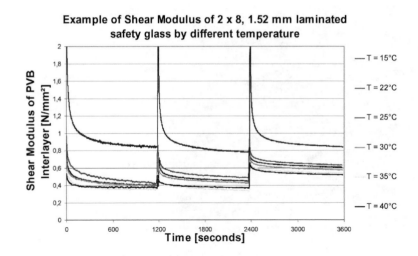

Figure 5. Shear modulus by temperatures between 15°C and 40°C

- rotation angle of the laminated safety glass by 2°, 4° and 6°
- decrease of the stiffness of the PVB-interlayer within 1200 seconds, Figure 4 and 5
- the stiffness of the shear modulus is dependent on temperature and loading time
- the stiffness of PVB-interlayer shows high scattering
- the thickness of the glass pane (2 x 6 mm, 2 x 8 mm or 2 x 10 mm) have no bearing on the stiffness of the PVB-interlayer

Examples

- Examples for possible field of application of glass load bearing elements in Construction Engineering
- Figure 6: Chamber of industry and commerce in Munich
    - i) Dimensions of the glass roof: 14 x 14 m
    - ii) Glass roof is consists of 5 main glass beams and crossbeams at intervals of 2 m
- Figure 7: Glass winter garden in London
    - i) Usage of transparent silicone
    - ii) No steel joints in visible area
    - iii) Beams and column are consist of 3 x 10 mm laminated safety glass

Figure 6. Chamber of industry and commerce (Munich)

Figure 7. Winter garden built completely in glass (London)

Codification

- DIN EN ISO 12543: *Glas im Bauwesen - Verbundglas und Verbund-Sicherheitsglas*
- DIN 1249: *Flachglas im Bauwesen*

References

- Knaack U.: *Konstruktiver Glasbau; Verlagsgesellschaft Rudolf Müller, Bau- und Fachinformationen GmbH & Co. KG, Köln 1998*
- Glas: *Architektur und Technik Ausgabe 1/2003 Transparente Dächer*
- Scarpino P.: *Berechnungsverfahren zur Bestimmung einer äquivalenten Torsionssteifigkeit von Trägern in Sandwichbauweise, Diplomarbeit, RWTH-Aachen, Lehrstuhl für Stahlbau, 2002*
- Langosch K.: *Experimentelle und numerische Untersuchung von Verbundglasscheiben unter dem Aspekt des visko-elastischen Materialverhaltens des Verbundmittels PVB, Diplomarbeit, RWTH-Aachen, Lehrstuhl für Stahlbau, 2004*
- Kasper R.: *Untersuchungen der Tragsicherheit von Verbundglasträgern, Dissertation, RWTH-Aachen, Lehrstuhl für Stahlbau, expected Herbst 2004*

Contacts

- Katharina Langosch
    Address: Institute of Steel Construction, RWTH Aachen,
    Mies-van-der Rohe Str. 1, D-52074 Aachen, Germany
    www.stb.rwth-aachen.de
    Fax. +49/ (0) 241 – 80-22140
    e-mail: katharina@stb.rwth-aachen.de
- Ruth Kasper
    Address: Institute of Steel Construction, RWTH Aachen,
    Mies-van-der Rohe Str. 1, D-52074 Aachen, Germany
    www.stb.rwth-aachen.de
    Tel. +49/ (0) 241 – 80-25181
    Fax. +49/ (0) 241 – 80-22140
    e-mail: ruth.kasper@stb.rwth-aachen.de

CONNECTION OF REPAIR SYSTEMS AND STEEL SUPPORT FOR STRENGHTENING AND INCREASING DURABILITY OF THE THIN EXTERNAL REINFORCED CONCRETE ELEMENTS

Description

- Key words: concrete, reinforced concrete, repair system of concrete, strengthening, durability,
- concrete repair systems based on cement mortar or concrete, modified with polymer, concrete repair system consists of:
  - ✓ connecting (bridging) layer
  - ✓ covering preparation for reinforcing steel
  - ✓ repair mortar or repair concrete
- in case very advanced degradation of reinforced concrete element, steel structure for supporting of damage element is required

Technical Information

- the thin elements in the buildings that are particularly exposed to weather are the balconies, the roofs under the entrance and the architectural grates
- different methods of structure repair corresponding to the extent of the damage were adopted, according to code [1]:
  - ✓ in the case of the balcony slabs of III and IV degree damage, the repair was preceded by hammering off the corroded concrete (Fig. 1), cleaning the corroded reinforcement up to degree II cleanness and cleaning and moisturizing the concrete.

Figure. 1. View of balconies ready for repair

- ✓ the extent of hammering was decided during the repair and the minimum was about 15 cm wide belts along the balcony's perimeter. In some cases all the corroded concrete was removed.
- ✓ two repair methods were applied with regard to the extend of the concrete loss and ADDIMENT POLYMENT and COMPAKTA systems [1, 2, 3, 4, 5] were used, small losses (down to 5 cm deep) were repaired using ready-made POLYMENT

preparations, bigger losses were filled with cement mortar or concrete mix modified with polymer Compakta Baudispersion emulsion,

✓ in every case, before connecting the old material with the new one, a bridging layer was made,

✓ in the case of V degree damage slabs an extra support steel structure was made (Fig. 2) adjusted each time to the local conditions.

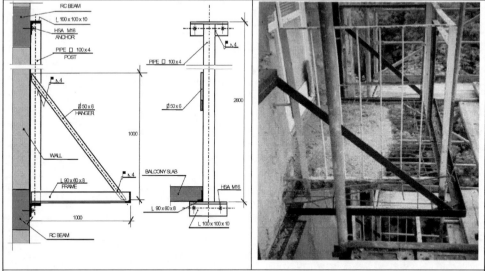

Figure 2. Supportive steel structures of balcony slabs

✓ the flashing and cement screet (also modified with polymer emulsion) was being done, the reinforced concrete structure of the balcony is subjected to the same factors that brought about its damage.

✓ in order to preserve the usability of the balcony, it is necessary to prevent destructive effect of the atmosphere, it can be done in a few ways, two the most frequently used are mentioned below (Fig. 3):

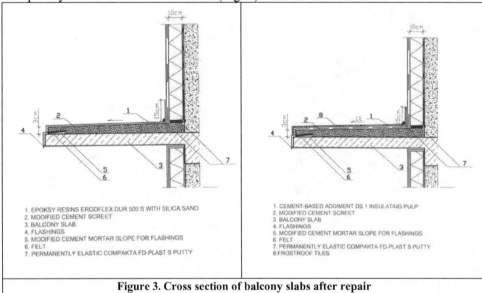

Figure 3. Cross section of balcony slabs after repair

Field of Application and End Users

- every kind of structures consists thin external reinforced concrete elements

Structural Assessment

- Advantages:
  repair systems and strengthening give the possibilities to restore the usability of RC elements and to increase durability of structures
- Disadvantages:
  repair is very expensive, time-consuming and laborious, when the degradation of element is very advanced

Economical Aspects

- giving possibilities to increase the period of structures usability significant

Robustness and Testing

every components of repair systems has been tested by Institute of Building Technology in Warsaw and have got adequate signature in 1997 – 2004, a preservation assessment was

Examples

Figure 4. Façade of the building before and after repair

carried out whose results were used to determine the extent of repairs, to prepare repair design and to properly supervise the repair of about 50 multi-floor apartment building in Rzeszów and town near Rzeszów.

Figure 5. Balcon after repair

Codification

- Standard: PN-88/B-01807: Protection against corrosion in construction industry. Concrete and reinforcement concrete structures. The rules of diagnosis structures.

References

1. K. Wróbel, W. Kubiszyn: Technical state estimation of residential building's balcony …….. in Rzeszow. Rzeszow 1998
2. K. Wróbel, W. Kubiszyn: Repairing design of reinforced concrete balconies and architectural grates of residential buildings ……. in Rzeszow. Rzeszów 1998.
3. K. Wróbel: Technical state estimation and repairing design of reinforced concrete balconies of living multi-flat building ……….. in Rzeszow. Rzeszów 2000.
4. K. Wróbel, W. Kubiszyn: Technical state estimation and repairing design of reinforced concrete balconies of living multi-flat building ……….. in Rzeszow. Rzeszów 2002.
5. Technical cards of systems ADDIMENT, POLYMENT i COMPACTA.

Contacts

- Name

  Address: Department of Building Structures,
  Faculty of Building and Environmental Engineering,
  Rzeszów University of Technology,
  35-959 Rzeszów, W. Pola 2
  email: wrobel@ prz.edu.pl
  Phone: +48 17 854 2974
  Fax.: +48 17 854 2974

COST C12 – WG1

Datasheet II.3.2.2.1

Prepared by: A. Mandara
[1] Second University of Naples
F.M. Mazzolani[2]
[2] University Federico II of Naples

CONFINEMENT OF MASONRY MEMBERS
BY STEEL ELEMENTS

Description

- The basic principle of confinement is the application of a compressive action in one or more directions transverse to that of the applied load, so to achieve conditions of multi-axial compressive stress (Figure 1). A quite rational structural system is obtained, in which *in-situ* materials are exploited in the most rational and effective way. This leads to both an increase of the member compressive strength compared with the case of uniaxial stress, and also to a remarkable improvement of the ductility properties at failure;

- Confinement interventions can be applied not only to single structural elements, like columns or walls, but also to a greater extent, for example encircling the whole building with a suitable system of tensioned members. Most of times, it is applied in order to increase the load bearing capacity in order to withstand new loads, or to make for structural modifications of the building (Figure 2). The most common way to apply confinement is by means of steel elements working in tension, such as tie-bars or tie-beams, fastened to the masonry by means of contrasting end-plates, as shown in. To this purpose, most adopted material in such interventions is steel, both mild, low carbon and high strength grade. Tension in the ties can arise either as a consequence of member lateral expansion, or due to external prestressing.

- Confinement offers ease of application and reversibility. A long-term adjustment of the confining force is also possible. In this case, due to the use of externally fastened plates, the intervention can be arranged in such a way to be easily controlled or removed, if necessary. This aspect complies with a commonly adopted policy, aiming at preventing existing buildings, in particular when they possess monumental value, from irreversible and/or inappropriate restoration operations.

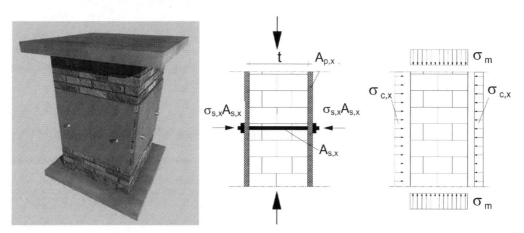

Figure 1. The basic principle of uniformly confined masonry

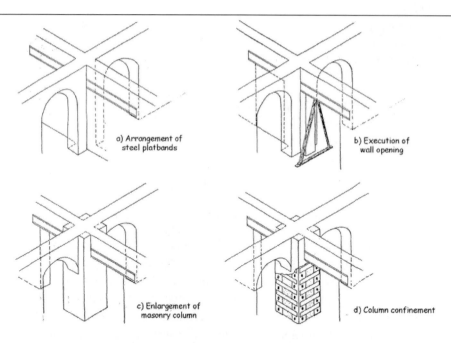

Figure 2. Typical situation demanding for confinement of a masonry member

Technical Information

Collapse in confined masonry can occur in three ways:

- bar yielding (Figure 3a);
- crushing (punching) of the masonry in the confined area (Figure 3b);
- shear-tension failure of masonry in unconfined areas (Figure 3c).

The main geometrical magnitudes influencing the collapse mechanism are the plate-to-bar cross section ratio (A_p/A_s) and the ratio between the unconfined wall length and the wall thickness (i/t). The use of small plates gives place to failure by punching of the masonry, whereas a comparatively great distance between plates involves collapse by shear-tension in the unconfined area. Relatively high values of A_p/A_s, instead, produce failure by bar yielding, which is also the collapse type providing the best performance in terms of ductility (Figure 4).

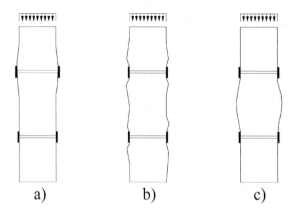

Figure 3. Collapse types of confined masonry: a) bar yielding; b) masonry crushing in the confined area; c) shear-tension of masonry in unconfined areas

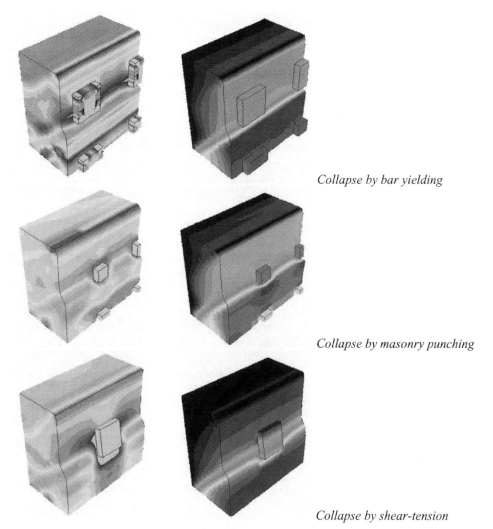

Collapse by bar yielding

Collapse by masonry punching

Collapse by shear-tension

Figure 4. Contour of Von Mises equivalent stress (left) and horizontal displacement (right) for the three collapse mechanisms obtained from a F.E.M. analysis.

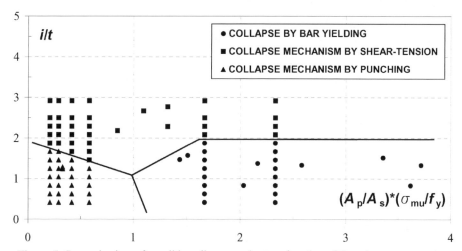

Figure 5. Synopsis view of possible collapse modes as a function of the relevant parameters: plate-to-bar cross section ratio (A_p/A_s); ratio between the unconfined wall length and the wall thickness (i/t); ratio between the unconfined masonry resistance and steel yield stress (σ_{mu}/f_y).

Collapse by bar yielding can be obtained with a close spacing between confining plates. The diagram of Figure 5 can be used for design the confinement intervention in such a way to have such a kind of failure. A method for the prediction of the load bearing capacity of confined masonry members, also accounting for different collapse mechanisms, is presented in Mandara & Mazzolani, 1998, Mandara & Scognamiglio, 2003 and Mandara & Palumbo, 2004.

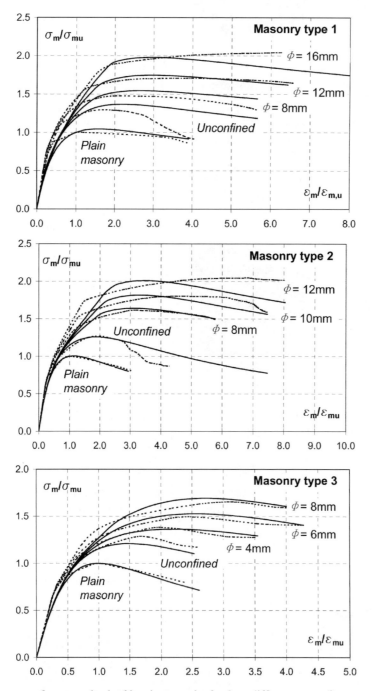

Figure 6. Increase of compressive load bearing capacity for three different types of masonry as obtained from theoretical (solid line) and F.E.M. model (dotted line) for uniform confinement of masonry walls as a function of steel bar diameter φ (Mandara & Palumbo, 2004).

Field of Application and End Users

Such technique can be profitably used in any strengthening operation of masonry buildings.

    i)   Increase of load bearing capacity of masonry elements loaded in compression in order to withstand new load or to undergo seismic retrofit;

    ii)   Recovery of existing damage in masonry members due to excess of axial load;

    iii)   Confinement of a whole building for improving its seismic performance;

Structural Assessment

- Confinement of masonry walls and columns
  - Advantages

    Increase of load bearing capacity with a negligible increase of weight;

    Increase of ductility at failure;

    Reduction of the risk of instability for slender elements;

    Possibility to recover cracks existing in the masonry;

    Inexpensive and reversible;
  - Disadvantages

    The degree of confinement can change over time due to anelastic deformation of masonry;

    The intervention cannot be easily hidden, in particular when adjustment is required;

- Building confinement
  - Advantages

    Increase of building resistance under seismic loads;

    Prevention of both in-plane and out-of-plane collapse mechanism of walls;

    Inexpensive and reversible;
  - Disadvantages

    Great effectiveness only in case of high-quality masonry buildings;

    Metal ties can undergo corrosion if not properly allocated and, if necessary, maintained;

- Potential

  A great enhancement of the structural performance can be achieved with the addition of few metal elements;

- Risk

  The long-term reliability of the system could be impaired by a poor maintenance and/or by an improper allocation;

Economical Aspects

- Such technique is by far the most convenient also from the economic point of view compared with others techniques;

Robustness and Testing

- Both effectiveness and reliability of this technique have been widely tested and verified in a millenary experience carried out on masonry buildings all over the world;

Examples

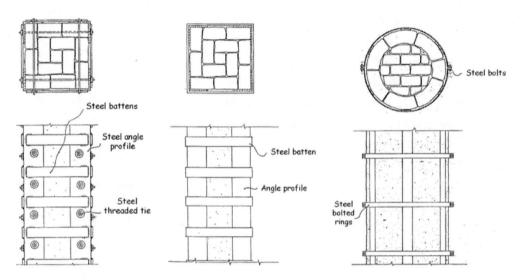

Figure 7. Confinement of square and circular columns

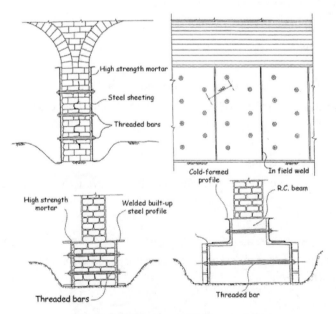

Figure 8. Confinement of masonry walls

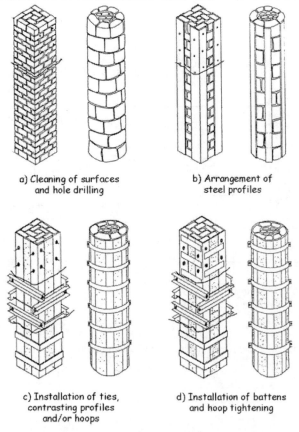

a) Cleaning of surfaces
and hole drilling

b) Arrangement of
steel profiles

c) Installation of ties,
contrasting profiles
and/or hoops

d) Installation of battens
and hoop tightening

Figure 9. Typical execution stages of column confinement by means of steel profiles

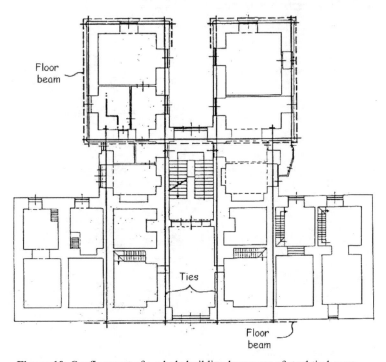

Floor beam

Ties

Floor beam

Figure 10. Confinement of a whole building by means of steel tie-beams

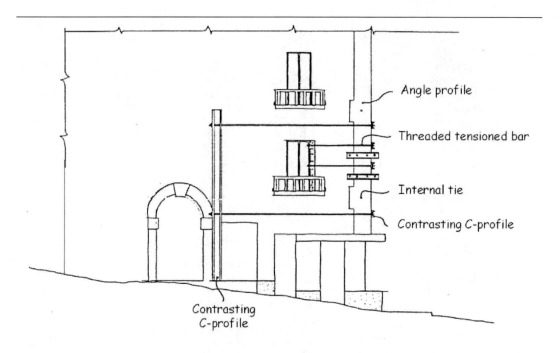

Figure 11. Detail of contrasting elements in building tying

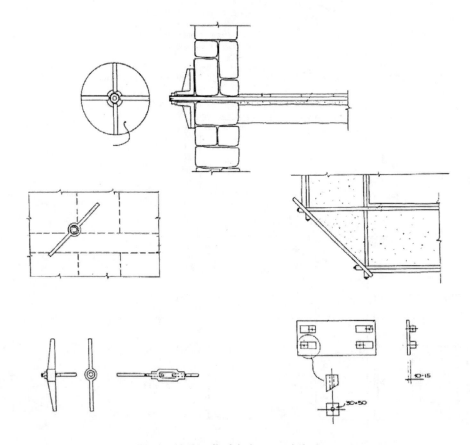

Figure 12. Detail of tie-beam end elements

Codification

- FEMA 356. (2000). Prestandard and Commentary for the Seismic Rehabilitation of Buildings. Report by the American Society of Civil Engineering (ASCE) for the Federal Emergency Management Agency (FEMA), Washington, D.C..
- FEMA 273 (1997), NEHRP Guidelines for the Seismic Rehabilitation of Buildings, Building Seismic Safety Council, Washington,D.C., USA (FEMA 274 Commentary).

References

- Ballio G. & Calvi G.M. 1993. Strengthening of Masonry Structures by Lateral Confinement, Proc. of IABSE Symp. *Structural Preservation of the Architectural Heritage*, Rome.
- Faella G., Manfredi G. & Realfonzo R. 1993. Stress-Strain Relationships for Tuff Masonry: Experimental Results and Analytical Formulations. *Masonry International*, Vol. 7, No. 2.
- Mandara A. 2002. Strengthening Techniques for Buildings, in Refurbishment of Buildings and Bridges (F.M. Mazzolani & M. Ivanyi Eds), Springer Verlag, Wien-New York.
- Mandara A. & Mazzolani F.M. 1998. Confining of Masonry Walls with Steel Elements, Proc. of Int. IABSE Conf. *Save Buildings in Central and Eastern Europe*, Berlin.
- Mandara A. & Palumbo G. 2004. Confined masonry members: a method for predicting compressive behaviour up to failure. Proc. of the *International Seminar on Structural Analysis of Historical Constructions*, Padova, Italy.
- Mandara A. & Scognamiglio D. 2003. Prediction of Collapse Behavior of Confined Masonry Members with ABAQUS, Proc. of the *ABAQUS Users' Conference*, Munich.
- Mazzolani F.M. 1996. Strengthening Options in Rehabilitation by means of Steelwork, Proc. of *SSRC International Colloquium on Structural Stability*, Rio de Janeiro.
- Mazzolani F.M. 2002. Principles and Design Criteria for Consolidation and Rehabilitation, in Refurbishment of Buildings and Bridges (F.M. Mazzolani & M. Ivanyi Eds), Springer Verlag, Wien-New York.
- Mazzolani F.M. & Mandara A. 2002. Modern Trends in the Use of Special Metals for the Improvement of Historical and Monumental Structures, *Engineering Structures 24*: 843-856, Elsevier.
- Sargin M. 1971. Stress-Strain Relationship for Concrete and Analysis of Structural Concrete Sections, *Study n. 4, Solid Mechanics Division, University of Waterloo, Canada.*

Contacts

- Alberto Mandara
  Address: Department of Civil Engineering, Second University of Naples, Via Roma 29, I-81031, Aversa (CE), Italy
  E-mail: alberto.mandara@unina2.it
  Tel. +39-081-5010216
  Fax. +39-081-5037370

- Federico M. Mazzolani
  Address: Department of Structural Analysis and Design, University of Naples Federico II, P.le Tecchio 80, I-80125 Napoli, Italy
  E-mail: fmm@cds.unina.it
  Tel. +39-081-7682443
  Fax. +39-081-5934792

STRENGTHENING OF TIMBER BEAMS
WITH CONCRETE IN AN ANCIENT MONASTERY

Description

- strengthening of an existing slab made of timber beams in an ancient building
- shear transfer between timber beams and the added concrete by means of screwed dowels
- composite slab in final state for better ULS and SLS

Figure 1. View of the building from outside

Technical Information

- adapting the existing ceiling in a monastery for new use
- future loads will cause further uses which are going to exceed the load capacity of the existing timber beams
- working out a solution in which, additionally to the timber beams, also the reinforced concrete beams with a above lying and running through concrete slab, have been put in
- composite action between timber and concrete with the help of screwed plugs
- increasing the stiffness of the ceiling in the final condition
- improving the load bearing capacity as well as the suitability of use
- the underside of the ceiling could not be destroyed because of the protection of historical buildings

- explaining all the works from above
- Working-steps:
  - taking down the surface and the rubble lying between the beams
  - relieving the ceiling
  - blank formwork and fixing the ceiling for that load-condition
  - direct use of the composite cross section because of all the following loads
  - plugging the beams and laying the reinforcement
  - concreting the ceiling

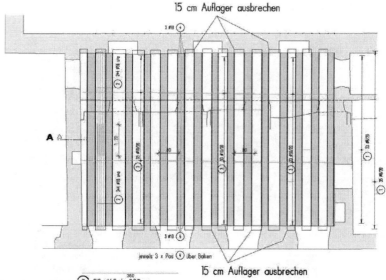

Figure 2. Plan view of slab

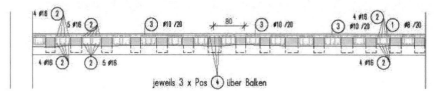

Figure 3. Cross section of slab

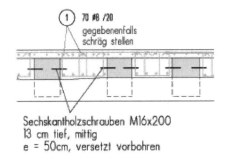

Figure 4. Detail of slab

Figure 5. Laying of reinforcement

Figure 6. Final state of the slab from below

Fields of Application

- solution for existing timber beams to create composite slabs with higher stiffness and improved ULS and SLS behaviour

Structural Assessment

- Advantages:

  making use of existing timber beams

  no need of destroying existing bearing structures

  higher stiffness leads to improved ULS and SLS

  slim floor depths
- Disadvantages:

  bringing in of screwed dowels and reinforcement between narrow arranged timber beams

Economical Aspects

- material savings
- fast erection
- only partly required interventions into the existing building structure

Codification

- EN-1995-1-1. *Design of timber structures*
- DIN 1052-1 Holzbauwerke

References

- Holz Beton Verbunddecke, Hartmut Werbner, Bauen mit Holz 1992
- Holz Beton Verbund im Hochbau, u. Meier, Schweizer Ingenieur und Architekt, 1994

Contact

- Dr. Christian Schaur

  Address: Dr. Ambros Giner Weg 30, A-6065 Thaur, Austria
  e-mail: schaur@tirol.com
  http://www.schaur.com

Prepared by: B. Vektaris[1], V. Barkauskas[1], R. Simkus[1]

[1] Institute of Architecture and Construction of Kaunas University of Technology
Kaunas/Lithuania, www.asi.lt

FRACTURE AND FINISHING OF HISTORICAL CERAMIC MASONRY

Description

- Physico – mechanical properties of the external surfaces of the walls of historical buildings depend on the environmental effects and the period of exploitation.

- Investigation of the material's physico – mechanical properties shows that an increase of decay, porosity and microporosity is the result of the alteration of the material's, structure as well as of the destruction of the crystal lattice [Miniotaite, 2001].

- With the use of renovation of historical buildings, new materials and new technologies it is urgent to find out what to do with the destructed masonry and how to guarantee structural integrity of the damaged and ageing masonry.

Methods and Structural Aspects

- Affected by the atmospheric and other external factors the construction material gradually gets mechanically weaker; its porosity and microporosity increase, its density decreases, water absorption increases, and so on.

- The general dependence of the structural alterations of the material during the process of its ageing is expressed by the ratio of the amount of micropores in the general porosity with the number of the years of exploitation [Barkauskas, 1958].

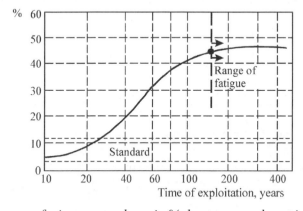

Figure 1. *Increase of micropore volume in % due to general part in total porosity*

- At the reach of a concrete limit point (a dot on the curve), the stabilization of the alteration of the material's microstructure takes place. It could be named as the limit of the material's fatigue ρ and expressed in the following way:

$\rho = p_m(1 - p_m)$, where p_m – porosity of the material by the parts of the unit.

- With the increase of micropores and microcracks in the material, their interior surface also increases. Water vapour condenses on their walls thus forming microfilms that have the properties of a hard body and therefore, when penetrating into the micropores, they crack the crystal – type structure of the material [Deriagin, 1937; Barkauskas, 2001].

- These statements are confirmed by the measurements of the ultrasonic speed along and across the layers of the material as well as by the data of the depth of the capillary water absorption. With the deterioration of the crystal structure of the material, the capillary gravity of the material decreases as well as intensity of the absorption (Figure 2).

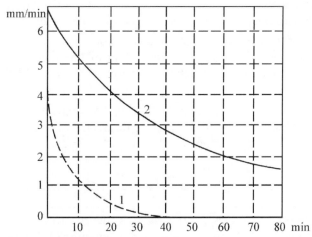

Figure 2. *The dependence of the intensity of the capillary absorption in new (Curve 1) and old (Curve 2) samples taken out of the walls on the time of soaking*

- By avoiding approvement and mathematical deduction and by relying on the equations offered by Grinsburg and Zalieskis [1950], the following equations of an average increase of the porous area ΔU, cm^2/g and the speed of material destruction K, m^2/g might be offered:

$$\Delta u = \delta \cdot \rho \cdot 10^{-2}(\omega_k - \frac{b}{S+c})$$

$$K = \frac{\Delta u}{u} = 0,6 \cdot 10^{-3} \frac{g \cdot \sqrt{\frac{\psi}{p}}}{\delta \cdot v \cdot p} (\omega_k - \frac{b}{b+c}),$$

where : δ – thickness of average film moisture, mm;
ρ – water density, g/cm^3;
ω_k – the upper limit of the increase of film moisture, cm^2;
c – number of the cycles of affection equivalent to the number of the years of exploitation;
S and b – the coefficients estimating the structure of the material and the experimental conditions;
ψ - air permeability of the sample, $\frac{cm^3}{cm \cdot s \cdot mm}$;
v – sample volume, cm^3;
p – sample porosity, %;
g – sample mass, g.

- An average pore surface u, cm^2 is calculated as follows: $u = \dfrac{4v \cdot p \cdot 100}{d \cdot g}$,

  where: d – average pore diameter, cm.

  The index of the speed of material destruction K characterizes the duration of its ageing up to the fatigue limit and shows both the thickness and area of the wall surface that should be either removed or strengthened.

Field of Application

- The conclusions regarding the investigation of the decay and ageing processes taking place in ceramic bricks might be widely applied in the estimation and reconstruction of the old masonry's technical state.

Realization of Application

- The Monumental Resurrection Church in Kaunas has been subject for the realization of the results of this given research.

Figure 3. The Resurrection Church in Kaunas, 2003

- The red brick construction of the church facades and the tower was deeply damaged by the atmospheric factors, rebuilding and the pulling down of constructions. In the process of the church renovation, the external surface strengthening, its restoration and finishing were planned to prevent them from moisture accumulation under the waterproof shell as well as from corrosion of inserted metallic details and could guarantee the aesthetic and long-term quality [Vektaris, Barkauskas, 2001].
- The prepared construction of the waterproof-finishing external shell consists of

several layers armed with fibre whose density (from the surface deep down) varies from $\rho \geq 1700$ kg/m^3 to $\rho \geq 2000$ kg/m^3; compressive strength from 5MPa to > 30 MPa, adhesion strength from 0,3 MPa to 0,6 MPa, and capillary water absorption from 3 kg/m^2 to 0,5 kg/m^2. The layers of dry mixtures have been made in Lithuania. The external surface was treated with the silicone liquid S-14.

- The scientific importance of the research lies in the discovery of the complex alterating polynomial function solution for the purpose of the church renovation, i.e. of climatic impact, stability of the tower construction, physical and mechanical properties of the materials, etc. [Barkauskas, Daunoravicius, Vektaris, 2003].

References

- Barkauskas V., 1958. Change of Physical-Mechanic Properties in Some Finish Materials under Climatic Effect. Proceedings of Academy of Sciences of LSSR, part B(22).
- Barkauskas V., 2001. The Content of the Resistance to Environmental Influences in the Active Layer of the External Surface. Civil Engineering. Vilnius, VII, Nr.1.
- Barkauskas V., Daunoravicius M., Vektaris B. Journal of Civil Engineering and Management, Vol. IX. Suppl. 2.
- Deriagin K., 1937. Works of Academy of Science USSR. Moscow. Russia.
- Ginsburg J., Zalieskij B. Works Institute of Geological Science of Academy Science. USSR. Vol. 122, № 137.
- Miniotaite R., 2001. Materials Science Vol. 7, № 4. Kaunas. Lithuania.
- Vektaris B., Barkauskas V., 2001. Designing and Realisation of the Plaster of Tower of the Resurrection Church. In: Proceedings of the Conference at Kaunas University of Technology "Concrete and Reinforced Concrete".

Contacts

- Dr. R.Simkus
  Address: Tunelio 60, Kaunas, 44405, Lithuania
  e-mail: rsimkus@asi.lt

Prepared by: Zivko Bozinovski[1] and Kiril Gramatikov[2]
  [1] Institut of Earthquake Engineering and Engineering Seismology Skopje
  [2] Civil Engineering Faculty, Skopje, Republic of Macedonia

REPAIR AND STRENGTHENING OF EXISTING MASONRY BUILDING STRUCTURES IN SEISMIC PRONE AREAS

Description

- Based on the performed synthesis of the results from analytical and experimental investigations of elements of RC and masonry systems in the world and in our country as well as the investigations performed in IZIIS - Skopje, proposed is a procedure for design and analysis of new structures and repair and strengthening of damaged RC and masonry systems exposed to static and dynamic effects.

- The proposed procedure assumes including of bearing elements in the nonlinear range of behaviour, but with a controlled and dictated mechanism of behaviour, up to ultimate strength and deformability states.

Technical Information

- For centuries masonry has been one of the main construction materials because of its quality as a thermal insulator, acoustic insulator, its hygroscopicity, etc. However, the experience from occurred earthquakes points to unfavourable behaviour of masonry structures since it is the masonry structures that have suffered mass failure and heavy structural damage during past earthquakes. Masonry structures are massive and have a high bearing capacity of walls under compression, insufficient bearing capacity under tension, low ductility capacity and inadequate connection of the structural elements into a whole, particularly improper interconnection of the bearing walls in both orthogonal directions and inadequate connection of the bearing walls with the floor structures.

- To satisfy the requirements of seismic design of high-rises in seismically active regions, it is necessary to apply a modern design concept that apart from the strength and deformability shall take into account the plastic excursions and the capability for seismic energy dissipation. Namely, it is necessary that elements that shall eliminate the unfavourable behaviour of masonry and enable its favourable behaviour be incorporated in masonry structures during their design or repair and strengthening. Most frequently, a solution is found in combining masonry with other materials that are characterized by a high ductility, tensile and shear strength.

- Design of new masonry structures or repair and/or strengthening of existing masonry structures is done by satisfying the requirements of the valid technical regulations, based on the most recent knowledge on seismic design and behaviour of this type of structures, controlling the strength, stiffness, deformability and capability of seismic energy dissipation of the bearing elements and the system as a whole.

Field of Application and End Users

Rehabilitation of existing masonry building structures.

Structural Assessment

- Advantages:
  i)   Strengthening is perform using materials with high strength, stiffness and deformability

- Disadvantages:
  i)   Masonry and reinforced concrete are materials with different characteristics and response for seismic effects.

Economical Aspects

- Proposed is a procedure for design and analysis of stable and economic systems exposed to static and dynamic effects.

Robustness and Testing

- Design Conditions

The design conditions contain several parameters connected with the type, geometrical, strength, stiffness and deformability characteristics of the bearing structural system, the capability of the elements to dissipate seismic energy by lineal and nonlinear behaviour, the purpose of the structure, the degree of required seismic protection, the location with the soil characteristics, the potential earthquake foci, the design criteria and other parameters in compliance with the regulations.

Definition of the strength and deformability characteristics of materials is done analytically or experimentally. The most appropriate is the experimental one in a laboratory or "in situ" on a particular bearing wall (Fig. 1). The seismic parameters on the considered site are defined on the basis of complex seismic, geological and geomechanical investigations of the wider location. For different return periods, the expected seismic effects are defined through different intensity and frequency content.

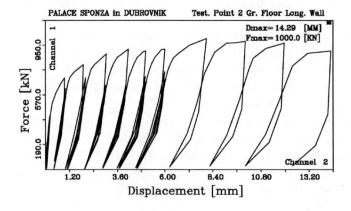

Figure 1. P-δ diagram, Palace Sponza, obtained during 'in situ' test.

- Static Analysis of the Structure Under Vertical Loads and Equivalent Seismic Forces

During the serviceability period, the structures are exposed to the effects of vertical and horizontal, static and dynamic, permanent and occasional, normal and abnormal loads. The vertical loads, dead or live, are mainly static and depend on the purpose of the structure, the material from which they are constructed and the finishing of the elements. The mathematical model of the structure is defined as a system with concentrated masses at the level of the floor structures, connected by springs and dampers. The system is fixed on a rigid or elastic base. The stiffness of the elements and the system as a whole is defined according to standard formulae for masonry structures and corresponding fixation conditions. Stiffness is more realistically defined on the basis of the working diagrams of elements, from the beginning of loading up to failure.

The intensity of the equivalent seismic forces depends on the category of the structure, the soil conditions in which it will be founded, the expected seismic effect at the considered location, the dynamic characteristics of the structure, the damping and the expected ductility capacity.

For the defined external effects on the bearing structural system in both the orthogonal directions, linear static analysis is performed for vertical loads and equivalent seismic forces.

- Analysis of the Elements up to Ultimate Strength and Deformability States

For the structural elements with geometrical characteristics, characteristics of materials and position in the structure, analysis of elements is done and hence analysis of the structure up to ultimate states of strength, stiffness and deformability. For several types of walls as are the stone walls, brick walls, walls of stone or brick with reinforced concrete jackets, confined brick masonry with reinforced concrete vertical and horizontal belt courses and stone walls with a concrete cladding, there is a simple, but sufficiently exact way of determining the capacities of strength, stiffness and deformability, in the linear and nonlinear range of behaviour (Fig. 2). The deformation in the same range is defined by the strength/stiffness ratio.

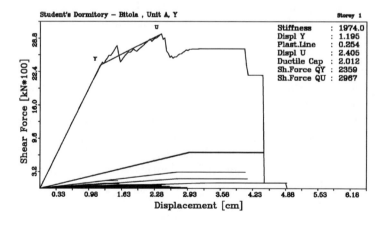

Figure 2. Shear force -relative displacement diagram

Generally speaking, for all the elements that are found in masonry structures, analysis of the strength capacity could be performed in a simple way. The bearing R/C elements are considered in repair and strengthening of masonry systems. For these elements, there is a more exact way of defining the ultimate strength and deformability value, by control of the mechanism of behaviour from the beginning of loading up to failure. Therefore, we are often forced to define parallelly the ultimate states of the masonry and reinforced concrete elements and arrive at storey capacities of strength, stiffness and deformability in both the orthogonal directions (Fig. 3) by their superposition.

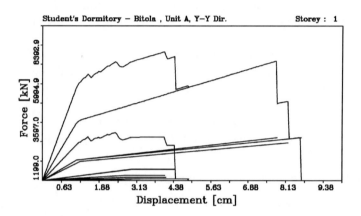

Figure 3. Force - displacement diagram for masonry and R/C elements.

- Nonlinear Dynamic Response

To predict the nonlinear dynamic response of a structure with discrete masses, the shear force - relative displacement relation should be established, i.e., the stiffness should be defined, known as hysteresis curve. The results from the experimental investigations of different types of structures point to extremely complex hysteretic behaviour of elements and the system as a whole. In practice, different authors have defined different idealized analytical hysteretic relationships and have developed several computer programmes for prediction of the nonlinear dynamic response of the structures.

- Repair and Strengthening

Based on the analysis of the existing state of the structure, comparison is made between the strength and deformability capacities of the bearing elements and the system as a whole in both the orthogonal directions with those required according to the regulations and those required for the analyzed response of the structure under expected earthquakes at the considered site, with intensity and frequency content. If the strength and deformability capacities are lower than the required, it is considered that the structure, it is concluded that the structure does not possess sufficient strength and deformability wherefore its repair by strengthening is necessary.

The solution for repair and/or strengthening depends on: the seismicity of the site, the local soil conditions, the type and age of the structure, the level and type of damage, time available for repair and/or strengthening, equipment and staff, restoration and architectonic conditions and requirements, economic criteria and required seismic safety. From the several analyzed possible solutions for repair and strengthening, selected is the most favourable from the aspect of stability, i.e., fulfillment of the design criteria according to the regulations, the possibility for realization of the solution, the available materials, the economic justification, the fulfillment of the sociological and aesthetic requirements. For the selected technical solution for repair and strengthening, performed is analysis of the stability of the structure (Fig. 4). For the repaired and strengthened structural system, the strength, stiffness and deformability capacities of the structure as well as the required strength, stiffness and deformability for the expected earthquakes with intensity and frequency content are defined.

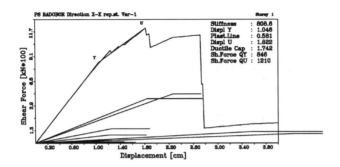

Figure 4. Storey P - Δ diagram, primary school Radobor, repaired state

■ Conclusion on the Stability and Vulnerability of the Structure

Based on the strength and deformability capacities of the bearing elements and the structure as a whole and on the basis of required strength and deformability for expected seismic effects with intensity and frequency content, conclusions are drawn regarding the stability of the structure and its vulnerability level.

Pilot structure

Out of several analyzed possible solutions for repair and strengthening of the main structural system, selected is the most favourable from the aspect of: stability, i.e., fulfillment of the design criteria according to regulations, possibility for realization of the solution, available materials, economic justification, fulfillment of social requirements and satisfying of aesthetic requirements (Fig. 5 to 6).

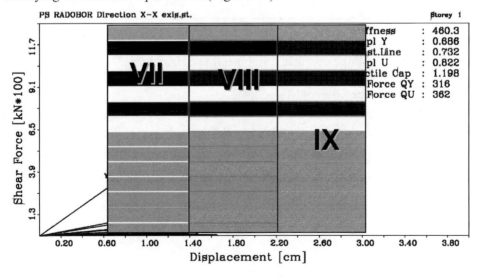

Figure 5. Storey P-Δ diagram, primary school Radobor, direction X-X, existing state

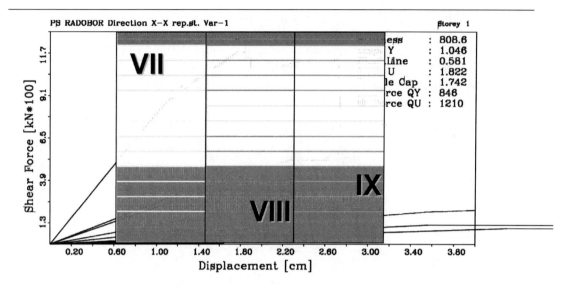

Figure 6. Storey P-Δ diagram, primary school Radobor, direction X-X, repaired state

Standards

- Code for technical regulations for the design and construction of buildings in seismic regions *(Official gazette of SFRJ No.31/81)*

References

- Bozinovski, Z., Stojanovski, B., and Petrusevska, R., "Repair and Strengthening of the Main Structural System by Analysis of the Stability of the Primary School Buildings in the Commune of Bitola, IZIIS Report 95-26 to 95-45, 1995.
- Bozinovski, Z., and R. Petrusevska, Repair, strengthening, reconstruction, adaptation and revitalization of the Elementary School 'Mosa Pijade' in Gostivar, Report IZIIS 96-38, Skopje, 1996.

Contacts

- Zivko Bozinovski

Address: Institute of Earthquake Engineering and Engineering Seismology, University "St. Cyril and Methodius", Salvador Aljende 73, P.O.Box 101, 1 000 Skopje, Skopje, Republic of Macedonia.
e-mail: zivko@pluto.iziis.ukim.edu.mk,
Phone:++389-2-3176 155, Fax: +389-2-3112-163

- Kiril Gramatikov

Address: Civil Engineering Faculty, University "St. Cyril and Methodius", Bul. Partizanski Odredi 24, P.O.Box 560, 1 000 Skopje, Republic of Macedonia,
e-mail: gramatikov@gf.ukim.edu.mk

STRENGTHENING OF A CONCRETE SLAB WITH FRP STRIPS

Description

- strengthening of an existing concrete flat slab in a shopping centre
- the application of FRP strips was necessary due to the cutting of different openings after concreting of the slab

Figure 1. View of the opening from inside

Technical Information

- some of the openings were placed in the near of the columns, therefore the most parts of the upper reinforcement had to be replaced by FRP cross sections
- the redistribution of moments leaded also to a necessary strengthening of the concrete section in the adjacent spans of the slab
- for the application of the FRP strips on the upper side of the slab the already laid screed had to removed again

Figure 2. Application of strip on the upper side of the slab

- special attention had to be given to the end anchorage of the strips at the edges of the opening
- for that purpose special L- shaped strips were used

Figure 3. Anchorage of strips

Figure 4. Detail of L- shaped strips

561

- the application of the strips on the ceiling had to take care on already installed transmissions

Figure 5. Application of strips on the ceiling

Fields of Application

- solution applicable for strengthening of concrete slabs
- substitution of removed or cut reinforcement sections

Structural Assessment

- Advantages:
  easy application of light weight strips with a small structural height
  no need of placing heavy steel girders under the slab
  usable also in narrow surroundings
  high load bearing capacity on the tension side
- Disadvantages:
  protection from fire and mechanical destruction needed

Economical Aspects

- material savings
- fast erection

References

- Federation for Structural Concrete (*fib*) Bulletin 14, 2001, *Externally bonded FRP reinforcement for RC structures*, ISBN 2-88394-054-1156292-9147-000-52
- ■

Contacts

- Dr. Christian Schaur
  Address: Dr. Ambros Giner Weg 30, A-6065 Thaur, Austria
  e-mail: schaur@tirol.com
  http://www.schaur.com

COST C12 – WG1

Datasheet II.3.2.5.2
Prepared by: T. Nagy-Gyorgy[1], V. Stoian[1], J. Gergely[2], D. Dan[1]

[1] Politehnica" University of Timisoara, Romania, http://www.ceft.utt.ro
[2] The University of North Carolina at Charlotte, USA, http://www.ce.uncc.edu/

REINFORCED CONCRETE WALLS STRENGTHENED WITH CFRP COMPOSITES

Description

- It was demonstrated that fibre reinforced polymer (FRP) composites are efficient materials to retrofit structural reinforced concrete elements, as beams or columns. These are non-ductil materials, but with high ultimate strength, low weight and thickness, and durability.

- Research on the behaviour of reinforced concrete walls strengthened with FRP composites is much reduced, although its use is more common.

- Objectives were to investigate the effectiveness of this strengthening technique. For that purpose five structural shear walls with staggered door openings were tested before and after composite retrofit.

Field of application

- Seismic retrofit of structural reinforced concrete walls
- Improve and supply the load bearing capacity of the walls before or after earthquake
- Restore the shear capacity of the walls

Research activity

- The research programme included test on a series of reinforced concrete walls, strengthening of the walls on one side with carbon fiber reinforced polymer (CFRP) composites and than retesting, comparing the experimental results.

- The model of the structural wall has the height of 260 cm, width of 125 cm, storey height of 60 cm and a thickness of 8 cm. The opening had the dimension of 25cm x 50 cm. The experimental model foundations were cast simultaneously with the concrete from the walls. There were studied five types of structural walls with different values of α angle (figure 1).

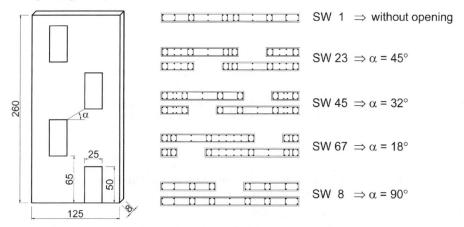

Figure 1. Geometrical parameters, configuration and notation of the walls

- Experimental models were provided with the same amount of steel reinforcement based on Romanian prescriptions (norm P85/2001).
- The elements were subjected to a constant vertical load V=50 kN at the top. The horizontal load (H) was applied monotonically for the wall without openings and cyclic for the rest of the elements into a displacement-controlled mode. The top displacements were increased with an average drift (horizontal top displacement divided by wall height).

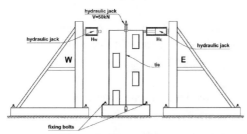

Figure 2. Test setup

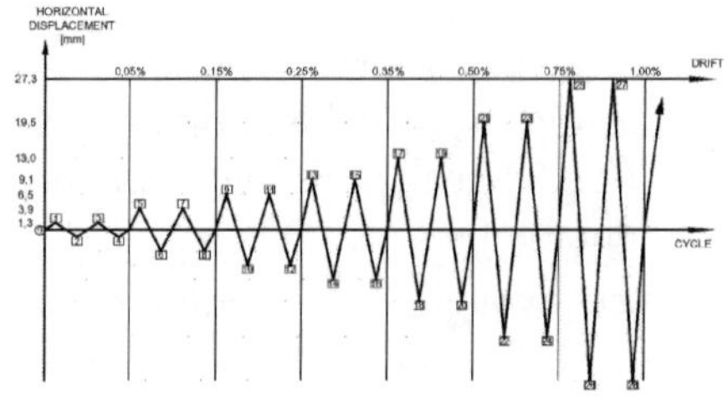

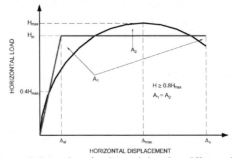

Figure 3. Loading diagram

- In the first phase, the simple wall specimens were tested up to failure (concrete crashing, steel reinforcement yielding), then the damaged and exfoliated parts of the walls were replaced with an epoxy based repair mortar and the existing cracks were filled with an epoxy resin. Followed the surface cleaning and the realization of the anchorage zone, which was very simple (figure 4), but as was demonstrated after the test, was very efficient and not presented any degradation.
- All the walls have been strengthened with CFRP composite fabric, on one side. The vertical sheets increased the bending capacity, while the horizontal sheets the shear capacity. The recorded data was the horizontal load, the horizontal displacement, the strain in the composite and the specimens' failure modes. The test setup and principles of the tests (loading schemes, value of the loads, cycles) was identical with the baseline specimens'.
- The procedure to evaluate the stiffness, the ductility and the equivalent elasto-plastic curve is shown in figure 5.

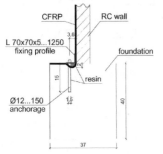

Figure 4. Principle of the anchorage zone

Figure 5. Procedure for determining the stiffness, the ductility and the equivalent elasto-plastic curve

Results of analysis

- The load bearing capacity of the damaged wall and than retrofitted on one side with CFRP increased considerably over the baseline value (practically, the load bearing capacity of the pre-tested walls was negligible).
- The strain in the CFRP composite indicates the major contribution in the load bearing

capacity of the walls, the average recorded value being 0.54-0.84%.
- The failure of the retrofitted wall was caused by gradually opening of the existing cracks, by debonding of the FRP in compression followed by tension, or in some cases by compression of the FRP.
- The maximum horizontal displacements of the retrofitted walls were generally higher, or at least equal with those of the baseline specimens.
- The results were highly dependent by the initial state of the retrofitted element (width and number of cracks, yielded steel reinforcement, rehabilitation method and material), respectively by the evaluation method. Based on the available method for the characteristics evaluation (such as the stiffness, ductility, elastic limit) the following results were obtained:
  - the stiffness of the elements decrease, in average, with 53%
  - the ductility of the elements decrease, in average, with 60%
  - the elastic limit of the walls increase, in average, with 47%
  - the failure load of the walls increase, in average, with 45%
- The anchorage system behaves excellently, without degradations.
- RC walls subjected to seismic forces had a ductile failure. Retrofitting such structural elements with composites maintain the characteristic ductile behaviour, but at the maximum load fails brittle.

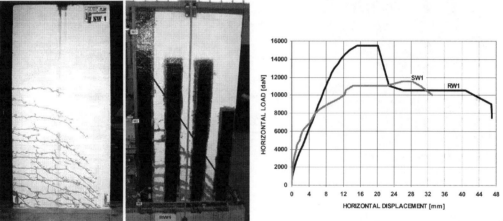

Figure 6. The SW1 and RW1 walls after the test, respectively the load-displacement diagrams
(monotonic test)

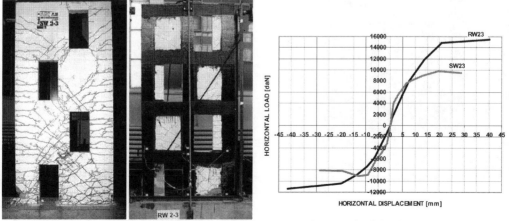

Figure 7. The SW23 and RW23 walls after the test, respectively the load-displacement diagrams
(cyclic test)

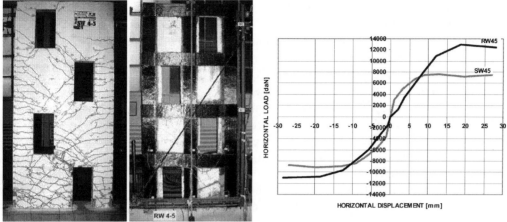

Figure 8. The SW45 and RW45 walls after the test, respectively the load-displacement diagrams
(cyclic test)

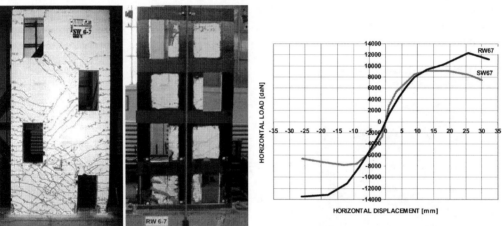

Figure 9. The SW67 and RW67 walls after the test, respectively the load-displacement diagrams
(cyclic test)

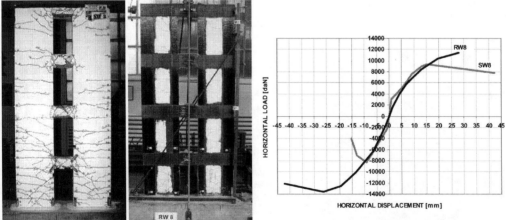

Figure 10. The SW8 and RW8 walls after the test, respectively the load-displacement diagrams
(cyclic test)

Table 1. Synthetic presentation of the experimental test results

Specimen		SW1	RW1	SW23	RW23	SW45	RW45	SW67	RW67	SW8	RW8
Max. horizontal load H_{max} [kN]	W	115	155	98	155	76	130	91	135	94	114
	E	-	-	93	114	91	108	78	123	83	136
Difference in capacity [%]	W	+ 35		+58		+ 71		+ 48		+ 21	
	E	-		+22		+ 19		+ 57		+ 63	
Max. displacement Δ_u [mm]	W	33	47	28	40	27	27.5	30	32	42	28
	E	-	-	28	41	27	27.5	26	26	15	42
Difference in displacement [%]	W	+ 42		+42		+ 1		+ 6		− 33	
	E	-		+46		+ 1		+ 0		+ 180	
Stiffness k_e [daN/cm]	W	348	159	309	109	298	91	199	105	154	89
	E	-	-	184	90	343	112	168	102	143	99
Difference in stiffness [%]	W	− 54		− 65		− 69		− 47		− 42	
	E	-		− 51		− 67		− 39		− 31	
Ductility D [-]	W	11.2	4.8	10.2	3.0	11.5	2.1	7.1	2.2	7.9	2.6
	E	-	-	6.6	3.6	11.3	3.2	6.3	3.1	3.2	3.4
Difference in ductility [%]	W	− 57		− 71		− 81		− 69		− 67	
	E	-		− 45		− 72		− 51		− 6	
Maximum strain in FRP [%]		-	0.54	-	0.63	-	0.79	-	0.58	-	0.83

Further developments

- Develop the design formulas for such type of strengthening applications
- Apply the results to retrofit buildings with RC wall structures

References

- Iso, M., Matsuzaki, Y., Sonobe, Y., Nakamura, H. & Watanabe, M. , *"Experimental study on reinforced concrete columns having wing walls retrofitted with continuous fiber sheets"*, CD -ROM Proceedings of the 12th World Conference on Earthquake Engineering, New Zealand,2000, Paper No 1865.

- Sugiyama T., Uemura M., Fukuyama H., Nakano K., Matsuzaki Y., *"Experimental study on the performance of the RC frame infilled cast-inplace non- structural RC walls retrofitted by using carbon fiber sheets"*, CD - ROM Proceedings of the 12th World Conference on Earthquake Engineering, New Zealand, 2000, paper No 2153.

- Lombard J., Lau D., Humar J., Foo S., Cheung M., *"Seismic strengthening and repair of reinforced concrete shear walls"*, CD - ROM Proceedings of the 12th World Conference on Earthquake Engineering, New Zealand, 2000, paper No 2032.

- Antoniades K. K., Salonikios T. N., Kappos A. J., *"Inelastic behaviour of frp-strengthened r/c walls with aspect ratio 1.5, subjected to cyclic loading"*, Concrete Structures in Seismic Regions - fib Symposium, Athens, Greece, 2003.

- Antoniades, K.K., Salonikios, T.N. and Kappos, A.J. *"Cyclic tests on seismically damaged R/C walls strengthened using FRP reinforcement"*, ACI Structural Journal, Vol. 100, 2003.

- Paterson J., Mitchell D., *"Seismic retrofit of shear walls with headed bars and carbon fiber wrap"*, Journal of Structural Engineering, May, 2003.

- AC125, ICC – ES, 2003. *Interim Criteria for concrete and reinforced and unreinforced masonry strengthening using fiber-reinforced polymer (FRP) composite systems.*

- Nagy-György T., Gergely J., Dan D., *"Reinforced concrete structural walls retrofitted with composites"*, International Conference in Civil Engineering and Architecture - EPKO 2004, Miercurea Ciuc, Romania, 2004.

Contacts

- Tamas NAGY-GYORGY
  Address: "Politehnica" University of Timisoara
  T. Lalescu 2, 300223 – Timisoara, Romania
  Tel/Fax: +40 256 403958
  e-mail: tnagy@ceft.utt.ro

- Valeriu STOIAN
  Address: "Politehnica" University of Timisoara
  T. Lalescu 2, 300223 – Timisoara, Romania
  Tel/Fax: +40 256 403958
  e-mail: stoian@fctm.dnttm.ro

- Janos GERGELY
  Address: The University of North Carolina at Charlotte
  9201 University City Blvd., Charlotte, NC 28223, USA
  Phone: 704 / 687-4166
  Fax: 704 / 687-6953
  E-mail: jgergely@uncc.edu

- Daniel DAN
  Address: "Politehnica" University of Timisoara
  T. Lalescu 2, 300223 – Timisoara, Romania
  Tel/Fax: +40 256 403958
  e-mail: ddan@ceft.utt.ro

MASONRY WALLS RETROFITTED WITH COMPOSITES

Description

- The seismic vulnerability of masonry buildings was evident during major seismic events across the world. A significant number of masonry buildings suffered extensive damage, and large number of partial or complete collapses were observed.

- Effective retrofit technique was needed, as fibre reinforced polymer (FRP) composites overlays, which increase the in-plan and out-of-plane strength and stiffness of the masonry walls

- Objectives were to investigate the behaviour of the unreinforced clay brick masonry walls subjected to in-plan shear loads strengthened with FRP composites on one side of the wall. For this reason have been performed several finite element analysis (FEA), and four specimens were built and tested

Field of application

- Seismic retrofit of structural masonry walls
- Improve the shear capacity of the masonry building before earthquake
- Restore the shear capacity of the walls

Research activity

- The research programme included the following steps:
  [1] Performing an analytical study
  [2] Building the walls
  [3] Testing, retrofitting and retesting the walls

[1] Performing an analytical study
- Have been performed an analytical study with a simplified (theoretical) model of the wall. The goal was to conceive a device in which the load system creates a pure in-plane shear of the wall, without much influence from the bending moment. This system is auto-equilibrant and, theoretically, the crack should form in the diagonal direction. The loads applied to the specimen were a constant vertical (V) and a constantly increasing horizontal (H) force (figure 1). With this test set-up a large number of finite element analysis (FEA) have been performed, by modifying the width to height ratio (d/h) of the elements (d/h=1, d/h=1.5 and d/h=2), the quality of the brick and the mortar, through the strength and the modulus of elasticity of the element, the horizontal load-steps, and finally by applying a constant vertical force of different magnitude.

- Theoretically, the application of the vertical force is not necessary in the case of homogenous materials, but the brick wall is composed of clay brick units and mortars, which have different characteristics. Therefore, to prevent a sliding failure mode, a vertical force was applied.

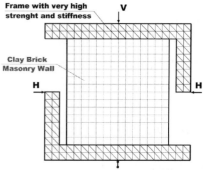

Figure 1. Structural model of the wall for FEA

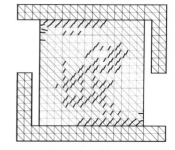

Figure 2. Crack distribution of the wall specimen

[2] Building the walls

- The specimens were 150 cm wide and 150 cm high, build of solid clay bricks with dimensions 6.3 x 24.0 x 11.5 cm and unit strength of $9.0 \div 10.0$ N/mm^2, the mortar strength being $13 \div 16$ N/mm^2. At the top and at the bottom were placed reinforced concrete beam with 50 x 150 x 25 cm.

[3] Testing, retrofitting and retesting the walls

- The wall was tested in a special device (figure 3), composed by a pair of L-shaped steel elements attached to a massive reinforced concrete block, which was part of the wall at the top and the bottom. The forces have been applied with hydraulic jacks. The vertical force was applied on the top of the specimen, acting through the reinforced concrete bond beam. The horizontal (shear) force has been applied through a series of steel bolts embedded in the reinforced concrete block, and mounted to the L-shaped steel elements at the top as well as at the bottom.
- The displacement of the wall was measured with displacement transducers. Six of the transducers were placed along the height of the wall, three on left and the other three on the right, measuring the specimen's horizontal displacement. The other four transducers measured the vertical displacements, two on each side of the specimen. Two were placed at the first mortar bed joints and other two measured the displacement of the steel frame.

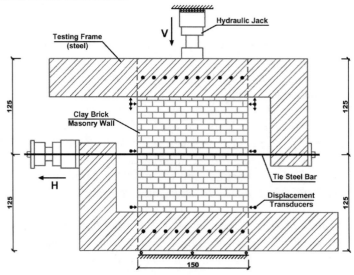

Figure 3. The specimen test set-up

- The specimens were tested in as-built condition up to failure and then retrofitted on one side with FRP composite layers and retested. The recorded data was the horizontal load, the horizontal and vertical displacement, the strain in the composite and the specimen failure modes.

i) Experimental tests of the UM1 and RM1 elements

- The UM1 (UM – Unreinforced Masonry) wall was initially tested in the as-built condition. A constant vertical force was V=200 kN and the monotonically increased horizontal force H was applied by an increment of 5 kN up to failure, who generated the required in-plane shear forces in the specimen. The failure mechanism of the wall was produced through a diagonal crack from the top-right to bottom-left corner, as was expected. The load - displacement diagrams at the top of the wall are typical for unreinforced masonry. The specimen behavior is close to linear. The load at the specimen failure was 190 kN and the maximum horizontal displacement reached 7 mm.

- To rehabilitate the precracked wall, three carbon FRP sheets were applied to one side. The FRP was applied to just one side, because in many situations the modification of the facades is not permitted, or it is very expensive to perform. Therefore, in these cases only the inside surfaces of the walls are accessible

- The test set-up for RM1 (RM – Reinforced Masonry) wall specimen was identical to the baseline test set-up (UM1), additionally strain gages were attached to the composite in the maximum stress zones and were aligned in the direction of the carbon fibres. The retrofitted wall reached a peak lateral load of 145 kN and a maximum horizontal displacement of 19 mm. The peak tensile stresses in the composite laminates reached approximately 33% of the ultimate value, which corresponded to an ultimate strain of 0.5%. This is a very good result, demonstrating that this solution worked really well with clay brick masonry. The dominant wall failure mode was extensive brick masonry cracking, followed by composite debonding at the cracks.

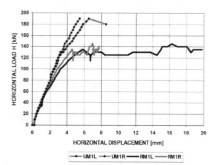

Figure 4. RM1 retrofitted wall test | **Figure 5.** Load-displacement diagram of UM1 and RM1 (L – left side, R – right side)

ii) Experimental tests of the UM2 and RM3 elements

- Another set of experiments have been performed on the element UM2 and RM3. The applied constant vertical force was V=300 kN and the monotonically horizontal force H was increased by an increment of 5 kN. The wall failure was very brittle through a diagonal crack. It can be mentioned that the constant vertical force was increased with 100 kN, correlating with the previous case, because the quality of the wall was superior to the first element and was necessary to avoid the sliding in horizontal bed joint. The horizontal force (H) at the failure was 300 kN and the maximum horizontal displacement exceeded 8 mm.

- The retrofitting was performed by applying on one side a carbon fibre fabric which covered the whole surface of the element.

- The retrofitted wall (RM3) reached a peak horizontal load (H) of 370 kN, the maximum horizontal displacement being 17 mm. The peak tensile stresses in the composite laminates reached approximately 10% of the ultimate value, which corresponded to an ultimate strain of 0.15%. This demonstrates that the composite had high reserves in the moment of the specimen failure. It can be mentioned, that the failure was produced through developing a new crack and through the extensive opening of this. The composite has been delaminated just in the crack zone, but it was not broken.

Figure 6. The tested RM3 wall

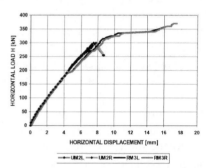

Figure 7. Load-displacement diagram of UM2 and RM3
(L – left side, R – right side)

iii) Experimental tests of the UM3 and RM4 elements

- In the next set of experiments have been tested at first the UM3 specimen. The constant vertical force was V=300 kN and the monotonically applied horizontal force H had an increment of 5 kN up to failure. The wall failure was also very brittle through a diagonal crack, which opened approximately 1.5 cm. The horizontal force (H) at the failure was 325 kN and maximum horizontal displacement did not exceed 3 mm.

- The retrofitting has been performed also by applying a carbon fibre fabric on one side which covered the whole surface of the element. The differences compared with the first two elements were the need of the crack injection and filling, which has been performed with cement mortar.

- The retrofitted wall (RM4) reached a peak horizontal load (H) of 270 kN, the horizontal displacement being 9 mm. The maximum tensile stresses in the composite laminates reached just 8% of the ultimate value, which corresponded to an ultimate strain of 0.12%. The failure has been produced through the extensive opening of the existing crack. The composite has been delaminated just in the crack zone, but not broken.

Figure 8. The RM4 wall before the test

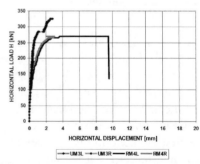

Figure 9. Load-displacement diagram of UM3 and RM4
(L – left side, R – right side)

iv) Experimental tests of the UM4 and RM5 elements
- In the last set of the experiments have been tested the UM4 and RM5 specimens. The applied forces were a constant vertical force V=300 kN and the monotonically increased horizontal force up to failure. The wall failure was also very brittle through a diagonal crack, which opened approximately 1.0 cm. The maximum horizontal force (H) was 320 kN and maximum horizontal displacement reached 16 mm.
- The retrofitting has been performed by applying a glass fibre fabric on one side which covered the whole surface of the element, the crack also was filled with cement mortar.
- The retrofitted wall (RM5) reached a peak horizontal load (H) of 335 kN, the horizontal displacement being over 38 mm. The maximum tensile stresses in the composite laminates reached 1.78 % of the ultimate value, which corresponded to an ultimate strain of 48 %. The failure has been produced through the extensive opening of the existing crack, concomitant with the composite debonding in the crack zone and without it's rupture.

Figure 10. The tested RM5 wall

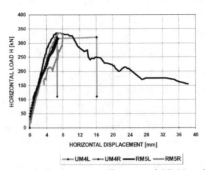

Figure 11. Load-displacement diagram of UM4 and RM5 (L – left side, R – right side)

Results and conclusions of the analysis

- The correction and injection mortars had an important role in restoring the load bearing capacity. The width of the initial crack is decisive in the evolution of the final capacity of the strengthened wall. If the crack was tight the capacity increased significantly over the reference value, while if the crack was wide, the ultimate load capacity was approximately equal to the initial value.
- A considerable capacity increase was observed for the precracked shear walls retrofitted with FRP composites (practically, the load bearing capacity of the cracked walls was negligible). The most advantageous strengthening system seems to be the composite with glass fibres, because it uses up to 50% of the load bearing capacity of the fibres.
- The failure of the retrofitted walls was caused by extensive cracking followed by FRP delamination and not due to tensile or shear failure of the FRP.
- The maximum horizontal displacements increased two times compared with the displacements of the baseline specimens that demonstrate the increase of the ductility and energy absorbing capacity of the retrofitted walls.
- Strengthening with FRP composites, using unidirectional fabric, placed in vertical direction on one side of the wall, has an important contribution in increasing or restoring the shear capacity of the structural masonry walls, in spite of the opinion of some researchers, who recommend neglecting the contribution of vertical FRP reinforcement, due to the dowel action effect.
- Unreinforced masonry walls subjected to shear forces behave in a very brittle way and

fail with or without warning. By strengthening such a non-ductile structural element with composites the characteristic behavior became rather ductile than brittle.

Further developments

- Wall tests with composites applied horizontally, to eliminate the formation of so-called dowel action
- Develop the design formulas for such type of strengthening applications
- Apply the results to retrofit buildings with masonry structure

References

- Triantafillou T., *„Composites: A New Possibility for the Shear Strengthening of Concrete, Masonry and Wood"*, Composite Science and Technology 58, Elsevier Science, 1998.
- Gergely J., Young D.T., Hooks J.L., Alchaar N., *"Composite Retrofit of Unreinforced Masonry Walls,"* SSRS-2 Conference, Fullerton-CA, March 2000.
- Stoian V., Gergely J., Nagy-Gyorgy T., *"Seismic Retrofit of Masonry Structures"*, COBASE Research Report, USA, 2002.
- Stoian V., Nagy-György T., Dan D., Gergely J., *„Retrofitting the shear capacity of the masonry walls using CFRP composite overlays"*, The SE 40EE International Conference in Earthquake Engineering, Skopje, Macedonia, 2003.
- AC125, ICC – ES, 2003. *Interim Criteria for concrete and reinforced and unreinforced masonry strengthening using fiber-reinforced polymer (FRP) composite systems.*

Contacts

- Tamas NAGY-GYORGY
    Address: "Politehnica" University of Timisoara
        T. Lalescu 2, 300223 – Timisoara, Romania
        Tel/Fax: +40 256 403958, e-mail: tnagy@ceft.utt.ro

- Valeriu STOIAN
    Address: "Politehnica" University of Timisoara
        T. Lalescu 2, 300223 – Timisoara, Romania
        Tel/Fax: +40 256 403958, e-mail: stoian@fctm.dnttm.ro

- Janos GERGELY
    Address: The University of North Carolina at Charlotte
        9201 University City Blvd., Charlotte, NC 28223, USA
        Phone: 704 / 687-4166
        Fax: 704 / 687-6953, E-mail: jgergely@uncc.edu

- Daniel DAN
    Address: "Politehnica" University of Timisoara
        T. Lalescu 2, 300223 – Timisoara, Romania
        Tel/Fax: +40 256 403958, e-mail: ddan@ceft.utt.ro

COST C12 – WG1
Datasheet II.3.2.6.2
Prepared by: Leo g.w. Verhoef (Assoc.Prof.)
   Renovation & maintenance techniques
   Delft University of Techniques

MASONRY AND FRP

Description

- restrenghtening of brickwork with strips of carbon fibre reinforced plastic to derive earthquake resistance

- restrenghtening of brickwork with strips and rods of carbon fibre reinforced plastic to derive limitation of crack widths for buildings with bearing brickwork facades and internally a half skleton concrete structure

- restrenghtening of brickwork with strips of carbon fibre reinforced plastic for strenghtening of foundations

Technical information

- the bond of brickwork does not perform well under seismic loads. To increase the reliability of the masonry structure under seismic load, it is possible to strenghten it with (C)FRP. The fibre strips are glued to the masonry wall and anchored into the masonry slab.

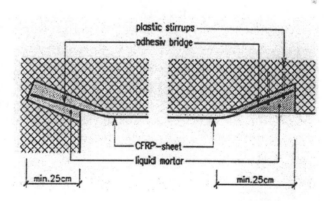

Fig. 1. strips in truss form   Fig. 2. anchor system in masonry

- the bond of brickwork does not perform well under restrained shrinkage, due to the behaviour of the stiff concrete floors resting on the bearing masonry facades. Renovation with eventually internal thermal insulation will cause larger thermal gradients, which will lead to larger crack widths. Flat strips glued on the inside above and below windows while rods are placed and glued in the joints on the outside of facades. With the effect that cracks will be limited to acceptable, low widths at the recurrence of restrained shrinkage.

Application

■ Application on masonry walls in earthquake endangered areas. The substrate has to be freed of all loose or unsound particles, such as wall paper, paint and rendering. In order to

Fig. 3. flat strips on the inside of facades to limit crack width

obtain a straight final position, all protruding surface elements have to be removed. Parts of the wall that lay deeper has to be filled up e.g. with epoxy resin.

■ Application on masonry walls for restrained shrinkage.

On the inside of the façade, horizontal strips have to be glued in a straight final position. The substrate has to be freed of all loose or unsound particles, such as wall paper, paint and rendering.

On the outside of the facades, a number of bed joints partly have to be cut out till at least 25 mm. The rods have to be glued into the backside of the opened bed joints. Afterwards the rest of the open bed joint has to be filled up with the correct mortar material.

■Two weeks after the repair, SIKA CARBODUR S 1012 CFRP strips of 100 mm width, 1.2 mm thickness and 2 m length can be glued to the rear side of the wall. Standard procedure was followed to apply the strips. Firstly the wall parts where the strips were to be placed were brush-cleaned. Simultaneously, the CFRP strips where de-greased. Next, the two-component SikaDur 30-Normal epoxy was applied to a thickness of up to 5 mm, due to the wall rear side unevenness. Finally, each strip was placed manually and evened out with a hand-held roller.

Fig. 4. the flat strips are placed manually and evened out with a hand-held roller

■ Calculations in advance is necessary in advance of the application for restrained shrinkage. For analysis a rate dependent, anisotropic Rankin-Hill continuum plasticity model has to be employed.

■For earthquake resitance: the major horizontal earthquake forces act at the floor levels.

The floor slabs have to be stiffened and strong enough to transfer these forces into the shear walls, from where they are carried back to the the foundation. In the transverse and longitudinal direction of the building, the lateral resistance is provided by the shear walls restrenghtened by CFRP. Wooden floors can be stiffened by covering them with a concrete layer. The (C)FRP strips glued on the shear walls have a vertical struss position.

■ values for glue, Dutch masonry and CFRP

Glue	Masonry	CFRP
k_n=3035 N. mm^{-3}	E=4 kN.mm^{-2}	E=165 kN.mm^{-2}
k_s=1265 N. mm^{-3}	v=0.2	v=0.2
-	-	-
ρ=900 kg.m^{-3}	ρ=1900 kg.m^{-3}	ρ=1900 kg.m^{-3}
f_t =1.0 N.mm^{-2}	f_{tx} =0.5 N.mm^{-2}	f_t =1500 N.mm^{-2}
$c = 1.0 + 0.1\sigma_n$	f_{ty} =0.25 N.mm^{-2}	
G_f^I = 0.02 N.mm^{-1}	g_{fx} =0.0007 N.mm^{-2}	
G_f^{II} = 0.05 N.mm^{-1}	g_{fy} =0.00035 N.mm^{-2}	
	m = 100 000 N.s.mm^{-2}	

■ For restrained shrinkage cases repairing of cracks has to be token in advance

■ Risks: long-term performance in relation to wet/warm environments are presently unknown.

Economical aspects

■ CFRP for earthquake resistance is easy to apply as an efficient and economical method.

It prevents to add to existing buildings new elements or layers such as shotcrete. New layers have adiddiculties in acceptance form an architecturel, technical and economical point of view.

- CFRP for restrained shrinkage is an efficient and economical method. Cracks with large widths will not occur but replaced by more cracks but be limited to acceptable, low widths at the recurrence of restrained shrinkage.

Robustness and Testing

- Earthquake resistance has been tested in full scale at the Swiss Laboratories for Materials Testing and Research. EMPA. These tests showed a significant increase in strength and ductility of the masonry wall elements.
- Restrained shrinkage resistance has been tested in a near full scale wall of typical Dutch clay brick masonry in the STEVIN laboratories of Delft University of Technology.

References

[1] Ernst basler und Partner AG (1995). Konkret 1995, Unternehmenspublikation 95. Ernst Basler und Partner AG Ingenieurunternehmen, Zollikon ZH (Switserland)

[2] Muttoni A., Schwarz j., Thuelimann B. (1988) Bemessung und Kostruktion von Stahlbetontragwerken mit spannungsfeldern. Vorlesung ETZH (Institute of Technology Zurich)

[3] Schwegler G. (1994a) Verstarken von Mauerwerk mit Hocchleistungsfaserverbund-werkstoffen. Dissertation: Eidgenossische Materialprufungs-und Forschungsanstalt Dubendorf, EMPA-Bericht Nr.229.

[4] Schwegler G. 1994b. Masonry construction strenghtened with fiber composites in seismically endangered zones. 10th European Conference on Earthquake Engineering. Vienna, Austria

[5] Van Zijl, G.P.A.G. & Verhoef, L.G.W.; Restrengthening of brickwork to reduce crack witdth; In Advances in Engineering Software vol 33/1, Elsevier Science,2002,p. 49-57

[6] Van Zijl, G.P.A.G. *Computational modelling of masonry creep and shrinkage.* Dissertation, Delft University of Technology, Delft, The Netherlands, 2000.

[7] Van Zijl, G.P.A.G. & Verhoef, L.G.W. Collaboration between brickwork and concrete. *Proc. Symp. On Maintenance and restrengthening of materials and structures: Brick and Brickwork* (eds. L.G.W. Verhoef and F.H. Wittmann), Zürich, pp. 83-96, 2000.

[8] Van Zijl, G.P.A.G., De Vries, P.A., Verhoef, L.G.W. and Groot, C.J.W.P. "Experimental confirmation of predicted restrained shrinkage damage in masonry walls". This proceedings.

[9] Van Zijl, G.P.A.G. and Verhoef, L.G.W. CDROM-Proc. 9th Canadian Masonry Symposium (2001).

[10] Lourenço, P.B. Dissertation, Delft University of Technology, (1996).

[11] Van Zijl, G.P.A.G., De Borst, R. and Rots, J.G. Int. J. Solids and Structures, 38(30-31), (2001) 5063-5079.

[12] Van Zijl, G.P.A.G. and Verhoef, L.G.W Double side restrenghtening of historic brickwork with rods and strips of carbon fibre reinforced plastic (CFRP)

Chapter III

Urban Design

COST ACTION C12

III.5 Building physics aspects

Introduction to Chapter III

1. Objectives of Working Group 3 – Urban Design

In the last century, especially in the last 50 years, the evolution observed in buildings construction technology was extraordinary when compared with the evolution occurred during the past centuries. Although there are spread around the world emblematical buildings that can be considered true milestones of high technology, these buildings does not constitute the standard for the construction and the majority of people continues to live in and to use buildings that do not possess a level of quality compatible with the contemporary development.

In European countries the recognition that there is a special need for higher quality in the field of urban buildings led to the development of several national and international research projects. Under the auspices of the European Commission and lately of the European Science Foundation, by way of COST projects, several funded Actions have been undertaken with the aim of developing new and suitable strategies for architects, sociologists, urban planners, local authorities and engineers.

The main objective of the COST Action C12 was to develop, combine and disseminate new engineering technologies to improve the quality of urban buildings, to reduce the disturbances of the construction process in urban areas and finally to improve the quality of life inside buildings, the quality of the urban environment and the quality of living in the urban habitat.

In the scope of COST C12, the urban buildings issues were approached by the Working Group 3 – Urban Design. The establishment of links between civil engineers involved in the development of new building techniques and those persons, as architects and urban planners, whose objective is a harmonious development of the city, was of primary importance in this Working Group.

The aim of WG3 was the integration of buildings built, modernized or rehabilitated using Mixed Building Technology in the design and planning of urban buildings. To achieve this goal, WG3 developed studies in several wide fields of knowledge in building science such as Aesthetics, Sustainable Development, Performance Management and Building Physics. Each of these fields was set as an objective for the WG3 study, where the following topics were developed:

- For Aesthetics the aspects selected were the buildings relation to site and adjacent buildings, the parameters for the aesthetic façade design and the sustainable perspectives for architectural design.
- In the research towards the Building Sustainable Development were included themes as sustainability assessment, life cycle assessment of building products, rational use of energy, site issues and climate change impacts on buildings.
- Management of the performance, the adaptability, the accessibility, the security and social demands of buildings was also defined as important issues to be developed by WG3.
- Building physics aspects as thermal performance, day lighting use and shading, and sound insulation were also set as specialities to be studied in order to identify the potentiality contribution of MBT for the building performance.

The future of the existing buildings was also set as an important issue to be performed by WG3, where the study of the technologies leading to the modernization of multi-storey prefabricated panel buildings was defined as a major objective.

2. Presentation of WG3 datasheets

In order to make easier the dessimination of the WG3 outputs the report of the work carried out was condensed in 12 Datasheets. Each one is independent from the others and deals with a specific issue. It is not necessary to read all the datasheets to collect the relevant information.
The content of each datasheet is briefly summarized in the next paragraphs.

The Future of the Existing Buildings
Two complementary perspectives of the foreseen future use of the existing housing stock is given in two datasheets: the first one focus the multi storey panel buildings with load bearing cross-walls and the second one exemplifies the modernization that can be performed in existing multi-storey family large panel building.

- **Multi-storey panel buildings with load bearing cross-walls:** The estates with multi-storey panel buildings from the second half of the 20th century are in general criticised. That criticism is partly connected to their size and physical appearance uniformity (sites and buildings) and partly connected to the tenants' social composition. As the problems associated with panel buildings cannot be solved by just demolishing and new building construction this datasheet shows some renovation solutions, carried out recently, that transformed old prefabricated buildings in fairly good buildings.
- **Example of modernization of existing multi-storey family large panel building:** The existing situation in large panel dwellings houses does not fulfil tenants' demands. Reorganisation of residential buildings and areas are needed. This datasheet shows some possible ways of adaptation of large panel buildings like: the redistribution of internal space (joining adjacent flats), the structural solutions for new habitable surface, adding of new loggia and balconies, the installation of additional lifts, the casing of balconies and loggia changing of architectural view of buildings.

Aesthetics Aspects
The aesthetics aspects are treated in two datasheets: the first deals with the parameters for the aesthetic of New Building Technologies and the second gives a sustainable perspective of the buildings aesthetics.
- The datasheet Parameters for the Aesthetic of New Building Technologies shows the aesthetics as the science of sensible cognition. Aesthetics' ambivalence explains itself in a physical dimension, concerned to senses, to perception of colours and superficial weavings of lights; and in an emotional dimension vivified by physical subjective perception. This also shows how new materials and new technologies may arise new aesthetic aspects.
- The datasheet Aesthetics – Sustainable Perspective shows how environmental technology does not need to be perceived as aesthetically disturbing. It also shows that the environmental efficiency improvement can be designed and organised from a combination of techniques and behavioural change of humans, stressing the idea of sustainability as the creation of technology that decreases environmental pressure, thus ensuring that future generations can use the same environmental qualities as we do today.

COST ACTION C12

Building Sustainable Development

In the field of building construction, sustainability of construction may be defined as a constant search for planning processes, design, construction, exploitation and reconstruction/replacement of buildings, maximizing a sustainable development.

Buildings last longer in the ecosystem than any other product made by our society. Therefore buildings sustainability is perhaps one of the most important issues that has been discussed in the scope of WG3 and that will be developed in the future in international and national forum.

In this chapter the building sustainable development is approached by 3 datasheets. Two datasheets treats the sustainability of Mixed Building Technology solutions (Sustainability Assessment and Towards Building Sustainability - Life Cycle Assessment of Building Products and the datasheet) and another one refers to the Climate Change (Climate Change Impacts on Built Environment)

- The datasheet for the Towards Building Sustainability - Life Cycle Assessment of Building Products shows how building sustainability assessment has to be taken into account not only in the construction phase but also the behaviour of the building over its service life and its final destination (demolition/recycling) including any change of its functionality that might happen during this period.

- The datasheet for the Sustainability Assessment shows a methodology to adopt in the assessment of the sustainability of construction solutions. For this task several parameters must be analysed, being most part of them independent and impossible to correlate. As the way that each parameter influences the sustainability is neither consensual nor constant along the time the presented methodology could constitute a first step to conveniently help the design teams in choosing more sustainable technologies in detriment of others less sustainable.

- The datasheet for the Climate Change Impacts on Built Environment reveals needs to re-evaluate the safety and durability of the built environment. The climate change takes place globally and inevitably according to the common opinion among the climate experts. Future changes in climate are expected to include additional warming, changes in precipitation patterns and amounts, sea-level rise and changes in the frequency and intensity of extreme winds and rains. The most widespread serious impacts are flooding, landslides, mudslides and avalanches. All these risks have consequences on the built environment that should be recognized as new planning and design conditions in many countries.

Buildings Performance

The management of the performance of Mixed Building Technology solutions are treated in two datasheets: Management of Building's Performance and the Framework for Adaptability:

- The datasheet Management of building's performance presents the modern approach to building's performance analysis that aims at managing design and construction process from the viewpoints that take into account the performance needs of the occupants and society. The focus of the performance analysis is the completed building in use. Decision-making can be supported by a toolkit in which the performance objectives are a part of evaluation of the cost-efficiency and eco-efficiency.

- The datasheet Framework for adaptability shows the short-term and long-term benefits of the possibilities to make changes in buildings. Adaptability has become an essential issue for the building and construction sector for two main reasons: pace of business requires rapid alterations in commercial and office buildings and demographic changes require more individually tailored residences. Improvement of adaptability responses these topical needs but it also helps to extend the service life of buildings. The methodology of Open Building can be applied both in new and old buildings to find solutions in practice.

Building Physics Aspects

The building physics assessment of Mixed Building Technology solutions are treated in three datasheets: Thermal Performance, Acoustics and Daylighting.

- The datasheet for the Thermal Performance shows that buildings are systems that interact with its surrounding environment and, in order to avoid excessive energy consumption and maintain the thermal comfort, shows some strategies for buildings design in a way that they can respond to the environment by the intelligent use of materials and a minimal reliance on machinery.
- The Acoustics datasheet shows how noise insulation (airborne sound insulation and impact sound insulation) can be evaluated. This datasheet also shows the acoustic performance of some MBT solutions and some technical solutions to improve the sound insulation in buildings.
- The Daylighting datasheet shows how light can be perceived by the human eye and the relation between light and building materials. This datasheet also presents the design factors and discusses the related design problems.

3. Conclusions

WG3 was a forum of exchanges between experts in separate disciplines which complement each other in the integration of Mixed Building Technology and Urban Design in view of higher living quality in urban environment. The results achieved can have direct positive impact on the environment through the savings on material and energy consumption and the possibilities for dismantling and recycling.

Expertise on each of WG3 topics exists in almost all European countries, but until now no integrated knowledge was available. In this context the COST scheme was the ideal platform to gather a multidisciplinary team to integrate new technologies in the field of urban buildings and to disseminate understanding and knowledge.

Luís Bragança   Heli Koukkari
WG3 Chairman   WG3 Vice-chairwoman

COST C12 – WG3

Datasheet III.1.1

Prepared by: Adam Rybka

Department of Civil Engineering
Rzeszow University of Technology
Poland

THE FUTURE OF EXISTING BUILDINGS
Example of modernization of existing multi-storey, family large panel building

Description

- In EU (15) countries exist 150 million dwellings, this means 400 dwellings per 1000 inhabitants. However, in some new Member States like Poland there are only 300 dwellings per 1000 inhabitants. Over half of the stock is built after 1946, every third dwelling is built between 1946-1970.

- Prefabricated building technologies were used all over in Europe in order to meet the social dwelling requirements, together with the cost reduction. Constructions of them were characterized by: arrangement of bearing elements, size and solution of prefabricated elements, solutions of connections as well as use of materials.

- Demolish of this buildings is alternative for modernization. Yet in view of lack of flats as well as limited possibilities of financial state as well as small affluence of society, it is not accepted in general. Maintenance and modernization of existing large slabs buildings is well founded.

- The existing situation in large panel dwellings houses does not fulfil tenants' demands. Reorganisation of residential buildings and areas are needed.

- Improvement of thermal comfort in large panel houses is needed. Due to climate conditions, change of composition of the existing roof's covers is needed. In the low buildings change of roofs shape seems to be rational solution. While changing of the roof's shape at the same time we can varies building shape and make its renovation or we have possibility of increasing number of flats in existing building.

- There is a need to increase the number of commercial and communal services and common spaces (e.g. retail shops, handicraft) in these residential areas. This can be obtained in changing first floors more open.

- Another thing is the problem of changing aesthetics of these houses.

Method of analysis

- Problems of load bearing structures of residential buildings are related with:
  - lack of habitable surface connected with demographic changes and needs of new generation of occupants
  - adaptation of existing functional solutions to new needs
  - tendencies of limiting of new terrains under housing building from economic regards and using of existing infrastructure (rule of balanced development).

- Modernization creates new possibilities of architectural view of building. Realization of full modernization together with adding new balconies, loggia and tilted roofs, creates change in overall appearance. General conception of structures should fulfil requirements and standard decisions.

- Order of operation requires the following stages:
  - study of technical evaluation of chosen object
  - study of architectural conception
  - study of constructional conception

- study of conception of using of materials.
- Conception of superstructure usually foresees technical solution for an additional storey. First step is to check, whether load bearing walls and foundations can safely transfer the additional loads. After finding positive result of the analysis it is possible to develop solution for the remaining requirements.
- A problem in the vertical transport stays in medium high multi-storey buildings. In this case new external lifts may be possible to add. Convenient access to individual flats gets itself into this way.
- Possible ways of adaptation of large panel buildings are:
  - redistribution of internal space (joining adjacent flats)
  - structural solutions for new habitable surface
  - adding of new loggia and balconies
  - installation of additional lifts
  - casing of balconies and loggia
  - change of architectural view of buildings.

Example of application: Modernization of multi-storey residential large panel building

- In Poland, about 1.5 million new apartments are needed in the near future. About 0.80 million dwellings are needed to replace the dwellings, which will be lost due to demolishing the buildings that have technically poor condition level.
- In Poland, large blocks and large panel residential buildings were built using prefabricated technologies between 1945-1990.
- In *large blocks buildings* lengthwise arrangement of bearing walls was applied. On lengthwise load-bearing walls, channel ceiling plates were put. Lengthwise arrangement of load-bearing walls causes considerable difficulties to do functional changes.
- First *large panel buildings* were built at the end of 50ies. Crossways arrangement of bearing walls was used. Ceiling plates were crosswise too. Internal walls were made from concrete. The possibilities of changes that type of multi-family buildings is very limited.
- Open dwelling system with large-size slabs and walls elements, *Szczecin system,* was used in Poland started from 1967. Crosswise configuration of load bearing walls was the main arrangement used in this system. This arrangement of load bearing walls was used in several systems: *W – 70 system, Wk – 70 system, OWT – 67 system* and *OWT – 75 system.* MBT can help to do functional changes in this type of building.
- The share of residential buildings built in large panel technology in Poland in years 1950-1990 is about 50%. Shape of existing building is presented in figures 1, 2, 3.

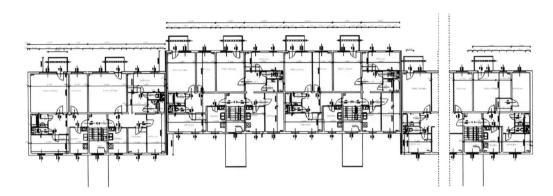

Figure 1. Repeatable level of existing building.

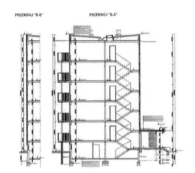

Figure 2. Section of existing building.

Figure 3. Facades of existing building.

- Changes are done by adding new elements to existing building. New level of flats and new roof was added. Also new loggias and balconies were added. New lifts help to better communication in the building. Project show haw these changes can be done in existing building and what are final results. (see fig. 4, 5, 6)

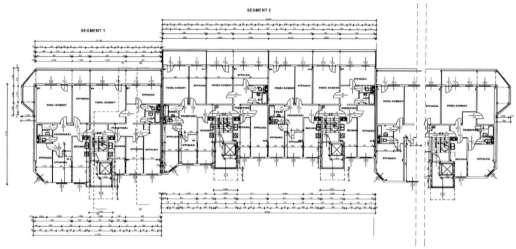

Figure 4. Project of changes in repeatable level of existing building.

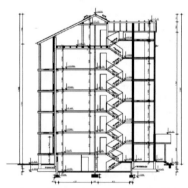

Figure 5. Project of changes in section of existing building.

Figure 6. Project of changes in facades of existing building.

Conclusions:

- The use of new technologies, now aiming at new instruments available at planner, allows more precise checks when develops the idea of project.
- Using Mixed Building Technologies multi-storey family houses can be change and their shape can be totally new.

References

- Collective work, 1998, *Concrete building*, Arkady, Warsaw
- Węglarz M., 1972, collective work, *Dwelling construction systems*, Arkady, Warsaw
- Concise Statistical Yearbook of Poland 1995, 1999. Polish Statistical Office, Warsaw 1996, 2000.
- Januszaniec B., 1985, *Concrete building for architects*, Białystok University of Technology, Białystok
- Kobiak J., 1991, *Ferro-concrete constructions*. Arkady, Warsaw
- Krol W., 1969, *Ferro-concrete constructions*. PWN, Warsaw
- Lewicki B., 1961, *Big size elements for dwelling buildings*, Arkady, Warsaw
- Parczewski W., 1995, *Concrete building for architects: finishing work elements*, Warsaw University of Technology Publishing House, Warsaw
- Rybka A., 1995, *Central Industrial District and the Polish awant-garde town-planning between wars*, Rzeszow University of Technology, Rzeszow
- Rybka A., 2001, *Evaluation of MBT buildings – Modernisation Aspects and an estimation of a heat insulation of multi-story houses made of large-sized prefabricated units*, COST C12, Bled
- Rybka A. 2003, *Large-panel dwelling building, currant stage – way of construction and architectural modification*, 207-216, Improvement of building structural quality by new technologies, Proceedings of the international seminar Lisbon, 19 and 20 April 2002, Universidade de Coimbra and Universidade de Minho, Portugal, COST C12
- Starosolski W., 1985, *Ferro-concrete constructions*, PWN, Warsaw
- Voivodships from 1995 till 2000, www.stat.gov.pl
- Zaleski S., 1997, *Remonty budynków mieszkalnych*, Arkady, Warsaw

Contacts *(include at least one COST C12 member)*

- Associate Professor Adam Rybka
  Address: Department of Civil and Environmental Engineering,
  Rzeszow University of Technology, W. Pola 2, 35 959 Rzeszow, Poland,
  e-mail: akbyr@prz.rzeszow.pl - Phone, 48 17 8651624

THE FUTURE USE OF THE EXISTING HOUSING STOCK
Multi storey panel buildings with load bearing cross-walls

Description

- The estates with multi-storey panel buildings from the second half of the 20th century are in general critiqued - a criticism that partly is connected to their size and uniformity, partly to the physical appearance of the sites and the buildings and partly to the social composition of the tenants. The last mentioned problem is of a nature not necessarily connected to the first ones described, but often strongly underlining them.

- The problems associated with panel buildings cannot be solved by just demolishing and new-building, due to both the number and the age of these buildings.

- In some countries the technical quality of the dwellings is fairly good, as renovation has been carried out recently in major parts of these estates. In such cases, the use is therefore locked for the next decades. However, one could foresee a greater rate of success for these estates and buildings if they are redesigned and rebuild taking the criticism seriously (Engelmark, 2003).

- Till now the buildings are supposed to be impossible to be rearranged because of the structural system that consists of transverse load-bearing concrete walls placed rather close together, in general giving rather strong limitations in the size of the rooms .

- Improvement of the existing multi-storey panel buildings may be done by (radical) rearrangement of inside spaces, but also by extensions of buildings horizontally and/or vertically, by adding new staircases, lifts, balconies, extra stories etc. and thereby also changing the architectural features of facades. Many different structural solutions can be developed (Engelmark, 2003; Rybka, 2003).

Methods of analysis

- The Danish multi-storey panel buildings were analyzed with regard to some factors that affect rearrangement possibilities (Sardemann, 2000; Schultz, 1999; Sørensen, 2000). Usually changes in the load-bearing structures do not cause the most difficult technical tasks – in non seismic areas at least (Northern Europe).

- The structural system of a panel building often consists of transverse load-bearing concrete walls placed rather close together, in general giving the maximum size of the rooms and also limiting the freedom in arrangement of flats (Fig. 1).

- Solutions may include taking up holes in load-bearing walls (to the limit of their load capacity) - replacement of walls with beams could also be a possibility - thereby giving flexibility in the size and different arrangements of flats.

- The facades are mostly non-load bearing and of a lightweight construction - easy to dismantle and therefore also easy to replace. Buildings like these can be compared to bookshelves, where the interior of the apartments in principle can be arranged - and replaced again and again - as easy as books on shelves (fig. 2 and 3).

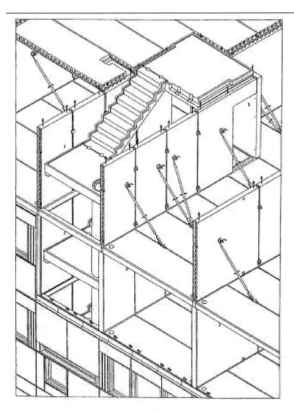

Fig. 1. The Jespersen System, the overall used prefab. system in Danish panel buildings. It is a so called open system with no limitations in suppliers of components.

Fig. 2. Typical Danish panel building with load bearing concrete cross walls (under erection 2003, but built just as done half a century before).

Fig. 3. "Placing books on the shelves and closing afterwards"

- The uniform standard height of rooms (in DK 2.5 m) gives limitation in developing other and more spacious apartments. Removing parts of the slabs will give opportunities in redesigning that are seen and have been in contemporary housing projects since the dawn of functionalism (Fig. 4).

Fig. 4. Possible future interior in panel buildings – dreamed of in a century as being standard (Le Corbusier, Pavilion de l'Esprit Noveau, Paris 1925).

- The above described alterations can be done with technology as known, but use of mixed building technology in the sense: not seen before, might ease the jobs and also bring solutions that fit more than just one purpose - e.g. adding to sound insulation. New combinations of materials and/or constructions might also open for solutions not described ahead nor foreseen at present.

- If furthermore additions outside the existing construction is taking in consideration, the possibilities are of cause manifold bigger (Fig. 5 and 6).

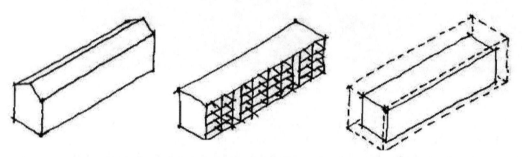

Fig. 5. A given building (Fig.5 A) and its load-bearing structure (Fig. 5 B): one can imagine a zone around it -"the addition zone" (Fig.5 C).

- The addition zone can of cause be totally filled with additional members - acting as extension of the usable floor area/building volume, at the same time as it can offer extra structural strength if needed.

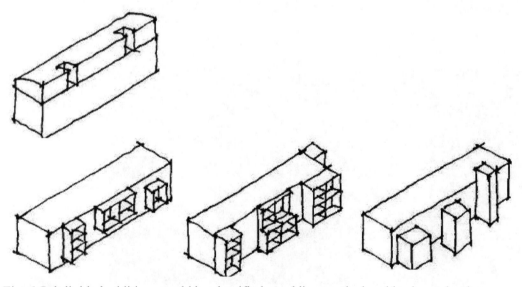

Fig. 6 Subdivided additions could be classified as adding vertical and horizontal, where horizontal parts could be seen as more or less connected to the original structure.

- Seeing the existing buildings as - what could be described like - "three dimensional building sites", also the site itself should be included: all the visible and hidden parts that give the base for the building (Fig. 7).

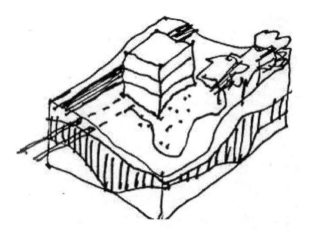

Fig. 7. The total "three dimensional building site".

Example of application: Danish research/innovation projects

- Additions: giving new access and extra floor area horizontal and vertical.

Fig.8. New lifts and staircases (Danish Ministry of Housing and Building, 1998)

Fig. 9. New access by lift and glazed balcony (Svensson & Wittchen, 1998)

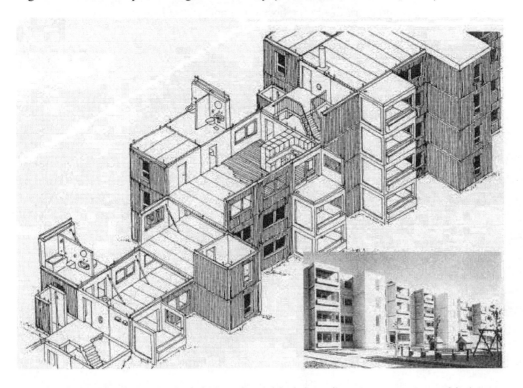

Fig. 10. Additions can in principle be of any kind: here in a newer estate and being part of the original building (Bertelsen, 1997).

Fig.11. A new projecting façade giving additional floor area and also internal staircase in a two storey flat (Danish Ministry of Housing and Building, 1997).

Fig. 12. Additions made individually (Danish Ministry of Housing and Building, 1997).

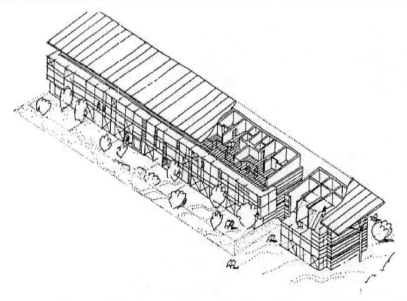

Fig. 13. Addition of flats/new roofing (Danish Ministry of Housing and Building, 1997).

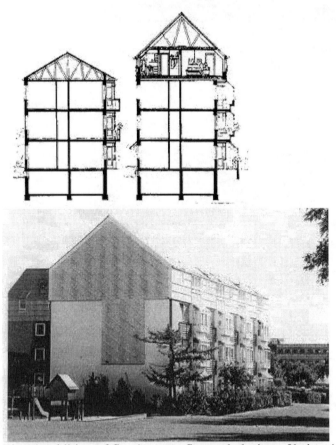

Fig.14.Addition of flats/new roofing and glazing of balconies (Danish Ministry of Housing and Building, 1997).

References

- Bertelsen, S., 1997. *Bellahøj – Ballerup – Brøndby Strand.* Hørsholm: Statens Byggeforskningsinstitut.
- Danish Ministry of Housing and Building, 1997. *10 Overcoats, Façade renovations – ten examples.* Copenhagen: Ministry of Housing and Building.
- Danish Ministry of Housing and Building, 1998. *Den lodrette forbindelse, forslag til nyindretning af trappeopgange I 50'er byggeriet.* Copenhagen: Ministry of Housing and Building.
- Engelmark, J., 2003. The existing buildings in the future. In: *Proc. Intern. Sem., Improvement of buildings' structural quality by new technology, COST Action C 12, Lisbon, 19-20 April 2002.* European Commission, Luxembourg: Office for Official Publications of the European Communities.
- Rybka, A., 2003. Large-panel dwelling building, current state – ways of construction and architectural modifications. In: *Proc. Intern. Sem., Improvement of buildings' structural quality by new technology, COST Action C 12, Lisbon, 19-20 April 2002.* European Commission, Luxembourg: Office for Official Publications of the European Communities.
- Sardemann, K., 2000. *Rebuilding cross-wall buildings 2.* Master thesis, Dept. of Civil Engineering, Technical University of Denmark. Lyngby: DTU.
- Schultz, M., 1999. *Rebuilding cross-wall buildings.* Master thesis, Dept. of Civil Engineering, Technical University of Denmark. Lyngby: DTU.
- Sørensen, G., 2000. *Analysis of cross-wall buildings.* Master thesis, Dept. of Civil Engineering, Technical University of Denmark. Lyngby: DTU.
- Svensson, O. & Wittchen, K.B., 1998. *Nye glastilbygninger I ældre etageboligbyggeri, SBI-rapport 286.* Hørsholm: Statens Byggeforskningsinstitut.

Contacts *(include at least one COST C12 member)*

- Associate professor Jesper Engelmark
   Address: BYG.DTU.Department of Civil Engineering, Technical University of Denmark, DTU – Building 118, DK-2800 Kgs. Lyngby, Denmark.
   e-mail: je@byg.dtu.dk, Phone: + 45 45251932/45251700, Fax : + 45 45883282
- Associate professor Adam Rybka
   Address :Dept. Of Civil and Environmental Engineering Rzeszow University of Technology, W Pola 2, 35 959, Poland
   e-mail : akbyr@prz.rzeszow.pl – Phone : +48 178651624
- Dr.-Ing. Frank Vogdt, Assistant Director of IEMB
   Address : Institut für Erhaltung und Modernisierung von Bauwerken Salzufer 14, D-10587 Berlin, Germany
   e-mail : zentrale iemb.de, Phone : +49 (0) 30-39921-6 Fax : +49 (0) 30-39921-850

COST C12 – WG3

Datasheet III.2.1
Prepared by: Marina Fumo
Department of Civil Engineering, University of Naples
Federico II - Italy

AESTHETICS ASPECTS

Description

- Science of sensible cognition (A. G. Baumgarten);
- aesthetics' ambivalence explains itself in a phisical dimension, concerned to senses, to perception of colours and superficial weavings of lights; and in a emotional dimension vivifyed by phisical subjective perception;
- new materials and new technologies may arise new aesthetic aspects.

Method of analysis

- In projectual step the aesthetics' care bases itself on site's study and on relationship's verification between the work and urban-social context;
- it's not possible to apply a scientific method to subjective dimension of aesthetic judgement, but it'spossible to specify some parameters that influence the physical perception as colour, superficial weaving, lights, shapes and materials that mark the relationship with the urban context;
- new materials and technologies' introduction push forward the conception of new formal expressions.

Field of application

- Generally aesthetics' care can be pursued in every projectual ambit. Particularly in architectonic project it's interesting both the works to build and ones to recover or mantain;
- in the project of "new" the advanced technologies allow to avoid known problems that deteriorate in the long run the aesthetics of architecture;
- in the project of recover the new technologies and new materials allow to erstore and preserve in the long run the formal expression of architectonic elements, or to insert contrast's elements;
- the technological culture of planner conditions the aesthetic result of architectonic work. The use of new technologies, now aiming at new struments available at planner, allows more precise checks when develops the idea.

Example of application

- The improvement of aesthetics in project of recover can be get by eliminating the troubles that infer the defacement of the built and the environment. For example to avoid the anti-aesthetic sistem of conditioning of houses can profit by materials that reduce the loss of heat and/or regulate the sun light that crosses the glass surfaces. In this field besides using active struments, as the crome-elettric glasses, it's possible to use passive sistems as the holographic films and the TIMs.The holographic films allow to direct the light towards precise points of internal ambients and to shield the undesired rays while the TIMs allow to reduce the conductance of a window maked with double glasses to 1.2 W/mq°K reducing it, therefore, to about ¾. With such materials it's possible, therefore, to improve the brigth performance and to regulate the climatic conditions of a ambient avoiding the installation of cumbersome devices of conditioning that deteriorate the aesthetics of external front of the building.

- a particularly interesting experimentation for its aesthetic development is the Guggenheim Museum in Bilbao planned by Frank O. Gehry. The museum is covered whit titanium plates, a material whit the excellent physical, mechanical and corrosion-proof characteristics. Titanium's mechanical resistance is comparable with one steel, but the mechanical resistance/density ratio is greater. Its resistance in the long run is due to a oxide-veil that covers it and protect it against corrosion, and to ability to oppose the development of mycro-organism making it inalterable to other forms of superficial attacks. In such way the architect Frank O. Gehry using a material coming from the naval sector and using a computer technology taking in loan from the sector aerospaziale, putting the attention particularly to the process of production, gets to realize an innovative work that also it integrates with the river Nervion, the bridge and the historical city, preesistenze of the site, becoming at the same time a point of symbolic reference and view (Figures 1, 2, 3).

Figure 1. Bilbao – Guggenheim Museum

Figure 2. Bilbao – Guggenheim Museum

In fact the hulling of the plates, gotten through a treatment that makes the bright material and of color silvered in comparison to the brown opaque, his natural color, becomes mirror of the surrounding one, assuming, besides also humoral tones among the various according to the climatic condition and of the sun.

Figure 3. Bilbao – Guggenheim Museum

The start-up of covering on the front of the Bilbao's museum makes use of advanced technologies.

The plates, in fact, are been singly dimensioned on a computer model and start-up with the help of laser layers to trace out the rigth placing, so that to exactly check the geometry of the solid and the dimension and disposition of the plates (Figures 4, 5).

Figure 4. Guggenheim Museum 's 3D model

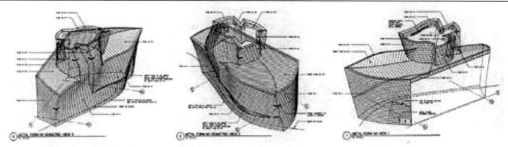

Figure 5. Guggenheim Museum 's 3D model

This has surely allowed the possibility of building an innovative architectural experience. The mass in work of the covering has happened then with the aid of laser pointers to trace its correct position.

References

- Berlage, Bergeijk, 1985. *Architettura, urbanistica, estetica*, Zanichelli, Bologna, Italia.

- Menna, 1968. *Profezia di una società estetica : saggio sull'avanguardia artistica e sul movimento dell'architettura moderna*, Lerici, Milano, Italia.

- Panza, Franzini, 1996. *Estetica dell' architettura*, Guerini, Milano, Italia.

- Soto, Pons, Cuito, 1998. *Guggenheim,* Gribaudo, Cavallermaggiore, Italia.

- Università degli studi di Roma "La Sapienza", 1996. *Plinio Marconi e l'estetica dell'architettura,* Rivista semestrale su temi monografici di architettura e urbanistica, Bollettino della Biblioteca della Facoltà di Architettura n. 54-55, Roma, Italia.

Contacts

- Prof. Marina Fumo
  Address: Department of Civil Engineering, University of Naples Federico II, P.le Tecchio, 80, 80125 Napoli, Italy
  e-mail: mfumo@unina.it - Phone, 0039.081.7682144

- Ing. Marialuigia Naponiello
  Address: Department of Civil Engineering, University of Naples Federico II, P.le Tecchio, 80, I-80125 Napoli, Italy.
  e-mail: naponiello@hotmail.com - Phone: +393478265211

COST C12 – WG3

Datasheet III.2.2

Prepared by: C.M. Ravesloot[1,2]
   [1] Delft University of Technology
   [2] Institution Cobalt Create Company Delft

AESTHETICS – SUSTAINABLE PERSPECTIVE

Description

- Sustainability is defined as the creation of technology that decreases environmental pressure, thus ensuring that future generations can use the same environmental qualities as we do.

- Environment is the set of conditions with which humans, animals, plants and minerals can develop optimally. In short environment is the set of conditions for life. If the set of conditions is not optimal there is an environmental problem.

- The techniques of environmental improvement can be qualified in three categories:

- Energy neutral building, minimum of fossil energy with maximum of sustainable energy

- Material neutral building, closed cycles of water, waste and material use, with a minimum of environmental damage, possibly with environmental benefits.

- Healthy and comfortable building, minimum of health risks for humans, animals and plants, maximum change of survival and optimal development.

- The environmental efficiency improvement can be designed and organised by engineers from a combination of techniques and behavioural change of humans (Lovins, L. Hunter-Lovins, E.A. von Weizsäcker, 1995).

- Environmental technology does not need to be perceived as aesthetically disturbing.

Method of Analysis

- Optimal conditions for life, as developed in new sustainable technology can have aesthetical impacts on peoples' minds. To improve our future potential and to clean up environmental problems from the past we need to improve sustainable technology with a factor 20 (Ehrlich, Speth, 1993).

- Decision-making is done on basis of arguments in conditional order (deJong, 1996).

- The first condition is technical feasibility. If a technique is not working efficiently yet, it is no use on a larger scale.

- The second condition is economical feasibility. Economical feasibility includes organisational, legal and process barriers before financing can become a problem. If there is no finance or return on investment, the change of realising large scale technology for sustainable development is very small.

- The third condition is social feasibility. Not always a positive argument from a sound technical and economical feasibility is followed by a positive decision making at the highest level of leadership. In practice also emotional and social-cultural arguments can have big influence on decision making. Aesthetics in architecture can be categorised under social-cultural arguments (Berlyne, 1971). If a very profitable and environmentally safe technique is ugly, so to say, it will not be used.

- Conditions for life can be put into a diagram called ecological tolerance diagram

- In this diagram the probability of survival of a certain ecosystem, like for instance a group of water plants in a swamp or a social coherent group of people in a city, is set out against the quality of a condition for life. For instance the presents of water or the aesthetical quality of a city.

- If conditions are too bad, too dry or too much repetition of similar architectural forms, the change of survival is limited. On the other hand if the conditions are too good, to much water in a swamp or too many explicit architectural forms in a city, survival changes are also limited.

- The area between too little and too high is the zone were optimal conditions for life can be found.

- Assuming we can reduce the aesthetical perception of urban areas and architecture to the simple counting of repetitive and variating objects and forms, we can put this in the diagram as well.

- According to design research of de Jong and Ravesloot, this way of argumenting can be viable under a certain set of conditions (de Jong, Ravesloot, 1995).

- By accepting the simplification of aesthetical impressions into counting repetition and variation, aesthetics in architecture can be put into an analysis method. This was first done in a case study by de Jong and Ravesloot in 1995 in the Amterdam quarter De Baarsjes.

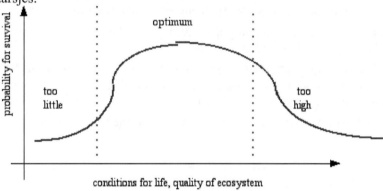

Figure 1. Ecological tolerance curve (de Jong, Ravesloot, 1995)

Field of application

- The introduction of sustainable building techniques in the practice of architecture is a daring enterprise. On the one hand we see a strong focus on technical feasibility and the challenge to improve efficiency, on the other we have to take into account the tendency in our society to obstruct or slow down innovations. Improving the efficiency in architecture is not only a technical problem. We build zero-fossil-energy houses. We are developing fully prefabricated and fully portable buildings, and we are capable of recycling many building materials. New techniques enable us to design houses that do not harm our health or create discomfort. Although technically we have the answer, the legal, financial and social structure of our society does not support the introduction of sustainable architecture.

- However, is it possible to give sustainable architecture such significance that we might counter the destructive trend in society and in architecture in particular. The imperative argument is that beauty might contribute to the success of sustainable architecture and make it acceptable and contributing to a trend-break.

- The most powerful cultural aspect of architecture is its pleasing quality. It is the architect's challenge to provide for artful buildings that like all art reflect society and may also express new architectural meanings and ideologies. From the beginning of this century especially the modern architects tried to design buildings that were an answer to social changes, being directly or indirectly the result of industrialisation in the building process. Architects interpreted this as a road toward more wealth and comfort. By now we are well aware of the fact that industrialisation also had less positive effects on architecture. This can be seen in the massive repetition of prefabricated building elements that cause dull architecture, especially in housing.

- From the analysis of aesthetical quality by de Jong and Ravesloot in the Amterdam quarter De Baarsjes we can conclude that indeed it is possible to count repeating and varying elements in several scale's of urban design and architecture. We can however not make clear that visual quality can be substituted by beauty. The visual quality of 'De Baarsjes' has only been rough. It might be an unhallowed simplification. But for the sake of the argument it can be assumed that beauty can be partly represented in its simplified form of visual quality. In the above assumed context of visual quality, the experiment to substitute visual quality for presented form in terms of repetition and variation, is more familiar to the vocabulary of environmental psychologists such as Michael Stephan (Stephan, 1994).

- The argumentation for the importance of beauty for architecture could be as follows:

- In ecology all physical conditions for life follow a similar rule expressed in the tolerance-curve (T.M. de Jong 1996).

- In psychology variations in form, related to time and scale, do arouse potential positive and negative valuation (Berlyne 1972, Gombroich 1982, Hekkert 1994).

- This provides the only justification for the assumption of relating visual quality and beauty.

- In the history of philosophy, art and architecture have been expressing (spiritual) knowledge and moral (Murdoch, 1993).

- This means that architecture and the built environment are strong enough to carry some cultural significance. So I put forward the hope and the possibility that sustainable architecture, seen in this tradition, could carry the message of a trend-break toward sustainable living.

- According to McAllister beauty has been playing a major role in accepting new paradigm's in science and technology (McAllister 1996).

- According to several insights and theoretical arguments, brain functions do react on something like an aesthetical imprint. (Sacks 1995; Stephan, 1994).

- The most interesting part of Stephan's 'Transformational Theory of Aesthetics' is that variations in the set of known experiences would arouse some emotional reaction. Variation is a trigger for brain activity. Architecture can do this too, by providing a proper set of repeating and varying elements and would supposedly be able to arouse some emotional tension and reaction.

- The variation, then, would be to design equivocal aesthetic experience, i.e. architectural images which contain a mixture of surprising variation in a sequence of recognisable repetition. Practically this would enhance prefabrication of differing elements. Buildings would become compositions of repeating elements with surprising variations in the building details and building knots perhaps (Ravesloot, Thöne 1996).

- Many questions remain unanswered, but the most important one would be: would a proper equilibrium of repeating and varying elements help people accept sustainable architecture, because it makes them think or just makes them feel at ease?

Example of application

- The optimum aesthetical is biases by cultural context and by personal opinion and perception. However, the concept of contemporary architecture is often characterised by a high number of repetitive elements in different parts of the buildings. Within scientific acceptable tolerances it can be put that the excessive use of repeating form elements a boring not appealing image might occur that might in most cases not be seen as aesthetical pleasing (Gombrich 1982).

Figure 2. Analysis of repetition in New York   **Figure 3.** Analysis of repetition in Berlin

- Example of an analysis of the sky line of down-town New York and High-rise housing in Berlin.

- To calculate our count the repetition and variation in buildings, it is absolutely imperative that the counting is done within each scale separately. According to de Jong and Ravesloot this is the only way to make valid conclusions. The so called scaleparadox can otherwise mess up all conclusions.

- As an example of a possible analysis a building is shown the Figure 4.

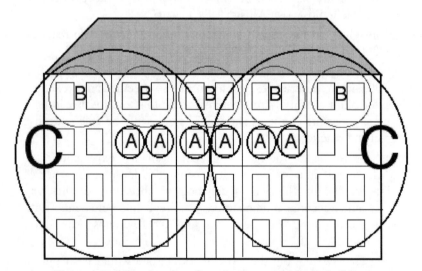

Figure 4. Building elevation of a multy-story multi family building.

- The analysis can result in the conclusion that within the scale of circle A only one form is seen, a window. However within the scale of circle B two similar window can be seen. Within the scale of circle C even more similar window can be counted, although this time in the same grouping of pairs like within scale B was seen.

- This analysis was tested in a case study in Amsterdam. The Quarter De Baarsjes was built in the period 1920-1940. The urban design and architecture is characterised by repetitions in the scales of 1000 m, 100 m, 10 m 1 m und 10 cm. The scales in between of 300 m, 30 m, 3 m, und 30 cm have more variations. The detailing of the connection between elements and components was often elaborated by architects, craftsmen and even artists.

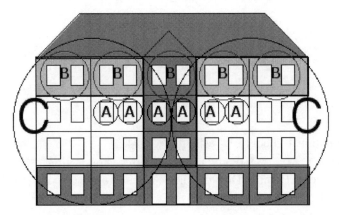

Figure 5. New design with colour that creates variations in the elevation.

- With this knowledge, sustainable architecture can, in the long run, profit from repetition from industrialisation, leading to cost reductions without endangering aesthetics. As long as enough variation in different scales is ensured a mix of variation and repetition, within the optimal zone, can be realised.

- In the example in the image below the elevation was altered with some more variation on the scale level of circle D, one higher than C. Also the connection between two C scales was varied by bringing in some colour. The B components were changed an used to make the connection between scales. The repetition of the window, as building components, however was not altered, leaving the potential for industrialisation in tact.

- Also in product development and industrial design it is accepted that beauty of men made things can be influenced by designing, as well as that the aesthetical perceptions of products can be anticipated by designers (Hekkert, 1995) (McAllister, 1996).

- A design for a future approach for industrialised prefabricated timber framing in housing based on maximal environmental gain and maximum aesthetical potential was put forward by Thöne and Ravesloot (Ravesloot, Thöne, 1996) (Thöne, Ravesloot, 1996).

References

- Berlyne, D.E. 'Aesthetics and Psychology', Appleton Century Crofts, New York 1971;

- Ehrlich, S. 'Can the world be saved', Ecological Economics Volume 2, 1993;

- Gombrich Ernst H., 'Ornament und Kunst, Schmucktrieb und Ordnungssinn in der Psychologie des dekorativen Schaffens', Klett Cotta Stuttgart, 1982;

- Hekkert, P., 'Artful Judgements, a psychological inquiryinto aesthetic preference for visual patterns', Delft University of Technology, 1995;

- Jong, T.M. de, Ravesloot, C.M., 'Beeldkwaliteitsplan De Baarsjes, Stadsdeel De Baarsjes', Amsterdam 1995;

- Jong, T.M. de, 'Essays over Variatie', Publicatiebureau Bouwkunde, Delft 1996;

- Lovins, L. Hunter-Lovins, E.A. von Weizsäcker, 'Faktor 4, doppelter Wohlstand, halbierter Naturverbrauch', Droemer Knaur, München 1995;

- McAllister, James W., 'Beauty and Revolution in Science', Cornell University Press, Ithaca, 1996;

- Ravesloot, C.M., Thöne, V., 'Modularer Holzbau unter esthetischen, ökologischen und ökonomischen Aspekten', Das Bauzentrum, Verlag Das Beispiel, Darmstadt, 1996;

- Schneider, R., 'Architektur die sich fühlt, Christian Hunzicker, Le Schtroumpf in Genf', Edition Fricke Köln, 1986;

- Sacks, O., 'PM', 1995;

- Stephan, M., 'Transformational Theory of Aesthetics', 1994;

- Thöne, V., Ravesloot, C.M., 'Method for designing connections between components', Conference, 'Detail Design in Architecture', Nene College Northampton 1996;

Contacts

- drs.ir. Christoph Maria Ravesloot

  Address: Delft University of Technology, Faculty of Architecture, Berlageweg 1, NL-2628 CR Delft, Netherlands.
  e-mail: C.M.Ravesloot@bk.tudelft.nl.

  Cobalt Create Company, ChristophMaria@Ravesloot.nl
  PO Box 2952 NL-2601 CZ Delft
  fax- voice mail: +31848753089

Prepared by: H. Gervásio[1], L. Simões da Silva[2], L. Bragança[3]

[1] GIPAMB, Ltd. – Lisbon, Portugal
[2] University of Coimbra – Coimbra, Portugal
[3] University of Minho – Guimarães, Portugal

TOWARDS SUSTAINABILITY: LIFE CYCLE ASSESSMENT OF BUILDING PRODUCTS

Introduction

- In building construction, sustainability may be defined as a constant search for planning processes, design, construction, exploration and reconstruction/replacement of buildings, minimizing environmental impact by limiting the consumption of non renewable fossil fuels and non renewable materials, minimizing pollutant emissions, and minimizing waste production.

- Buildings in general last longer in the ecosystem than any other product made by our society. Therefore sustainability might imply the consideration of a minimum lifetime of about 100 years so that renovation/replacement cycles underlying the concept of sustainability can be considered.

- Any assessment of a building's sustainability has to take into account not only the construction phase but also the behaviour of the building over its service life and its final destination/disposal (demolition/recycling) including any change of its functionality that might happen during this period.

- Due to the necessity of assessing the life cycle of a building, particularly the issue of environmental impacts, specific methodologies are being developed - the so-called Life-Cycle Assessment Methods.

- Life-cycle design of building products is a "cradle-to-grave" analysis, from the gathering of raw materials to their ultimate disposal, and each step is examined for its environmental impact.

- A building product's life-cycle can be organized into three phases: pre-building, building and post-building:

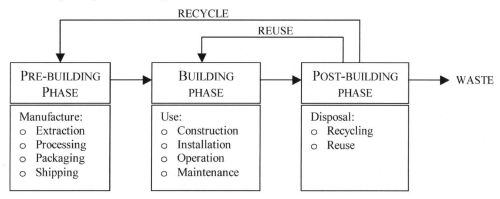

Figure 1. Phases of the building material life cycle

- o The pre-building phase includes the production and the transportation to a construction site, and has probably the most potential for causing environmental damage. A close understanding of the environmental impacts in this phase will lead to a wise selection of building products and materials.

- o The building phase refers to a building material's service life.
- o The post-building phase refers to the building materials at the end of their service life. At this point a material may be completely reused, partially recycled or dispose of.
- ▪ In view of a sustainability assessment and as a result of the definition of sustainable development of building construction, several features should be taken into consideration:
  - o Social aspects (health, hygiene, security and wellbeing)
  - o Economic aspects
  - o Functional aspects (flexibility to change the functionality)
  - o Technical aspects (durability, reliability and service behaviour)
  - o Environmental aspects (consumption of natural resources such as energy, raw materials, water and air; pollution problems and waste production)

Methodology

- ▪ A study conducted by the University of Michigan defined criteria that lead to the selection of sustainable building materials. In the table below, those criteria are grouped according to the affected building life-cycle phase. The presence of one or more of these "green-features" in a building material can assist in determining its relative sustainability.

Green Features		
Pre-building Phase	**Building Phase**	**Post-building Phase**
Waste reduction	Energy efficiency	Biodegradable
Pollution prevention	Water treatment & conservation	Recyclable
Recycled	Nontoxic	Reusable
Embodied energy reduction	Renewable energy source	Others
Natural materials	Durability	

- o Waste reduction: measures to reduce waste in the manufacturing process allow to increase the resource efficiency of building materials.
- o Pollution prevention measures taken during the manufacturing process can contribute significantly to environmental sustainability.
- o Recycled content in a product means that it was produced from post-industrial or post-consumer waste. By recycling materials, the contained embodied energy is preserved. The energy used in the recycling process for most materials is far less than the energy used in the original manufacturing.
- o Embodied energy: in most cases the greater a material's embodied energy, the greater the amount of energy required to produce it, implying more severe ecological consequences
- o Natural materials are generally lower in embodied energy and toxicity than man-made materials. They require less processing and are less damaging to the environment.
- o The aim of using energy-efficient materials is to reduce the amount of generated energy that must be brought to a building site. The long-term energy costs of operating a building are heavily dependent on the materials used in its construction.
- o A system for water treatment/conservation allows to increase the quality of water

and/or reduce the amount of water used on a site.

- o Non-toxic materials are less hazardous to construction workers and building's occupants.

- o Renewable energy systems can be used to supplement or eliminate traditional heating, cooling and electrical systems through the utilization of the natural energy available in nature (wind, solar radiation and geothermal heat)

- o Materials that last longer will, over a building's useful life, be more cost-effective than materials that materials that need to be replaced more often. Durable materials that require less frequent replacement will require fewer raw materials and will produce less landfill waste over the building's lifetime.

- o Very durable materials may have many useful years of service life left when the building in which they are installed reach the end of its service life, and may be easily extracted and reinstalled in a new site.

- o Recyclability measures a material's capacity to be used as a resource in the creation of new products.

- o The biodegradability of a material refers to its potential to naturally decompose when discarded.

- The International Standards Organization has developed a series of standards dealing with environmental management and life cycle assessment in general.

- According to ISO standards series 14040-14049, the general LCA methodology involves four steps: the goal and scope definition, the inventory analysis, impact assessment and interpretation step.

  - o The goal and scope definition step spells out the purpose of the study and its breadth and depth;

  - o The inventory analysis step identifies and quantifies the environmental inputs and outputs (inventory flows) associated with a product over its entire life cycle.

    In this step the functional unit, which is the subject of the study, is translated into a product system. The scope and boundaries of the system and the involved processes as well as the level of detail have to be defined.

  - o The impact assessment step characterizes these inventory flows in relation to a set of environmental impacts. Environmental Impacts are classified in different groups of impacts each with their own characteristics. Normalisation makes it possible to weigh the different impacts and compare them.

  - o The interpretation step combines the environmental impacts in accordance with the goals of the LCA study. Results of the LCA calculations need to be analysed and all the choices and assumptions of the analyses need to be evaluated. A number of tests to check the validity of the results are prescribed in the standards.

- Selecting environmentally preferable building products is a fundamental step towards sustainability. However the environmental performance must be balanced against economic performance, so that the methodology can appeal to everyone involved: manufactures, designers and final customers.

- Integrated life cycle design aims to improve on methods to incorporate and balance the most influential aspects: Human conditions, Economic conditions, Culture and Ecologic considerations.

Applications

- The U.S. National Institute of Standards and Technology developed a software program – BEES (Building for Environmental and Economic Sustainability), which implements a methodical technique for selecting environmentally and economically balanced building products. The environmental performance of building products is measured using the environmental life-cycle assessment approach specified in ISO 14040 standards, and the economic performance is measured using the American Society for Testing and Materials standard life-cycle cost method (E917). Environmental and economic performances are combined into an overall performance measure using the ASTM for multiattribute decision analysis E1765.
  - The purpose of this project is to provide a useful tool for the residential sector which will allow a cost-effective reduction in building-related contributions to environmental problems.
  - However, for the time being, this tool has some limitation as the building products are compared as individual components, not taking into consideration the entire building or component assemblies.
- Green-buildings consist of a holistic building strategy making use of environmental planning, design, specification, labour management and technologies to reduce the negative impact development upon the planet. The main guidelines of green-buildings are:
  - Materials efficiency through:
    - Use of existing materials from the site
    - Use of recycled materials
    - Reduction of the quantity of materials and waste.
  - Energy efficiency and greenhouse gas reduction through:
    - Ventilation, heating & cooling
    - Effective management of lighting & equipment energy demands
    - High efficiency artificial lighting
  - Water efficiency and reduced water demand through:
    - Rainwater collection and use
    - Efficient water supply appliances
  - Occupant health, comfort and productivity through:
    - Openable windows
    - Significantly improved internal air quality

Future developments

- Methodologies for integrate life-cycle assessment. Some features such as the technical performance of a building material (durability, long-time behaviour, etc), or building physics should be included in the overall performance.
- New-technology construction materials bring new uncertainties to every project. New models should allow to characterize uncertainty in the underlying data (environmental, technical performance, costs, etc) and to see how these uncertainties affect the overall life-cycle analysis.
- Models integrating sensitivity analysis, in order to analyse how changes in individual parameters can affect the overall analysis.

Examples

- Successful cases study of sustainable building around the world:
  http://www.sustainable.doe.gov/buildings/gbsstoc.shtml

References

- American Society for Testing and Materials, *Standard Practice for Measuring Life-cycle Costs of Buildings and Building Systems*, ASTM Designation E 917-94, West Conshohocken, PA, March 1994.

- American Society for Testing and Materials, *Standard Practice for Applying the Analytic Hierarchy Process to Multiattribute Decision Analysis of Investments Related to Buildings and Building Systems*, ASTM Designation E 1765-95, 1995
  http://www.60lgreenbuilding.com

- International Standards Organization (ISO), *Environmental Management – Life-Cycle Assessment – Principles and Framework*, International Standard 14040, 1997.

- ISO, *Environmental Management – Life-Cycle Assessment – Goal and Scope Definition and Inventory Analysis*, International Standard 14041, 1998.

- ISO, *Environmental Management – Life-Cycle Assessment – Life Cycle Impact Assessment*, International Standard 14042, 2000.

- Lippiatt, B., *BEES 2.0. Building for environmental and economic sustainability. Technical Manual and User Guide. NISTIR 6520*, NIST, Building and fire research laboratory, National Institute of Standards and Technology, Gathersburg, 2000.

- Lippiatt, B., *Building for environmental and economic sustainability (BEES)*, CIB/RILEM Symposium: Materials and Technologies for Sustainable Construction, Gavle, 1998.

- Public Technology Inc. and U.S. Green Building Council, *Sustainable Building Technical Manual – Green Building Design, Construction and Operations*, Public Technology Inc. and U.S. Green Building Council, Washington DC, 1996.

Contacts

- Luís Simões da Silva
  Address: Department of Civil Engineer, University of Coimbra, Polo II, Coimbra, Portugal.
  e-mail: luisss@dec.uc.pt, Phone: +351239797216, Fax: +351239797217

- Helena Gervásio
  Address: GIPAMB Consulting, Ltd., Av. Elias Garcia, nº76, 3A, Lisbon, Portugal.
  e-mail: helena@lisboa.gipac.pt, Phone: +351217938502, Fax: +351217938460

- Luís Bragança
  Address: Department of Civil Engineer, University of Minho, Guimarães, Portugal.
  e-mail: braganca@civil.uminho.pt, Phone: +351253510200, Fax: +351253510217

SUSTAINABILITY ASSESSEMENT DATASHEET

Introduction

- The construction industry in general and the buildings sector in particular, contributes to the degradation of the environment through the depletion of natural resources. Building construction consumes 40% of the raw stone, gravel, and sand used globally each year, and 25% of the virgin wood. Building also account for 40% of the energy and 16% of the water used annually worldwide (Roodman, 1995).

- New materials and constructive solutions tend to be more sustainable than the conventional ones. Construction systems with light gauge steel framing structures are one of the solutions that appeared in response to the thrive for sustainable construction.

- In the assessment of the solutions' sustainability, several parameters could be analysed, some of them not correlated and/or not expressed in the same units. On the other hand, the way that each parameter influences the sustainability is neither consensual nor unalterable in time. So, it is difficult to express a solution' sustainability in absolute terms, through an indicator that integrates all of the analyzed parameters and that allows the quantitative classification of a solution' sustainability. In this way, the sustainability is a relative subject that should be evaluated comparatively and relatively to the most widely used solution – conventional /reference solution - in a certain country/local.

- In a constructive solution sustainability assessment process, the first step consists in gathering the most relevant functional and technical data about the constructive solution. The second step consists in selecting an appropriate method that allows the quantitatively assessment of the sustainability.

- The methodology to adopt should be simple and flexible, to conveniently help the design teams in choosing a certain technology in detriment of others less sustainable.

Sustainability Assessment Methods and Tools

- The most important systems and tools for the sustainable assessment:

    i) Building Research Establishment Environmental Assessment Method (BREEAM) (BRE, 2004);

    ii) Leadership in Energy & Environmental Design (LEED) (USGBC, 2004);

    iii) Green Building Challenge (GBTool) (Greenbuilding, 2004).

- These methodologies aim the evaluation of the overall sustainability of a building. Their application is complex and needs the previous knowledge of some data. The sustainability assessment tools have datasheets, although the data is related with the particular aspects of the country of origin, which makes its application in a different country very difficult. These systems focus on the building environmental impact assessment, mainly in a global perspective. The sustainability of the constructive solutions is one of the analysed aspects in the assessment of the buildings global sustainability.

- In this perspective one good methodology is the Methodology for the Relative Assessment of the Constructive Solutions Sustainability (MARS-SC[1])(Mateus, 2004). In MARSC-SC the evaluation of the sustainability is accomplished relatively to the most applied solution – conventional/reference solution - in a certain place. Three groups of parameters are approached: **environmental, functional** and **economical**. The number of parameters analyzed on each group can be adjusted depending on the specific characteristics of each constructive solution, on its functional demands, on the evaluation objectives and on the available data. The table 1 shows some of the most important parameters that could be analysed in this methodology. The methodology follows the following steps:

  i)   **Selecting the parameters to analyse**;

  ii)   **Calculation of the comparison indexes**. The comparison between the solution under analysis and the reference solution is accomplished at the level of each parameter through a comparison between indexes. These indexes express the relationship between the value of a certain parameter in the solution under analysis and the same parameter in the conventional solution that allows verifying, relatively to each analyzed parameter, if the solution is better or worse than the conventional constructive solution. The indexes are calculated by the equation 1:

$$I_X = \frac{V_X}{V'_X}$$ [Equation 1]

with,

I_X – Index of the parameter X;

V_X – Value of the parameter X in the solution in analysis;

V'_X – Value of the parameter X in the conventional/reference solution.

  iii) **Graphical representation of the indexes**. The indexes are represented graphically in a geometric illustration with the same number of sides as the number of parameters in analysis. The graphical representation of the indexes is the Sustainable Profile. The smaller the area of the sustainable profile is, the more sustainable is the solution.

  iv)   **Sustainable score calculation**. This sustainable score (SS) of the constructive solution is calculated from equation 2.

$$SS = W_1 x \sum_{i=1}^{m} \frac{I_{Env. P.(i)}}{m} + W_2 x \sum_{i=1}^{n} \frac{I_{F. P.(i)}}{n} + W_3 x \sum_{i=1}^{o} \frac{I_{Ec. P.(i)}}{o}$$ [Equation 2]

with,

$W_1+W_2+W_3=1$;

SS – Sustainable score of the solution in analysis;

W_1 – Weighting factor of the environmental performance;

---

[1] From the Portuguese "Metodologia de Avaliação Relativa da Sustentabilidade de Soluções Construtivas"

W_2 – Weighting factor of the functional performance;

W_3 – Weighting factor of the economical performance;

$I_{Env. P. (i)}$ – Environmental parameters index i;

$I_{F. P. (i)}$ – Functional parameters index i;

$I_{Ec. P. (i)}$ – Economical parameters index i;

m – number of environmental parameters in analysis;

n – number of functional parameters in analysis;

o – number of economical parameters in analysis.

There is no consensus on the way as each group of parameters influences the sustainability of a constructive solution. However, it's a current consideration that in terms of turning the artificial environment more compatible with the natural one, without forgetting the functionality of the solutions, the weight of the environmental and functional parameters should be higher than weight of the economical parameters. Through the sustainable score value it is possible to evaluate if the solution in analysis is worse or better than the conventional/reference solution. The more close to zero the sustainable score is, the more sustainable is the solution.

Table 1 – *Parameters that could be analysed in the MARS-SC methodology*

Parameter Group		
Environmental	**Functional**	**Economical**
• Mass	• Air born sound insulation	• Construction cost (first cost)
• Primary energy consumption (PEC)	• Percussion sound insulation	• Utilization cost
• Recycled content	• Thermal insulation	• Maintenance cost
• Reuse potential	• Durability	• Rehabilitation cost
• Recycling potential	• Fire resistance	• End use value
• Raw materials reserves	• Constructability	• End use treatment cost
• Distance of transportation	• Flexibility	• (…)
• Global warming potential	• Aesthetics and innovation	
• Eutrophication potential	• (…)	
• Water intake		
• (…)		

Table 2 – Classification of the constructive solution sustainability throw the Sustainable Score (SS)

Sustainable Score (SS) value	Classification
< 1	The solution in analysis sustainability is better than the conventional/reference solution one
≈ 1	The solution in analysis sustainability is equivalent to the conventional/reference solution one
> 1	The solution in analysis sustainability is worse than the conventional/reference solution one

Example of application

- **Applied methodology**: In this example the sustainability was assessed through the methodology MARS-SC (Mateus, 2004). The sustainability of the constructive solution was evaluated relatively to the conventional solution through the comparison of two environmental parameters (mass -m- and primary energy consumption -PEC-), three functional (air born sound insulation -$D_{n,w}$-, thermal insulation -U- and occupied space –os-) and one economical (construction cost –cc-). In this way there are six indexes: i) mass index (I_m); ii) primary energy consumption index (I_{PEC}); iii) air born sound insulation index ($L_{Dn,w}$); iv) thermal insulation index (I_U); v) construction cost index (L_{cc}); vi) occupied space index (I_{os}). The weighting factors of each group of parameters considered in this study were: i) environmental parameters (W_1) = 0.45; ii) functional parameters (W_2) = 0.45; iv) economical parameters (W_3) = 0.10.

- **Description of the solution in study**: The structure of the wall is formed by 14cm height lightweight steel profiles. The interior covering is composed by two plasterboard layers, with a total thickness of 2,5cm. The external covering is composed by the structural covering and by the external thermal insulation with rendering. The structural covering is formed by 1,2cm thick OSB panels. The external continuous insulation, composed by 1cm thick expanded polystyrene plates, is fastened to the OSB panels. The final covering is a 1cm thick layer of render. The cavity between the exterior and interior coverings is filled out with 14 cm thick rock wool blankets (figure 3).

- **Description of the conventional/reference solution**: The conventional/reference solution is one of the most applied technologies in exterior walls in Portugal. The solution is a double (15+11 cm) hollow brick wall with a 2 cm thick extruded polystyrene layer placed on the air gap. Each surface of the wall is covered by a 1,5cm thick layer of render.

- **Local of study**: The presented data and the results refer to the present construction context in the North of Portugal.

Technology Description

- The light constructive technologies are one of the possible answers to the aims of the sustainable construction through: the decrease of the raw material consumption in the construction, the use of more ecological constructive materials and the superior industrialization level of the constructive processes.

- The structure of a LGSF wall is composed by a system of horizontal and vertical light weight steel profiles (Figure 1). The average thickness of this type of walls varies between 95 and 146 mm.

- The parts of this type of wall are shown in Figure 3.

Figure 1. Structural elements of a light gauge steel framing exterior wall.

Figure 2. Exterior view of a lightweight exterior wall.

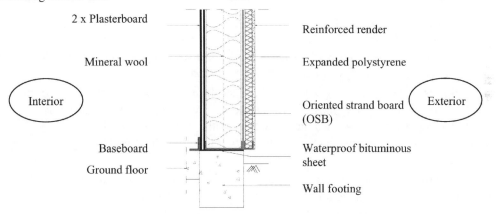

Figure 3. Sketch of a typical light gauge steel framing exterior wall.

Functional Aspects

- **Structural capability**: the steel is one of the construction materials that have the best relation resistance/weight. So, it is possible to build with slender structural elements and consequently with smaller mass. With this technology it's possible to build buildings with 2 to 3 floors with load walls.

- **Thermal mass**: The mass per unit of surface in this type of walls is significantly inferior to the mass of the conventional masonry walls (Almeida, 2002). This situation can negatively influence the buildings thermal behaviour, unless the mass of the remaining constructive elements is enough for the thermal storage.

- **Insulation**: The application of high thickness insulating materials between the two covering faces turns the global coefficient of thermal transmission quite lower than the conventional masonry walls one (Almeida, 2002). The application of the external continuous insulation contributes to the high insulation of this solution while corrects the thermal bridges at the level of the structural elements

- **Sound insulation**: In spite of the lower mass of this solution, experimental results showed that the use of great thickness absorbent materials, as the rock wool, turns, in equality of thickness, the air born sound insulation of this technology higher than the conventional masonry walls one (Bragança, 2002a).

- **Fire resistance**: The structural elements of this technology are sensitive to the fire.

Although they are protected with low thermal conductivity materials with a great fire resistance (rock wool and plasterboard). Studies, made in agreement with the EC3 methodology, showed that the fire resistance of this technology is about 60 minutes (Gervásio, 2002).

Environmental Sustainability

- **Construction mass**: The mass of this constructive solution is about 27% of the mass of a conventional masonry wall with the same thermal insulation. In this way, this solution is more maintainable under the point of view of the natural resources preservation (Mateus, 2004).

- **Primary energy consumption (PEC)**: In spite of its quite inferior mass, the primary energy consumption in this solution is just slightly inferior to the conventional masonry walls one. This situation is justified by the use of a big quantity of steel, material that needs great quantity of energy to be processed.

- **Waste production**: This technology is characterized by a great industrialization of the constructive process, being most of its elements prefabricated. The extremely controlled productive process allows the reduction of the residues. The elements are produced with the exact dimensions needed to accomplish their functions, not giving place to wastes.

- **Recycled content**: The steel could be 100% recycled, so, in this solution, the recycled raw materials content could be higher than the conventional walls one.

- **End use and recycling potential**: In this solution the constructive materials/elements are punctually joined, being easy to separate during dismantling. On the other hand, the biggest amount of the materials can be easily reused or recycled: the steel profiles can be easily reused in another building or 100% recycled; the mineral wool and the plasterboards can be directly reused; the OSB panels can also be reused or if its conservation do not allow, they could be energetically valued as combustible.

Buildability, Availability and Cost

- This constructive technology is new in Portugal, so the availability of specialized workmanship for this technology is still limited, what makes its buildability low. The small number of companies specialized in metallic structure technology also increases its cost and reduces its competitiveness

- In Portugal, the importation of the biggest amount of the consumed steel in association to the abundance of raw materials needed for the conventional construction, makes this solution less competitive in terms of construction cost.

- In a global economical analysis of the materials life cycle, the higher construction cost could be compensated by the biggest end value of this solution (most materials can be directly reused or recycled) and by the lower energy cost during the utilization phase (due to the superior thermal insulation of this solution).

Sustainable Profile and Sustainable Score

- Table 1 lists the results obtained in the sustainability assessment. The results were obtained through numerical evaluation methods.

Table 1 – *Results obtained in the sustainability assessment*

Parameter group	Parameter	Conventional solution	Studied solution	Index	
Environmental	Mass (kg)	279.00	76.00	$I_m =$	0.27
	PEC (k.W.h/m^2)	197.00	171.00	$I_{PEC} =$	0.87
Functional	$D_{n,w}$ (dB)	51.00	51.00	$I_{Dn,w} =$	1.00
	U (W/m^2.°C)	0.70	0.23	$I_U =$	0.33
	Thickness (cm)	33.00	19.60	$It =$	0.59
Economical	Cost (€/m^2)	46.68	133.40	$I_{cc} =$	2.85

- **Sustainable profile**: The labels graphical representation results in the sustainable profile presented in figure 4.

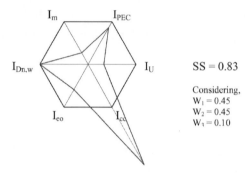

SS = 0.83

Considering,
$W_1 = 0.45$
$W_2 = 0.45$
$W_3 = 0.10$

Figure 4. Sustainable profile and sustainable score (SS) of the solution in analysis.

- **Comments**: The results show that, relatively to the parameters in analysis, the constructive solution in study is more sustainable than the conventional one. The factors that contribute most to this result are: the lower mass, the lower thickness and the biggest thermal insulation. Although, the constructive cost of the analysed solution is about 185% higher than the cost of the conventional solution. However, in a global economical analyses that approaches the several phases of a building's life cycle, this difference could be lower, justified mainly by the following factors: the smaller mass allows important savings at the level of the footings; the smaller thickness allows the maximization of the marketable interior areas; the superior thermal isolation potentially decreases the energy consumption necessary for the maintenance of the interior temperature; in the end of the building's life, the deconstruction processes are simple, because the biggest amount of the elements are linked mechanically, which allows the reuse of elements that are in good conservation and the recycling - the steel is 100% recyclable.

References

- Almeida, M. & Bragança, L. & Silva, S & Mendonça, P., 2002. *Thermal Performance of a MBT Solution – a case study*. Proceedings of the XXX IAHS

World Congress on Housing – Housing Construction, an Interdisciplinary Task, University of Coimbra, Portugal.

- Bragança, L. & Almeida, M. & Silva, S. & Mendonça, P., 2002a. *Acoustic Performance of a MBT Solution – a case study*. Proceedings of the XXX IAHS World Congress on Housing – Housing Construction, an Interdisciplinary Task, University of Coimbra, Portugal.

- BRE. 2004. *Building Research Establishment Environmental Assessment Method – BREEAM, Building Research Establishment Ltd, England.*

- Roodman, D. M. & Lenssen N. 1995. *A Building Revolution: How Ecology and Health Concerns are Transforming Construction*, Worldwatch Paper 124, Worldwatch Institute, Washington, DC, March.

- Gervásio, H. & Santiago, A. & Silva, L. & Bragança, L. & Mendes, J. 2002. *Safety and Functional Assessement of MBT Building Solutions in View of Sustainability*. Proceedings of the XXX IAHS World Congress on Housing – Housing Construction, an Interdisciplinary Task, University of Coimbra, Portugal.

- Greenbuilding. 2004. *Green Building Challenge – GBTool*.

- Mateus, R. 2004. *New Building Technologies for Sustainable Construction*. Thesis for the master in sciences degree, Engineering School, University of Minho, Portugal.

- USGBC. 2004. *Leadership in Energy & Environmental Design – LEED, U.S. Green building council.*

Contacts

- Prof. Luís Bragança

    Address: Department of Civil Engineering, University of Minho, Azurém, Guimarães, Portugal.

    e-mail: bragança@civil.uminho.pt, Phone: +351 253 510 200, Fax: +351 253 510 217

- Eng. Ricardo Mateus

    Address: Department of Civil Engineering, University of Minho, Azurém, Guimarães, Portugal.

    e-mail: ricardomateus@civil.uminho.pt, Phone: +351 253 510 200, Fax: +351 253 510 217

CLIMATE CHANGE IMPACTS ON THE BUILT ENVIRONMENT

Description

- Climate Change has been identified as one of the most important environmental challenges to the mankind.
- Climate Change according to United Nations Framework Convention refers to a change which is attributed directly or indirectly to human activity that alters the global atmosphere and which is in addition to natural climate variability.
- The average air temperature of the globe is projected to be 1.4-5.8 °C higher in 2100 than in 1990 according to climate model-based scenarios completed by the Intergovernmental Panel on the Climate Change IPCC. This estimation includes the natural variability.
- Future changes in climate are expected to include additional warming, changes in precipitation patterns and amounts, sea-level rise and changes in the frequency and intensity of some extreme events.
- The most widespread serious impacts are flooding, landslides, mudslides and avalanches.
- The impacts of the climate change in Europe vary. Southern Europe and the European Arctic are more vulnerable that other parts of Europe.
- Climate change impacts on buildings and living surroundings are related primarily to the structural safety, human health and durability. Especially the variability and extremes cause the main concerns.

Method of analysis and case study

- Risk analysis and vulnerability analysis are the common methodologies to evaluate the effects of climate change on the buildings and built environment. Risk analysis at the global level on different branches of economy and infrastructure has been made by the international experts of IPCC.
- The regional scenarios for Nordic hemisphere countries were studied and one moderate scenario was selected as a basis to a qualitative assessment of the possible impacts.

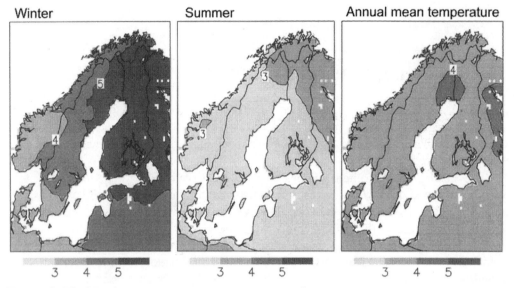

Figure 1. The North is warming more in wintertime than in summertime.

- A review of potential impacts on the built environment was produced based on the projected global changes in climate phenomena and their impacts that have high confidence of occurrence (IPCC) as well as the engineering knowledge concerning the climate-related requirements of the buildings and construction processes.

- The effects of the foreseen changes on the building and planning methods were assessed qualitatively. The effects were assessed also with comparisons with the current and well-known risks of failures and damages caused by the similar environmental hazards as expected to be caused by the climate change..

- Quantitative analysis was made about the consumption of thermal insulations in basement and soil structures and buildings envelopes.

- Together with industrial and administrative experts, a review was produced on the essential effects on building processes and solutions.

The impacts of the global climate change on planning and building in Finland

- The average temperature is foreseen to increase 4 °C in whole country in the next 100 years, according to the Swedish Regional Climate Modelling Programme SWECLIM.

- The most important climate changes that may have influence on the construction sector, were listed as follows:
  - Extreme phenomena, such as strong winds and heavy rains as well long dry or rainy periods, will occur more often.
  - The probability of flooding in riverside and coastal as well as urban areas will increase.
  - The changes in precipitation will be significant, up to 10%. The changes in precipitation are most significant in the autumn
  - Warming will shorten the duration of snow and ice cover and decrease the ground frost depth about 0.5–1.0 meters
  - Due to increasing rain the water level in rivers and lakes raises and erosion and landslides may occur at higher levels than before.

Outcome

- The probabilistic models for environmental loads (snow, wind and ice) are difficult to base on climate change models as so far the prediction is based on recorded data. Thus prediction of the change of structural safety or on the other hand the need to change the regulations is difficult to present. Further research has been initiated.

- Drains for rainwater will be overloaded more often, particularly in wintertime when the ground is frozen. It will be necessary to renew and renovate drains for rainwater in order to avoid the urban flooding risks.

- Risks on public health and environment may be caused by flooding in the areas, where toxic or noxious materials are stored or handled. New planning and design regulations are needed in order to avoid these imminent risks of contamination of waters and soil.

- The deterioration risk of buildings envelopes (walls, roofs and windows) increases with driving rains. The grainy fouling and change of colour are the typical aesthetic damages, which can be caused by driving rain. Consequently, the life-time of structures and/or the maintenance cycle gets shorter.

- The thermal behaviour of the structures of the buildings envelopes will depend more on the repeated cold and warm periods.

- When winters get warmer, more working days on site are available and the construction process and logistics are easier to manage.

- Due to more rainy winters, protective measures in transport and storage of construction products and drying of wet structures will become more demanding. Also, building materials are subjected to weather conditions and may be damaged by high humidity.

- In general planning, i.e. regional land use plans and master plans, it is essential to determine the flood, land slide and other risk areas and to leave them outside building areas or to restrict construction in them.

- In detailed planning, i.e. town plans, it is important to define sites and building areas to control location of buildings, networks and other structures. The minimum construction level near waterways should be regulated and special attention should be paid to microclimate, topography and soil. New recommendations for planning as well as specifications and complements to regulations may be necessary.

References

- http://www.wmo.ch/ (Programmes, WPC, IPCC)
- Ala-Outinen T. et al. 2004. *Impacts of climate change on the built environment* VTT Tiedotteita - Research Notes 2227. VTT Building and Transport, Espoo. 83 p. + App. 6 p. (in Finnish)

Contacts

- Senior Research Scientist Tiina Ala-Outinen, VTT Building and Transport, P.O.Box 1805, FIN-02044 VTT. Tel +358-9-456 6688, e-mail tiina.ala-outinen@vtt.fi
- Senior Research Scientist, Dr. Lasse Makkonen, VTT Building and Transport, P.O.Box 1805, FIN-02044 VTT. Tel. +358-9-456 4914, e-mail lasse.makkonen@vtt.fi

COST C12 – WG3

Datasheet III.4.1
Prepared by: Pekka Huovila & Heli Koukkari
      VTT Building and Transport
      Finland

MANAGEMENT OF BUILDING'S PERFORMANCE

Description

- Building's performance is a set of characteristics with which a concept can be qualitatively described.
- In the performance-based design, the focus is the use rather than construction of a building. The contractor identifies the market goals and makes decisions, how the users' needs are gathered.
- Analysis of performance with hierarchy of characteristics and requirements offers an integrative platform for different specialists to make common decisions.
- Software should be effectively used in gathering and managing the performance requirements of a building and its surrounding
- The essential requirements of construction works presented in the Construction Products Directive (CPD) are a part of the requirements.
- Transforming performance-based demands into appropriate technical solutions will benefit new business models.

Method of analysis

- Performance (or "functionality") requirements are usually organized in a systematic manner and accompanied by a set of specifications. For buildings, there exist several examples of analysis dating back to the well-known state-of-the-art of architecture and engineering by Roman Marcus Vitruvius Pollio in 27 BC.
- Recently, an EU funded Thematic Network PeBBu - Performance Based Building - and the working group W060, Performance Concept in Building, of the Int. Council for Research and Innovation in Building and Construction CIB, have produced models for analysis.
- The procedure to present a general criteria of a building from the point of view of fitness for use or fitness for purpose consists of the following:
  - Analysis of characteristics of a building and its surroundings
  - Qualitative description of the characteristics
  - Interpretation of the qualitative requirements into technical specifications
  - Classification of the specifications
  - Selection of verification methods and requirements

        TITLE OF THE MAIN LEVEL
        Requirement of the main level
           rationale: description, why the requirement is presented
           Basic information about the requirement
              Requirement of the sublevel
               Requirement: description
               Value: value or classification
               Verfication: methods like calculation, experiments, standards etc.

- The Construction Products Directive (89/106/EEC) of the European Community presents an overall attempt to control the basic quality of the built environment. It constitutes both the general and specific criteria with which construction works must comply. Performance requirements on works are described with six essential requirements (ER's) and durability that is considered in relation to each ER.
- The construction products are evaluated based on their ability to fit with the essential requirements. A lot of harmonization work has been done in the European Organisation for Technical Approvals EOTA and in the European Standardisation Organisation CEN in order to organise the technical requirements in a hierarchy and present the detailed verification methods.
- The essential requirements of the Construction Products Directive and their interpretation (Interpretative Documents -ID's) are

ID no 1 - mechanical resistance and stability (ER n°1)
ID no 2 - safety in case of fire (ER n°2)
ID no 3 - hygiene, health and the environment (ER n°3)
ID no 4 - safety in use, including durability (ER n°4)
ID no 5 - protection against noise (ER n°5)
ID no 6 - energy economy and heat retention (ER n°6)

- Incorporation of the systematic approach and technical information in a software application, a tool for analysis and decision making can be developed.
- Performance analysis is used in product development work to handle the market demands and official approval procedures, too.

Outcome

- Characteristics of a building is presented by the aid of the following main titles and subtitles in a functionality analysis developed at VTT (VTT Prop®)

A CONFORMITY	B PERFORMANCE
A1 Location	**B1 Indoor Conditions**
A1.1 Site characteristics	B1.1 Indoor climate
A1.2 Transportation	B1.2 Acoustics
A1.3. Impacts on surroundings	B1.3 Illumination
A2 Spatial Systems	B1.4 Vibration conditions
A3 Services	**B2 Service life and deterioration risk**
	B3 Adaptability
C COST AND ENVIRONMENTAL PROPERTIES	**B4 Safety**
C1 Life Cycle Costs	B4.1 Structural safety
C1.1 Investment costs	B4.2 Fire safety
C1.2 Operation costs	B4.3 Safety in use
C1.3 Maintenance costs	B4.4 Intrusion safety
C1.4 Demolition and disposal costs	B4.5 Natural catastrophes
C2 Environmental Pressure	**B5 Comfort**
C2.1 Biodiversity	**B6 Accessibility**
C2.2 Resources	**B7 Usability**
C2.3 Emissions	

- A software tool has been developed at VTT for management of performance requirements, EcoProP. It has a database of performance requirements and easy-to-use interfaces to the database. The application provides estimates the life cycle costs of the building.
- The EcoProP has the following features:
    - A number of prerequirements definition sets, which correspond to the possible requirements of different project types
    - User may select from one to five pre-set performance levels for each requirement and then add own comments
    - Description of the verification methods, e.g. reference to generally accepted calculation methods, experimental methods, standards or methods of expert evaluation
    - LCC- analysis is based on the cost factors associated with different performance levels and the baseline information of the project.
    - The essential requirements of the Construction Products Directive and their interpretation are incorporated in the functionality requirements

References

- Huovila P. & Leinonen J. Managing performance in the built environment. CIB World Building Congress 2001 - Performance in Product and Practice. Wellington, NZ, 2 - 6 April 2001. 8 p.
- Huovila P. Decision Analysis Support in Sustainable Building Projects. Int.Conf. on Sustainable Building 2002. Oslo, Norway, 23 - 25 Sept. 2002. CIB; iiSBE; Norwegian Research Institute. 6 p.
- Bramwell Jack, Importance of Construction Products Directive and Energy Performance in Buildings Directive to PeBBu. Report on the meeting held in Brussels 24 June 2003. http://www.pebbu.nl/
- Prior Josephine J. & Szigeti Francoise: Why all the fuss about Performace Based Building. Thematic Network - PeBBu - Performance Based Building, News Article July 2003. http://www.pebbu.nl/
- Requirements management - EcoProP. VTT Technical Research Centre of Finland. Brochure.

Contacts

- Senior Research Scientist Pekka Huovila, VTT Building and Transport, P.O.Box 1800, FIN-02044 VTT. Tel +358-9-456 5903, e-mail pekka.huovila@vtt.fi
- Senior Research Scientist Heli Koukkari, VTT Building and Transport, P.O.Box 1805, FIN- 02044 VTT. Tel +358-9-456 6816, e-mail heli.koukkari@vtt.fi

FRAMEWORK FOR ADAPTABILITY

Description

- Adaptability is regarded as an offer for the users to take part in design or as a possibility to later change spaces and service systems in a building - or both. Adaptability has become a competitive advantage for developers.
- Adaptability and flexibility of all kinds of buildings improve their capability to serve business, communities and inhabitants in the era of rapid changes.
- A growing proposition of a building's construction cost go into high-value, factory-made, electronics-loaded, soft-ware programmed components and subsystems; a correspondingly decreasing proposition will go into on-site construction of the structure and cladding (Mitchell, 2000).
- Adaptability of the technical systems is over-riding compared with structural frame as their technical and economical lifetime is shorter.
- Adaptability of structural frame is seldom possible and instead it should be such that different user needs concerning spaces and technical systems can be taken care.
- Adaptability enhances the possibilities to realize "ageing-at-home" policy accepted world-wide when the proposition of older persons is increasing.
- Adaptability is a step forward in creation of built environment suitable for all to use. Design-for-All is defined as the designing of products, services and systems that are flexible enough to be directly used, without assistive devices or modifications, by people within the widest range of abilities and circumstances as is commercially practical (Porrero & Ballabio 1998).
- The products and processes in relation to the Open Building form the physical basis of the response to the market needs of adaptability.

Method of analysis

	Fixed, unique	Changeable
Universal	**Minimum level** e.g. basic requirements stated by finanzers Problems: determination of basic requirements	**Adaptable product** e.g. adaptable furniture Problems: product development, marketing
Individual	**Customized** e.g. individually designed one-family house Problems: price, later inflexibility	**Adaptable system** e.g. Open Building system Problems: design codes, creation of product markets, price of adaptability

Markku Norvasuo: Evaluation of universal and individual design (1999).

Concept of Open Building

- The concept of Open Building recognizes the structural frame as a permanent part of a building ("support") and all other structures and technical systems as removable parts ("infill")
- The number and shape of structures in a load-bearing frame should allow user-oriented design and adaptability of the "infill"

Adaptability of common areas of a residential quarter

- In the project "Accessible Residential Quarter", recommendations and solutions were search for usability and functionality of pedestrian routes and common spaces (Koukkari et al. 2001)
- The moving of the primary users (inhabitans of different ages and capabilites) is related to personal life and household. Safety and convenience are important.
- The secondary users (maintenance, postal delivery, e-shopping, waste disposal and home aid) are usually in business. The cost-effectiveness of these services depends on the speed and accessibility.
- Minimum requirements for accessibility are given in planning and building codes of several states, and they usually refer to moving with a wheelchair
- Adaptabality is a means to save in the building costs without prohibition later usability and functionality by imroving accessibility features
- Adaptability of moving routes means provision for later changes like space for an elevator or storage areas and structures in which ramps and handrails can be fixed

Repair Concept

- In the project "Repair Concept - With Repair Construction Apartment for All Life Phases", tools for the evaluation of the quality and costs of a renovation work were developed (Tiuri, Sarja & Laine, 2001).
- The concept is an adaption of the principle of the open system building for a renovation project, especially in the prefabricated housing stock of the 1960s and 1970s.
- The emphasis is on the kinds of refurbishments and reconstruction methods that increase the independence of the technical systems from the supporting frame and the independence of one apartment from the others.
- The basis of the concept is the division of a building into the common and private spaces: the common spaces are to be designed suitable for all and the private spaces are to be designed adaptable.

Outcome

Accessible residential quarter

- Classification of the characteristics of an accessible and safe route between an apartment and busstop as well as those of other daily routes in a home environment. Classification concerns also common spaces and courtyards of a residential building
- Each aspect in classification is described in such a way that five levels can be defined based on convienence, safety and security in use. The minimum level given in codes was taken as the middle level. Adaptability improves the level of functionality.
- The qualification is described for lay-out and surfaces, structures, air quality, lighting, use of waste disposal and delivery of postal and other services

- User-oriented approach is regarded as a necessary condition for an accessible entrance and courtyard of a multi-storey apartment building

Repair Concept

- The model solutions of repair and refurbishment work are developed to be used as tools in evaluation of the functional goals and methods.
- The models are alternate or completing methods of the realization of the accessibility, adaptability and convenience.
- The models for improvement are categorized in seven main classes:
  - accessibility in the staircases (I)
  - common service spaces (II)
  - division and total shape of the flats (III)
  - outer spaces and holes of the flats (IV)
  - accessibility inside the flats (V)
  - versatility of the flats (VI)
  - inside climate of the flats (VII)
- In each main class, several methods as alternates or completing ones are presented for decision making. The models are described with drawings and cost estimations, too.
- For example, the methods for Improvement of versatility of a flat (VI) are adaptability of the lay-out (VI-1), adaptability of kitchen furniture (VI-2), adaptability of bathroom fittings (VI-3) and improvement of sound-proofing between flats (VI-4)
- For each option, more technical information is given as alternative or completing methods. For example, Improvement of adaptability of a flat (VI-1) may need movable partitions (VI-1.1), new solutions for electric and telecommunication circuits (VI-1.2), new routing for heating pipelines (VI-1.3) new routing for water and sewage pipelines (VI-1.4) and flexible floor heating system (VI-1.5) that allow for later changes.

References

- Koukkari H. et al. 2001. *Accessible residential quarter.* Espoo: Technical Research Centre of Finland, VTT Research Notes 2090. 112 p. + app. 68 p. (in Finnish)
- Koukkari H. & Laine J. 2002. *Repair concept - study on two building projects.* Espoo: VTT Technical Research Centre of Finland. Project Report. (in Finnish)
- Mitchell, W.J. 2000. e-topia. 2nd ed., Cambridge, MA: Massachusetts Institute of Technology. 184 p.
- OECD. 1999. *Conference on ageing, housing and urban development*, Oslo 21-23 May 2000, Background Paper. Paris: OECD Room Document No 1. 48 p.
- Tiuri U., Sarja A. & Laine, J. 2001. *Repair Concept. With repair construction apartment for all life phases.* Espoo: Technical Research of Finland, VTT Research Notes 2082. 45 p + app. 126 p. (in Finnish)

Contacts

- Research Professor Asko Sarja, VTT Building and Transport, P.O.Box 1803, FIN-02044 VTT. Tel +358-9-456 4120, e-mail asko.sarja@vtt.fi
- Senior Research Scientist Juhani Laine, VTT Building and Transport. P.O.Box 1803, FIN- 02044 VTT. Tel +358-9-456 4752, e-mail juhani.laine@vtt.fi
- Senior Research Scientist Heli Koukkari, VTT Building and Transport, P.O.Box 1805, FIN- 02044 VTT. Tel +358-9-456 6816, e-mail heli.koukkari@vtt.fi

COST C12 – WG3

Datasheet III.5.1
Prepared by: M. Almeida & L. Bragança
University of Minho
Portugal

THERMAL ASSESSMENT OF A MIXED BUILDING TECHNOLOGY SOLUTION

Introduction

- Thermal comfort can be defined has a state where a person does not feel any sensation of unpleasantness or irritation that can distract him from his activities in that moment. However, since this state of comfort is the result of different sensations, where many subjective factors interfere, this state won't be easy to define.

- In the design phase there are some climatic actions to consider in order to obtain a state of thermal comfort:
    - exterior temperature;
    - solar radiation;
    - wind.

- Buildings are systems that closely interact with their surrounding environment but some twentieth century buildings deny that interaction and their thermal comfort is obtained by the application of expensive heating, cooling and lightning equipments. In order to avoid this excessive energy consumption and maintain the thermal comfort standards, the buildings must be designed in a way that they can respond to the environment by the intelligent use of materials and a minimal reliance on machinery, as observed in Figure 1 and Figure 2 (Goulding, 1993).

Figure 1. Design without regarding the surrounding environment

Figure 2. Design regarding the surrounding environment

Thermal Performance

The buildings thermal performance can be characterized by the balance between the heat losses and heat gains taking into account their heat storage capacity. In this balance the three fundamental parameters are the insulation level, thermal inertia use and solar radiation control.

Heat Losses

The heat losses can occur by conduction through the external surfaces, infiltration and ventilation. Heat losses can be controlled with the application of insulation in the external envelope as well as controlling the air infiltration rates and the ventilation mechanisms.

- **Insulation**

The insulation level of the external envelope is related to the resistance that this construction element offers to the heat flow (Figure 3). The following expression gives the heat flow through one element:

$$Q = U \cdot S \cdot (\theta_i - \theta_e)$$ (1)

Where:

U – overall heat transfer coefficient (W/m^2.°C);
S – surface area of the element (m^2);
θ_i – interior air temperature (°C);
θ_e - exterior air temperature (°C).

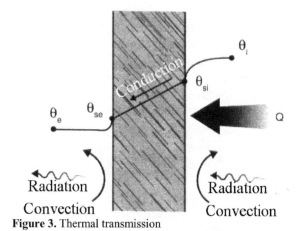

Figure 3. Thermal transmission

A side-effect of the use of insulation in the external envelope is the increase of the superficial temperature of the envelope. This increase results in a higher mean radiant temperature, which increase the comfort level. The most effective mean of increasing this comfort level by this side-effect is the use of low-U glazing.

With the increase of insulation one must be very careful in order to avoid thermal bridges (points with higher U-values). One measure used to avoid thermal brides is the application of continuous thermal insulation on the exterior façade of the building.

- **Infiltration and Ventilation**

In order to maintain a healthy environment inside buildings it's needed a certain amount of fresh air. To control the heat losses by infiltration and ventilation there are many points that must be controlled:

- Type of ventilation - natural or mechanical;

- Air flow through openings and cracks in the external envelope;

- Infiltration - the rate of air infiltration in a house depends of the wind velocity, pulsation effects or temperature differences;

- Occupants use of doors and windows.

Heat Storage

The heat storage his controlled by the thermal inertia of the building elements. To increase the heat storage, the envelope material must be heavyweight and with high thermal diffusivity.

- **Thermal inertia**

Thermal inertia is responsible for the reduction of inside air temperature peaks and for the delay between the accumulation of energy and its respective release. Then, the thermal inertia can be very useful in winter because the accumulation of heat occurs when there is solar radiation and therefore the heat is less needed. The release of heat, due to the delay, occurs only at night when it is most needed (Figure 4 and Figure 5).

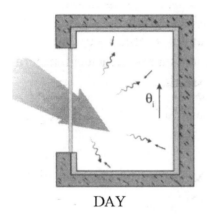
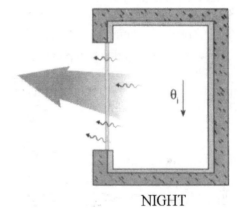

DAY   NIGHT

Figure 4. Accumulation of the energy   **Figure 5.** Realese of the energy

Heat Gains

Heat gains can be obtained either by solar radiation or by internal gains.

- **Solar Radiation**

In order to obtain heat gains by solar radiation there must exists a great area of south oriented windows. But, to convert the solar radiation into useful heat gains, the building must also possesses thermal inertia.

In Southern countries, it is also necessary to take care with excessive solar gains, which are responsible for frequent overheating situations, making mandatory the use of shading devices in all windows. Solar radiation is an essential source of heat gains that can balance the heat losses but its use and control must be carefully thought and defined since the very beginning of the project, in the design phase.

- **Internal gains**

Internal heat gains are other kind of gains that can occur inside buildings that are related with the normal use of the habitation (Achard, 1989):

     1. heat supply from occupants;

     2. heat supply from artificial lightning;

     3. heat gains due to appliances.

Thermal Performance Evaluation

- There are many ways to evaluate the thermal performance of buildings. The method used was the one recommended by the EN ISO 13790: 2003. This method was based on the determination of the anual energy needs.

- This quantification was based on the thermal load concept defined as the calorific power required to be supplied to (in Winter) or to be removed from a space (in Summer) so that the comfort conditions, such as the temperature (20°C in the heating season and 25°C and 50% relative humidity in the cooling season), remain constant.

- The thermal load is obtained from the energy balance of the space under consideration. Four separate components are found in this balance:
    - Heat exchange by conduction through the opaque envelope, comprising roofs, floors and walls;
    - Heat exchange by solar radiation and by conduction through glazed areas;
    - Heat exchange associated with the indoor air renewal, which can be caused by forced ventilation, natural ventilation or by infiltration;
    - Heat released by internal heat sources, such as equipment, lighting and occupants;

- The annual heating needs (HN) result from the algebraic sum of three components integrated for the heating season divided by the useful floor area (A_p):
    - Heat loss by conduction through the external envelope, taking also in consideration the losses through elements in contact with the ground and losses through thermal bridges, Q_t;
    - Heat loss associated with the indoor air renewal, Q_v;
    - Heat gains by internal heat sources and by solar radiation, Q_{gu}.

$$HN = \left(Q_t + Q_v - Q_{gu} \right) / A_p \qquad (2)$$

- The methodology used to calculate the annual cooling needs (CN) it's complementary to the one used for the annual heating needs and are calculated by the following equation:

$$CN = Q_g \cdot (1 - \eta) / A_p \qquad (3)$$

Where:

   Q_g – are the building total gains;

   η - is the factor of gains use;

- The building total gains are given by integrating the algebraic sum of the following portions, for the cooling season:
    - Heat gains by conduction through the outer envelope, Q_1;
    - Solar heat gains through glazed areas, Q_2;
    - Heat gains through indoor air renewal, Q_3;
    - Heat gains from indoor heat sources, Q_4.

Guidelines and/or codification

- Directive 2002/91/CE of the European Parliament of 16 December 2002 –*Thermal Performance of Buildings*.

- EN ISO 13790: 2003 - Thermal performance of buildings - *Calculation of energy use for space heating*.

Example of application: The thermal performance of a MBT building

- Building description

  MBT buildings combine different structural construction methods in one building. The majority of materials used in this solution are lightweight. Only for structural or energy (e.g. thermal storage) proposes are used heavyweight materials.

  In order to show the thermal performance of a MBT solution it was selected an office building belonging to the Portuguese Electricity Company (EDP), located in Coimbra - Portugal.

  The building analysed (Figure 6) was rehabilitated using a MBT strategy only in the last floor whilst the first two floors kept the original conventional construction characteristics.

  The internal layout of the MBT floor is an open space (Figure 7).

Figure 6. Front and lateral view of the building

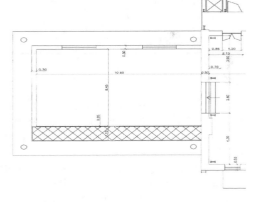

Figure 7. Schematic plan of the MBT building

The MBT building was compared with a virtual building, with the same geometry, but designed in a conventional way in what concerns the structure, walls, windows and roof.

The exterior walls in the MBT building have an overall conductance of 0.20 W/m^2.ºK, the floor 0.90 W/m^2.ºK, the roof 0.65 W/m^2.ºK and the windows 3.30 W/m^2.ºK.

The exterior walls in the conventional building were considered to have an overall conductance of 0.85 W/m^2.ºK, the floor 1.00 W/m^2.ºK, the roof 0,90 W/m^2.ºK and the windows 4,20 W/m^2.ºK.

The considered climatic data, for both seasons, was the one correspondent to sequences of typical winter and summer days corresponding to the building's location

- Coimbra, denoted in the Portuguese thermal regulations as zone I1 and zone V2 for winter and for summer, respectively. The reference data for these zones considers an average heating degrees-day, in an 18°C base, of 1460°C/year, in winter, and an exterior design temperature of 33°C for summer with temperature amplitude of 13°C.

- **Thermal Evaluation**

The energy consumption in both buildings (MBT and conventional one) was calculated according to the referred methods. Table 1 summarizes the annual heating and cooling needs, partial and total values, estimated for the analysed buildings.

Table 1 - Thermal performance of the MBT and Conventional buildings

		MBT Solution	Conventional Solution	MBT / Conventional
Heating Needs *[kWh/year]*				
- Opaque Exterior Envelope		1250	3185	↓ 61%
- Glazing		3788	2237	69%
- Air Renewal		2323	2323	0%
Useful Solar Gains [kWh/year]		3900	3514	11%
HN [kWh/year]		3461	4231	↓ 18%
Cooling Needs *[kWh/year]*				
- Opaque Exterior Envelope		327	484	↓ 32%
- Glazing	Without roller screen	3233	1491	117%
	With roller screen	1655		11 %
CN [kWh/year]	Without roller screen	3560	1975	80%
	With roller screen	1982		0%
HN + CN [kWh/year]	Without roller screen	7021	6206	13%
	With roller screen	5443		↓ 12%

- **Comments**

This study shows that along a typical year, the overall energy needs in the MBT solution are 13% higher than the energy needs in the Conventional solution. However, in winter the MBT solution has a better performance since the heating needs are 18% lower than the ones of the Conventional solution. This best thermal behavior in the heating season is due to a better thermal insulation level of the building's opaque envelope, which makes possible a significant reduction on the heat losses through the building skin. Due to its solar strategy (South glass curtain), the MBT building has more useful solar gains (11% more) than the Conventional one. However, due to the same reasons, it also shows a greater glazing area, leading to a considerable increase in the heat losses (69% more).

In summer the behavior is the opposite. The MBT building has cooling energy requirements 80% greater than the Conventional solution. In the final balance, the MBT solar strategy has a very negative impact. This worst behaviour is also due to the lack of thermal inertia in the MBT building. As previously referred, the MBT solution is a lightweight one leading to a light inertia building. The thermal inertia is responsible for the delay and amplitude decrease of the heat wave that enters into the building. The greater the thermal inertia the better indoor environment is achieved, preventing the risk of an overheating of the indoor air.

About the solar exposure, the MBT building has a glass curtain wall in the South facade, two windows in the North facade and no openings in the East and West facades. The shading device placed on the roof is not enough to protect the south glass curtain in summer leading to large cooling needs.

Only through a better solar strategy exposition it is possible to reduce the cooling energy requirements in the MBT building. Table 1 also shows the results obtained in the case of having an interior roller screen in the MBT building activated during all the cooling season. In this case, the study shows that the cooling needs in the MBT building become equivalent to the ones in the Conventional building. Along a typical year, the overall energy needs in the MBT solution would be 12% lower than the energy needs in the Conventional solution.

References

- Canha da Piedade, A; Moret Rodrigues, A; Roriz, L., 2000. *Climatização em Edifícios – Envolvente e Comportamento Térmico*. Edições Orion, Amadora.

- Goulding, J.; Lewis, J.; Steemers, T., 1993. *Energy Conscious Design – A Primer for Architects*. B.T. Batsford, London.

- Achard, P.; Gicquel, R., 1989. *European Passive Solar Handbook*. Directorate-General XII for Science, Research and Development.

- Almeida, M.; Bragança, L.; Silva, S.; Mendonça, P. 2002. *Thermal Performance of a MBT solution - a case study*. Proceedings of the XXX IAHS - World Congress on Housing, Coimbra, Portugal, 9-13 September, 2002.

Contacts

- Prof. Manuela Almeida
    Address: Department of Civil Engineering, University of Minho, Campus de Azurém, Guimarães, Portugal.
    e-mail: malmeida@civil.uminho.pt
    Phone: +351 253 510 200, Fax: +351 253 510 217
- Prof. Luís Bragança
    Address: Department of Civil Engineering, University of Minho, Campus de Azurém, Guimarães, Portugal.
    e-mail: braganca@civil.uminho.pt
    Phone: +351 253 510 200, Fax: +351 253 510 217

COST C12 – WG3

Datasheet III.5.2
Prepared by: L. Bragança[1], J. Patrício[2]

[1] University of Minho
[2] National Laboratory for Civil Engineer
Portugal

ACOUSTIC ASSESSMENT OF MIXED BUILDING TECHNOLOGY SOLUTIONS

Introduction

- Sound can be defined as the propagation of pressure and depression waves the human ear can detect.

- If a wall is submitted to pressure and depression waves it will vibrate emitting a sound with a frequency equal to that emitted by the source. This transmission of sound depends on the acoustic energy that hits the wall and on its interior structure (Meisser, 1978).

- In building acoustics two main types of sound propagation can be considered: the airborne sound and the structure borne sound (normally known as impact sound).

- In what respects airborne sound, there are three main transmission paths between adjacent rooms (Figure 1):
    - 1. Direct transmission through joints, discontinuities and cracks;
    - 2. Direct transmission due to the vibration of the partition itself; and
    - 3. Flanking transmission.

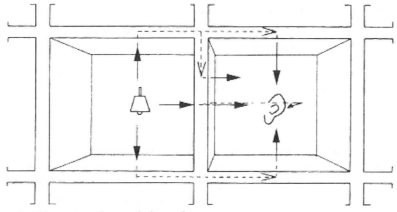

Figure 1. Airborne sound transmission paths

- The reduction of transmission of airborne sound between the source room and the receiving room is the so-called airborne sound insulation.

- Concerning the structure borne sound, which is caused by people walking, sliding of chairs, dropping of objects, etc., i.e., the sound energy that reaches the receiving room due to an impact hit. In the same way, the reduction of the radiation process in the receiving room, related to impact sound, is called the impact sound insulation.

- In order to get a good acoustic performance, three fundamental aspects should be considered at the design stage:
    - Type of construction (single leaf, double leaf, etc.);
    - Mass per unit of area of the partition, internal losses, existence (or not) of absorption material partially filling the air cavities;
    - Air tightness and sealing.

Flanking transmission

The flanking transmission is determined by the type of element and edge conditions. If there is a high flanking transmission the sound insulation between two adjacent rooms can be much lower than expected.

Regarding airborne sound, the flanking transmission can only be negligible in the case of façades, for its sound insulation is very poor when compared with that of the flanking elements.

In the case of impact sound, to minimize the flanking transmission a floating floor self-contained within each room should be used. Each partition should neither be built on top of the floating floor nor reach the ceiling where a special sealing treatment must be used. Regarding movable partitions, floating floors should be separated by joints at all possible positions.

Impact sound

Concerning the impact sound insulation and in order to get a good acoustic performance some fundamental aspects should be taken into consideration:
    - the use of resilient layers separating the floor beneath the partitions;
    - the use of floating floors which should be cut off beneath the partitions;
    - the use of resilient floor coverings;
    - the use of suspended ceilings made of boards with small stiffness.

Massive walls

Massive walls, as brick or concrete block masonry walls, are not usually used in MBT buildings because they are very heavy. The sound insulation of this type of partitions is almost entirely influenced by their mass per unit area. Other factors with less influence are the loss factor, the stiffness, and the bending conditions. In the case of the use of these types of walls in a steel construction it's important to insulate the steel profiles from the walls and enclose them with a resilient separating lining in order to avoid the vibration propagation to the steel structure.

Lightweight board walls

Lightweight board walls are the most commonly used solutions to make partitions and separating walls in MBT buildings. This type of walls can be made with a single frame or a double frame supporting single leaf or double leaf boards.

For lightweight board walls, normally, less mass is required to achieve the same sound insulation as in the case of massive walls. But the elastic characteristics of the board leafs has to be carefully chosen in order to locate the coincidence frequency below the

100 Hz frequency band.

The sound insulation of lightweight board walls is not influenced by the steel supporting frames provided the boards are effectively separated from the construction as shown in Figure 2 (Cremer, 1973).

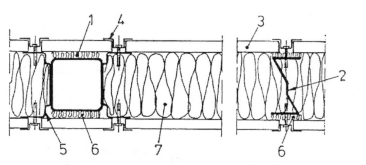

1 – 50 x 60 mm steel support
2 – Z-shaped steel sheet support
3 – 13.5 gypsum board with plastic coating
4 – Omega profile and rubber profile
5 – Metal sheet profile
6 – Intermediate layer of foamed plastic strips
7 – 73 mm mineral wood matting

Figure 2. Cross section of a lightweight double leaf wall

A new separating wall type comprising steel studs and sheathing boards has been developed in co-operation between Rautaruukki Ltd and VTT (Salmi, 2002). The work has resulted to a single-leaf wall with so called acoustic steel studs RAN AWS-studs (see Figure 3). Compared to traditional double-framed wall systems, the use of the new RAN AWS wall system brings significant benefits in constructability and economics. Erection of single stud system is swift and efficient because of the amount of studs reduced to a half and simplified installing of insulation.

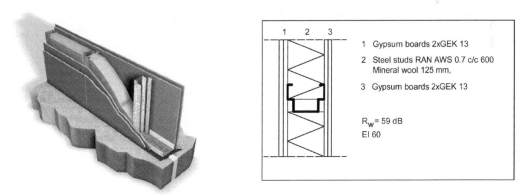

1 Gypsum boards 2xGEK 13

2 Steel studs RAN AWS 0.7 c/c 600 Mineral wool 125 mm,

3 Gypsum boards 2xGEK 13

R_w= 59 dB
EI 60

Figure 3. The new RAN AWS lightweight separating wall for load bearing and non-load bearing applications (Salmi, 2002)

Massive floors

For this type of bare floors, good impact sound insulation is not an easy task. Normally, to accomplish national regulations the use of acoustically efficient floor coverings is a basic need. Besides, the influence of flanking transmission is not negligible. Notwithstanding, a good impact sound insulation can only be achieved by using a resilient floor finish.

Double leaf floors

This type of floors can be used in lightweight constructions. However, they must be designed with care in order to minimize flanking transmission as well as good air tightness (Josse, 1977). To ensure good impact sound insulation a resilient floor finish can do the job.

Building's Acoustic Performance Assessment

- Due to the inaccuracy of the existing numerical estimation methods (Bragança, 2001), the only effective way to evaluate building's acoustic performance is by measuring each building partition noise insulation.

- In order to measure the airborne sound insulation between rooms a sound source is placed inside the emission room. The sound pressure levels in the source room (Figure 4) and in the receiving room (Figure 5) are measured on a 1/3-octave band. The sound pressure levels should be spatially averaged.

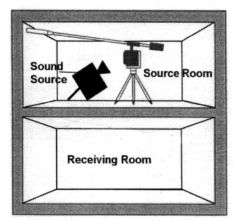

Figure 4. Sound pressure level in the source room   **Figure 5.** Sound pressure level in the receiv. room

- <u>Sound insulation of façade elements and façades</u>

  A method to evaluate the sound insulation of façade elements, $D_{2m,n}$, considers a loudspeaker as the sound source and the measurement of sound pressure levels at a distance from the façade equal to 2 m, and inside the reception room on the assumption that the sound field is diffuse. The $D_{2m,n}$ is given by the following equation:

$$D_{2m,n} = L_{1,2m} - L_2 + 10 \log\left(\frac{A}{A_0}\right) \text{ dB} \tag{1}$$

Where:

$L_{1,2\,m}$ - is the average sound pressure level on the surface of the façade, measured at 2 m distance from the façade;

L_2 - is the average sound pressure level in the receiving room;

A - is the equivalent sound absorption area of the element;

A_0 - is the reference equivalent sound absorption area in the receiving room.

- <u>Airborne sound insulation between rooms</u>

  The recommended method to evaluate the sound insulation of building partitions between rooms, D_n, on the assumption that the sound field in both rooms is diffuse. D_n is given by the following equation:

$$D_n = L_1 - L_2 + 10\log\left(\frac{A}{A_0}\right) dB \qquad (2)$$

Where:

L_1 - is the average sound pressure level in the source room;

L_2 - is the average sound pressure level in the receiving room;

A - is the equivalent sound absorption area of the receiving room;

A_0 - is the reference equivalent sound absorption (equal to 10 m^2).

- Impact sound insulation of floors

To measure the impact sound insulation of floors a standardised tapping machine should be used in order to exert a standard impact in the source room floor. Then the sound pressure levels in the receiving room on a 1/3-octave band should be measured (Figure 6). The sound pressure levels should be also spatially averaged.

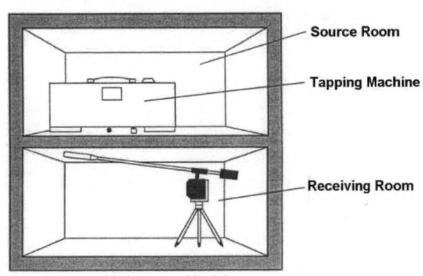

Figure 6. Measure of the impact sound insulation

The recommended method to evaluate the impact sound insulation of floors considers the normalized impact sound pressure level, L'_n, which is given by the following equation:

$$L'_n = L_2 - 10\log\left(\frac{A}{A_0}\right) \qquad (3)$$

Where:

L_2 - is the average sound pressure level in the receiving room;

A - is the equivalent sound absorption area of the receiving room;

A_0 - is the reference equivalent sound absorption (equal to 10 m^2).

- The last step is the fitting of a standard reference curve to the measured sound reduction/transmission curve (D_n and L'_n). The resulting values are the weighted sound reduction index ($D_{n,w}$) and weighted normalized impact sound pressure level index ($L'_{n,w}$).

Guidelines and/or codification

- NP EN ISO 140-4: 2000. Acoustics - *Measurement of sound insulation in buildings and of building elements - Part 4: Field measurements of airborne sound insulation between rooms.*

- NP EN ISO 140-5: 2000. Acoustics - *Measurement of sound insulation in buildings and of building elements - Part 5: Field measurements of airborne sound insulation of façade elements and façades.*

- ISO 140-7: 1998. Acoustics - *Measurement of sound insulation in buildings and of building elements - Part 7: Field measurements of impact sound insulation of floors.*

- EN ISO 717-1:1996. Acoustics - *Rating of sound insulation in buildings and of building elements - Part 1: Airborne sound insulation.*

- EN ISO 717-2: 1996. Acoustics - *Rating of sound insulation in buildings and of building elements - Part 2: Impact sound insulation.*

Example of application: The acoustic performance of a MBT building

- Building description

    MBT buildings combine different structural construction methods in one building. The majority of materials used in this solution are lightweight. Only for structural or energy (e.g. thermal storage) proposes are used heavyweight materials.

    In order to show the acoustic performance of a MBT solution it was selected an office building belonging to the Portuguese Electricity Company (EDP), located in Coimbra - Portugal.

    The building analysed (Figure 7) was rehabilitated using a MBT strategy only in the last floor whilst the first two floors kept the original conventional construction characteristics.

    The internal layout of the MBT floor (Figure 8) is an open space.

Figure 7. Front and lateral view of the building

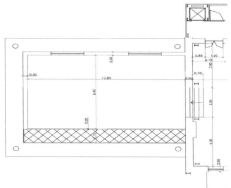

Figure 8. Schematic plan of the MBT building

- Acoustic Evaluation

Numerous "in situ" measurements where made in order to characterize the sound reduction index of façade ($D_{2\,m,n}$), the weighted sound insulation of building partitions between rooms ($D_{n,w}$) and the weighted impact sound insulation of floors ($L'_{n,w}$).

These experimental values were compared with the values experimentally obtained in the conventional part of the building with the same geometry in order to better show the acoustic performance of the MBT solutions. Table 1 summarizes the results of the acoustic evaluation.

Table 1 - Buildings acoustic performance

Element type	$D_{n,w}$	$L'_{n,w}$	$D_{2\,m,n}$
MBT Solution	Measured		
South façade (90% glass + 10% opaque)	-	-	30
East/West façade (0% glass + 100% opaque)	-	-	50
North façade (19% glass + 81% opaque)	-	-	40
Floor	53	70	-
Conventional Solution	Experimental Database		
South façade (26% glass + 74% opaque)	-	-	33
East/West façade (9% glass + 91% opaque)	-	-	35
North façade (19% glass + 81% opaque)	-	-	34
Floor	48	77	-

- Comments

Analyzing the results obtained in the acoustic evaluation of the MBT building it is possible to conclude that in almost all cases the MBT solution has a better acoustic performance than the conventional building, in spite of having less mass. This fact can be explained by the better quality of the glazing and the higher level of insulation of the exterior wall. One of the few cases where the MBT solution has a worst acoustic performance is in the south façade. Such is due to the large area of fenestration of the MBT solution.

The reason for the better airborne sound insulation of the MBT building floor lies in the floor finishing and in the suspended ceiling used, which with small stiffness and backed with mineral wool quilts, in spite of the higher mass of the conventional building floor, provide this performance.

The same reason can be pointed out to explain the better impact sound insulation of the MBT building floors. Although the conventional building floor has a higher mass, in the MBT floor, the air gap and the mineral wool quilts between the slab finishing and the suspended ceiling, increases the airborne sound insulation and reduces the impact sound pressure level.

References

- Bragança, L.; Patrício, J. 2004. Case study: Comparison between the acoustic performance of a Mixed Building Technology building and a conventional building. Building Acoustics, Volume 11, London, "Accepted and to be published in N.º 3, June,

2004".

- Bragança, L. et al. 2001. Acoustical Performance of Lightweight and MBT Buildings. Working Paper presented in the WG3-Urban Design meeting, Bled, Slovenia, 16-17 September, 2001.

- Cremer, L. and Heckl, M.1973. Structure-borne sound: structural vibrations and sound radiation at audio frequencies. Berlin, Springer-Verlag.

- Salmi, P. et al. 2002. New AWS steel stud for partition walls. International Iron and Steel Institute World Conference IISI. Luxembourg 15-17 May 2002 Steel in sustainable construction. International Iron and Steel Institute. Brussels, p. 197 - 202.

- Josse, R. 1977. A l'usage des architectes ingénieurs urbanistes. Éditions Eyrolles, Paris.

- Meisser, M. 1978. La Pratique de L'Acoustique dans le Batiment. Éditions Eyrolles, Paris.

- Patrício J. 2004. Acústica nos edifícios (Acoustics in buildings) Author's Edition, Lisbon.

- Patrício, J.; Bragança, L. 2004. Some Considerations about the Contribution of Roller Shutters Positions to Noise Insulation of Façades. Submitted to Building Acoustics Journal, in March, 2004 (situation: under revision).

Contacts

- Prof. Luís Bragança

  Address: Department of Civil Engineering, University of Minho, Azurém, Guimarães, Portugal.
  e-mail: braganca@civil.uminho.pt
  Phone: +351 253 510 200, Fax: +351 253 510 217

- Dr. Jorge Patrício

  Address: National Laboratory for Civil Engineer, Av. Do Brasil, 101, 1700-066 Lisboa, Portugal.
  e-mail: jpatricio@lnec.pt
  Phone: +351 218 443 273, Fax: +351 218 443 028

DAYLIGHTING ASSESSEMENT OF MIXED BUILDINGS TECHNOLOGY SOLUTIONS

Introduction

- Light is the part of the electromagnetic radiation spectrum which can be perceived by the human eye.
- Natural light has two main components: Direct light and diffuse light
- Light and building materials: Transmission, reflection, absorption, refraction

Technology Characterization

- <u>General</u>

  MBT buildings combine different structural construction methods in one building. The majority of materials used in this solution are lightweight. Only for structural or energy (e.g. thermal storage) purposes are used heavyweight materials.

  Opening sizes are usually required to be larger for daylighting purposes than what would be desirable in terms of heating/cooling loads. The use of thermal mass can help to dampen some of the effects of the larger glazing areas. On the other hand, innovative daylighting and shading systems often require lightweight solutions, calling for a Mixed-Building Technologies approach.

- <u>Relations to consider</u>

  -Natural light and thermal behavior

    Summer – shading, selective use of thermal mass

    Winter – solar gains, selective use of thermal mass

  -Natural light and visual comfort

    Discomfort glare

    Disability glare

    Working environments (Offices, schools) – VDU's

  -Natural light and acoustical control

  -Natural light and views out

  -Natural light and security

  -Natural light and aesthetics

- <u>Design issues</u>••

  Natural light can be brought into the space through:

  Vertical, horizontal or tilted planes

  Direct light through openings in opaque planes

  Reflected light through openings in opaque planes

Diffuse light through openings in opaque planes

Diffuse light through building elements built of translucent materials

••

- Design Factors••

  Openings design (windows, skylights)

  Lightshelves, reflectors, etc.

  Types of glazing materials

  Advanced glazing materials – light redirecting materials, selective materials, intelligent glazings, etc.

  Shading elements (exterior/interior, movable/fixed)

  Interior finishes

- Urban design issues••

Recommendations / Codes

- Recommended Daylight factor levels
- Recommended illuminance levels
- Codes

Technology Characterization

- Software

  Radiosity

  Ray tracing

  Daylight Factor calculations

- Daylight studies with models

  Artificial sky (diffuse light)

  Heliodon (direct light)

  Sky dome (direct + diffuse light)

  Measurements / Photography / Video

- Daylight studies in existent buildings

  Type:

  Illuminance measurements

  Luminance measurements

  Daylight Factor measurements

  Glare studies

  Time duration:

  Spot measurements (hand-held devices, instant data acquisition kits)

Long term measurements (data loggers)

Video photometry
Picture histograms
Occupant surveys

References

- Murdoch, J., 1994, Illumination engineering--from Edison's lamp to the laser, Macmillan Pub. Co
- The Chartered Institute of Building Services Engineers, Daylighting and Window Design, UK
- Baker, N., Fanchiotti, A., Steemers, K., Daylighting in Architecture: A European Reference Book, James&James

Contacts

- Prof. Luisa Caldas
  Address: Department of Civil Engineering and Architecture, Technical University of Lisbon, Av. Rovisco Pais, 1049-001 Lisbon, Portugal.
  e-mail: luisa@civil.ist.utl.pt, Phone: +351 218418346,
  Fax: +351 218418344

COST ACTION C12

General conclusions

All along the COST C12 Action, expertise on Mixed Building Technology, Structural Integrity under Exceptional Actions and Urban Design has been gained through extensive contacts and exchanges amongst researchers from more than twenty European countries.

Beyond its high scientific interest, the so-acquired knowledge brings to designers, architects, urban planners and authorities an integrated set of technical information, which is likely to positively influence the construction of new urban structures meeting the imposed structural requirements and the highest non-structural demands of the people using them. The integrated character of gathered and analysed material presented in the present book should be underlined what, at the authors' knowledge, may be considered as a "première" in the technical field.

Because of the extremely wide range of topics to be covered, the question has been raised during the final stage of the COST C12 Action to know which format should be adequately adopted to transfer to scientists and practitioners the main results of the cooperative work within COST C12, in terms of basic knowledge, design and practical use.

To achieve this goal, a so-called "datasheet format" has been followed in which the reader will find, for each item covered by the final report, the key information allowing him to become familiar with the topic, to understand its bases and implications and to learn about the technical solutions, advices or recommendations proposed by the authors.

It has so been well understood that datasheets have not been drafted to gather the whole available knowledge on each individual topic, but more as a guide reviewing the main aspects and issues of clearly identified technical problems. But in the datasheets, the possibility is also given to the reader to go deeper in its investigations through a list of appropriate references to be consulted and/or specialists to be contacted.

Obviously, this final report should in no way be considered as a sum of the whole knowledge available nowadays throughout the world. It reflects the experience of those who have been involved in the Action for four years and who tried to be as complete as possible in reporting on the various topics addressed in the publication.

Finally the authors hope that the present book will be of help in the implementation of new technologies in the construction field, leading to an increase of buildings' structural quality and to a direct positive impact on the environment through the savings on material and energy consumption and the possibilities for dismantling and recycling.

Printed and bound by CPI Group (UK) Ltd, Croydon, CR0 4YY

01/11/2024

01782640-0003